Applied Hydrogeology of Fractured Rocks

Second Edition

Applied Hydrogeology
of Fractured Rocks

Second Edition

B.B.S. Singhal
Formerly Professor, Department of Earth Sciences,
Indian Institute of Technology Roorkee, Roorkee, India

R.P. Gupta
Professor of Earth Resources Technology,
Department of Earth Sciences,
Indian Institute of Technology Roorkee, Roorkee, India

 Springer

Dr. B.B.S. Singhal
Formerly Professor
Indian Institute of Technology Roorkee
Department of Earth Sciences
Roorkee, India
brijbsinghal@gmail.com

Dr. R.P. Gupta
Professor
Indian Institute of Technology Roorkee
Department of Earth Sciences
Roorkee, India
rpgupta.iitr@gmail.com

This is a revised and updated edition of *Applied Hydrogeology of Fractured Rocks,* Kluwer Academic Publishers (now Springer), 1999.

ISBN 978-94-007-9019-3   ISBN 978-90-481-8799-7 (eBook)
DOI 10.1007/978-90-481-8799-7
Springer Dordrecht Heidelberg London New York

Cover illustration: Fig. 2.1 in this book and a schematic of triple porosity model in carbonate rocks: groundwater occurs in the inter-pore spaces, along fractures and joints and in larger subsurface conduits (by Ravi Gupta and Mukesh Kumar Singh).

Cover design: deblik, Berlin

Printed on acid-free paper

Springer is part of Springer Science+Business Media (www.springer.com)

Verily, all this is water. All the created beings are water. The vital breaths in the body are water. Quadrupeds are water. Edible crops are water. Ambrosia is water. The luminaries are water. Vedic formulae are water. Truth is water. All deities are water.

– from Mahanarayan Upanishad

Those who take managerial action to conserve and use water from glaciers, mountains, rivers, wells and also rainwater prosper all the time.

– from Atharva Veda

Note: Upanishads & Vedas are ancient Indian scripture originally in Sanskrit.

Preface

Preface to the Second Edition

The main objective in producing the second edition of "Applied Hydrogeology of Fractured Rocks" is to incorporate in the book recent advances in this field. In this new edition, topics such as tracer and isotope techniques, groundwater contamination and groundwater modeling have been enlarged into full-length chapters. Besides, additional information has been incorporated in nearly all the chapters particularly for updating latest techniques and information in geophysical exploration, satellite remote sensing sensors, hydrogeology of crystalline and carbonate rocks, well hydraulics, and groundwater conservation and management, with numerous examples. The critical suggestions from colleagues and esteemed reviewers of the first edition have been taken into consideration to the extent possible.

We are thankful to a number of persons for their suggestions, inputs and contributions in this edition, particularly to D.R. Galloway, USGS, Thomas Hahmann, German Aerospace Center, O. Batelaan, University of Brussels, Jiri Krasny, Charles University Prague, M. Thangarajan, National Geophysical Research Institute, A.K. Saha, Delhi University, Sudhir Kumar, National Institute of Hydrology, and G.C. Mishra, Ashwani Raju, D.C. Singhal, R.K. Tewari and A.K. Singh of IIT Roorkee. We appreciate the cooperation extended by the faculty of the Department of Earth Sciences, IIT Roorkee. Mukesh K. Singh assisted in designing the cover page sketch. Thanks are particularly due to Sarvesh Kumar Sharma for softcopy preparation of the manuscript.

Roorkee   B.B.S. Singhal
February 2010   R.P. Gupta

Preface to First Edition

In recent years, particular attention has been focused on the hydrogeology of low permeability rocks as is evidenced by a large number of seminars, symposia and workshops held on the above theme, throughout the world. The hydrogeological aspects of such rocks have attracted greater attention of scientists and engineers, both as a source of water supply, mainly in developing countries, and also for the potential repositories for the safe disposal of high-level radioactive waste, particularly in Europe and North America. Development of geothermal resources is another area of interest.

While teaching graduate and post-graduate students of Earth Sciences and Hydrology at the University of Roorkee, Roorkee, and in several other specialist courses, we realized the need of textbook devoted to fractured rock hydrogeology—to emphasize various aspects of exploration, development, water quality, contamination and assessment, including the application of newer tools remote sensing and geographical information systems etc. to the problem. With this in view, we have endeavoured to all earth scientists and engineers engaged in the field of fractured rock hydrogeology.

Scientific tools and methods of study in fractured rock hydrogeology include a number of aspects, viz. structural mapping, remote sensing, geophysical exploration, geographical information systems, field and laboratory hydraulic testing, including drilling, pumping tests, modeling, and assessment etc. Each of these is a topic in itself, such that separate books are available on individual topics. We have, however, endeavoured to strike a balance so that the reading material is suitable for a graduate/post-graduate level study.

Parts of the manuscript were reviewed by a number of colleagues—A.K. Bhar, S. Balakrishna, D. Kashyap, G.C. Mishra, B. Parkash, A. Prakash, G. Ramaswamy, R.G.S. Sasrty, D.C. Singhal and B.S. Sukhija. We are greatly obliged to them for their help and comments in arriving at the final presentation. We are specially indebted to C.P. Gupta who has contributed Sect. 17.7 of the book. The financial support to one of us (BBSS) received from Council of Scientific and Industrial Research, Government of India, during 1993–1994 and from Association of Geoscientists for International Development (1994–1997) is gratefully acknowledged. We are grateful to the University of Roorkee, for extending the facilities. We also appreciate the assistance provided by Yash Pal and N.K. Varshnay in typing and drafting of the work.

Finally, we are also indebted to our families for enduring four years of our preoccupation with this book.

Roorkee   B.B.S. Singhal
                                                                              R.P. Gupta

Contents

Abbreviations

AGU	American Geophysical Union
ASCE	American Society of Civil Engineers
AWRC	Australian Water Resources Council
BGS	British Geological Survey
CGWB	Central Ground Water Board, Govt. of India
CSC	Commonwealth Science Council
EPA	Environmental Protection Agency, USA
GSI	Geological Survey of India
IAH	International Association of Hydrogeologists
IAEA	International Atomic Energy Agency
IAHS	International Association of Hydrological Sciences
IGC	International Geological Congress
IGW	Proceedings International Workshop on Appropriate Methodlogies for Development and Management of Ground Water Resources in Developing Countries, NGRI, India.
NGRI	National Geophysical Research Institute, Hyderabad, India.
NRC	National Research Council, USA
UN	United Nations
UNESCO	United Nations Educational, Scientific and Cultural Organisation
WHO	World Health Organisation

Introduction and Basic Concepts

1.1 Need and Scope of the Book

More than half of the surface area of the continents is covered with hard rocks of low permeability. These rocks may acquire moderate to good permeability on account of fracturing and hence are broadly grouped under the term fractured rocks, in the context of hydrogeology.

The importance of systematic hydrogeological studies in fractured rocks was realised by the international community about 50 years back when an international symposium was held at Dubrovink in the erstwhile Yugoslavia (UNESCO-IAHS 1967). Such studies have gained greater importance in recent years as is evidenced by several special publications and international seminars and workshops sponsored by IAH, IAHS, IGC[1] and other organisations, world-over (e.g. Wright and Burgess 1992; Sheila and Banks 1993; NRC 1996; IAH 2005; Ronka et al. 2005; Witherspoon and Bodvarsson 2006; Krasny and Sharp 2008). US Geological Survey has been carrying out detailed field and model studies on the characterisation of fluid flow and chemical transport in fractured rocks (Shapiro 2007; also see websites: http://water.usgs.gov/nrp/proj.bib/hsieh.html; http://toxics.usgs.gov; http://water.usgs.gov/ogw/bgas/).

Hard rocks including igneous, metamorphic and strongly cemented sedimentary rocks and carbonate rocks cover about 50% of the earth's land surface i.e. approximately 30 million km^2 including extensive areas in several continents. In several developing countries of Asia, Africa and Latin America, greater emphasis is being given to supply safe drinking water to the vast population living in hard rock terrains.

This has necessitated evolving efficient and economic methods of groundwater exploration, assessment and development. In this context, integrated geological, remote sensing and geophysical methods have proved very successful. Further, disposal of solid and liquid wastes is an important problem in both developing and developed countries which has threatened water quality, and for this purpose low-permeability rocks form the targets. Therefore, special attention is required to locate suitable sites for the safe disposal of such waste, especially high-level waste (HLW) (Witherspoon and Bodvarsson 2006).

Thus, studies on the hydrogeology of fractured rocks are being pursued for different purposes and with widely differing objectives, such as:

1. Development of safe groundwater supplies for domestic and irrigation purposes.
2. Contaminant migration studies, in order to estimate the movement of pollutants through fractures etc.
3. Tapping of geothermal resources involving estimation of extractable amount of hot fluids from the natural geothermal gradients.
4. Development of petroleum and gas reservoirs.
5. Underground disposal of nuclear waste.
6. Construction of underground rock cavities for storing water, oil and gas etc. and underground passages like tunnels etc.
7. In several other geotechnical problems, e.g. hydromechanical effects (hydromechanical coupling) on stability of rock slopes, seepage from dams and tunnels, land subsidence and triggering of earthquakes etc.

The various above aspects require a clear understanding and proper description of the hydrogeological char-

---

[1] For abbreviations/acronyms see list of Abbreviations.

B. B. S. Singhal, R. P. Gupta, *Applied Hydrogeology of Fractured Rocks*,
DOI 10.1007/978-90-481-8799-7_1, © Springer Science+Business Media B.V. 2010

Table 1.1 Comparison of granular and fractured-rock aquifers

Aquifer characteristics	Aquifer type	
	Granular rock	Fractured rock
Effective porosity	Mostly primary	Mostly secondary, through joints, fractures etc.
Isotropy	More isotropic	Mostly anisotropic
Homogeneity	More homogeneous	Less homogeneous
Flow	Laminar	Possibly rapid and turbulent
Flow predictions	Darcy's law usually applies	Darcy's law may not apply; cubic law applicable
Recharge	Dispersed	Primarily dispersed, with some point recharge
Temporal head variation	Minimal variation	Moderate variation
Temporal water chemistry variation	Minimal variation	Greater variation

acteristics of fracture systems. In this work although, the main emphasis is on the fractured rocks as source of water supply, the other aspects such as water quality, tracer and isotope studies, contaminant transport and development of geothermal resources are also included.

Earlier, the principles of groundwater occurrence and movement were studied mainly for homogeneous porous medium, due to simplicity and widespread groundwater development in such formations. The characteristics of fluid flow in fractured rocks differ from those of homogeneous medium, primarily due to their heterogeneity (Table 1.1).

1.2   Hydrological Cycle

The hydrological cycle can simply be defined as the circulation of water between ocean, atmosphere and land (Fig. 1.1). It is best to describe the start of the cycle from the oceans as they represent vast reservoir of water although there is no definite start- or end-point of the cycle. Water evaporates from the ocean surface and goes to the atmosphere where in vapour form it undergoes circulation depending upon the distribution of temperature and wind velocity. Under suitable atmospheric conditions, the water vapour condenses resulting in precipitation in the form of rainfall or snow. The main interest of hydrologists is on the distribution of precipitation in time and space and its subsequent disposal after reaching the land surface. Some of the water which reaches the land surface drains as surface flow or runoff, some seeps down into the ground by infiltration. Part of this infiltration flows down to be added to groundwater reservoir while part of the moisture from shallow depths returns to the atmosphere by the process of evapotranspiration. A part of the water which is added to the groundwater reservoir, flows laterally under its hydraulic gradient and is discharged to the rivers, lakes or oceans as baseflow from where part of it may return to the atmosphere by evaporation. This will form the longest cycle and it is of interest to a hydrogeologist. A part of the precipitation, before fall-

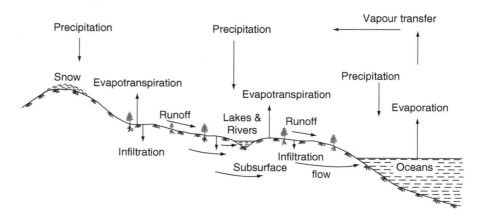

Fig. 1.1 Schematic diagram of the hydrological cycle

ing on the land surface, may return to the atmosphere by evaporation or interception by vegetation, thereby representing the shortest duration of the hydrological cycle.

The estimate of world's water balance in terms of various components of the hydrological cycle is attempted by several workers, e.g. Berner and Berner (1987). It is estimated that the total precipitation on the earth is of the order of 0.49×10^6 km³ per year, out of which 0.110×10^6 km³ falls on the continents and the remaining 0.386×10^6 km³ on the oceans. The subsurface and surface runoff to the oceans is about 37 000 km³ and the remaining 73 000 km³ is lost by evapotranspiration from the continents. The evaporation exceeds precipitation over the oceans; the difference is made up by runoff from the continents.

The distribution of water between land and sea has varied with time. For example, during the last Pleistocene glaciation, about 18 000 years ago, sea level was lowered by around 130 m due to a transfer of about 47×10^6 km³ of water (equal to about 3.5% of the oceanic volume) from the oceans to the land (Berner and Berner 1987).

Recent changes in atmospheric temperature due to various human activities, particularly the burning of fossil fuels and deforestation are also bound to affect the hydrological cycle. Records show that due to continuous increase in the atmospheric concentration of CO_2 and other greenhouse gases world over, the earth-surface temperature may rise by about 2 °C by 2050 and more than 3 °C by 2100 (Dingman 1994) which is likely to disturb the hydrological balance. This will require major adjustments in the demand and supply of water at different locations (see Sect. 20.10 in this book).

The various components of the hydrological cycle are described briefly in the following paragraphs.

Precipitation: The major types of precipitation are drizzle, rain, snow and hail. Atmospheric precipitation is a result of condensation of water vapour around hygroscopic nuclei in the atmosphere. Rainfall is measured with the help of rain gauges and is expressed in terms of depth of water in millimetre or centimetre. In the case of snow, generally the water equivalent is used as a measure of the precipitation; roughly 10–12 mm depth of snow equals 1 mm depth of water. Precipitation intensity, i.e. precipitation per unit of time, influences the recharge–runoff relation as rainfall of moderate intensity will be effective for groundwater

recharge than short spells of high intensity. The latter will cause greater runoff leading to floods.

The areal distribution of rainfall in a basin for a given storm or period is expressed in terms of isohyets which are drawn by joining points of equal precipitation in an area. The average depth of precipitation over a basin can be computed by (a) arithmetic mean method, (b) isohyetal method, and (c) Thiesen polygon method. The reader may refer to any textbook on hydrology for the details of these methods, e.g. Dingman (1994). The isohyetal method is more accurate as it takes into consideration the influence of topography on precipitation.

Evapotranspiration (ET) consists of two components—evaporation and transpiration. Evaporation is the loss of water to the atmosphere from water and land surface due to vapour pressure gradient between the evaporating surface and the air. Solar radiation is the principal energy source for evaporation. Evaporation rate is also affected by wind velocity as it brings unsaturated air to the evaporation surface. Evaporation from free water surface in arid and semi-arid regions may exceed the rainfall. Evaporation from soil surface depends on depth to water-table and type of soil (Fig. 1.2). It will be equal to evaporation from a free water surface if the soil surface remains saturated with water.

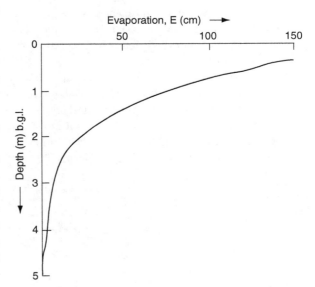

Fig. 1.2 Relation between depth of water table and rate of evaporation based on data from an experimental site in Shangqiu, PR China. The aeration zone is composed of clayey sand. (After Chen and Cai 1995)

Transpiration is the loss of water from vegetation when the vapour pressure in the air is less than that in the leaf cells. The amount of transpiration depends on the type, density and size of plants. Phreatophytes cause heavy loss of water due to transpiration while xerophytes which have shallow root system cause minimal transpiration loss.

In field conditions, it is difficult to separate evaporation from transpiration. The total loss of water from the soil surface and plants is termed evapotranspiration (ET). It is also known as 'water loss' as this part of water is not available for water supply. The terms evapotranspiration and consumptive use are frequently used interchangeably by agricultural scientists as it indicates the amount of water in raising plants. Potential evapotranspiration (PET), a term introduced by Thornthwaite (1948), is the maximum amount of water that will be removed by evapotranspiration if the soil has sufficient water to meet the demand. The actual evapotranspiration (AET) is the amount of water evapotranspired under the existing field conditions. AET is usually less than PET being about 50–90% of the potential value. However, when soil is saturated with water as in waterlogged or swampy areas, the AET may equal PET value. Losses due to evapotranspiration depend on depth of water-table and type of vegetation (Fig. 1.3).

ET losses are quite significant in water limited environments (WLE) i.e. areas where the ratio between yearly precipitation and potential evapotranspiration is less than 0.75 viz. arid, semi-arid and some subhumid regions (Lubczynski 2009). Evaporation from free water surface is measured by evaporation pans. Lysimeters are used for estimating ET losses by performing experiments under various soil moisture and vegetative cover conditions. Empirical formulae are also developed for estimating ET losses from some

readily available meteorological data. The commonly used formulae are summarized by de Marsily (1986); Dingman (1994).

Under shallow water-table conditions, the following simple relation (White's formula) can be used to estimate ET losses:

$$E = S_y \, (24 \, h \pm s) \qquad (1.1)$$

where E = daily rate of evapotranspiration, S_y = specific yield, h = water-table rise between midnight and 4 a.m., and s = rise or fall of the water-table during the 24 h period.

Infiltration is the process of absorption of water from rainfall or other surface water bodies into the soil. Infiltration is an important component of hydrological cycle. It supplies water for the growth of plants, contributes to base flow of streams and recharges the groundwater reservoir. *Infiltration capacity*, f_p is defined as the maximum rate at which water can penetrate into the soil in a given condition. *Infiltration rate*, f is the actual rate at which water enters into the soil. The relation between f and f_p is $0 \leq f \leq f_p$. The actual rate of infiltration will be less than infiltration capacity unless the rainfall intensity is equal or more than infiltration capacity. The rate of infiltration decreases exponentially with time due to swelling of clay particles and growth of algae. The maximum amount of water that soil can hold is called *field capacity*. Infiltration can be estimated either by using infiltrometers or by hydrograph analysis. Infiltration mainly depends on the properties of the soil/rock and vegetative cover.

The infiltration process in fractured rocks is not very well understood. With continuous recharge, an increase in infiltration in fractured rocks is observed which could be explained due to removal of infilling material from fractures (Salve et al. 2008).

Runoff is the part of precipitation which reaches the stream by several routes. Runoff from a basin consists of mainly four components namely, surface runoff, interflow, channel precipitation and groundwater (base) flow (Fig. 1.4 & 1.5). The part of precipitation which infilters downward is added to the groundwater reservoir. A part of this moves laterally under the existing hydraulic gradient and is discharged into the stream with a greater time lag by effluent seepage.

After the storm on account of direct runoff, water-level in the stream rises as compared with the groundwater level causing thereby a reverse flow from the stream into the groundwater reservoir resulting in bank

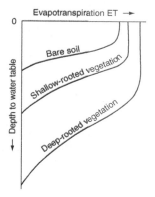

Fig. 1.3 Relation between evapotranspiration and water-table depth for different terrain conditions. (After Bouwer 1978)

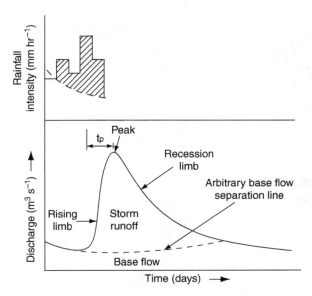

Fig. 1.4 Rainfall intensity histogram and the resultant river hydrograph

storage (Fig. 1.5). As the stream stage falls, the water from bank storage flows back to the stream causing a temporary rise in stream flow.

In actual practice, the total stream flow is divided into two parts—direct runoff and baseflow. Direct runoff consists of overland flow (surface runoff), interflow and channel precipitation, whereas base flow is mainly from groundwater source. In water management studies, it is important to estimate the groundwater component of a stream discharge, which can be achieved by the separation of the hydrograph. In some

areas the contribution from baseflow is quite significant, e.g., in Britain, streams flowing over Cretaceous Chalk and Tertiary sandstones receive about 75% of their total discharge from base flow annually (Inseson and Downing 1964).

A typical stream hydrograph shows the variation in stream discharge with time as a result of a storm. A concentrated storm rainfall will result in a single-peaked hydrograph while multiple peaks are formed where there are a number of successive storms or a variation in rainfall intensity. A single-peaked hydrograph consists of three parts—rising limb, crest (peak) segment and falling or recession limb (Fig. 1.4). The shape of the rising limb depends upon the character of storm i.e. duration and intensity distribution of rainfall. The crest or the peak of the hydrograph represents the maximum concentration of the runoff. It occurs after a certain time lag with respect to rainfall intensity. The recession limb represents withdrawal of water from storage within a basin after all the inflow to the channel has stopped. It is mainly a function of geomorphological characteristics i.e. topography, drainage pattern, soil and geology of the drainage basin.

The recession curve, also called as depletion curve, is described by a recession Eq. 1.2.

$$Q_t = Q_0 \, e^{-\alpha t} \qquad (1.2)$$

where, Q_t is discharge at time, t after the recession started, Q_0 is initial discharge at the start of the

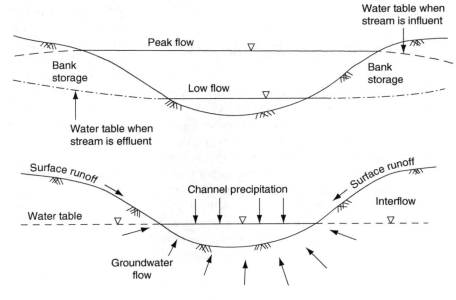

Fig. 1.5 Variation in groundwater conditions due to rise and fall in the river stage

recession, α is recession constant dependent on the basin characteristics, and t is time since the recession began.

The procedure of separation of base flow (groundwater flow) is given in several books on Hydrology (e.g. Linsley et al. 1982; Dingman 1994). There are several modifications of this approach. All these methods are of arbitrary nature. The influence of hydrogeology in selecting a suitable method of hydrograph separation is discussed in UNESCO (1972). Groundwater flow component (Q_{GW}) can also be estimated from the chemical composition of stream water, groundwater and surface water using Eq. 1.3 (Dingman 1994).

$$Q_{GW} = Q \left(\frac{C - C_{SW}}{C_{GW} - C_{SW}} \right) \qquad (1.3)$$

where Q is stream flow, C is the concentration of chemical species in the stream water, and C_{SW} and C_{GW} are its concentration in surface water and groundwater respectively. For this purpose, a variety of dissolved constituents (Cl, SO_4, HCO_3), environmental isotopes (^2H, ^3H and ^{18}O) and electrical conductivity have been used (Freeze and Cherry 1979; Dingman 1994). Radon (^{222}Rn) activities in groundwater and river water can also be used to estimate groundwater discharge to rivers (Cook et al. 2006).

In areas where field data are scarce or unavailable, remote sensing—GIS technology can be used to generate surrogate hydrogeological data (e.g. Meijerink et al. 1994; also see Chap. 4).

1.3 Classification of Subsurface Water

Water which occurs below the groundsurface is termed as subsurface water, to distinguish it from surface water. A simple classification of subsurface water with respect to its depth of occurrence and the extent to which it saturates the soil is given in Fig. 1.6. Depending on the degree of saturation, two depth zones can broadly be identified, i.e. the zone of aeration (vadose zone) and the zone of saturation (phreatic or groundwater zone). In the vadose zone the intergranular space is only partly filled with water, the remaining space is occupied by air. Therefore in this zone processes of

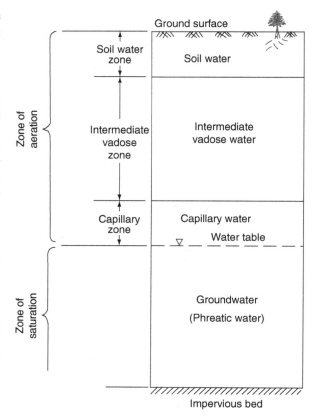

Fig. 1.6 Classification of subsurface water

oxidation and leaching are more prominent. The zone of saturation as it is named is saturated with water with the exclusion of air. This forms the zone of reduction and deposition of minerals. In unconfined aquifer, water-table represents the upper surface of the zone of saturation. As the water-table fluctuates depending on recharge and discharge conditions, the thickness of the two zones also changes seasonally. Being the main source of water supply to wells and springs, phreatic water or groundwater is of major interest to the hydrogeologists.

The zone of aeration is divided into three zones from top to bottom—soil water zone, intermediate vadose zone and capillary zone. There are no sharp boundaries between the various zones. The soil water zone is of interest to agricultural scientists as it provides water for the growth of vegetation. Water in this zone occurs either as hygroscopic water which remains adsorbed or as thin film by surface tension. The moisture content in soil water zone changes as a result of loss of water due to evapotranspiration and therefore it shows diurnal variations.

Water in the intermediate zone is termed as intermediate vadose water or suspended subsurface water as it is held due to intermolecular forces against the pull by gravity. The thickness of this zone may be zero when water-table is close to the ground surface or it may be even more than 100 m under deep water-table conditions as in the arid regions.

The *capillary zone*, also termed as *capillary fringe* extends above the water-table upto the height of the capillary rise which mainly depends on the size of the intergranular openings. The height of the capillary rise, h_c for water of specific weight (γ) $1\,g\,cm^{-3}$ and at 20°C is given by the relation

$$h_c = 0.15/r \qquad (1.4)$$

where r is the radius of the capillary opening which can be taken to be equivalent to pore size. The height of the capillary rise varies from about 2.5 cm in gravel to more than 100 cm in silt.

As the pore sizes in natural material will vary within short distances, microscopically the upper surface of the capillary zone will be uneven.

The zone of aeration and zone of saturation is separated by water-table or phreatic surface, which is under atmospheric pressure. The water-table may be either very close to the ground surface in areas of intensive recharge or may be several hundred metres deep in arid regions. Fluctuations in water-table indicate changes in groundwater storage, either due to natural reasons or by man's activity. Therefore, the monitoring of water-table fluctuation is of importance for the management of water resources. We return to this subject in Chap. 20.

1.4   Classification of Water with Respect to Origin

The relationship between various genetic types of water is shown in Fig. 1.7. Most of the water is of meteoric type as it is a result of atmospheric precipitation being a part of the present day hydrologic cycle. It is the main source of water to wells and springs. The

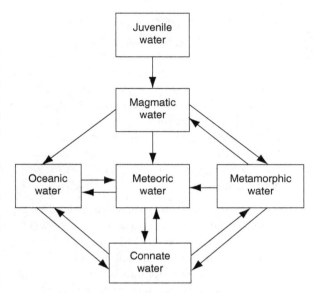

Fig. 1.7 Relationships between different genetic types of water. (After White 1957, reproduced with permission of the Geological Society of America, Boulder, Colarado, USA)

other types of water which are more of academic interest are connate water, juvenile or magmatic water and metamorphic water.

Juvenile water is also known as *new* water, as it is introduced in the hydrosphere for the first time. Magmatic water is mainly of juvenile origin derived from either deep seated magma or may be of shallow volcanic origin.

Connate water is the remnant of ancient water retained in the aquifers and is not in hydraulic continuity with the present day hydrological cycle. It is therefore, also known as fossil water although this term for water is a misnomer. It may be either of marine or fresh water origin. Connate water is commonly associated with oil and gas where it is usually of marine origin. In arid regions, it represents the past pluvial climate as in the Sahara desert.

Metamorphic water or rejuvenated water is the term used for water derived from hydrous minerals like clays, micas, etc. due to the process of metamorphism. It is more of academic interest rather than a source of water supply. The various genetic types of water can be distinguished, to some extent, on the basis of hydrochemical and isotopic data (White 1957; Matthess 1982).

1.5   Hydrological Classification of Geological Formations

The occurrence and movement of groundwater depends on the geohydrological characteristics of the sub-surface formations. These natural formations vary greatly in their lithology, texture and structure which influence their hydrological characteristics. The geological formations are accordingly classified into the following three types depending on their relative permeabilities:

1. Aquifer: Aquifer is a natural formation or a geological structure saturated with water which has good hydraulic conductivity to supply reasonable quantity of water to a well or spring. Unconsolidated sedimentary formations like gravel and sand form excellent aquifers. Fractured igneous and metamorphic rocks and carbonate rocks with solution cavities also form good aquifers. The hydraulic conductivity of an aquifer should be generally more than $10^{-6}\,\mathrm{m\,s^{-1}}$.

2. Aquitard: Aquitard is a formation having insufficient permeability to make it a source of water supply but allows interchange of groundwater in between adjacent aquifers due to vertical leakage. Therefore, aquitards serve as semi-confining layers. Examples are that of silt, clay, shale and *kankar* (calcrete).

3. Aquiclude: It is a confining formation which is impermeable like unfractured crystalline rocks, clays and shales. In nature, truly impermeable formations are rare as every geological unit has some hydraulic conductivity.

1.5.1   Types of Aquifers

The lateral continuity and vertical boundaries of aquifers are often not well defined. The aquifers may be of localised occurrence or may extend over distances of several hundred kilometres. Aquifers in the great Australian basin and in the Sahara desert (Nubian sandstones) have been traced over lateral distances of several hundred kilometers. Based on hydraulic characteristics of confining layers, aquifers are classified into the following types (Fig. 1.8)

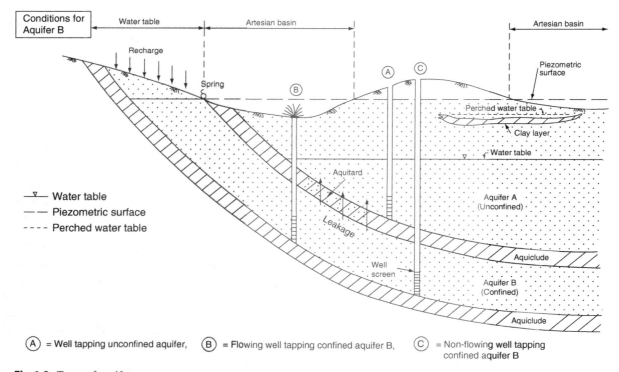

Fig. 1.8 Types of aquifers

1. Confined aquifers

    A confined aquifer, also known as artesian aquifer, is overlain and underlain by a confining layer. Water in a confined aquifer occurs under pressure which is more than the atmospheric pressure. The piezometric (potentiometric) surface, which is an imaginary surface to which water will rise in wells tapping confined aquifer, should be above the upper surface of the aquifer, i.e. above the base of the overlying confining layer. Confined aquifers are mainly recharged at the outcrops which form the intake areas. Groundwater movement in a confined aquifer is similar to a conduit flow. A confined aquifer may change to unconfined aquifer either with the time or space depending on the position of the potentiometric surface which inturn depends on the recharge and discharge from the aquifer. Confined aquifers, with potentiometric surface above the land surface, support flowing wells. In India, for example, flowing well conditions exist in alluvial aquifers of the *Terai belt* in the sub-Himalayan region and in Tertiary sandstone aquifers in South India.

    Confined aquifers may be formed under the following three types of geological conditions:

    (a) Stratiform multilayered formations: They occur either as gently dipping beds, monoclines or synclines. They usually form high pressure systems, viz. Dakota sandstones of USA, Great Artesian Basin in Australia, Nubian sandstone in North Africa, and Cuddalore sandstone in South India.

    (b) Fractures and joints: In igneous and metamorphic rocks groundwater may occur under confined conditions in joints and fractures. In volcanic rocks, like Deccan basalts of India, vesicular horizons and interflow spaces may form confined aquifers at some places.

    (c) Solution cavities: Groundwater in soluble rocks, like limestones may also occur under confined condition, viz. the Roswell basin in New Mexico, USA.

2. Leaky or semi-confined aquifer

    In nature, truly confined aquifers are rare because the confining layers are not completely impervious. In leaky aquifers, aquitards form the semi-confining layers, through which vertical leakage takes place due to head differences across it (Fig. 1.8).

3. Unconfined aquifers

    An unconfined or phreatic aquifer is exposed to the surface without any intervening confining layer but it is underlain by a confining layer. It is partially saturated with water. The upper surface of saturation is termed water-table which is under atmospheric pressure. It is recharged directly over the entire exposed surface of the aquifer. In unconfined aquifer, gravity drainage due to pumping is not instantaneous. They show delayed drainage which is more in clay and silty formation as compared to coarse grained material.

4. Perched aquifer

    Perched aquifer is a type of unconfined aquifer separated from the main regional aquifer by localised clay lens or any other impervious material in the zone of aeration. The thickness and lateral extent of perched aquifer is controlled by the shape and size of the clay layer. Being of limited extent, perched aquifers are only a source of limited water supply. Perched aquifer can be distinguished from the main unconfined aquifer, by sudden fall of water-level in the borehole during drilling as it cuts across the underlying clay layer. Therefore the borehole will become dry until the regional water-table is reached.

5. Double porosity aquifer

    A double or dual porosity aquifer, viz. fractured rocks consists of two parts—the matrix blocks and the fractures. The blocks have low permeability but high storativity while fractures have high permeability but lower storativity. The hydraulic behaviour of a double porosity aquifer is described in Chap. 9.

6. Triple porosity aquifer

    Karst aquifers are characterised by triple porosity consisting of matrix blocks, solution cavities and large solution conduits (Chap. 15).

The distinction between different types of aquifers is often difficult. It is necessary to have data about subsurface lithology, water-levels and hydraulic parameters of aquifer and confining layers for identifying a particular type of aquifer. The response of different aquifer types to pumping is discussed in Sect. 9.2.3.

Further, the distinction between unconfined and confined aquifer is also a matter of scale. An aquifer which locally appears to be of confined type but on a regional scale the different aquifers may be interconnected forming an unconfined aquifer system.

1.6 Methods and Stages of Investigations

The exploration of groundwater resources should be done in stages as depicted schematically in Fig. 1.9. At the first instance a search should be made for existing data available in published or unpublished reports of the federal and state departments of Geology and

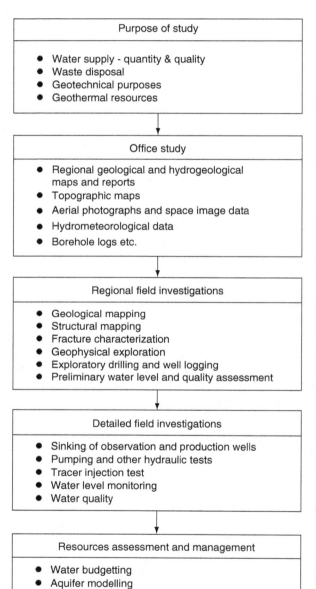

Water Resources, academic theses and dissertations. Additional information can be obtained from local land owners, construction firms and highway departments. The following information is of particular relevance in this context:

1. Surface and subsurface data about the rock types and their attitude. In case of hard rocks, data about depth to bed rock and structural aspects like folds, faults and other rock discontinuities is of more relevance;
2. Geomorphological and drainage characteristics: satellite images, aerial photographs and topographic maps should be used for this purposes;
3. Borehole logs about subsurface geology;
4. Hydrometerological data on rainfall, snowfall, evapotranspiration, runoff etc.;
5. Groundwaer levels and their seasonal variation;
6. Hydraulic characteristics, viz. hydraulic conductivity, transmissivity, storativity and well yield, etc.;
7. Water quality;
8. Groundwater withdrawals.

The existing data could be compiled and reinterpreted wherever necessary. In the absence of required information, dedicated investigations, depending on the nature of the problem, ought to be planned.

Summary

In the earlier part of the human civilisation, most of the groundwater development was from porous alluvial formations located adjacent to source of recharge. Therefore the basic principles of the occurrence and movement of groundwater were developed as applicable to homogeneous formations. In view of greater demand of water in the last half a century, interest in heterogeneous formations viz. fractured rocks has gained importance not only as a source of water supply but also for other purposes viz. disposal of waste material, development of petroleum reservoirs, tapping of geothermal power, and several other geotechnical problems.

The circulation of water on the earth planet is described in terms of hydrological cycle consisting of precipitation, evapotranspiration, runoff and infiltration. The occurrence and movement of groundwater depends on the lithology, and structure of the geological formation.

Fig. 1.9 Steps in exploration, assessment and management of groundwater resources

---

Figure content (Fig. 1.9):

Purpose of study
- Water supply - quantity & quality
- Waste disposal
- Geotechnical purposes
- Geothermal resources

Office study
- Regional geological and hydrogeological maps and reports
- Topographic maps
- Aerial photographs and space image data
- Hydrometeorological data
- Borehole logs etc.

Regional field investigations
- Geological mapping
- Structural mapping
- Fracture characterization
- Geophysical exploration
- Exploratory drilling and well logging
- Preliminary water level and quality assessment

Detailed field investigations
- Sinking of observation and production wells
- Pumping and other hydraulic tests
- Tracer injection test
- Water level monitoring
- Water quality

Resources assessment and management
- Water budgetting
- Aquifer modelling
- GIS analysis
- Resources planning and management

Further Reading

Chow VT (1964) Handbook of Applied Hydrology. McGraw-Hill Book Co. Inc., New York, NY.

IAH (2005) The future of hydrogeology. (ed. C Voss), J. Hydrol. 13(1), p. 341.

Krasny J, Sharp JM (eds) (2008) Groundwater in Fractured Rocks. IAH Selected Papers Series, volume 9, CRC Press, Boca Raton, FL.

Linsley RK et al. (1949) Applied Hydrology. McGraw-Hill Book Company Inc, New York, NY.

Fractures and Discontinuities

2

2.1 Introduction

From the hydrogeological point of view, fractures and discontinuities are amongst the most important of geological structures. Most rocks possess fractures and other discontinuities (Fig. 2.1) which facilitate storage and movement of fluids through them. On the other hand, some discontinuities, e.g. faults and dykes may also act as barriers to water flow. Porosity, permeability and groundwater flow characteristics of fractured rocks, particularly their quantitative aspects, are rather poorly understood. Main flow paths in fractured rocks are along joints, fractures, shear zones, faults and other discontinuities.

There is a great need to understand hydraulic characteristics of such rocks, in view of: (a) groundwater development, to meet local needs; and (b) as depositories for nuclear and other toxic wastes.

There could be multiple discontinuities in fractured rocks along which groundwater flow takes place. A number of factors including stress, temperature, roughness, fracture geometry and intersection etc. control the groundwater flow through fractures. For example, fracture aperture and flow rate are directly interrelated; non-parallelism of walls and wall roughness lead to friction losses; hydraulic conductivity through fractures is inversely related to normal stresses and depth, as normal stress tends to close the fractures and reduce the hydraulic conductivity.

It has also been noted that fracture permeability reduces with increasing temperature. As temperature increases with depth, thermal expansion in rocks takes place which leads to reduction in fracture aperture and corresponding decrease in permeability. Further the permeability is also affected by cementation, filling, age and weathering (see Chap. 8).

Parallel fractures impart a strong anisotropy to the rock mass. On the other hand, greater number of more interconnected fractures tends to reduce anisotropy. Further, larger fracture lengths, greater fracture density and larger aperture increase hydraulic conductivity.

Therefore, summarily, for hydrogeological studies, it is extremely important to understand and describe the structure of the rock-mass and quantify the pattern and nature of its discontinuities (van Golf-Racht 1982; Sharp 1993; Lee and Farmer 1993; de Marsily 1986).

2.2 Discontinuities—Types, Genetic Relations and Significance

Discontinuity is a collective term used here to include joints, fractures, bedding planes, rock cleavage, foliation, shear zones, faults and other contacts etc. In this discussion using a genetic approach, we group discontinuities into the following categories:

1. Bedding plane
2. Foliation including cleavage
3. Fractures (joints)
4. Faults and shear zones, and
5. Other geological discontinuities.

2.2.1 Bedding Plane

Primary bedding and compositional layers in sedimentary rocks form the bedding plane. Usually, it is the most significant discontinuity surface in all sedimentary rocks such as sandstones, (Fig. 2.1b) siltstones, shales etc., except in some massive sandstones or

Fig. 2.1 Examples of fractured rocks; **a** Metamorphic rocks (meta-argillites) in Khetri Copper Belt, India. Several sets of fractures including shear planes are developed; some of the fractures possess infillings. **b** Sandstones of Vindhyan Group, India; bedding planes constitute the dominant discontinuity surfaces. (Photograph (b) courtesy of A.K. Jindal)

limestones. Bedding plane can be readily identified in the field owing to mineralogical-compositional-textural layering.

Bedding plane, being the most important discontinuity, imparts anisotropy and has a profound influence on groundwater flow in the vadose zone. The groundwater flow is by-and-large down-dip (Fig. 2.2).

Folds are flexures in rocks formed due to warping of rocks. Although a wide variety of folds are distinguished, the two basic types are anticlines (limbs dipping away from each other) and synclines (limbs dipping towards each other). Folding leads to change or reversal in dip directions of beds, and this affects groundwater flow. Further, folding is accompanied by fracturing of rocks. In an anticline, the crest undergoes higher tensional stresses and hence develops open tensile fractures, which may constitute better sites for groundwater development.

2.2.2 Foliation

Foliation is the property of rocks, whereby they break along approximately parallel surfaces. The term is restricted to the planes of secondary origin occurring in metamorphic rocks. Foliation develops due to parallel-planar alignment of platy mineral grains at right angles to the direction of stress, which imparts fissility. The parallel alignment takes place as a result of recrystallisation during regional dynamothermal metamorphism, a widespread and common phenomenon

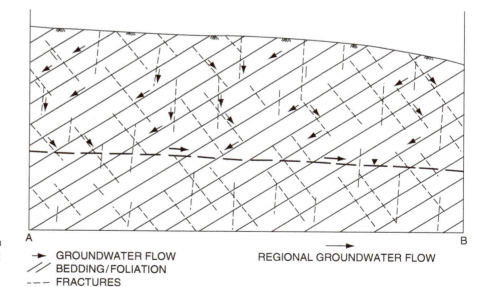

Fig. 2.2 Schematic diagram showing the role of bedding planes and fractures on groundwater movement in the vadose zone

A   B

→ GROUNDWATER FLOW
// BEDDING / FOLIATION
- - - FRACTURES
⊥ WATER TABLE

→
REGIONAL GROUNDWATER FLOW

in crystalline rocks. *Rock cleavage* is almost a synonymous term. It is also used for planes of secondary origin along which the rock has a tendency to break in near-parallel surfaces. Some terms are used for specific metamorphic rocks. Thus, the term *slaty cleavage* is used for rock cleavage in slates; *schistosity* is used for schists and *gneissosity* for gneisses. Foliation planes may or may not be parallel to bedding. Foliation that is parallel to the bedding is often referred to as bedding foliation. Fracture cleavage is produced by closely-spaced jointing. In many schistose rocks, shear cleavages are developed due to closely spaced shear-slip planes, known as slip-cleavage. In a folded region, the foliation often developed parallel to the axial plane of folds is called the axial plane foliation.

Foliation in metamorphic rocks has a profound influence on groundwater movement, possessing quite the same role as bedding in the sedimentary rocks, both being the most significant discontinuities in the respective rock categories (e.g. see Fig. 2.2).

2.2.3 Fractures and Joints

2.2.3.1 Introduction and Terminology

Fractures, also called joints, are the planes along which stress has caused partial loss of cohesion in the rock. It is a relatively smooth planar surface representing a plane of weakness (discontinuity) in the rock. Conventionally, a fracture or joint is defined as a plane where there is hardly any visible movement parallel to the surface of the fracture; otherwise, it is classified as a fault. In practice, however, a precise distinction may be difficult, as at times within one set of fractures, some planes may show a little displacement whereas others may not exhibit any movement. Slight movement at right angles to the fracture surface will produce an open fracture, which may remain unfilled or may get subsequently filled by secondary minerals or rock fragments.

'Fracture zones' are zones of closely-spaced and highly interconnected discrete fractures. They may be quite extensive (length > several kilometres) and may even vary laterally in hydraulic properties.

Fracture-discontinuities are classified and described in several ways using a variety of nomenclature, such as: joints, fracture, fault, shear, gash, fissure, vein etc.

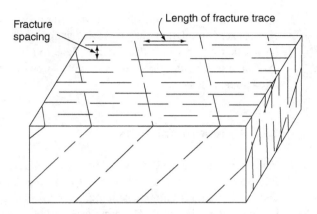

Fig. 2.3 Two sets of fractures are schematically shown in the block. An individual fracture has limited spatial extent and is discontinuous in its own plane. Fracture spacing and fracture trace length are indicated for one set

Generally, the term fracture is used synonymously with joint, implying a planar crack or break in rock without any displacement. The terms fault and shears are used for failure planes exhibiting displacement, parallel to the fracture surfaces. Gash is a small-scale open tension fracture that occurs at an angle to a fault. Fissure is a more extensive open tensile fracture. A filled-fissure is called a vein.

An individual fracture has a limited spatial extent and is discontinuous in its own plane (Fig. 2.3). On any outcrop, fractures have a certain trace lengths and fracture spacings. By mutual intersection, the various fracture sets may form interconnected continuous network, provided that the lengths of the joints in the different sets are much greater than the spacings between them (see Fig. 2.18). The interconnectivity of fractures leads to greater hydraulic conductivity.

2.2.3.2 Causes of Fracturing

Although fractures are extremely common and widespread in rocks, geologically they are still not well-enough studied (Price and Cosgrove 1990). Complex processes are believed to be involved in the origin of fractures, which are related to geological history of the area. Fractures are created by stresses which may have diverse origin, such as: (a) tectonic stresses related to the deformation of rocks; (b) residual stresses due to events that happened long before the fracturing; (c) contraction due to shrinkage because of cooling of magma or dessication of sediments; (d) surficial

movements such as landslides or movement of glaciers; (e) erosional unloading of deep-seated rocks; and
(f) weathering, in which dilation may lead to irregular
extension cracks and dissolution may cause widening
of cavities, cracks etc.

2.2.3.3 Types of Fractures

Firstly, fractures may be identified into two broad types:
(a) systematic, which are planar, and more regular in
distribution; and (b) non-systematic, which are irregular and curved (Fig. 2.4). The non-systematic fractures
meet but do not cross other fractures and joints, are
curved in plan and terminate at bedding surface. They
are minor features of dilational type and develop in the
weathering zone. Curvilinear pattern is their general
characteristic. Parallel systematic fractures are treated
as a set of fractures.

Geometric classification—Considering the geometric relationship with bedding/foliation, the systematic fractures or joints are classified into several types.
Strike joints are those that strike parallel to the strike
of the bedding/foliation of the rock. In dip joints, the
strike direction of joints runs parallel to the dip direction of the rock. Oblique or diagonal joints strike at
an angle to the strike of the rocks. Bedding joints are
essentially parallel to the bedding plane of the associated sedimentary rock.

Depending upon their extent of development, fractures may be classified into two types: first-order and
second-order. First-order fractures cut through several
layers of rocks; second-order fractures are limited to a
single rock layer. Further, depending upon the strike
trend of fractures with respect to the regional fold
axis, fractures are designated as longitudinal (paral-

lel), transverse (perpendicular) or oblique ones (see
Fig. 2.7 later).

Genetic classification—Genetically, the systematic
fractures can be classified into three types:

1. *Shear fractures*, which may (or may not) exhibit
   shear displacement and are co-genetically developed
   in conjugate sets with a dihedral angle 2i > 45°.
2. *Dilational fractures*, which are of tensile origin,
   commonly, developed perpendicular to the bedding
   plane, and are open fractures with no evidence of
   shear movement.
3. *Hybrid fractures*, which exhibit features of both
   shear and dilational origin. They may occur in conjugate sets with a dihedral angle 2i < 45°. They are
   open (extension!), may be partly filled with veins,
   and may also exhibit some shear displacement.

The physical stress conditions under which these three
types of fractures develop are illustrated by the Mohr
diagram in Fig. 2.5. The curve ABC is a Mohr envelope. The stress circles touching the Mohr envelope
at A, B and C points indicate different failure conditions of the rock. In condition 'A', the principal maximum compressive stress is negative, i.e. extensional,
and therefore it leads to a dilational failure. In condition, 'C', a typical conjugate shear failure takes place,
such that the dihedral angle 2i > 45°. 'B' represents a
condition that there is a positive maximum principal
compressive stress and a negative minimum principal compressive stress, i.e. the effective normal stress
perpendicular to the fracture plane is negative (extensional). This can be attributed to high fluid pressure
conditions at depth. Hence, there is a tendency for such
shear fractures to open and also get filled with minerals. Typically, in such hybrid shear-extension fractures,
the dihedral angle is 2i < 45°.

Conjugate shear fractures developing at greater
depths are of ductile nature and possess a large 2i
(~90°). On the other hand, conjugate brittle shears
develop at a shallower depth and possess a smaller 2i
(~60°). Further, brittle deformation causes derivative
shears of several orders to form successively deviating
trends, which cause a spread in the trends of conjugate shears (Ruhland 1973). The process of shearing is
also accompanied by tensile deformation. Thus, brittle
deformation may produce fractures of different magnitude and direction in successive orders. In a rock mass
fractured by three orders of brittle deformation, tensile

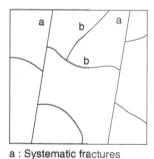

Fig. 2.4 Systematic and non-
systematic types of fractures

a : Systematic fractures
b : Non-systematic
fractures

Fig. 2.5 Basic genetic types of fractures: *A* extension fracture; *B* hybrid extension-shear fracture; *C* shear fracture. The Mohr diagram indicates the stress conditions for these failures. σ_1 and σ_3 are the maximum and minimum principal compressive stresses respectively

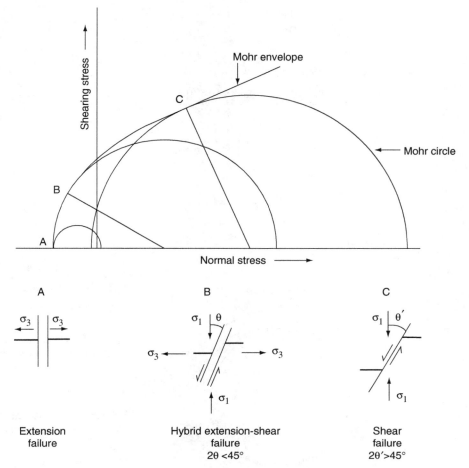

fracturing may spread over a range of about 75° and shear fracturing over a range of nearly 135° (Fig. 2.6).

2.2.3.4   Discrimination Between Shear and Extension Fractures

The rheological principles indicate that there is no sharp categorization between extension and shear fractures. In fact, all gradations from one category to the other take place. However, from hydrogeological point of view, it is important to distinguish between shear and extension joints as dilational joints are more open and possess greater hydraulic conductivity than shear joints. Discrimination between shear and tension joints may be difficult, particularly in complexly deformed areas. However, the following features may help in their discrimination:

1. Shear joints may exhibit displacement parallel to the plane of the joints, which is absent in the case of extension joints.
2. Shear joints commonly occur in conjugate sets which may be indicated by a statistical analysis.
3. In field, slickensides and other criteria of relative movement may be observed in the case of shear joints.
4. Generally, extension joints are open and shear joints are tight.
5. The orientation of the joints with respect to the bedding/foliation and or fold-axis can provide information on shear vs. tensile origin of fractures, as shear joints occur in oblique conjugate sets whereas extension joints occur as longitudinal and transverse joints forming an orthogonal pair (Fig. 2.7).
6. The cumulative trend diagram of fractures may also provide information on the related stress field,

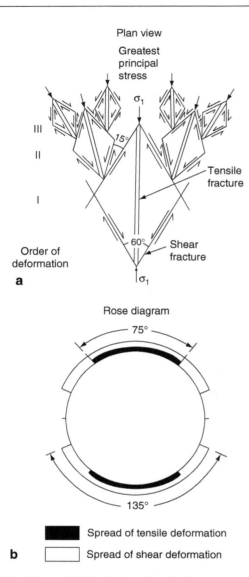

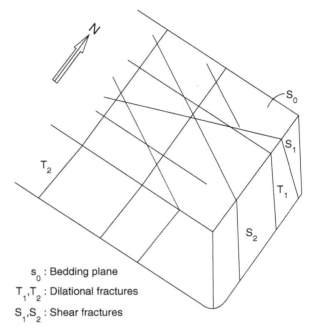

Fig. 2.7 Development of four sets of fractures in the case of simple dipping strata (see also Fig. 2.9)

Fig. 2.6 Scheme of brittle deformation of a homogeneous rock mass. The rose diagram illustrates the ranges of orientations of tensile and shear fractures due to three orders of deformation. (After Ruhland 1973)

2.2.3.5 Orientation of Fractures vis-à-vis Regional Structure

Ideally, in the case of simple-dipping strata, four sets of fractures (systematic) develop (Fig. 2.7). S_1 and S_2 form a conjugate set of shear fractures and T_1 and T_2 are extension fractures. All these fractures are perpendicular to the bedding plane and contain the intermediate principal compressive stress σ_2.

Figure 2.9 shows a simplified ideal relationship between fractures and folds. σ_1 is the maximum principal compressive stress perpendicular to the fold-axis (b). A conjugate set of oblique trending right-lateral and left-lateral shear fractures develops. There are two sets of extension fractures, one longitudinal and the other transverse to the fold-axis, both being mutually orthogonal.

During folding, bending of a bed causes extension on the convex side and compression on the concave side (Fig. 2.10). This results in extension fractures and normal faults on the crests of anticlines. Less commonly, thrust faults also develop in the inner areas of compression.

and therefore the likely trends of shear and extension fractures; the maximum principal compressive stress bisects the dihedral angle of conjugate shear fractures and is parallel to the tensile fracture.

A field example of large-scale tensional and shear fractures extending for several kilometres is given in Fig. 2.8 where tensional fractures appear as wide open and shear joints are characterized by relative displacements.

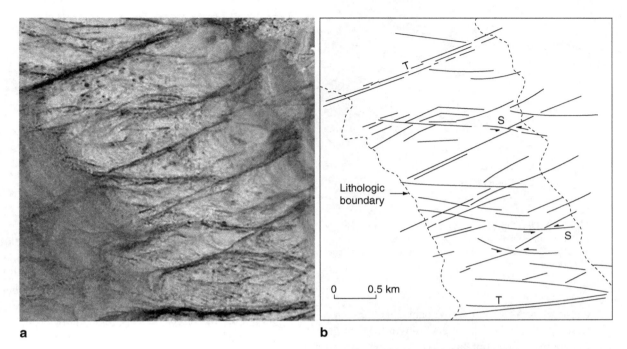

a   **b**

Fig. 2.8 a Example of large-scale tensional and shear fractures extending for several kilometres in a part of the Precambrian Cuddapah basin, India; black-and-white image generated from GoogleEarth. **b** Interpretation map of the above image; fractures marked *T* are wide open tensional fractures that are vegetated implying groundwater seepage, *S* are shear fractures exhibiting lateral relative displacement at places

2.2.3.6 Other Types of Fractures

Sheeting joints: These joints are generally flat, somewhat curved and nearly parallel to the topographic surface, often developed in granitoid rocks. They are closely developed near to the surface, and their spacing increases with depth. They are generated due to release of overburden stress.

Columnar joints: Joints of this type are tension fractures generated due to shrinkage in rocks. Shrinkage may occur due to cooling or dessication. Igneous rocks contract on cooling. Mud and silt shrink because of

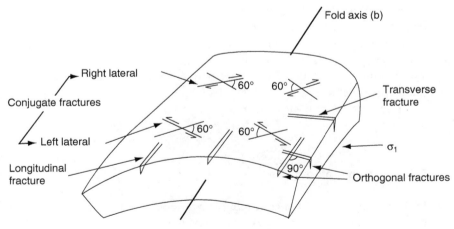

Fig. 2.9 Ideal relationship between major joint sets in a folded bed. There are two sets of conjugate shear fractures and two sets of mutually orthogonal dilational fractures. All the fractures are shown vertical

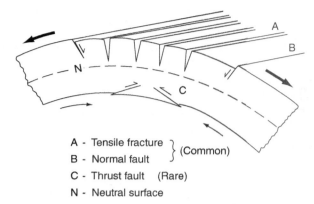

A - Tensile fracture ⎫
B - Normal fault   ⎬ (Common)
C - Thrust fault   (Rare)
N - Neutral surface

Fig. 2.10 Development of extensional fractures and normal faults on the crest and the upper axial zone of an anticline. Thrust faults may also develop, occasionally, in the inner area of compression

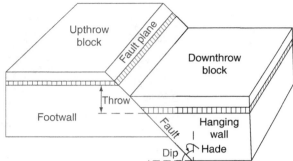

Fig. 2.11 Common terms used in describing a fault

dessication. As a result, polygonal and columnar joints develop. The columns are generally a few centimetres to a metre in diameter, and several metres long (high). Frequently, the columns are four, five or six sided in shape (see Fig. 14.5).

2.2.4   Faults and Shear Zones

Rupture and shear movement due to stresses leads to faulting. The stress in rocks is mostly a result of mountain building tectonic activity. From a hydrogeological point of view, faults and shear zones constitute very important types of discontinuities in rocks. Faults are planes and zones of rupture along which the opposite walls have moved past each other, parallel to the surface of rupture. The orientation of a fault plane is defined in terms of strike and dip, as that of any other plane in structural geology. Faults vary in dimension from a few millimetres long with minor displacement to several hundred kilometres in strike lengths with movement of several tens of kilometres.

2.2.4.1   Terminology

In describing faults, a range of terminology is used; only some of the more important terms are introduced

here (see e.g. Billings 1972; Price and Cosgrove 1990). The fault block above the dipping fault plane is called *hanging wall*; the block below the faults plane is called *footwall* (Fig. 2.11). The angle which a fault plane makes with the vertical plane parallel to the strike of the fault is called *hade*; it is complement of the dip. In many instances the displacement is distributed through a zone, called the fault zone, which may be a few centimetres to hundreds of metres wide. Faults may exhibit simple translational or rotational movements. Slip is the relative displacement as measured on the fault surface. Strike-slip and dip-slip are the displacements along the strike direction and dip direction respectively on the fault plane. *Throw* is the vertical displacement caused by the fault. The blocks which have moved up and down are called *upthrow* and *downthrow blocks*, respectively.

Shear zones are generally filled with broken and crushed rocks, which may be embedded in clay matrix. Shear zones tend to be more extensive and continuous than joints.

2.2.4.2   Classification

Faults are classified in various ways in the literature. Some classifications are based on geometrical relations between the fault plane and country rocks. Genetically, faults are classified into three types: (a) normal, (b) reverse and (c) strike-slip. They are related to the stress conditions (Fig. 2.12).

In the case of a normal fault, hanging wall moves relatively downward. Normal faults are generally high-angle faults caused when σ_1 is vertical. Reverse faults are generally low-angle (gently dipping) faults

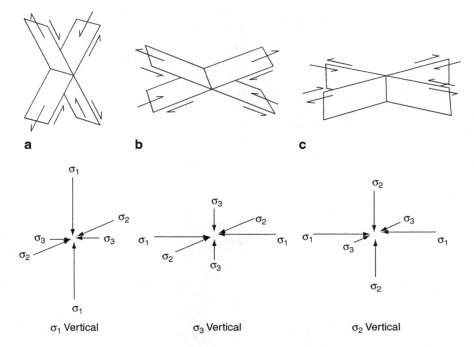

Fig. 2.12 The basic genetic types of faults: **a** normal, **b** reverse and **c** strike-slip. Orientations of principal compressive stresses are also shown

caused when σ_3 is vertical. It is characterised by the relative upward movement of the hanging wall. Strike faults are vertical faults marked by movement only in the strike direction of the fault. These are caused when σ_2 is vertical.

2.2.4.3 Recognition of Faults in the Field

A number of criteria are used to decipher the presence of faults, though in a specific case only some of the features may be present. Some of the more important criteria include: (a) displacement of key beds; (b) truncation of beds and structures; (c) repetition and omission of strata; (d) presence of features indicating movement on fault surface such as slickensides, mylonite, breccia, gouge, grooving etc; (e) evidence of mineralisation, silicification, along fault zones; (f) physiographic features such as fault scarps, offset ridges, etc; (g) alignment such as springs alignment, pond alignment, vegetation alignment, rectilinearity of a stream; (h) indication of sudden anomalous changes in river course, such as knick points, offset of streams, anomalous or closed meanders etc.; (i) erosional features such as triangular facets, unpaired terraces etc.

Investigations for faults may be made in outcrops, road cuttings, mines or other excavations, where smaller faults could often be directly observed. A larger fault, on the other hand, may be identified on stratigraphic and physiographic evidences, and particularly on remote sensing images, as only small segments of the fault may be exposed in field, and the feature may be largely covered under soil, debris or vegetation (see Sect. 4.8.11).

2.2.4.4 Effect of Faults on Groundwater Regime

Faults may affect groundwater regime in numerous ways, some of the more important being the following:

1. It is well known that faults may have such effects as truncation, displacement, repetition or omission of beds. In this light, the distribution and occurrence of aquifers may be affected by faults as locally an aquifer unit may get displaced/truncated/omitted (Fig. 2.13a, b).
2. A fault may bring impervious rock against an aquifer, which would affect groundwater flow and distribution (Fig. 2.13a).

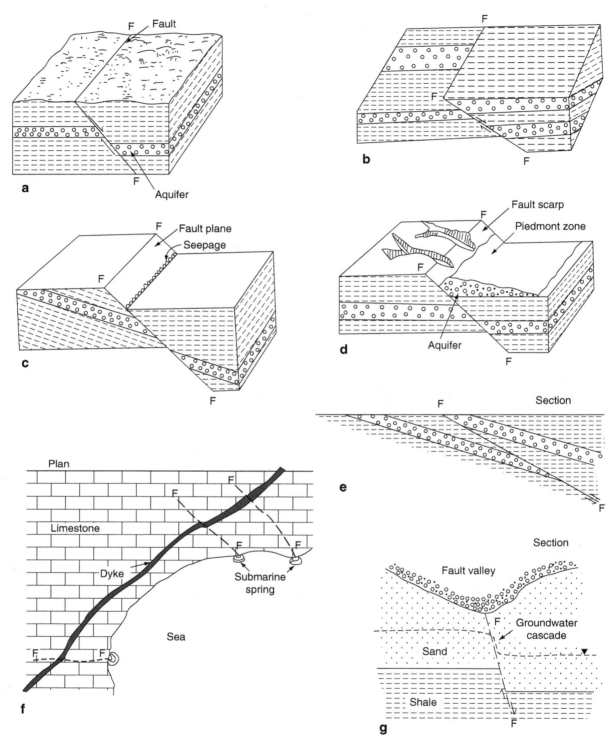

Fig. 2.13 Effects of faults on aquifers (for details see text)

3. Truncation of an aquifer by a fault may lead to seepage and formation of a spring line along the fault (Fig. 2.13c).
4. A fault may lead to a scarp; intensive erosion of the upthrow block and deposition of extensive piedmonts on the downthrow block may follow; the piedmont deposits may serve as good aquifers (Fig. 2.13d).
5. An aquifer may get repeated in a borehole due to thrust faulting; further it may also get re-exposed on the surface for recharge (Fig. 2.13e).
6. Vertical dykes, veins etc. which generally act as barriers to groundwater flow, may be breached by faults and this may produce local channel-ways across the barrier (Fig. 2.13f).
7. A fault may lead to a groundwater cascade (Fig. 2.13g).
8. Faults create linear zones of higher secondary porosity; these zones may act as preferred channels of groundwater flow, leading to recharge/discharge.
9. A fault may lead to inter-basinal subsurface flow.
10. A fault zone, when silicified, may act as a barrier for groundwater flow.

Figures 2.14 and 2.15 give field examples of extensive faults with strike length of kilometres, showing displacements of beds and marked by preferential alignment of vegetation indicating groundwater seepage.

2.2.5 Other Geological Discontinuities

In addition to the above structural features, there could be other geological boundaries such as unconformities and intrusive contacts which may act as discontinuities.

Unconformity is a surface of erosion and nondeposition separating overlying younger strata from the underlying older rocks. Conglomerate beds and palaeosols usually occur along the unconformity surface which often forms good aquifers. An unconformity implies that a hydrologeological unit may get laterally pinched out and spatially replaced by another unit (Fig. 2.16).

Intrusive contacts are other geological boundaries of significance in the context of hydrogeology. Intrusive bodies occur in a variety of shapes and sizes, such

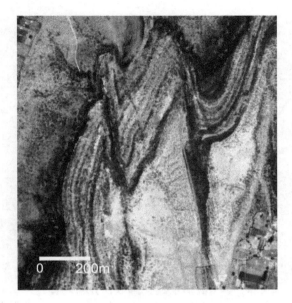

Fig. 2.14 Faults displacing the sedimentary layers of sandstones and shales (Vindhyan Super Group, near Chittaurgarh, India). The terrain has a semi-arid climate. Note the preferential growth of vegetation along fault zones related to the seepage of groundwater. Sedimentary layering is also marked by vegetation banding. Black-and-white image from GoogleEarth

as batholiths, dykes, sills etc. Their relation with the host rocks could be concordant, transgressive, or discordant. The igneous plutonic bodies crystallize under high pressure and temperature; they are devoid of primary porosity. Therefore, hydrogeological characters

Fig. 2.15 Large-scale parallel faults running for several kilometres displacing the sedimentary layers of Cuddapah basin, India. Note the vegetation alignment along the southern parts of fault zones related to the seepage of groundwater. Black-and-white image from GoogleEarth

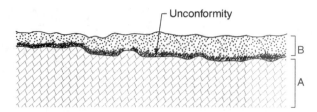

Fig. 2.16 Development of an aquifer along an unconformity between two impervious beds *A* and *B*

of the host rocks and intrusive rocks may be vastly different from each other. Hence, igneous contacts act as regional boundaries from a hydrogeological point of view.

2.3   Fracture Characterization and Measurements

A fractured rock mass can be considered to be made up of three basic components: (a) fracture network, (b) matrix block and (c) infillings along the fractures, if present (see Fig. 2.1). A single fracture or discontinuity plane is characterised by its orientation, genetic nature (shear/tensile), persistence and aperture etc. Several fracture planes of the same type create a fracture set. They have certain spacing (frequency). Several intersecting intercommunicating fracture sets create a fracture network which facilitates fluid flow. Thus, it is extremely important to characterize discon-

tinuities and make their measurements, for a meaningful application. The various important parameters are summarised in Table 2.1.

The concept of Representative Elementary Volume (REV) is very important and may be introduced here. REV is the minimum rock mass volume which has the hydraulic and or mechanical properties similar to those of the rock mass. For mechanical properties, a sample size of a few cubic meters may be sufficient for approaching a REV; however, in case of hydraulic flow, REV may be substantially larger, and in some cases, it may not even exist due to strong anisotropy and spatial variability of rock characters.

2.3.1   Number of Sets

Several sets of discontinuities are often developed in a rock mass, three to four sets being most common. Number of sets of discontinuities in a exposure can be statistically determined by contouring the pole-plots (see Fig. 2.18). Relevant data as orientation, spacing, length, aperture etc. has to be collected for each set of discontinuity.

2.3.2   Orientation

Orientation is the parameter to define a single fracture plane in space, using angular relationships, as for any

Table 2.1 Parameters for discontinuity characterization

Parameter	Description
1. Number of sets	Number of sets of discontinuities present in the network
2. Orientation	Attitude of discontinuity present in the network
3. Spacing	Perpendicular distance between adjacent discontinuities of the same set
4. Persistence	Trace length of the discontinuity seen in exposure
5. Density	
– linear	Number of fractures per unit length
– areal	Cumulative length of fractures per unit area of exposure
– volumetric	Cumulative fractured surface area per unit bulk rock volume
6. Fracture area and shape	Area of fractured surface and its shape
7. Volumetric fracture count	Number of fractures per cubic metre of rock volume
8. Matrix block unit	Block size and shape resulting from the fracture network
9. Connectivity	Intersection and termination characteristics of fractures
10. Aperture	Perpendicular distance between the adjacent rock-walls of a discontinuity, the space being air or water-filled
11. Asperity	Projections of the wall-rock along the discontinuity surface
12. Wall coatings and infillings	Solid materials occurring as wall coatings and filling along the discontinuity surface

geological planar surface. It is defined in terms of dip direction (angle with respect to north) and dip amount (angle with horizontal). The orientation is expressed in terms of a pair of numbers, such as 25°/N 330°, implying a plane dipping at 25° in the direction 330° measured clock-wise from the north. In field, inaccuracies often creep-in the measurements, and therefore statistical analysis is desirable.

Rose diagram is a method of displaying the relative statistical prevalence of various directional trends, e.g. strike direction of fractures, lineaments etc. It can be prepared for parameters such as number or length, i.e. number of joints direction-wise, or length of joints direction-wise. Frequently, the directions are grouped in 10° interval. Frequency in a group-interval is represented along the radial axis, the length of petals becoming a measure of relative dominance of the trend. The strike petals possess a mirror image about the centre of the rosette. Data on the magnitude of dip cannot be incorporated in the rosette, and may however be shown outside the circumference (Fig. 2.17). *Histogram plot* is another way to represent the relative prevalence of the trends.

Spherical projection: For representing orientation of geological planar surfaces, the method of stereographic equal-area projection is frequently employed, as it accurately shows the spatial distribution of data. Basic concepts on great-circle plots and π-pole plots to represent planes can be found in any standard text on structural geology (e.g. Billings 1972; Price and Cosgrove 1990). The method of plotting pole has a relative advantage over the great-circle method in that clusters of poles and their relative concentrations can be readily ascertained

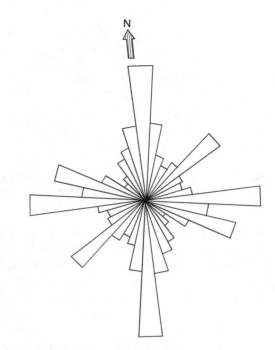

Fig. 2.17 Rose diagram of strike trends of discontinuities showing their relative prevalence

on such plots by contouring. Schmidts-net is often used for density contouring to provide information on highest concentration, i.e. the most dominant fracture plane. Figure 2.18 gives an example.

It may be important to find the over-all effect of various discontinuities. The mean direction of a group of poles can be represented by a simple vector-sum of all the constituting poles, following Fisher distribution. Similarly a resultant vector can be calculated

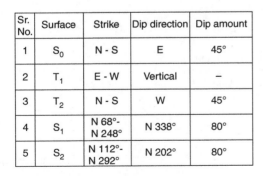

Sr. No.	Surface	Strike	Dip direction	Dip amount
1	S_0	N - S	E	45°
2	T_1	E - W	Vertical	–
3	T_2	N - S	W	45°
4	S_1	N 68°- N 248°	N 338°	80°
5	S_2	N 112°- N 292°	N 202°	80°

a   Fracture orientation data   **b**   Great circle diagram   **c** π-Pole diagram

Fig. 2.18 **a** Orientation data of discontinuities. S_0 is bedding plane, T_1 and T_2 are tensile fractures and S_1 and S_2 are shear fractures. Their great-circle and π-pole diagrams are shown in figures (**b**) and (**c**) respectively

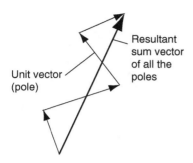

Fig. 2.19 Principle of determining the mean pole direction as the resultant vector sum of all the vectors (poles), following Fisher method

by summing all the clusters, to give a net directional effect of all the sets of discontinuities (Fig. 2.19).

It may be mentioned here that only selected and not all of the discontinuities present may play a significant role in fluid movement in the rock mass. Therefore, selection and data integration ought to be done judiciously. Further, Sharp (1993) gave a more useful concept for integration of discontinuity trend, frequency and aperture data to make hydraulic zonation maps (see Sect. 7.2.5).

2.3.3 Spacing (Interval)

Systematic joints are roughly equidistant and possess parallelism, and therefore, the parameter statistical

spacing has significance. It describes the average (or modal) perpendicular distance between two adjacent discontinuities of the same set. It has a profound influence on rock mass permeability and groundwater flow. Fracture spacing is reciprocal of the fracture frequency or linear fracture density. It also controls fracture intensity and matrix block size.

Fracture separation (f_s) is related to lithology and thickness of the bed (b), and is given as (Price and Cosgrove 1990):

$$f_s = Y \cdot b \qquad (2.1)$$

where Y is a constant related to lithology. Modelling and theoretical approaches also show that fracture spacing and bed thickness should have a linear relationship, for a given lithologic material.

By spreading a tape in any convenient direction on an outcrop face, average apparent spacing (f_{sa}) between fractures of a set can be measured. This measurement has to be corrected for angular distortion (θ) to give the value of true fracture interval, perpendicular to the fracture orientation. The correction angle (θ) equals the angle between the direction of tape alignment and the pole to the fracture plane, and can be easily computed using a stereographic net (Fig. 2.20). The true fracture spacing (f_s) can be obtained from the measured fracture spacing (f_{sa}) as:

$$f_s = f_{sa} \cdot \cos\theta \qquad (2.2)$$

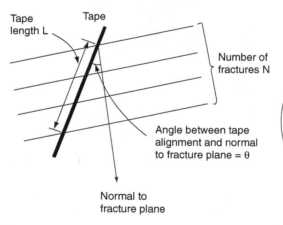

Fig. 2.20 Spacing of fractures and computation of true fracture spacing. **a** Measurements in field for apparent fracture spacing. **b** Angle of correction, i.e. the angle between the line of tape alignment and pole to the fracture plane, computed by stereographic method

Apparent fracture spacing = L / N = dm
True fracture spacing = dm cos θ

a

Normal to
fracture plane

S–S = Great – circle of the outcrop face orientation

T = Tape alignment

P = Pole to the fracture plane

b θ = Correction angle

Further, it has been reported that fairly reliable estimates of fracture spacing can also be given by the P-wave velocity using seismic refraction techniques (see Sect. 5.6).

2.3.4 Persistence (Fracture Length)

Fracture persistence or length is a measure of the extent of development of discontinuity surface (Fig. 2.21). This carries the notion of size and controls the degree of fracturing. It is a crude measure of the penetration length of a fracture in a rock mass. Fracture trace length is also related to fractured surface area. As some of the discontinuities are more persistent and continuous than others, it becomes a very important parameter in controlling groundwater flow.

Persistence is rather difficult to quantify, as it would differ in the dip and strike directions. It can be measured by observing the discontinuity trace length in an exposure, in both dip and strike directions.

The observed trace length may be only an apparent value of the true trace length due to various types of bias creeping in the data during measurements in exposures, drifts, excavations, benches etc. For example, the biases could be like: (a) inability to recognize fracture traces shorter than a certain threshold length will lead to a bias (truncation of the histogram); (b) inability to measure full length of the traces owing to incomplete exposures in drift-walls, excavations etc. will lead to recording of censored length data; (c) the observed length of fracture trace depends on the relative orientation between the fracture plane and the exposure face; (d) in the sampling area or scanline, a stronger discontinuity is more likely to appear than a weaker one. Considering such aspects, methods for estimating the true trace length are discussed by a few workers (e.g. Pahl 1981; Laslett 1982; Chiles and de Marsily 1993).

2.3.5 Fracture Density

Fracture density is measured for each set of fracture set separately and corresponds to the degree of rock fracturing. It can be described in three ways: linear, areal and volumetric, depending upon whether the measurement/computation corresponds to length (1D), area (2D) or volume (3D) aspect, respectively.

1. Linear fracture density (1D fracture density, d_1) is the average number of fractures of a particular set, per unit length measured in a direction perpendicular to the fracture plane. It equals fracture frequency (F_f) and is the reciprocal of fracture spacing.
2. *Areal fracture density* (2D fracture density, d_2) is a way to quantify persistence of the discontinuity. It is the average fractured length (of traces) per unit area on a planar surface.
3. Volumetric fracture density (3D fracture density, d_3) is the average fractured surface area per unit rock volume, created by all the fractures of a given set.

All types of fracture densities, d_1, d_2, and d_3 have the same dimension (L^{-1}). The volumetric density (d_3) is independent of direction and is a static parameter, like porosity. On the other hand, areal and linear densities are directional parameters and have bearing on fluid flow.

Both d_1 and d_2 depend on the orientation of the fractures vis-à-vis that of the scanline/exposure face. However, d_3 is independent of direction and can be estimated from a survey with boreholes or scanlines, with the help of proper weighting of the observed fractures (Chiles and de Marsily 1993). For computing the correct weighting factors, consider first the case of a borehole or a scanline survey (Fig. 2.22). The surveyed straight line can be considered as a cylinder of length L with a small section p, as in the case of a borehole. If n fractures intersect the survey line and i_i is the acute angle made by the *ith* fracture plane with the borehole, then the fracture surface within the

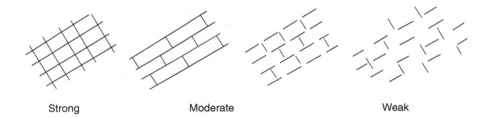

Fig. 2.21 Influence of persistence of discontinuity on the degree of fracturing and interconnectivity

Strong   Moderate   Weak

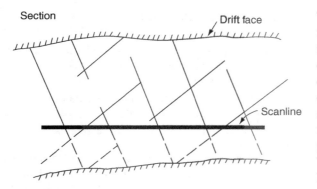

Fig. 2.22 Scanline method of discontinuity survey

cylinder is $p/\sin i_i$, for the *ith* fracture. Hence, 3D fracture density d_3 is:

$$d_3 = \frac{1}{L \cdot p} \sum_{i=1}^{n} \frac{p}{sin\,\theta_i} = \frac{1}{L} \sum_{i=1}^{n} \frac{1}{sin\,\theta_i} \qquad (2.3)$$

Thus, the weighting factor is related to the acute angle between the fracture plane and the scanline.

Similarly, considering the case of an areal survey, the exposure can be considered as a layer of area S and a small thickness e. Within the surveyed rectangle, if n fractures are traced on the exposure (Fig. 2.23), and *ith* fracture has a trace length l_i and makes an angle i_i with the exposure plane, then the fractured surface area for the *ith* fracture is $e \cdot l_i/\sin i_i$. Hence, 3D fracture density is:

$$d_3 = \frac{1}{S \cdot e} \left(\sum_{i=1}^{n} \frac{e \cdot l_i}{\sin\,\theta_i} \right) = \frac{1}{S} \left(\sum_{i=1}^{n} \frac{l_i}{\sin\,\theta_i} \right)$$
$$\qquad (2.4)$$

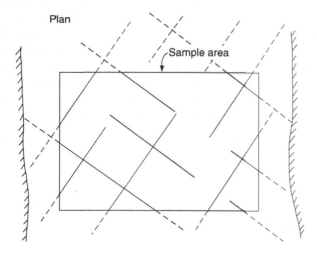

Fig. 2.23 Areal method of discontinuity survey

Thus, the weighting factor is related to acute angle and fracture trace length. If all the fractures have the same trace length, then l is constant. For parallel fractures, i_i can be replaced by i. For example, in an area in Bundelkhand granites (India), the 3D fracture density was computed using the scanline method at 64 observation sites. It is observed that d_3 in the area varies generally from about $6\,\mathrm{m}^{-1}$ to $21\,\mathrm{m}^{-1}$, whereas there are smaller pockets of higher values of d_3, of the order of $31\,\mathrm{m}^{-1}$. The variation in d_3 across the study area is shown in Fig. 2.24, where the magnitude of d_3 is plotted as a circle of appropriate radius.

With simplifications and assumptions, d_1, d_2 and d_3 can be interrelated; if fracture orientations are purely random, then (Chiles and de Marsily 1993):

$$d_1 = 1/2 \cdot d_3 \qquad (2.5)$$

$$d_2 = \pi/2 \cdot d_3 \qquad (2.6)$$

2.3.6 *Fracture Area and Shape*

Fracture area can be estimated from the strike trace length and dip trace length, assuming that the fractured surface has a certain regular shape, e.g. circular, square, elliptical, rectangular or polygonal. Out of these the case of circular discs is the simplest. Disc diameter D can be related to fracture surface area A as:

$$A = (\pi/4) \cdot \left(D^2 + S_D^2 \right) \qquad (2.7)$$

where S_D is the standard deviation of disc diameter distribution. Statistical aspects on the bearing of fracture shape on area estimation are discussed by a few workers (e.g. Lee and Farmer 1993). The 3D density of disc centres τ, average disc surface area A and the 3D fracturation density d_3 are interrelated as:

$$d_3 = \tau \times A \qquad (2.8)$$

2.3.7 *Volumetric Fracture Count*

Volumetric fracture count (V_f) is the total number of fractures per cubic meter (m^3) of rock volume and is determined from the mean fracture spacing as:

$$V_f = 1/f_{s1} + 1/f_{s2} + 1/f_{s3} \cdots + 1/f_{si} \qquad (2.9)$$

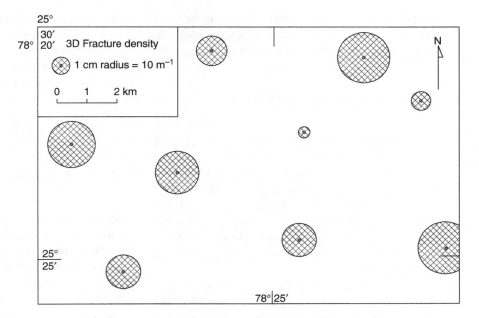

Fig. 2.24 Map showing variation in 3D fracture density in a part of Bundelkhand granites, Central India

where f_{si} is the mean fracture spacing of the ith fracture set in metres. This also carries the notion of fracture intensity which is defined as the number of discontinuities per unit length, measured along a line, area or volume. Volumetric fracture count has a direct bearing on the size of matrix blocks and the representative elementary volume (REV).

2.3.8   Matrix Block Unit

The rock block bounded by fracture network is called matrix block unit. Each matrix block unit can be considered to be hydrogeologically separated from the adjacent block. The shape of the matrix block unit could be prismatic, cubical or tabular, as governed by the orientation of fractures and their distribution (Fig. 2.25). For example, predominantly vertical fractures produce columnar and parellelopiped blocks (e.g. columnar joints in basalts); dominantly horizontal joints lead to plates and sheets (e.g. sheeting joints in granitoid rocks). These features impart hydraulic anisotropy to the geologic unit.

Consider an ideal case where beds are horizontal and fractures only vertical. It is known that fracture spacing and bed thickness are directly related (Eq. 2.1). It follows that a particular lithology has a tendency to develop block units of a certain shape, the block unit volume being dependent upon the bed thickness.

Block size is also related to the volumetric fracture count V_f. The maximum number of matrix blocks N_{bmax} can be expressed as (Kazi and Sen 1985):

$$N_{bmax} = \left(\frac{V_f}{3} + 1\right)^3 \qquad (2.10)$$

Fractal concepts are also used to define fragmented rocks. It is found that for fragmented materials including rocks, there is a size-frequency relationship of the form:

$$N(r) \propto \left(r^{-D}\right) \qquad (2.11)$$

where $N(r)$ is the number of fragments with a characteristic linear dimension greater than (r) and D is the

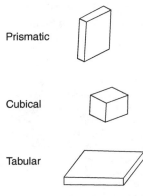

Fig. 2.25 Shape of matrix block units: prismatic, cubical and tabular

fractal dimension. It is believed that in future, fractal dimensions could be very useful in defining rock mass characteristics (e.g. Mojtabai et al. 1989; Ghosh 1990).

2.3.9 Fracture Connectivity

Discontinuities may exhibit differing termination and connectivity characteristics (Fig. 2.26a). Intersection of discontinuities is important as groundwater flow takes place through multiple fractures. Greater continuous inter-communication among the fracture network is provided by a higher degree of fracturing. Fracture connectivity increases with increasing fracture length and fracture density, as the chance of fracture intersection increases.

For evaluating connectivity it is necessary to study how the fractures terminate. Barton et al. (1987) classified fractures into three categories: abutting, crossing, and blind. The fractures of blind type do not intersect other fractures and remain unconnected. Laubach (1992) suggested that in many cases fracture connectivity may be gradual and that many fractures earlier classified as abutting, were really diffuse (interfingering type). He grouped fracture terminations into blind, diffuse and connected (which includes abutting). The data can be plotted in a ternary diagram to represent the bulk condition of fracture intersection in the rock mass (Fig. 2.26b). As an example, it is shown in the figure that most the joints in Bundelkhand granites (BG) are of connected type.

2.3.10 Rock Quality Designation (RQD)

RQD is a semi-quantitative measure of fracture density which can be estimated from core recovery data.

RQD is defined as the ratio of the recovered core more than 4 in. (about 10 cm) long and of good quality to the total drilled length and is expressed as a percentage. Although RQD is mainly used in assessing the geomechanical properties of rocks, it is also considered to be an important parameter in assessing relative permeability.

2.3.11 Aperture

Aperture is the perpendicular distance separating the adjacent rockwalls of an open discontinuity, in which the intervening space is air or water-filled. Aperture may vary from very tight to wide. Commonly, subsurface rock masses have small apertures. Tensile stress may lead to larger apertures or open fractures. Often shear fractures have much lower aperture values than the tensile fractures.

Aperture may increase by dissolution, erosion etc. particularly in the weathered zone. It may decrease with depth due to lithostatic pressure, and there fracture wall compression strength is an important parameter governing aperture as lithostatic pressure tends to close the fracture opening. Table 2.2 gives aperture ranges as usually classified in rock mechanics.

Fracture aperture can be measured by various methods which include feeler gauge, flourescent dyes, impression packer, tracer test, hydraulic test etc. Readers may refer to Indraratna and Rajnith (2001) for details of various methods used for measurement of fracture aperture. Often, measurement of aperture in surface exposures is made with a vernier caliper or gauge and the measured opening is termed as the mechanical aperture. In the laboratory, fracture aperture can be estimated by impregnating rock samples

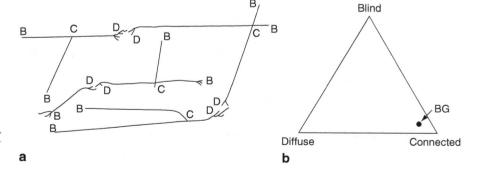

Fig. 2.26 Fracture connectivity. **a** Different types of fracture terminations: *B* blind; *C* crossing; *D* diffusely connected. **b** Ternary diagram of fracture terminations (After Laubach 1992); the point *BG* corresponds to data from Bundelkhand granites indicating high degree of fracture interconnectivity

Table 2.2 Aperture classification by size. (After Barton 1973)

Aperture (mm)	Term
<0.1	Very tight
0.1–0.25	Tight
0.25–0.50	Partly open
0.50–2.50	Open
2.50–10.0	Moderately wide
>10.0	Wide

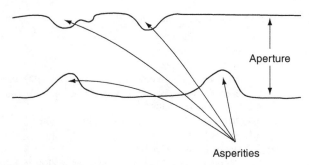

Fig. 2.27 Asperities in fracture walls

with dyes or resin and by studying the thin sections under the microscope. This method can even be used in soft sediments viz. clay till (Klint and Rosenbem 2001). Lerner and Stelle (2001) have suggested two field techniques to estimate the in-situ spatial variation of fracture aperture: one is the conventional slug hydraulic testing using packers and in the second technique NAPL (sunflower oil) is injected into isolated fractures in a borehole.

The term 'equivalent aperture' is introduced to account for the variation in fracture which can be estimated from tracer test and hydraulic tests. The terms 'tracer aperture' and 'hydraulic aperture' are introduced by Tsang (1999) depending on the method of estimation. The hydraulic aperture is estimated from hydraulic tests based on the Cubic Law:

$$T_f \propto a^3 \tag{2.12}$$

where a is the fracture aperture and T_f is the transmissivity of the formation (also see Sect. 7.2.1).

The data on discontinuity sets with corresponding apertures is to be recorded. Asperities affect the aperture size and also render its measurement difficult in field. Therefore, when considering fluid flow, apertures are defined in terms of flow properties, as volumetric flow rate is governed by the cube of aperture. Aperture can be integrated with fracture density to give an integrated function representative of hydraulic conductivity (see Sect. 7.2.5).

2.3.12   Asperity

Fracture walls are not flat parallel smooth surfaces but contain irregularities, called asperities (Fig. 2.27). The asperity reduces fluid flow and leads to a local channelling effect of preferential flow. This reduces the

effective porosity and makes flow velocities irregular. Observations on asperities should be made for each type of fracture surface and measurement made. Mean height of asperities together with Reynolds number Re has a direct influence on flow regime, i.e. laminar vs. turbulent flow (Sect. 7.1.1).

2.3.13   Wall Coatings and Infillings

It is the solid material occurring between the adjacent walls of a discontinuity, e.g. clay, fault gouge, breccia, chert, calcite, etc. Filling material could be homogeneous or heterogeneous, and could partly or completely fill the discontinuity. The material may have variable permeability, depending upon mineralogy, grain size, width etc. The net effect of wall coatings and infillings is a reduced aperture.

2.4   Methods of Field Investigations

Methods of field investigations can be classified into two broad types (Jouanna 1993): 2D and 3D (Table 2.3). The 2D methods are based on observations made at rock surface, at surface or subsurface levels. They include scanline surveys, borehole surveys, and different types of areal surveys (Fig. 2.28). These methods give an idea of the hydrogeological properties at and around the site of observation.

The 3D methods are aimed at gathering information on the bulk volumetric properties involving inner structure of the fractured rock mass. There can be direct or indirect 3D methods. Brief descriptions of the various 2D and 3D methods are given below.

Table 2.3 Methods of field investigations

1. 2D Methods—Based on rock surface observations on lithology, structure, fractures, and their characteristics; made at surface or subsurface levels
    1.1 Scanline surveys
    1.2 Areal surveys—on outcrops, pits, trenches, adits, drift etc. including terrestrial geophotogrammetry and remote sensing
    1.3 Borehole surveys—including drilling, study of oriented cores, borehole logging, dipmeter, borehole cameras and formation microscanner methods
2. 3D Method—Investigations aimed at bulk volumetric properties of rock mass in 3D
    2.1 Hydraulic well tests
    2.2 Hydrochemical methods
    2.3 Geophysical methods including seismic, electrical, EM, gravity, magnetic and georadar

2.4.1 Scanline Surveys

Scanline surveys involve direct observation of rock features along a line on the rock surface, e.g. on an outcrop, drift face, excavation, adit etc. (Fig. 2.22). Scanlines are usually horizontal; however, vertical scanlines are preferred where fractures are mostly horizontal. Data on fractures obtained by sampling techniques such as along scanline (and also borehole) are strongly biased towards the fractures oriented perpendicular to the scanline/core and needs to be corrected for sampling bias by applying correction (Terzaghi 1965).

A suitable rock exposure or face is selected. A sample scanline is marked on the face, and its orientation (rake on the face) is recorded. Fractures intersecting the line are collected. Each fracture is represented by its trace which can be measured. Observations are made for various parameters, like: location of the fracture trace intersection with the scanline; orientation of the fracture and angle made with the scanline; termination type if seen and connectivity; alternatively, whether the

fracture extends beyond the top of face/batter; fracture type and other relevant fracture characteristics.

2.4.2 Areal Surveys

Areal surveys can be treated as extension of the scanline surveys. They are used for surveying fracture characteristics on a rock surface area, e.g. on an outcrop, drift face, adit, tunnel etc. (Fig. 2.23). In field, an area is first demarcated on a rock surface for observation and statistical sampling. Detailed observations on fracture characteristics are made with-in the marked area where all the fractures data are collected.

Direct observations and field mapping at natural rock outcrops is an old established technique. Weathering, surficial cover, soil, vegetation etc. influence the accessibility and visibility of good outcrops. Excavations, pits and trenches are made to expose the fresh rocks at shallow depth for visual inspection. Subsurface direct observation can be made in adits and tun-

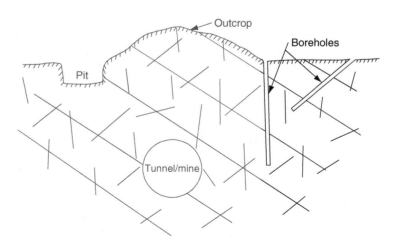

Fig. 2.28 The various 2D methods of field investigations

nels. Geological maps of rock faces exposed can be prepared and fracture characteristics measured.

Remote sensing includes study of photographs and images acquired from aerial and space platforms. This technique can give valuable information on geology, structure, fractures, lineaments etc. and forms an important mapping tool (Chap. 4). Further, stereophotographs of rocks exposed in outcrops, scarps, excavations, etc. can be taken from a ground-based (terrestrial) platform. These stereo pairs can be studied and measurements of fracture characteristics can be done in laboratory.

2.4.3 Borehole Surveys

These are the only tools for direct observations of rock surface and features occurring at depth. A number of methods are available. As drilling is expensive, optimum combination of methods is employed for getting maximum information from drilling. In the case of vertical and sub-vertical fractures, inclined bores are preferred to intercept a number of such fractures. Study of drill cores, particularly oriented drill cores provides data on orientation of structures, fractures, their apertures as well as infillings. Further, borehole walls can be studied in several ways. Geophysical well logging is a standard technique, including electrical, caliper, radioactivity, magnetic logging etc. These give information on lithology and structure. Borehole televiewer provides images of the borehole walls with joints and fractures. Dipmeter and formation microscanner help measure orientation of structural features at depth in-situ (for drilling and well observation techniques, see Chap. 5).

It may be mentioned here again that data on fractures obtained by sampling techniques such as along scanline and borehole are strongly biased towards the fractures oriented perpendicular to the core/scanline and needs to be corrected for sampling bias by applying a correction (Terzaghi 1965).

2.4.4 3D Methods

As mentioned above, 3D methods are aimed to provide information on bulk volumetric properties of the fractured rock mass. These methods include hydraulic well tests, hydrochemical methods and geophysical techniques. The hydraulic well tests comprise pumping tests of various configurations and types, and give bulk volumetric assessment. Slug tests will give a first hand dependable information about the hydraulic conductivity at much lower costs than pumping tests (see Chap. 9). In hydrochemical methods various types of geochemical tracer studies and solute transport studies are carried out for bulk volumetric hydrogeological characterization (see Sect. 10.3). Further, a number of geophysical methods are used such as seismic, electrical, EM, gravity, magnetic and georadar. They are briefly described in Chap. 5 from a hydrogeological investigation point of view.

Summary
Most rocks possess fractures, broadly termed as discontinuities here, which facilitate storage and movement of fluids through the medium The discontinuities may be formed by planar surfaces such as bedding plane, foliation, fractures, faults shear zones etc. The common systematic fractures are of three main genetic origins: extensional, shear and hybrid. Faults can cause truncation or repetition aquifers and may lead to formation of springs and interbasinal subsurface flow. Discontinuities are characterized in terms of a number of parameters such as orientation, spacing, persistence, fracture and shape, connectivity, aperture coatings etc., and these data may be collected from field surveys by scanline method or areal surveys or in borehole observations.

Further Reading

Billings MP (1972) Structural Geology. 3rd ed., Prentice Hall, New Delhi, 606 p.

Lee CH, Farmer I (1993) Fluid Flow in Discontinuous Rocks. Chapman and Hall, London, 169 p.

Marshak S, Mitra G (2006) Basic Methods of Structural Geology. 2nd ed., Prentice Hall, New Jersey, 446 p.

van Golf-Racht TD (1982) Fundamentals of Fractured Reservoir Engineering. Elsevier, Amsterdam, 710 p.

Hydrogeological Investigations

3

The purpose of most hydrogeological investigations is to locate potential sites for development of adequate quantity of reasonably good quality groundwater for a particular use: domestic, irrigation or industrial etc. The quantity and quality criteria would depend on the local needs and the socio-economic conditions of the people. Alternative possible sources of water supply, e.g. import of surface water or groundwater from adjoining areas may also be considered. Further, hydrogeological investigations in fractured rocks are also taken up for the selection of suitable sites for waste disposal including radioactive waste, tapping of geothermal power and in several other geotechnical problems, viz. tunnelling, mining, hill slope stability etc. These include geological, geomorphological, geohydrological studies given in this chapter. Besides, remote sensing and geophysical studies are described in Chaps. 4 and 5, hydraulic properties and methods of aquifer characterisation are discussed in Chaps. 8–9, and tracer techniques are described in Chap. 10.

3.1 Geological Investigations

Geological investigations include surface mapping of different lithological units and their structural and geomorphological features on a scale of about 1:20 000–1:200 000. Aerial photographs and satellite images are preferred as base maps for geological and geomorphological mapping.

In this text, from a hydrogeological point of view, the various rock types are broadly classified into four groups (Table 3.1). Their detailed hydrogeological characters are described in Chaps. 13–16.

In fractured rocks, special emphasis has to be provided for mapping of lineaments, fractures and other rock discontinuities (Chap. 2). Field mapping should also include information about the orientation and density of fractures, although their subsurface distribution may be different which can be deciphered from subsurface investigations (Chap. 5).

Data about the thickness and composition of the weathered zone (regolith) is particularly important especially in crystalline rocks (Chap. 13). In volcanic rocks, in addition to lineaments, attention should also be paid to the characters of individual flow units and interflow formations (Chap. 14). In carbonate rocks, mapping of various solution (karst) features and springs are of special importance (Chap. 15).

Based on exploratory drilling and well log data (Sect. 5.12), subsurface maps and sections, viz. fence diagrams, isopach maps, structural contour maps etc. are prepared to project the subsurface distribution and configuration of aquifers, aquitards and aquicludes (e.g. Walton 1970; Erdelyi and Galfi 1988). Besides, subsurface geophysical exploration and well log data can be used to extract valuable information on fracture characteristics.

3.2 Geomorphological Investigations

Geomorphological investigations include delineation and mapping of various landforms and drainage characteristics. These contribute significantly in deciphering areas of groundwater recharge and their potentiality for groundwater development. Geomorphic mapping can best be done from remote sensing data using satellite images and aerial photographs on suitable scales (say, e.g. 1:20 000–1:50 000) (Chap. 4). Preferably, a river basin should be taken as a unit for geomorpho-

B. B. S. Singhal, R. P. Gupta, *Applied Hydrogeology of Fractured Rocks*,
DOI 10.1007/978-90-481-8799-7_3, © Springer Science+Business Media B.V. 2010

Table 3.1 Hydrogeological classification of rocks

Rock group	Rock types	Mode of occurrence	Main features important for groundwater occurrence
Crystalline rocks	Non-volcanic igneous and metamorphic rocks, viz. granites, gneisses, schists, slates and phyllites, etc.	Large size massifs and plutons; regional metamorphic belts	Weathered horizon, fracture and lineaments
Volcanic rocks	Basalts, andesites and rhyolites	Lava flows at places interbedded with sedimentary beds	Fractures, vesicles and interflow sediments
Carbonate rocks	Limestones and dolomites	Mostly as chemical precipitates with varying admixtures of clastics in a layered sedimentary sequence	Fractures and solution cavities
Clastic rocks	Consolidated: sandstones and shales; unconsolidated: gravel, sand, clay etc.	Interbedded sedimentary sequence	Intergranular pore spaces and fractures

logical data analysis, though, it may be also realized that surface water basin may not always coincide with the groundwater basin.

3.2.1 Landforms

Landform is an end product of natural weathering. It depends on three main factors: (a) present and past climatic conditions, (b) rock types and their structural features, and (c) the time span involved in weathering.

Genetically, the landforms are divided into two groups: (1) erosional landforms, and (2) depositional landforms. Erosional landforms are typically associated with the resistant hard rock terrains. They comprise: (a) residual hills, (b) inselbergs, (c) pediments, (d) buried pediments with weathered basement, and (e) valley fills. These are described in Sect. 13.2. Depositional landforms are developed by depositional processes of various natural agencies, e.g. river, glacier and wind etc. The depositional landforms are typically made-up of unconsolidated sediments and may occur in the regional setting of hard rock terrains. Therefore, they may play an important role in the groundwater development for local needs (Chap. 16).

3.2.2 Drainage Characteristics

Qualitative and quantitative drainage characteristics of a basin provide an indirect clue to the hydrogeological characteristics of the area and therefore are useful in

the assessment of groundwater resources. The important characteristics are drainage pattern and drainage density. These are related with the lithology, structure and permeability of the bedrock.

Drainage is said to be internal when few drainage lines are seen on the surface and drainage appears to be mostly subsurface, e.g. commonly in limestone and gravel deposits (Fig. 3.1). External drainage is the one in which the surface drainage network is seen to be well developed. Drainage channel-ways are characterized by the typical serpentine-sinuous shape and pattern. Their manifestation depends upon dimensions of the channel, scale of study and also on physical condi-

Fig. 3.1 Internal drainage in limestone. (Aerial photograph courtesy of Aerofilms Ltd.)

Table 3.2 Common drainage patterns in fractured and other hard rock formations and their geological significance

Type	Description	Geological control
Dendritic	Irregular, branching of streams, resembling a tree	Homogeneous materials and crystalline rocks; horizontal beds; gentle regional slope
Rectangular	Streams having right-angled bands	Jointed/faulted rocks, e.g. sandstones, quartzites etc.
Rectilinear	Straight line pattern	Strong linear structural control
Angulate	Streams joining at acute angles	Joints/fractures at acute angles to each other
Parallel	Channels running nearly parallel to each other	Steep slopes; also in areas of parallel elongate landform
Trellis	Main streams running parallel, and minor tributaries joining the main streams nearly at right angles	Dipping or folded sedimentary or low-grade meta-sedimentary rocks; areas of parallel fractures
Radial	Streams originating from a central point or region	Volcanic cones, domes and residual erosional features
Annular	Ring-like pattern	Structural domes

tions of the channel, which may change from season to season, viz. due to flooding, drought, change in sediment load and vegetation cover etc.

Drainage pattern is the spatial arrangement of streams, and is, in general, very characteristic of rock structure and lithology. The drainage patterns can be studied and mapped on topographic maps, aerial photographs as well as satellite images, on scales as per requirements. Common types of drainage patterns, characteristic of fractured and other hard rock terrains are listed in Table 3.2 and illustrated in Fig. 3.2.

Among various other drainage characteristics of a basin, drainage density, D_d is an important parameter. Drainage density (D_d) is the ratio of total channel

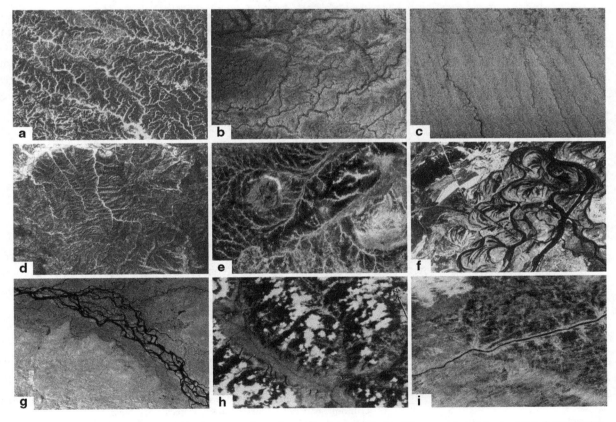

Fig. 3.2 Drainage patterns: **a** dendritic, **b** rectangular and angulate, **c** parallel, **d** trellis, **e** annular and sub-radial, **f** meandering, **g** braided, **h** barbed, **i** rectilinear. (Modified after Gupta 2003)

lengths of streams within a basin to the area of the basin. It has the dimensions of L^{-1} and can be expressed as

$$D_d = \sum L/A \qquad (3.1)$$

where D_d is drainage density in per kilometre, ΣL is the total channel length in kilometre, and A is the basin area, in square kilometre. The accuracy of measurement of the total lengths of all streams would depend on the scale of the map, as all the streams may not be shown on a map of small scale. Large scale aerial photographs therefore provide a better idea of drainage network for computing D_d.

Drainage density exhibits a very wide range of values in nature depending upon the relief, climate, resistance to erosion and permeability of rock material (Table 3.3). In general, low drainage density is characteristic of regions of highly resistant or highly permeable surface and low relief. High drainage density is found in regions of weak or impermeable subsurface materials, sparse vegetation and mountainous relief (Strahler 1964). Although low drainage density is considered to indicate regions of high permeability, but this criterion should be used with care as other factors like relief, climate and resistance to erosion may outweigh the influence of permeability of surficial material in a certain area. In areas of low relief, D_d may be more indicative of permeability of surface material and therefore could be used as a criterion for selection of suitable sites for shallow wells. Drainage density also influences runoff pattern and thereby infiltration capacity of the rock material. For example, high drainage density removes surface runoff rapidly, decreasing lag time and increasing the peak of the hydrograph. Among other drainage network parameters, bifurcation ratio, stream frequency and basin shape are also of hydrological interest. Readers may further refer to for example Strahler (1964), Gregory and Walling (1973) and Meijerink et al. (2007).

3.3 Geohydrological Investigations

These include groundwater level measurements, estimation of hydraulic properties of aquifers and assessment of groundwater quality. Study of springs including their location, variability of discharge, chemical composition and temperature also provide useful geohydrological information, especially in carbonate and other fractured rocks.

An observation well network is established for monitoring groundwater levels and groundwater quality. For estimating hydraulic characteristics of aquifers, pumping tests and tracer injection test are carried out. Tracer tests help in not only estimating hydraulic conductivity but also flow mechanism which is important from the point of view of evaluating contaminant transport. In fractured rocks, packer (Lugeon) test and slug test are preferred, at the first instance, as these do not require observation wells. Cross-hole hydraulic test and pumping test are necessary for detailed estimation of hydraulic parameters (Chap. 9). As pumping tests are expensive and require time and manpower, one has to be very selective in implementing such a programme. From groundwater development point of view, groundwater resources, both static and dynamic, are to be computed, and a management strategy for groundwater development has to be worked out (Chap. 20).

3.3.1 Water-level Measurements

3.3.1.1 Observation Wells—Design and Networks

The purpose of groundwater observation well network is to install observation wells or piezometers to monitor the water-levels and quality of groundwater. Keeping

Table 3.3 Drainage density (D_d) values for some terrains in USA. (After Strahler 1964)

D_d (km^{-1})	Geological terrain
1.9–2.5	Resistant sandstone strata of the Appalachian Plateau
5.0–10.0	Rocks of moderate resistance in humid central and eastern USA
12.5–19.0	Strongly fractured and deeply weathered igneous and metamorphic rocks in southern California under dry summer and sub-tropical conditions
125–250 and at places 688–812	In bad land developed on weak clays, barren of vegetation

Table 3.4 Types of observation well networks. (Modified after Heath 1976; UNESCO 1977)

Types of networks	Objectives	Products
Hydrogeologial	(a) Status of storage (b) Areal extent and interconnection of aquifers (c) Demarcation of recharge and discharge areas and groundwater flow regime	Regional water-table and/or potentiometric surface map showing water-level fluctuations over a selected period
Water management	(a) Effect of stresses on recharge and discharge conditions (b) Degree of confinement (c) Hydraulic characteristics of aquifers	(a) Local water-level maps (b) Hydrographs showing changes in water-level with time (c) Time-drawdown curves
Baseline	To determine effect of climate, topography and geology on groundwater storage	Hydrographs showing effect of climate, topography and geology
Water quality	(a) Water rock interaction (b) Contaminant transport	Water quality maps and sections
Special purpose	Effect of mining, irrigation and drainage projects	Water-level and quality maps

in view the regional and local requirements, the planning and design of such a network should depend on hydrogeological situations, purpose of investigations, stage of development, as well as political and social demands (UNESCO 1972, 1977). Observation wells can be drilled by using any conventional method of water-well drilling (see Chap. 17). In unconsolidated formations, wells are provided with screens tapping the zone of interest. In consolidated rocks, open-end wells can serve the purpose. The design of piezometers for measurement of water-levels should be different than those installed for collection of water samples for chemical analysis. The piezometers for water-level measurements should have small diameter to accommodate the water-level measuring device and without filter pack (to minimize the effect on groundwater flow) and an effective seal at the top of the screen to check the ingress of water from the surface (Houlihan and Botek 2007). The various types of observation networks are listed in Table 3.4.

The water-level fluctuations in the fractured zones and matrix are likely to be different due to differences in their hydraulic properties. Therefore water-level fluctuations in observation wells in fractured rocks should be interpreted with care.

The magnitude of water-level fluctuation also depends upon the relative location of observation well in the basin with respect to groundwater recharge/discharge zones. Wells located close to surface water divide will exhibit greater fluctuation of water-table than those in the midregions or near streams, due to rapid drainage from the upstream areas.

The purpose of hydrogeological network is to allow preparation of water-table or piezometric surface maps. Such maps provide information on the direction of groundwater movement, effect of stresses and changes in groundwater storage as well as areal extent of aquifers and nature of boundaries. In heterogeneous formations, it is necessary to construct a group of observation wells (well nests) tapping different aquifer zones, in order to determine the possible aquifer interconnections and comparison of water-levels in various aquifers (Fig. 3.3). In fractured rocks, the observation wells should be put in relation to rock discontinuities tapping different fracture sets for assessing extent of interconnection and the role of different fractures as conduits for groundwater flow. Lineaments and intersection of vertical fractures are potential sites for monitoring wells. In situations where the matrix blocks have appreciable porosity, it may be necessary to monitor the blocks as well as the high permeability zones (NRC 1996).

The objective of water management network is to obtain information about the effect of recharge and withdrawal on groundwater system. Such wells should be located in proximity to pumping well fields in clusters tapping the pumped aquifer as well as the overlying and underlying aquifers. In order to determine the extent of interconnection, separate observation wells should be installed in unconfined and confined aquifers (Fig. 3.3).

The baseline network consists of observation wells located in areas which are not significantly affected by withdrawals or other man-made stresses. The purpose

Fig. 3.3 Network of observation wells tapping different aquifers in the vicinity of a pumping well

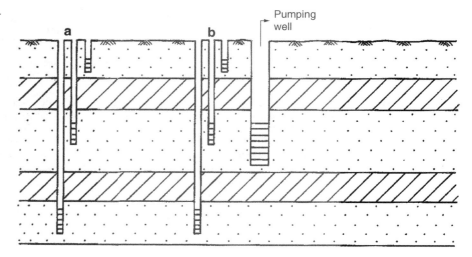

of this network is to determine effect of climate, geology and topography on groundwater levels.

The water quality network is used to determine changes in water quality due to rock-aquifer interaction in the direction of groundwater flow and also with depth, and the effect of man-induced contaminants on water quality. Water quality characteristics can be extremely variable in fractured rocks and karst aquifers. This is particularly true during and after recharge. In order to study the effect of contamination from a point source, at least one observation well should be located upstream from the possible source of pollution, in order to know the background quality. The other observation wells are installed in the down-gradient side from the source of contamination at suitable distances, depending on the rock permeability. In fractured rocks, the monitoring well network should be designed considering the fracture network to enable initial detection of the contaminant (Meyer et al. 1994). In addition to observation wells, data from springs and caves etc. can also provide useful information.

The special purpose network is designed to determine effect of mining and other engineering structures like dams, canals and drainage channels on groundwater levels. In case of open cast mines, the observation well network is put around the periphery of the excavation. In order to determine the effect of rivers and canals on groundwater regime, the observation wells should be put in line across the surface water bodies. If an area is bounded on both the sides by rivers or canals, the observation wells should be located not only near these hydraulic boundaries, but also in the interfluve area.

3.3.1.2 Timing and Frequency of Measurements

The frequency of waterlevel measurements in wells and discharge from springs depends on the nature of aquifer and the purpose of study, the broad guidelines being as follows:

Type of aquifer	Frequency of measurement
Confined aquifer	One per month
Unconfined aquifer	Three per month
Wells adjacent to rivers	Daily during flood times
Snowmelt areas	Daily
Irrigated areas	Related to time of irrigation

In drainage studies and also in the study of effect of external forces, viz. changes in barometric pressure, ocean tides, earthquakes, etc., continuous recording of water-level using automatic water-level recorders is necessary.

Spring discharge measurements are of importance for water resources assessment, especially in fractured and karstic rocks. Plots of spring discharge (or stage) and water quality in relation to time, known as spring hydrograph and chemograph respectively, have been used extensively in karst aquifer studies (see Sects. 15.3 and 15.6). Spring discharge in the range of 0.005–$0.01 \, m^3 s^{-1}$ is measured by volumetric methods while higher discharges could be measured using weirs. Sampling frequency would depend upon the trend of variability of the parameter. Samples should be collected through at least one major recharge event, as such an event is likely to cause rapid changes in the spring discharge, hydraulic head in wells and also in water quality. Samples must also be collected during base-flow (lean period).

3.4 Hydrogeological Maps

Data obtained from field and laboratory studies is plotted on hydrogeological maps and sections. It is necessary to know the purpose of preparing such maps to decide the type of information to be provided and the manner of its presentation. Unlike geological maps, hydrogeological/hydrological maps deal with essentially transient phenomena, viz. groundwater resources, groundwater levels, water quality etc. Therefore, hydrogeological maps should also indicate the dates of surveys, reliability of data and methods of survey.

3.4.1 Regional Maps

Hydrogeological maps depict a number of parameters including the geology of the area, major rock discontinuities, extent of aquifers, hydraulic characteristics, contours of water-table and piezometric surface, water quality, surface water bodies and important meteorological characteristics. Hydrogeological maps may be international, national, regional or local. They usually vary from 1:1 000 000 to 1:250 000 in scale. Considerable variation may be there in the amount of information included in different maps depending on the chosen scale, size of the area and the purpose of the map. For example, a hydrogeological map of the Czech Republic was prepared on the scale 1:1 000 000, showing variations in transmissivities and yield characteristics (Krasny 1996). Further, Russian hydrogeologists (Zektser and Dzhamalov 1988) compiled a World Groundwater Flow map on a 1:100 000 000 scale depicting data of specific groundwater discharge.

For geothermal studies, geothermal resource maps are prepared showing temperature and pressure distribution of a hydrogeological unit which are of practical importance for development of geothermal power.

In the earlier years, different workers and organisation used differing legends and colour codes to depict various groundwater features. This created difficulties in proper interpretation, comparison and compilation of data. To overcome this problem, a common legend including colour code for hydrogeological maps has been suggested for international adoption (UNESCO 1983; Castany et al. 1985). Hydrogeological maps of several countries have been published using the above code (Castany et al. 1985; Verma and Jolly 1992; Struckmeier 1993).

One of the earliest attempts to prepare hydrogeological map of India was made by Taylor (1959). Subsequently, the Geological Survey of India published a geohydrological map of the country on scale 1:2 000 000 (GSI 1969). In 1976, the Central Ground Water Board, compiled and published a hydrogeological map of India on 1:5 000 000 scale which was updated in 1989. The revised map (CGWB 1989) depicts major lithological units, their groundwater potential in terms of well yields and groundwater quality. A map of India showing major groundwater provinces is given in Fig. 20.3.

3.4.2 Map Media—Hard Copy and Electronic

Hard copy maps use paper, photographic film or other similar materials for map record and display in colour or black-and-white. Most of the hydrogeological maps are available as hard copy prints throughout the world. However, with the transformation in information technology, recently, computer coded maps in electronic media are now becoming available in some countries. These maps may use raster or vector format in GIS, and may be restructured and re-formatted as per requirements. Further, such maps can depict different hydrogeological parameters and also use the standard internationally accepted legend or colour scheme.

The computer coded maps have the advantage of digital compilation and processing in the GIS environment (Chap. 6). Further, the original resolution of data may be retained in such maps in comparison to the hard copy medium where the resolution gets generally reduced. These maps and data may be available in electronic media such as cartridges, tapes and diskettes for transfer to other systems where they could be put as monitor (soft copy) for display, processing, editing etc. With the developments in information super-highways, it is not far that hydrogeological data on public domain may be available on networks like internet/world wide web etc. facilitating world wide super fast accessibility.

3.4.3 Special Purpose Maps

Special thematic maps can also be prepared giving certain specific information, e.g. water-table maps, isopiestic maps, water quality maps and other such parameters as spatial distribution of permeability, transmissivity, specific yield, well yields, and specific capacity etc.

3.4.3.1 Groundwater Level Maps

Based on water-level measurements, various types of contour maps, e.g. water-table maps, piezometric surface maps, depth to water-level and water-level fluctuation maps can be prepared. These maps depict the spatial and temporal variations of the groundwater resources of an area. Water-table maps represent the position of water-level in unconfined aquifer while piezometric surface represents the pressure conditions in a single or a group of confined aquifers. The construction of such maps assumes that the vertical hydraulic gradients are not significant and that the well intersects enough fractures. Measurements of water-levels over an area should be completed within

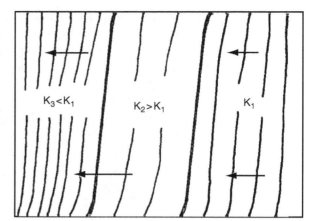

Fig. 3.4 Effect of hydraulic conductivity (K) on the spacing of water table contours. K_1, K_2 and K_3 are the hydraulic conductivities of the three layers from right to left

a short time interval. The contours represent the lines of equal potential and therefore flow lines can be drawn at right angles to these contours. The spacing of the contours depends upon flow rate, aquifer thickness and hydraulic conductivity (Fig. 3.4). The shape of contours and their values indicate recharge and discharge areas. The upward and downward swing of water-table contours along a stream would indicate effluent and influent condition respectively. Groundwater mounds represent areas of recharge while groundwater depressions are developed due to excessive withdrawal (Fig. 3.5). In unconfined aquifers, the water-table is regarded to be a subdued replica of the ground topography, though this may not be necessarily valid in all the situations.

Care should be taken in the interpolation of water-table contours across the hydrogeological boundaries as illustrated in Figs. 3.6 and 3.7. This is even of greater importance when the observation points are

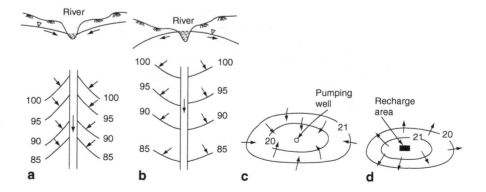

Fig. 3.5 Features exhibited by water table contour maps: **a** effluent seepage, **b** influent seepage, **c** groundwater depression, **d** groundwater mound

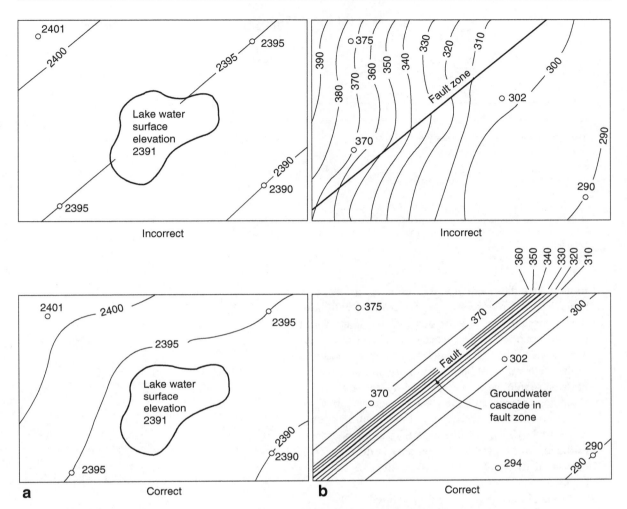

Fig. 3.6 Contouring water-table maps in areas of **a** topographic depressions occupied by lakes, and **b** fault zones. Incorrect and correct contour maps are shown. (After Davis and DeWiest 1966)

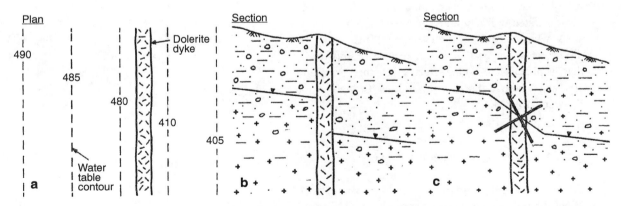

Fig. 3.7 Effect of dyke leading to groundwater compartments, inhibiting linear interpolation of water-table contours. **a** Plan, **b** and **c** are correct and incorrect interpolations respectively

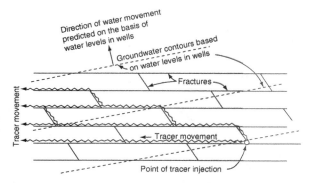

Fig. 3.8 Comparison of groundwater flow direction inferred from water-table contours with the actual flow direction determined from tracer studies in a fractured rock. (After Davis and DeWiest 1966)

limited and these are not uniformly distributed. In sparsely fractured rocks, the water-table will have the "stair-step" appearance due to large contrasts in hydraulic conductivity between matrix blocks and fractured zones. In these settings, water-levels measured in the wells may not be in continuity and the concept of "water-table" becomes less clearly defined. Therefore, maps prepared on the basis of such data may not give correct idea of groundwater flow. Figure 3.8 gives an example where the direction of groundwater flow estimated from water-table contours is different than the actual direction of flow determined from tracer studies.

Mountainous regions with fractured rocks pose special problems in the preparation and interpretation of water-table maps due to varying infiltration rates depending upon topography and intensity of fractures. The influence of topography, subsurface lithology, permeability and infiltration on piezometric surface is well demonstrated in fractured basalts region of Yucca Mt., USA, where in areas of high relief and fresh outcrops, the groundwater gradients are steep, as compared to that in plateau areas characterized by thicker weathered zones with higher transmissivities (Domenico and Schwartz 1998).

3.4.3.2 Depth to Water-level Maps

These provide information only about the spatial variation of depth to water-level, from the ground surface and therefore can not be used for determining direction of groundwater flow.

3.4.3.3 Water-level Fluctuation Maps

These maps indicate the change of water-level in wells during a given period of time. Such maps indicate rates of recharge and discharge to aquifers and changes in storage of water within the aquifer and aquitards. The reasons for such fluctuations are described in Sect. 20.4. Groundwater level trends can also be depicted by hydrographs where variation in groundwater level is plotted against time of observation (see Fig. 20.3). These are of importance in groundwater management studies.

Water-level fluctuation in granular aquifers are of smaller magnitude compared to fractured media as fractures and other conduits respond quickly to recharge events. Therefore, moderately fractured and karst aquifers may show complex response to recharge events (see Chap. 20).

Summary

Hydrogeological studies include lithological, structural, geomorphological and hydrological studies. This should be supported with surface and subsurface geophysical investigations. In hard rocks, fractures studies form and important components of field investigations. Geomorphological investigations include landform and drainage characteristics which in turn depend on the lithology and rock structures.

Groundwater level measurements in time and space form an important component of hydrogeological investigations which requires careful planning of the observation well network and also the time and frequency of water-level measurements. Interaction of surface water and groundwater should also be assessed. The groundwater level data are expressed as water-table maps or sections. The interpretation water level data needs proper integration with geological, structural and geomorphological information.

Further Reading

Davis SN, DeWiest RJM (1966) Hydrogeology. John Wiley & Sons, Inc., New York, NY.

Walton WC (1970) Groundwater Resource Evaluation. McGraw-Hill, New York, NY.

Remote Sensing

4

4.1 Introduction

Remote sensing, encompassing the study of satellite data and aerial photographs, is an extremely powerful technique for earth resources exploration, mapping and management. It involves measurements of electromagnetic (EM) radiation in the wavelength range of about 0.4 μm–1 m, from sensors flying on aerial or space platforms to characterize and infer properties of the terrain. Remote sensing has evolved primarily from the methods of aerial photography and photointerpretation used extensively in 1950s–1960s. The technique has grown rapidly during the last four to five decades. In the context of groundwater studies, remote sensing is of great value as a very first reconnaissance tool, the usual sequence of investigations being: satellite images—aerial photographs—geophysical survey—drilling. Geological interpretation derived from aerial photos/remote sensing has been extensively used for the purpose of identification of lineaments or fracture zones along which flow of groundwater may take place and for landform investigations suitable for groundwater prospecting particularly in hard rocks (e.g., Mabee and Hardcastle 1997; Moore et al. 2002; Salama et al. 1994).

Fundamental Principle The basic principle involved in remote sensing is that each object, depending upon its physical characteristics, reflects, emits and absorbs varying intensities of radiation at different EM wavelength ranges. The curves depicting relative intensity of light reflected/absorbed/emitted by the objects at different wavelengths are called spectral response curves (Fig. 4.1). Using information from one or more wavelength ranges, it is possible to discriminate between different types of ground objects, e.g. water, dry soil, wet soil, vegetation, etc., and map their distribution on the ground. A generalized schematic of energy data/flow in a typical remote sensing programme is shown in Fig. 4.2.

Advantages and Limitations The chief advantages of remote sensing technique over other methods of data collection are due to the following:

1. Synoptic overview: remote sensing permits delineation of regional features and trends.
2. Feasibility: in some inaccessible areas, remote sensing may be the only way to get the information.
3. Time saving: The technique saves time and manpower as information about a large area is quickly gathered.
4. Multidisciplinary applications: the same remote sensing data can be used by workers in different disciplines of natural sciences.

Though remote sensing provides direct observations and inputs to all other components of hydrological cycle, viz. oceanic, atmospheric and surface water, there are limitations in remote sensing applications to groundwater, chiefly arising from the fact that the EM radiation have a limited depth of penetration—say fraction of a millimeter in the visible range to a couple of meters (in dry desert conditions) at the most in the microwave region. Therefore, interpretations on subsurface hydrogeology have to be based on indirect surface evidences and features such as landform, lithology, structure, vegetation, soil, drainage, land use, surface anomalies etc. (see Sect. 4.7).

The implementation of satellite remote sensing techniques to hydrogeological problems in developing countries has faced constraints in the past due to several factors such as high costs of satellite sensor data, expensive software tools, and lack of proper expertise/training etc. (Jha and Chowdary 2007). However,

B. B. S. Singhal, R. P. Gupta, *Applied Hydrogeology of Fractured Rocks*,
DOI 10.1007/978-90-481-8799-7_4, © Springer Science+Business Media B.V. 2010

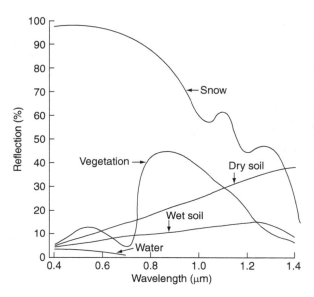

Fig. 4.1 Spectral response curves of selected objects

the above issues now appear to be resolved, at least in part, by the availability of free downloadable data (e.g. Landsat TM and ETM), free software packages (e.g. ILWIS), and several tutorials available on the web.

4.2 Physical Principles

EM Spectrum and Sources of Radiation Electromagnetic spectrum is the ordering of EM radiation according to wavelength, or frequency, or energy. The nomenclature used for different parts of the EM spectrum is shown in Fig. 4.3c.

For remote sensing purposes, the sensors may utilize either naturally available radiation from the Sun or the Earth or artificial radiation. The technique involving artificial illumination is called active, in contrast to the one utilizing naturally available radiation, which is termed passive. Solar reflected radiation dominates in the ultraviolet—visible—near-infrared parts of the spectrum (Fig. 4.3a), which, therefore, is called solar reflection region. The Earth-emitted radiation dominates in the 3–20 μm wavelength region, and this spectral region is therefore called thermal IR. Besides, artificial illumination using radar is frequently used in the microwave region.

Atmospheric Interactions The EM radiation while passing through the Earth's atmosphere interacts with atmospheric constituents and is selectively scattered, absorbed and transmitted. Raleigh scattering is the most important type of scattering and leads to haze and low-contrast pictures in the UV-blue parts of the EM spectrum. Further, selective absorption of the EM radiation takes place by atmospheric gases such as H_2O-vapour, CO_2 and O_3 etc. The spectral regions of least atmospheric absorption are called atmospheric windows, as these can be advantageously used for looking at the Earth's surface from aerial/space platforms. Major atmospheric windows occur in the visible (VIS), near-IR (NIR), shortwave-IR (SWIR), thermal IR (TIR), and microwave regions (Fig. 4.3b).

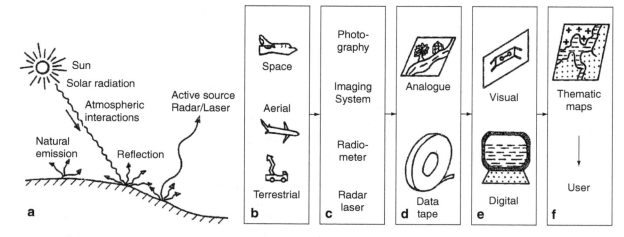

Fig. 4.2 Schematic data flow in a typical remote sensing programme

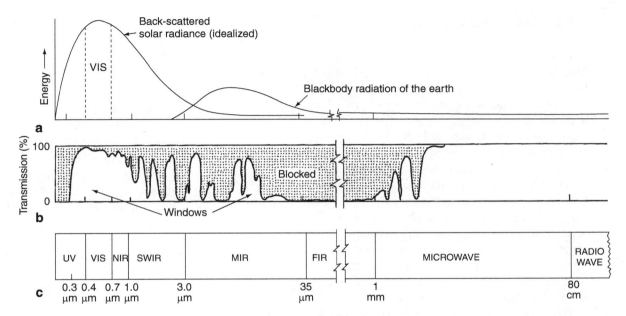

Fig. 4.3 a Backscattered solar radiation (idealised) and black body radiation of the earth, **b** atmospheric windows, **c** EM spectrum and nomenclature

Platforms Sensors for EM radiation can be operated from airborne, spaceborne or ground based platforms (Fig. 4.2b). Some of the typical airborne platforms are helicopters, aircrafts and balloons, whereas satellites, rockets and space shuttles constitute the space platforms. The hydraulic platforms and hand-held field instruments are used for collecting ground truth, to help better interpret aerial and space remote sensing data.

4.3 Remote Sensors

A sensor is a device to measure radiation intensity. It comprises optical components and detectors. A wide variety of sensors operate in various parts of the EM spectrum. For a comprehensive discussion on remote sensors, the reader is referred to other texts (Sabins 1997; Lillesand et al. 2007; Gupta 2003). Keeping in mind the operational aspects for groundwater investigations in fractured rocks, we consider here four main types of remote sensors: (a) photographic systems, (b) line scanning systems, (c) digital cameras, and (d) imaging radar systems.

4.3.1 Photographic Systems

Perhaps, the most familiar remote sensing systems have been photographic. Main advantages of photographic systems have been good geometric accuracy and high resolution, whereas chief limitations include logistics related to the retrieval of films from spacecrafts and limited spectral range of operation. A standard film is panchromatic black-and-white and gives output in shades of gray. Other films are colour out of which colour infrared film (CIR) is most important for resources investigations. Filters permit sensing in selected wavelengths of light and form an integral part of modern photographic system.

The photographic systems can be deployed from space, aerial or terrestrial platforms. From aerial platforms, the vertical photography with an overlap of 70–75% for stereo viewing has been a standard technique in remote sensing (e.g. Wolf 1983) and its unique advantage of stereo viewing with good geometric accuracy has not been taken-over by any spaceborne remote sensing technique till date. Aerial photographs normally at of 1:20 000–1:50 000 scales are used, the larger scale photography being ideally employed for detailed site selections.

4.3.2   Line Scanning Systems

Multispectral line scanning systems have been the most important device for remote sensing from satellite platforms. A scanner is a non-photographic device and generates digital data on intensity of ground radiance. The entire scene is considered to be comprised of a large number of smaller cells. Radiation from each unit cell is collected and integrated by the scanning system, to yield a brightness value, or digital number (DN), which is ascribed to the unit cell (Fig. 4.4). This process is called scanning. There are several advantages of scanners over photographic techniques related to retrieval of data and sensing in extended wavelength range with high spectral, spatial and radiometric resolutions. Further, the digital information from scanners is amenable to computer processing enhancement etc., and for integrated interpretation in GIS.

Resolution of a sensor is an important parameter and is given in terms of spatial, spectral, radiometric and temporal aspects. Spatial resolution implies the ground resolution and corresponds to a pixel on the image. Spectral resolution means the span of wave-length range over which a spectral channel operates. Radiometric resolution generally implies the total number of quantization levels used in the sensor. Temporal resolution refers to the repetivity of observations over an area.

Types of Line Scanners There are two basic types of line scanners: 1. opto-mechanical (OM) line scanners and 2. charge-coupled device (CCD) line scanners. The OM line scanners have been extensively used from aerial and space platforms in multispectral mode at a wide range of wavelength from visible, near-IR, SWIR to thermal-IR. In this, the collector optics includes a moving plane mirror to reflect radiation from the ground which is collected and directed on-to the filter and detector assembly for quantization. The MSS, TM and ETM+ on Landsats are typical OM line scanners. The CCD line scanners are solid-state scanners and utilize photoconductors as the detector material. In a CCD line scanning system, a linear array of CCDs comprising several thousand detector elements is placed at the focal plane of a camera system to sense the radiation and convert the same into electrical signal. Typically, for each spectral band there is one CCD line array viewing the entire swath, simultaneously. Thus, there is no moving part in the sensor. The CCD line scanning system has been used in many earth-observation satellites sensors, such as SPOT-HRV, IRS-LISS, JERS-OPS etc.

4.3.3   Digital Cameras

Digital imaging camera constitutes a rather recent development in remote sensing technology and provides high-definition images at reduced costs. A digital camera employs a solid-state area-array (using CCDs or CMOS) with millions of photo-sensitive sites, which produces a two-dimensional digital image. Thus, the basic difference between a film-based photographic camera and a digital imaging camera is that in the latter the film is replaced by solid-state electronics. The incident light is focused by the optical device on the photo-sensitive area array that generates a signal which is quantized and recorded. The main advantages of digital imaging cameras arise from their direct digital output, fast processing, higher sensitivity,

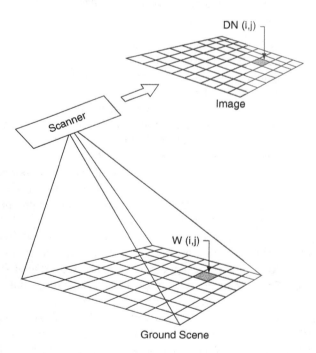

Fig. 4.4 Working principle of a scanner. The ground radiance $W_{(ij)}$ from a ground cell is converted into a digital number $DN_{(ij)}$ at the corresponding position in the image

better image radiometry, higher geometric fidelity and lower costs. A major limitation is imposed by detector technology that limits its usability in the wavelength range from visible to near-IR, presently. The development and growth of digital imaging technology by using area-arrays has been very fast. Space sensors using this technology include all the modern high-resolution satellite sensors such as IKONOS, QuickBird, CARTOSAT, GEOEYE etc.

4.3.4 Imaging Radar System

The electromagnetic spectrum range 1 mm–100 cm is commonly designated as microwave. The atmosphere is nearly transparent at these wavelengths. Basically, the microwave technique is of two types: (a) passive, and (b) active. The passive microwave radiometry has a major limitation of coarse spatial resolution (about 10–20 km^2). We therefore, in the context of groundwater applications, restrict this discussion to active microwave remote sensing, also called imaging radar.

Radar (Radio Detection and Ranging) basically operates on the principle that artificially generated microwaves sent in a particular direction collide with objects and are scattered. The back-scattered radiation is received, amplified and analyzed to determine location, electrical properties and surface configuration of the objects. The radar has all-time and all-weather capability, independent of solar illumination and atmospheric-meteorologic factors.

For imaging purpose, the radar is mounted in the configuration of Side-Looking Airborne Radar (SLAR), at the base of the sensor platform (Fig. 4.5). The radar transmits short pulses of microwave EM energy, illuminating narrow strips of ground, perpendicular to the flight direction. The radiation back-scattered from the ground is recorded to give information about the objects on the ground. As the sensorcraft moves forward, a two-dimensional image is generated.

Two basic types of imaging radars have been used: (a) real aperture and (b) synthetic aperture (e.g. Trevett 1986). Synthetic aperture radar (SAR) utilizes advanced data processing algorithms to yield higher spatial resolution and is employed in modern remote sensing programmes. Commonly, the SAR systems

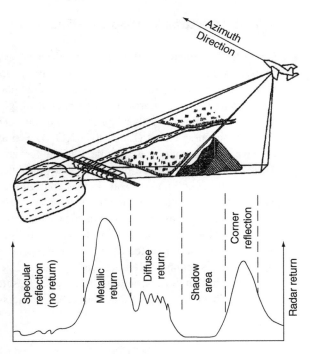

Fig. 4.5 Principle of side-looking airborne radar and typical radar returns. (Modified after Sabins 1997)

have a ground resolution of the order of 5–30 m, depending upon the altitude and sophistication in data processing. A number of SAR experiments have been conducted from aircrafts and spacecrafts. Important spaceborne SAR missions have been SEASAT, ERS-1/2, JERS (FUYO)-1, RADARSAT, ENVISAT-1 and DAICHI (ALOS).

Geometry of SAR imagery is fundamentally different from that of photographs and scanner images because of the oblique viewing configuration of the imaging radar system. Several serious geometric distortions arise due to image displacement, particularly in mountainous areas. Further, shadows and look-direction effects are also prominently manifested on SAR images.

The backscattered signal received at the SAR antenna is called "radar return" and carries information about the properties of the object. It depends upon numerous factors, such as radar wavelength, EM beam polarization, local incidence angle, target surface roughness and complex dielectric constant. Thus, it has to be kept in mind during interpretation that the radar response is governed by a complex interplay of radar system and ground terrain factors.

4.4 Important Spaceborne Sensors

LANDSAT Program Till date, the Landsat program of NASA has provided the most extensively used remote sensing data, the world over. Its chief plank has been in delivering unrestricted global data of good geometric accuracy at rather low costs. Starting 1972, till date, seven satellites (Landsat-1, 2, 3, 4, 5, 6 and 7) have been launched under the Landsat program. These satellites have been placed in near-polar, near-circular, sun-synchronous orbit. In this configuration, as the satellite orbits in the nearly north–south plane, the Earth below spins around its axis, from west to east (Fig. 4.6). Thus, different parts of the globe are 'seen' by the satellite during different north–south passes and entire globe is scanned. Remote sensing data are generally acquired in the descending node, i.e. as the satellite moves from the north to the south. The Landsats-1,

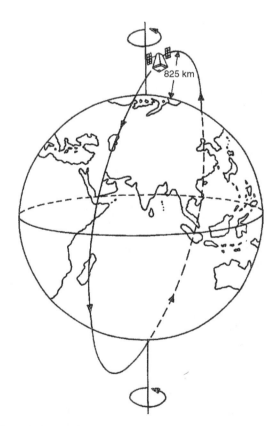

Fig. 4.6 Typical near-polar orbit of an earth observation satellite such as Landsat, IRS etc. As the satellite revolves in N–S orbit, the earth spins from W to E; therefore different parts of the earth are observed by the satellite in different N–S passes

2, 3 were placed at an altitude of 918 km with a repeat cycle of 18 days, and L-4, 5, 6 and 7 at an altitude of 705 km with a repeat cycle of 16 days.

For remote sensing, the Landsats have used OM-line scanners producing ground scenes of nominally 185 × 185 km size. Initially, the Multispectral Scanning System (MSS) was used which provided excellent multispectral data in four spectral bands and made the remote sensing experiment a tremendous success. The Thematic Mapper (TM) was an improved OM scanner yielding data in seven spectral bands with higher spectral, spatial and radiometric resolutions. Enhanced Thematic Mapper Plus (ETM+) being used on Landsat 7 is a further improved version yielding data in eight spectral bands (Table 4.1).

SPOT Program Under the French remote sensing program, a series of five satellites (SPOT-1, 2, 3, 4, and 5) have been launched since 1986, all being in near-polar sun-synchronous orbit. The SPOT program has used CCD linear array devices for remote sensing. The first sensor was called HRV (High Resolution Visible) used in SPOT-1, 2, 3 and yielded data in panchromatic and multispectral modes. Further, it could be tilted to acquire data in off-nadir viewing mode enabling more frequent repetitive coverage and stereoscopy. The HRV sensor was modified as HRVIR sensor on SPOT-4 improving on resolution and more number of spectral bands. The latest HRG (high resolution geometrical) sensor is a further improved version being used on SPOT-5. It provides high spatial resolution in panchromatic band and four multispectral bands in green, red, near-IR and SWIR bands (Table 4.1).

IRS Program Commencing 1988, the Indian Space Research Organisation (ISRO) has launched a series of land observation satellites, naming the series as Indian Remote Sensing Satellite (IRS) program. All these satellites have been placed in near-polar, sun-synchronous orbit. The sensors used have been typically CCD linear arrays to yield data in multispectral mode and called Linear Imaging Self Scanning (LISS-I, -II, -III, and -IV). LISS-I and LISS-II were used on initial IRS-1A and -1B platforms to yield data in blue, green, red, and near-IR spectral ranges. LISS-III is an improved version used on IRS-1C and -1D and provides data in green, red, near-IR and SWIR bands. LISS-IV is a multispectral sensor on RESOURCESAT platform and provides relatively good spatial resolution (5.8 m)

Table 4.1 Comparison of spatial resolutions of selected spaceborne remote sensors (all data in m)

		Panchro-matic	Blue	Green	Red	Near IR	SWIR I	SWIR II	Thermal-IR
LANDSAT	MSS*	–	–	79×79	79×79	79×79	–	–	–
	TM*	–	30×30	30×30	30×30	30×30	30×30	30×30	120×120
	ETM+	15×15	30×30	30×30	30×30	30×30	30×30	30×30	60×60
IRS	LISS I*	–	72×72	72×72	72×72	72×72	–	–	–
	LISS II*	–	36×36	36×36	36×36	36×36	–	–	–
	LISS III	–	–	23×23	23×23	23×23	70×70	–	–
	Pan	5.8×5.8	–	–	–	–	–	–	–
	WiFS	–	–	–	188×188	188×188	–	–	–
	LISS IV	–	–	5.8×5.8	5.8×5.8	5.8×5.8	–	–	–
CARTOSAT-1		2.5×2.5	–	–	–	–	–	–	–
CARTOSAT-2A		0.8×0.8	–	–	–	–	–	–	–
SPOT	HRV-Pan*	10×10	–	–	–	–	–	–	–
	Multi*	–	–	20×20	20×20	20×20	–	–	–
	HRVIR-HR*	–	–	–	10×10	–	–	–	–
	Multi*	–	–	20×20	20×20	20×20	20×20	–	–
	HRG	2.5×5	–	10×10	10×10	10×10	20×20	–	–
JERS-OPS*		–	–	18×24	18×24	18×24	18×24	18×24	
DAICHI (ALOS)									
	PRISM	2.5×2.5	–	–	–	–	–	–	–
	AVNIR	–	10×10	10×10	10×10	10×10	–	–	–
CBERS		20×20	20×20	20×20	20×20	20×20	–	–	–
TERRA	ASTER	–	–	15×15	15×15	15×15	30×30	30×30	90×90
IKONOS	Pan	1×1	–	–	–	–	–	–	–
	Multi	–	4×4	4×4	4×4	4×4	–	–	–
QuickBird-2	Pan	0.6×0.6	–	–	–	–	–	–	–
	Multi	–	2.5×2.5	2.5×2.5	2.5×2.5	2.5×2.5	–	–	–
FORMOSAT		2×2	8×8	8×8	8×8	8×8	–	–	–
KOMPSAT		1×1	4×4	4×4	4×4	4×4	–	–	–
WorldView-1		0.55×0.55	–	–	–	–	–	–	–
GEOEYE-1	Pan	0.4×0.4	–	–	–	–	–	–	–
	Multi	–	–	1.65×1.65	1.65×1.65	1.65×1.65	–	–	–

*Archive data may be available from these sensors

data in green, red and near-IR multispectral bands. A high resolution panchromatic band (Pan) with stereo capability has also been included in all but the initial programs (Table 4.1).

FUYO (JERS) and DAICHI (ALOS) Programs The Japanese earth resources remote sensing satellite programs, FUYO and DAICHI, both have included an optical sensor as well a SAR sensor on the same platform. The satellites have utilized near-circular, sun-synchronous and near-polar orbit. The Japanese Earth Resources Satellite-1 (Fuyo-1) (1992) employed an optical sensor (OPS) using CCD linear arrays that yielded data in green, red, near-IR and SWIR bands. The Fuyo-1 SAR operated at 23.5 cm wavelength

and provided data with a ground resolution of 18 m. The DAICHI (Advanced Land Observation Satellite, ALOS) carries two optical sensors—one called PRISM, that provides stereo data with 2.5×2.5 m spatial resolution, and another called AVINIR, that provides multispectral data in blue, green, red and near-IR bands with 10×10 m spatial resolution. In addition to the optical sensors, ALOS also carries an L-band SAR sensor called PALSAR.

TERRA-ASTER Sensor A very important sensor for geoscientific applications is the ASTER (Advanced Spaceborne Thermal Emission and Reflection) radiometer, launched by the NASA as a part of the Earth Observation Satellite program aboard TERRA platform

Table 4.2 Salient characteristics of selected spaceborne SAR systems

Spaceborne system Parameter	Altitude (km)	Wavelength (cm)	Band	Swath Width (km)	Azimuth resolution (m)	Range resolution (m)
Seasat* (USA)	795	23.5	L-band	100	25	25
SIR-A* (USA)	259	23.5	L-band	50	40	40
JERS-1* (Japan)	568	23.5	L-band	75	18	18
ERS-1/2* (ESA)	777	5.7	C-band	100	28	26
Radarsat* (Canada)	~800	5.7	C-band	10–500	9–100	10–100
Envisat-1 (ESA)	777	5.7	C-band	100	25	25

*Archive data available

in 1999. ASTER carries moderate resolution (15 m, 30 m, 90 m) imaging sensors in a wide range of 14 spectral bands lying in the visible—near infrared—SWIR—thermal infrared parts of the EM spectrum. The spectral bands are such that they are sensitive to broad chemical-mineralogic constituents (e.g. contents in terms of Fe–O, argillic, carbonates, silica, felsic vs. mafic etc.) and can be used to derive broad mineralogic compositional information (for more details, see e.g. Ninomiya et al. 2005; Hook et al. 2005; Rowan et al. 2003).

High Spatial Resolution Space Sensors A general trend in the evolution of spaceborne remote sensing systems has been the deployment of multispectral bands, one each in green, red and near-IR bands, with good spatial resolution to allow generation of high-resolution CIR composites. This may be with or without one still higher spatial resolution panchromatic band with stereoscopic capability. AB selected list is given in Table 4.1.

SAR Programs During the past three decades, a number of spaceborne SAR sensors have been launched by various agencies for resources investigations. Salient specifications of these sensors are listed in Table 4.2. The first spaceborne SAR sensor was the SEASAT (1978) which operated for about 100 days. A series of Shuttle Imaging Radar (SIR-A,-B,-C) yielded radar images of selected parts of the earth in different radar imaging modes on experimental basis. The European Resources Satellite (ERS-1/2) was a very important SAR mission that provided global data and has been particularly useful for SAR interferometry. JERS-1 and RADARSAT-1 also yielded useful SAR images from space.

Currently, data may be available from *ENVISAT-ASAR*, ALOS-PALSAR and TerraSAR-X. The ENVISAT-ASAR can operate in alternate polarization mode.

The *ALOS-PALSAR* is the first fully polarimetric L-Band spaceborne sensor. The *TerraSAR*-X is the first commercial SAR sensor to provide up-to 1 m resolution. It can operate in alternate polarization mode, provide polarimetry data and along track interferometry. Planned sensors include *RADARSAT-2*, *Sentinel-1* (follow-on to the ENVISAT), and *TerraSAR-L (Cartwheel)* and *Tandem-X*, that are novel concepts in SAR imaging to provide single pass along-track and cross-track interferometry.

4.5 Interpretation Principles

To commence a study, first, the remote sensing photographs and images are indexed to provide information on location, scale, orientation and extent of the area covered. Mosaicking provides a bird's-eye view of the entire area. Overlapping photographs and images may be studied stereoscopically. Products of a multispectral set can be combined in colour mode. Digital data can be processed for enhancement, classification and interpretation. It can also be combined with ancillary geodata in a GIS approach, for a comprehensive interpretation.

Several photo recognition parameters are used to study and interpret features on photographs and images. These are called elements of interpretation and geotechnical elements (Table 4.3) (for details see e.g. Colwell 1960; Mekel 1978). Convergence of evidence is a general underlying principle in photo and image interpretation. It implies integrating all the photo-evidences and analyzing where all the photo-parameters are leading to, collectively.

Ground truth is necessary for reliable interpretation of remote sensing data. It aims to provide a reference

Table 4.3 Elements of photo-interpretation and geotechnical elements

A. Elements of photo-interpretation	
1. Tone	It is a measure of the relative brightness of an object in shades of grey
2. Colour	Appropriate terms are used to describe the colour; use of colour dramatically increases the capability of subtler distinctions
3. Texture	It means tonal arrangements and changes in a photographic image; it is the composite effect of unit features too small to be discerned individually
4. Pattern	It refers to spatial arrangement of features, like drainage, vegetation, etc.
5. Shadow	Shadow helps in studying the profile view of objects, and is particularly useful for man-made objects
6. Shape	Outline of an object in plan
7. Size	Size of an object is related to scale
8. Site/association	Mutual association of objects is one of the most important guides in photo/image interpretation
B. Geotechnical elements	
1. Landform	Constructional and depositional landforms are associated with various natural agencies of weathering and erosion, and are extremely helpful in identifying hydrogeomorphological characteristics of the site
2. Drainage	It is one of the most important geotechnical element and includes the study of drainage density, valley shape and drainage pattern
3. Soil	Soil characteristics depend upon the bedrock material and agencies of weathering. They have characteristic hydrological properties which govern surface run-off vis-à-vis infiltration
4. Vegetation	Vegetation distribution and type may be related to lithology and subsurface seepage zone

base for interpretation, as also to verify anomalous responses, if any. Parameters for field survey may include: rock/soil type, geological structure, soil moisture, vegetation type and density, land use, groundwater level etc. Often, ground truth observations are made by field-checking. At times, adequately large-scale photographs and images may also serve as ground truth for smaller-scale satellite image data.

4.6 Interpretation of Remote Sensing Data

The discussion here is divided into the following:

1. Panchromatic photographs and images
2. Multispectral image data
3. CIR photographs and composites
4. Thermal IR image data, and
5. SAR imagery.

4.6.1 Panchromatic Photographs and Images

Panchromatic sensors are broad-band sensors operating in the visible range (0.4–0.7 μm wavelength) and carry a major advantage of high spatial resolution in comparison to that of multispectral sensors (Table 4.1). Panchromatic photographs and images may be obtained from a variety of platforms or sensors and may differ in their geometry and scale; however, they are quite similar as far as the spectral characters are concerned, and therefore, their interpretation must follow the same line of logic. The technique of air photo-interpretation has been described in numerous standard publications (e.g. Colwell 1960; Miller and Miller 1961; Ray 1965). Numerous factors such as atmospheric condition, topography, slope, aspect etc. affect the image data (Table 4.4). Target reflectance i.e. albedo and its variation across the scene is very important, and deciphering this parameter holds the key to remote sensing applications. Spectral characteristics and identification features of selected features are described below.

On panchromatic products, deep and clear water body appears dark; on the other hand, turbid and shallow water body appears in shades of light gray to gray. Vegetation appears dark gray to light gray, depending upon the density and type of foliage. Soils appear in various shades of gray, the variation being related to moisture content, organic matter and grain size of the soil. Cultural features such as cities, settlements, roads and railway tracks can often be recognized by their shapes and outlines.

Stereo aerial black-and-white photographs have been routinely used for the study of landforms,

Table 4.4 Salient factors affecting image data in the solar reflection region

Primary variables	Secondary variables	Comments
1. Atmospheric factors	Composition of the atmosphere leading to absorption; particulates and aerosols, leading to scattering; relative humidity, cloud cover, rain etc.	May vary within a scene from place to place
2. Topography and slope aspect	Goniometric aspects, landscape position and slope direction; sun-sensor look angle	Vary from place to place within a scene, depending upon sun-local topography relation
3. Target reflectance	Albedo of the object; surface texture and coating	Deciphering this attribute holds the clue in remote sensing

lithology, structure, lineaments, soil, vegetation etc. Identification of rock types is based on convergence of evidence derived from a number of parameters (Sect. 4.8.8). Interpretation of panchromatic data can provide valuable inputs in geohydrological studies, particularly in the context of runoff, evaporation and groundwater recharge (e.g. Colwell 1960).

4.6.2   Multispectral Image Data

Multispectral image data in the solar reflection region are available on a routine basis from a number of space-sensors such as LANDSAT-ETM+, SPOT-HRG, IRS-LISS-III/LISS-IV, TERRA-ASTER, IKONOS, QUICKBIRD, and GEOEYE etc. (Table 4.1). As spectral distribution of these sensors is quite comparable, interpretation of image products from these sensors must follow a common line of argument. Spectral characters of common objects on multispectral bands are briefly mentioned below (Table 4.5).

Water exhibits different characters, depending upon its suspended load and the depth of the water body. Clear and deep water bodies are generally dark, and turbid and shallow water bodies reflect in the shorter wavelengths (blue/green). In the NIR and SWIR, all water bodies appear black. Vegetation in general appears medium-dark in the VIS and bright in the NIR. Cropland is medium gray and is marked by the characteristic field pattern. Soils and fallow fields (dry) are light in the VIS and medium gray in the NIR—SWIR bands. Moist ground is medium gray in the VIS but very dark in the NIR and SWIR. Rocky terrain (bare) is usually brighter in the VIS than in the NIR and SWIR, and may be characterized by landforms and structures. Clayey soils are medium to light toned in the VIS, NIR and SWIR-I, but are very dark in the SWIR-II, due to strong absorption by the hydroxyl anions. Snow is bright in the VIS—NIR and exhibits typical absorption in the SWIR-I. Clouds appear bright in the all the bands due to non-selective scattering. The multispectral data have opened vast opportunities for mapping and monitoring of surface features for hydrogeological investigations and a multitude of other applications.

4.6.3   CIR Photographs and Composites

Colour infrared (CIR) photography has been carried out from aerial platforms world over, as also from selected space missions (e.g. Skylab, Metric Camera and Large Format Camera etc.). False colour composite (FCC) is an extensively used technique for combining

Table 4.5 Response characteristics of important multispectral bands

Blue	Green	Red	NIR	SWIR-I	SWIR-II
Very strong absorption by vegetation and Fe–O; good water penetration; high scattering by suspended particles	Some vegetation reflectance; good water penetration; scattering by suspended particles	Very strong absorption by vegetation; some water penetration; scattering by suspended particles	High reflectance by limonite and vegetation; total absorption by water	General higher reflectance; insensitive to moisture contained in vegetation or hydroxyl bearing minerals; absorption by water, snow	High absorption by hydroxyl-bearing minerals, carbonates hydrous minerals vegetation, water and snow

multispectral images of the same area through colour coding. In an FCC, three images of a multispectral set are projected concurrently, one each in a different primary colour, and geometrically superimposed to yield a colour composite. This optical combination carries information from all the three input images, in terms of colour variation across the scene. As these colours are not shown by the objects in nature, these products are called false colour composites (FCC's). Though, as such, any spectral band may be coded in any primary colour, a standard FCC uses a definite colour coding scheme as follows:

- Response in green wavelength range—shown in blue colour.
- Response in red wavelength range—shown in green colour.
- Response in NIR wavelength range—shown in red colour.

This is also called CIR composite as the colour coding scheme is analogous to a CIR photograph. On these products, vegetation appears in shades of red; deep and clear water bodies are dark, whereas shallow-silted water bodies are cyanish; dry sands are white-yellow while wet fields are cyanish-light grayish. Bare rocky slopes appear in shades of gray.

4.6.4 Thermal IR Image Data

The EM wavelength region of 3–25 μm is popularly called thermal infrared (TIR) region. Any sensor operating in this wavelength region would primarily detect the thermal radiative properties of ground materials. Out of this the 8–14 μm region has been of greatest interest, as this is characterized by the peak of the Earth's blackbody radiation and an excellent atmospheric window (Fig. 4.3). The thermally radiated energy is a function of two parameters: surface temperature, and emissivity. The surface temperature is governed by a number of physical properties, among which thermal inertia is most important in governing the diurnal temperature variations. If thermal inertia is less, diurnal temperature variation is more and vice-versa.

A typical thermal IR remote sensing experiment uses a set of pre-drawn (night) and day (noon) passes. Commonly, TIR image is a simple (radiant) tempera-

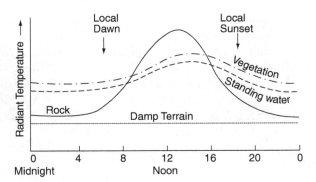

Fig. 4.7 Thermal response of rock, vegetation, standing water and damp terrain in a diurnal cycle

ture image. Other processed images, like a diurnal temperature difference image or thermal inertia image can also be generated through data processing and used for interpretation.

Topographic features are enhanced on day-time thermal images due to differential heating and shadowing. However, on night-time images, the topography gets subdued. All objects on the ground surface undergo cyclic heating and cooling in the day/night cycle, thus showing systematic temperature variations in a diurnal cycle (Fig. 4.7). Standing water appears brighter (relatively warmer) on the night-time image and darker (cooler) on the day-time image. Soils with higher moisture are cooler than those with lower moisture content. Shallow groundwater conditions are marked as zones of cooler temperatures on night-time thermal images (Heilman and Moore 1982). Soil moisture estimation has practical utility in agriculture and water budgeting. Temperature contrasts of the order 5–10°C could be produced by soil moisture variation. Effluent seepage would also lead to cooler ground surface. Drainages are also well shown on thermal data. Vegetation in general, is warmer on the night-time and cooler on the day-time image, in comparison to the adjoining unvegetated land.

Structural features like folds, faults etc. may be manifested due to spatial differences in thermal characters of rocks. Bedding and foliation planes appear as sub-parallel linear features due to thermal contrasts of compositional layering. Faults and lineaments may act as conduits for groundwater flow and may be associated with springs. This would lead to evaporative cooling along a line or zone producing linear features (Fig. 4.17a). Therefore, thermal data can be used for exploration of shallow aquifers and water-bearing fractures. In general, the spatial resolution of sensors is an important aspect in applications. The aerial thermal scanners have spatial resolution of the order of 2–6 m, quite suitable for such applications. In contrast, the available imagery from spaceborne thermal IR sensors is generally of coarser resolution (90 m for ASTER), which has put constraints the applicability of space-acquired TIR imagery for such applications.

4.6.5 SAR Imagery

As mentioned earlier, the radar return carries valuable information about the physical, geometric and electrical properties of ground objects. On a SAR image, intensity of radar return is shown in shades of gray, such that areas of higher backscatter appear brighter (Fig. 4.5). For common visual image interpretation, the black-and-white radar image is widely used, but the data can also be combined with other geodata sets in a GIS approach.

Commonly on a SAR image, most of the area is dominated by diffuse scattering caused by rough ground surface and vegetation (Fig. 4.5). Very strong radar return is caused by metallic objects and corner reflectors (hills and buildings). Specular reflection is produced by smooth surfaces, such as quiet water bod-ies, playa-lakes etc and there is little return at the SAR antenna.

Study of drainage network is often the first step in radar image interpretation. The waterways are generally black to very dark on the radar images, as water surface leads to specular reflection. However, in semi-arid to arid conditions, the valley bases are often vegetated, and appear bright on the radar images (Fig. 4.25). General terrain ruggedness is an important geotechnical element and orientation of individual landform facet in relation to look direction is very important for SAR image interpretation. Soil moisture is another important parameter as the radar return is governed by complex dielectric constant (δ) of objects. Water has very high δ, as compared to dry soil and rock. With increase in soil moisture content a regular increase of backscattering coefficient of the mixture takes place (Fig. 4.8), and this variation can be used to broadly estimate soil moisture on the ground.

Owing to the fact that minor details are suppressed, SAR images have value for regional landform studies. SAR images are found to be immensely useful for structural delineation. Lineaments are extremely well manifested on SAR images. The lithologic interpretation on radar image has to be based on indirect criteria like surface roughness, vegetation, soil moisture, drainage, relief, and special features like sinkholes etc.

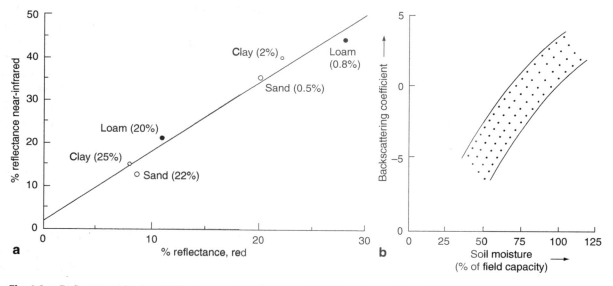

Fig. 4.8 a Reflectance of soils of different texture and moisture content (after Belward 1991), **b** Increase in radar backscattering coefficient with rise in soil moisture. (Redrawn after Bernard et al. 1984)

The radar image texture is an important parameter in deciphering lithological units.

Depth penetration is another important point in the context of geological-geohydrological applications. Ground penetration of the imaging radar depends upon the wavelength and increases with longer wavelengths as shorter wavelengths undergo a sort of "skinning effect". Further, the depth penetration is inversely related to complex dielectric constant (δ). Higher surface moisture inhibits depth penetration, whereas some limited depth information can be obtained in hyper-arid regions (Figs. 16.1, 16.2). It is generally accepted that in dry desertic condition, the depth of penetration of imaging radar is limited to <0.5 m for C-band and <2.0 m for L-band.

4.7 Groundwater Indicators on Remote Sensing Data Products

As mentioned earlier, remote sensing data provides primarily surface information, whereas groundwater occurs at depth, may be a few metre or several tens of metres deep. The depth penetration of EM radiation is barely of the order of fraction of a mm in visible part of the EM spectrum, to hardly a few metres in microwave region, at best. Therefore, remote sensing data is unable to provide any direct information on groundwater in most cases. However, the surface morphological—hydrological—geological regime, which primarily governs the subsurface water conditions, can be well studied and mapped on remote sensing data products. Therefore, remote sensing acts as a very useful guide and efficient tool for regional and local groundwater exploration, particularly as a forerunner in a cost-effective manner.

In the context of groundwater exploration, the various surface features or indicators can be grouped into two categories: 1. first-order or direct indicators, and 2. second-order or indirect indicators. The first-order indicators are directly related to groundwater regime (viz. recharge zones, discharge zones, soil moisture and vegetation). The second-order indicators are those hydrogeological parameters which regionally indicate the groundwater regime, e.g. rock/soil types, structures, including rock fractures, landform, drainage characteristics etc. (Table 4.6).

Table 4.6 Important indicators of groundwater on remote sensing data. (After Ellyett and Pratt 1975; Gupta 1991)

1. First-order or direct indicators
   - (a) Features associated with recharge zones; rivers, canals, lakes, ponds etc.
   - (b) Features associated with discharge zones; springs etc.
   - (c) Sol moisture
   - (d) Vegetation (anomalous)
2. Second-order or indirect indicators
   - (a) Topography
   - (b) Landforms
   - (c) Depth of weathering and regolith
   - (d) Lithology
   - (e) Geological structure
   - (f) Lineaments, joints and fractures
   - (g) Faults and shear zones
   - (h) Soil types
   - (i) Soil moisture
   - (j) Vegetation
   - (k) Drainage characteristics
   - (l) Special geological features, like karst, alluvial fans, dykes and reefs, unconformities, buried channels, salt encrustations etc. which may have a unique bearing on groundwater occurrence and movement

Image Data Selection Selection of remote sensing data for groundwater applications has to be done with great care as detection of features of interest is related to spatial resolution and spectral band width of the sensor as well as seasonal conditions of data acquisition. For example, small-scale image data are good for evaluating regional setting of landforms and regional structure, whereas large-scale photographs/images are required for locating actual borehole/structural sites. Similarly, an understanding of spectral response of objects is crucial for selecting spectral bands of remote sensing data. Further, temporal conditions (rainfall, soil moisture, vegetation etc.) may affect the manifestation of features on a particular data set. Figure 4.9 gives an example of the same area (granitic terrain in Central India) imaged by the same sensor (IRS-LISS-II) in two different seasons: post-monsoon and pre-monsoon. The post-monsoon image shows presence of a widespread thin vegetation cover, and therefore the distribution of various landforms and lineaments is not clear on this image. On the other hand, on the summer (pre-monsoon) image, the various landforms like buried pediments, valley fills and lineaments are clearly brought out.

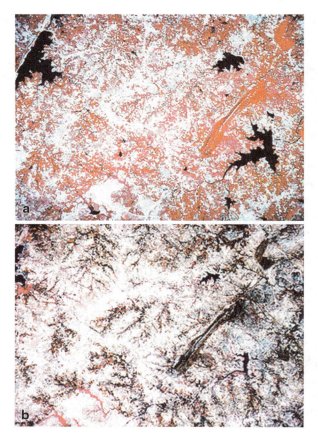

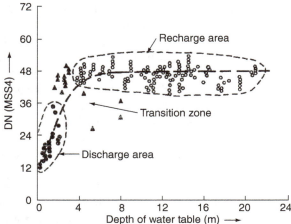

Fig. 4.10 Relationship between NIR reflectance (MSS4 DN values) and depth of water-table; the distribution of groundwater discharge and recharge areas is indicated. (Modified after Bobba et al. 1992)

Fig. 4.9 A set of **a** post-monsoon (dated 13 October 1996) and **b** pre-monsoon (7 April 1997) IRS LISS-II FCCs of a part of Bundelkhand granites in Central India. Note that the various landforms (buried pediments, valley fills etc.) and lineaments are better deciphered on the pre-monsoon (summer) image

types of optical and microwave image data. However, certain spectral bands such as near-infrared, thermal infrared and microwave are highly sensitive to surface moisture. Figure 4.10 is a plot of NIR reflectance against depth of water-table and shows that the groundwater discharge zones have shallow water-table and lower reflectance than the recharge zones.

Multispectral image data depicting spatial characteristics, patterns and shapes of surface water bodies can indicate whether they are recharging the groundwater. For example Fig. 4.11 is a Landsat image of

4.8 Thematic Applications

Application potential of remote sensing technique based on various groundwater indicators is briefly discussed in the following paragraphs, with examples.

4.8.1 Features Associated with Recharge Zones

Surface water bodies constitute an important source of groundwater recharge and hence their identification is useful. Water bodies such as streams, canals, ponds, reservoirs are often very well marked on all

Fig. 4.11 Features associated with groundwater recharge in the sub-Himalayas; the thinning of streams towards downstream indicates influent seepage. (Landsat MSS4 image)

a region marked by an overall low drainage density. The streams appear to get thinner and thinner towards downstream; evidently, they are losing water into the ground implying influent groundwater seepage.

4.8.2 Features Associated with Discharge Sites

Springs or effluent groundwater seepage may be detected on panchromatic images, multispectral image data and thermal IR data due to the fact that higher surface moisture may result in abnormal lower ground reflectance, anomalous local vegetation or temperature distribution. Quantitative estimates of groundwater discharge in lakes/streams etc. is difficult to obtain as remote sensing provides qualitative indicators of groundwater discharge.

In the area shown in Fig. 4.12, a number of springs and gradual building-up of streams southwards is due to effluent groundwater seepage. Another example of springs in coastal areas is shown in Figs. 15.13, 15.14 detected by thermal IR data. Effluent seepage may also be detected by SAR, as the microwaves are sensitive to surface moisture and vegetation. Remote sensing has been utilized to estimate spring discharge from areal extent of swamps and vegetation. When groundwater contains minerals, there may occur chemical deposits, staining or mineral alterations at points of water springs and seeps, which can be detected by high-spatial resolution remote sensing.

4.8.3 Soil Moisture

Soil moisture content influences the response of soil to EM radiation, throughout the EM spectrum. On the panchromatic band, a higher moisture in soils leads to darker photo-tones. As the infrared reflectance is highly sensitive to moisture content, the soil moisture variation can be detected using the NIR and SWIR bands. For example, very high reflectance in the visible and NIR bands and low drainage density point towards near dry surface conditions, absence of vegetation and deep water-table conditions. The waterlogged areas are also easily identified on NIR band. Figure 4.13 shows quartzite ridges and intervening waterlogged phyllite valleys. Reflectance of soil is influenced by soil texture and moisture content (Belward 1991) (Fig. 4.8a); therefore, reflectance data can hardly be used for quantitative estimation of soil moisture content in a general way. The thermal-IR region is also sensitive to surface moisture. The wet areas are sites of evaporative cooling and appear cooler (darker) than the adjacent dry areas.

As far as application of microwave part of the EM spectrum for soil moisture studies is concerned, imaging SAR response is also broadly related to soil

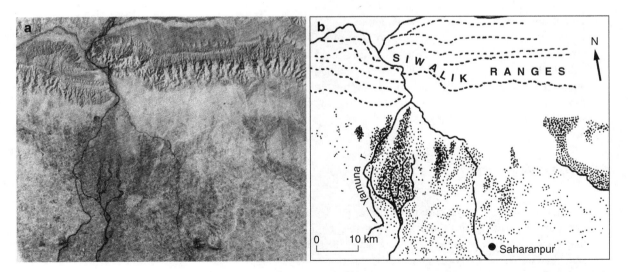

Fig. 4.12 Features associated with groundwater discharge: **a** Landsat MSS4 image, and **b** interpretation map. Note the building up of streams due to effluent seepage

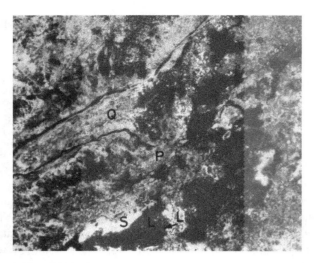

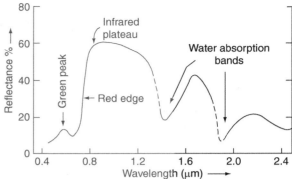

Fig. 4.14 Spectral reflectance curve of healthy vegetation in the solar reflection region

Fig. 4.13 IRS-LISS-II image of a part of northern Rajasthan, (India). The region is underlain by quatzite (*Q*) (ridges), and phyllits (*P*) (valleys). The valleys are waterlogged (*L*); white patches are salt encrustations (*S*)

moisture variation (Fig. 4.8b). However, precise estimation is made difficult by the fact that SAR response is affected by soil moisture as also by a number of parameters such as surface roughness, topography, and vegetation density and vegetation wetness. Passive microwave sensing has good scope for soil moisture studies but with coarse spatial resolution. Soil moisture estimates with spatial resolution of about 20 km are obtained from satellite sensor passive sensing in microwave frequencies of C-band (6.9 GHz) and X-band (10.7 GHz). The goal of such data is to obtain global soil moisture estimates with about 25–50 km spatial resolution and regional maps of soil moisture are available on the internet. A new satellite program called Hydros is under consideration for launch in about 2010 to estimate soil moisture in shallow (about 5 cm) soil zone, and is primarily based on microwave sensors (Entekhabi et al. 2004).

4.8.4   Vegetation

Vegetation forms an important general guide for groundwater investigations, and anomalous vegetation may serve as a direct guide for groundwater. Remote sensing data, particularly of dry seasons, may bring out vegetation anomalies as water-bearing fractures may sustain vegetation even under dry/drought conditions.

Vegetation can be studied on panchromatic and/or multispectral images, thermal-IR images and SAR data. The spectral response curve of healthy vegetation is shown in Fig. 4.14. In the visible part of the EM spectrum, the pigments in leaves govern the reflectance; therefore, vegetation is generally darker than the adjacent rock or soil on panchromatic, and multispectral blue, green and red bands. Healthy vegetation is strongly reflective in the NIR and this character helps in discriminating healthy vis-à-vis diseased/decaying vegetation. In the SWIR region, the reflectance is governed by leaf water content and this spectral band has importance in evaluating vegetation stress. On SAR images, dense canopy leads to a more intense scattering and a corresponding higher radar return (Fig. 4.23).

Suitably large-scale stereo pairs may yield information on the type of plants to indicate groundwater conditions in the area. For example, presence of xerophytic plants indicate dry arid conditions, whereas phreatophytic vegetation (plants taking water directly from the water-table) denotes shallow water-table. In such conditions, proliferation of water resistant plant species can also be observed on remote sensing data (e.g. Bacchus et al. 2003). Palm trees indicate shallow water-table but highly mineralized water, and mangroves flourish along the sea shore.

4.8.5   Topography and Digital Elevation Model

Surface topography has pronounced influence on groundwater occurrence. Topography also governs basin boundaries. The regional topography is best studied on stereo photographs/images. Besides, topographic maps, multispectral images and radar images are also useful for the study of topography.

Digital Elevation Model (DEM) forms a very basic and important input in any study of the earth's surface features. DEM can be obtained in many ways. Digitization of topographic maps followed by interpolation has been a widely used practice for generating DEM. However, the major short-comings of this are lower accuracy due to deformation of topographic map paper and many generalizations during topographic map preparation.

The shuttle radar topographic mapping (SRTM) mission has provided DEM of a large part of the land surface that can be downloaded. The SRTM global coverage has so far been restricted to 3-arc-sec (about 90 m) but for certain parts 1-arc-sec (about 30 m) DEM is available. In general, the absolute height accuracy is about 15 m and the relative accuracy is <10 m.

Digital photogrammetry developed for stereo aerial photography can also be applied to the stereo satellite image data and stereo data from space sensors such as SPOT, ASTER, IRS can be used. The principle of parallax (relief displacement) is applied and the amount of parallax is proportional to the relief. Various software packages are available for generating DEM. For ASTER data, the accuracy of DEM is normally considered to be around 15–40 m depending upon the terrain. High-resolution stereo-systems such as Cartosat, Quick-Bird, Ikonos etc. allow generation of DEM with higher accuracy ($\approx$1–2 m).

Elevation data can also be downloaded from GoogleEarth which provides point data up-to an accuracy of 1 m in relatively flat areas. This data if downloaded in grid-form can also be converted into DEM. LIDAR surveys yield vertical accuracy in the range of 10–30 cm; however there are problems in data processing due to vegetation cover.

4.8.6 Landforms

One of the widest applications of remote sensing image data, particularly of stereo photography, has been in the study of landforms, because: (a) remote sensing particularly with stereoscopic capability gives direct information on the landscapes, involving slopes, relief and forms, and (b) landforms can be better studied on a

regional scale provided by remote sensing data, rather than in the field. Some special landform features such as karst, buried channels, dykes, alluvial fans etc. are of particular interest in groundwater studies, and they are discussed separately at relevant places in this book. Genetically, the landforms are divided into two broad groups: 1. erosional landforms, and 2. depositional landforms.

1. *Erosional landforms:* Erosional landforms are typically associated with the resistant hard rock terrains and they are discussed in more detail in Chap. 13. From the point of view of groundwater occurrence, valley fill is the most important landform unit in such a geologic setting. Often, it may be difficult to distinguish between buried pediments and valley fills on panchromatic images and other raw data. However, dedicated image processing of multispectral data could lead to better discrimination of such landform features, as illustrated in Sect. 4.9.
2. *Depositional landforms:* Depositional landforms are a result of depositional processes of various natural agencies, viz. river, wind, and glacier etc. They differ in form, shape, size, occurrence, composition, association, and hence also in their groundwater potential. However, as they may also occur in the regional setting of fractured hard rock terrains, they play an important role for groundwater development for local needs. These are discussed in Chap. 16.

4.8.7 Depth of Weathering and Regolith

In hard rock terrains, the depth of weathering and thickness of regolith are important considerations in locating dugwells in rural areas. Spatial distribution of rock outcrops can give a reasonable idea of the likely thickness of regolith/overburden, and depth to bedrock. Thickness of regolith is also related to landform and is likely to be more at lower slopes and valleys. For this purpose, studies of stereo photographs and CIR composites of multispectral images are particularly useful (Fig. 4.19).

4.8.8 Lithology

Different lithologic units, i.e. rock types can be inferred indirectly based on the rock structures and landforms. The first task is to identify the terrain, i.e. fractured hard rocks vs. unconsolidated sediments. On photographs and images, the hard rock areas are characterized by the presence of a number of features such as compositional bandings, bedding, foliation, fractures, folds etc. Further, they have the characteristic erosional landforms, as described earlier. The unconsolidated sediments are marked by the absence of the rock structures and presence of various depositional landforms such as alluvial fans, point bars, moraines, dunes etc. From the point of view of hydrogeological investigations, the fractured hard rocks have been divided into four groups in this work: (a) crystalline rocks, (b) volcanic rocks, (c) carbonate rocks, and (d) clastic rocks. Their salient photocharacters are summarized in Table 4.7.

4.8.9   Geological Structure

Remote sensing images and photographs have found extensive applications for structural studies, to decipher planar discontinuities in the terrain, and understand their disposition and mutual-relations. The discontinuities may be visible on simple photographs and images, or remote sensing data may be processed specifically to enhance certain structural features. In general, remote sensors (photographic cameras, line scanners, digital cameras etc.) provide plan-like information, the data being collected while the sensor views the earth vertically from the above. Vertical and steeply dipping planar discontinuities are therefore prominently displayed on these images and photographs. On the other hand, gently dipping or subhorizontal

Table 4.7 Photocharacters of broad lithologic groups

Crystalline rocks

Widely different weathering characters are exhibited, depending upon composition, structure, and climate; more prone to weathering in humid, warm climates than in cold, dry climates; commonly occur as large size bodies (plutons) forming low-lying or undulating terrains; in arid conditions, sometimes show woolsack weathering in which isolated outcrops protrude through the weathered mantle. Drainage is generally medium density; massive rocks have dendritic pattern while jointed and foliated rocks exhibit angular-rectangular patterns; plutonic igneous rocks are massive; metamorphic rocks show foliation; lithological layering in gneisses leads to bandings. Several sets of joints may be present; granites also exhibit sheeting joints. Vegetated cover is variable. Spectral characters depend upon weathering, soil and vegetation; generally fresh rock surfaces are lighter toned. Also generally, if acidic they are lighter toned than basic and ultrabasic rocks.

Carbonate rocks

These are highly susceptible to solution by water; in humid climates, carbonate rocks show karst topography. In arid regions, the carbonates form ridges and hills. Drainage density in both arid and humid terrains is low; internal drainage is high in humid areas. Bedding may be weakly developed; intercalations of other lithologies may enhance lithological layering. Generally 3–4 sets of well-developed joints are observed; joint surfaces may become curved due to solution activity. Light coloured, thin calcareous soil is often observed in carbonate terrain; removal of carbonates may result in an iron oxide layer called terra-rosa. Vegetation is variable, depending upon weathering and climate. Carbonate rocks usually appear light toned; very often exhibit mottling due to variation in surface moisture.

Volcanic rocks

These are highly susceptible to weathering, which obliterates characteristic landforms and features in older flows. Drainage is highly variable; absent or very coarse due to high porosity in newer flows; often older flows display high drainage density and dendritic drainage. Foliation or bedding is absent in volcanic rocks; successive flows interbedded with weathered material and non-volcanic sediments may resemble coarse bedding (pseudo-bedding) shown on the scrap faces. Joints may be well developed; columnar joints are typically present in basaltic flows. Acidic flows have light toned soil cover and basaltic flows have darker toned soil cover. Vegetation is sparse on younger flows; weathered areas may support vegetation and cultivation. Tonal characters are related to composition, age, weathering, soil moisture and vegetation.

Clastic rocks

Sandstones are resistant to weathering, whereas shales are incompetent and easily eroded. Sandstones from hills, ridges, scarps, and topographically prominent features; shales form valleys and lower hill slopes. Sandstones have low to medium drainage density, and rectangular-angular pattern; shales have high drainage density and dendritic drainage; intercalated sequences of sandstone-shale produce trellis pattern. Bedding is often very prominent. Sandstones often exhibit compositional layering; massive and pure varieties lack compositional layering. Shales are less commonly exposed, as they are usually covered under weathered debris. Several sets of joints may be developed. Soil cover is variable. Shales usually possess thick soil cover. Vegetation bandings may be present, sandstones being vegetated and shales used as agricultural grounds. Tone is highly variable, depending upon vegetation and soil moisture.

structural discontinuities are comparatively subdued, and are difficult to delineate, especially in the rugged mountainous terrain.

Geological structures include bedding, foliation, folds, joints, lineaments, faults and shear zones etc. In this section, the discussion is confined to bedding/foliation and folds. Bedding is the primary discontinuity in rocks and is due to compositional layering. On remote sensing images, bedding appears as regular and prominent linear features, which may be marked by differential weathering, tone, texture, soil and vegetation; the linear features due to bedding appear longer, even-spaced and more regular, in contrast to those produced by foliation and joints.

Orientation of bedding or foliation is important. The strata may be sub-horizontal, inclined or vertical, and may form segments of larger fold structures. The outcrop pattern on photographs and images is influenced by structure and relief in the area. Folds can be delineated by tracing the bedding/marker horizon along the swinging strike and recognition of dips. Broad, open, longitudinal folds are relatively easy to locate on satellite images than tight overturned isoclinal folds, owing to the small areal extent of hinge areas, which may provide the only clues of their presence. Therefore, such folds need to be studied on appropriately larger scales.

4.8.10 Lineaments Including Joints and Fractures

Lineament studies are most important for groundwater investigation in hard rock terrains. The technique of mapping fracture traces and local lineaments from aerial photographs for locating zones of higher permeability in hard rock terrain was developed by Parizek (1976); also Lattman and Parizek (1964), and the same has now been extended to satellite imagery. It has been shown that wells located on or close to fracture traces yield many times more water than the wells away from the fracture traces, and they are more consistent in their yield when located on lineaments than under other conditions.

What Is a Lineament? The photolinears, i.e. linear alignment of features are one of the most obvious features on aerial photos and satellite images. The term lineament is now used, by and large, in a geomorphological sense, as "a mappable simple or composite linear feature of a surface whose parts are aligned in a rectilinear or slightly curvilinear relationship, and which differs distinctly from the pattern of the adjacent features, and presumably reflects a subsurface phenomenon" (O'Leary et al. 1976). Hence, this category includes all structural topographical vegetational, soil and lithological alignments, which are very likely to be the surface expression of buried fractures and structures. This definition seems to be the most practical in the context of remote sensing image interpretation.

Manifestation of Lineaments Lineaments occur as straight, curvilinear, parallel or en-echelon features (Fig. 4.15). Generally, lineaments are related to fracture systems, discontinuity planes, faults and shear zones in rocks (Fig. 4.16). The term also includes fracture traces described from aerial photographic interpretation by some workers. Dykes and veins may also appear as linear features but these can be identified in field or large-scale photographs. In some cases, fractures may be covered under regolith. The pattern of a lineament is important on the image. Lineaments with straighter alignments indicate steeply dipping surfaces and by implication, they are likely to extend deeper below the ground surface.

Ground element corresponding to a lineament would depend on the scale of remote sensing image. Both small-scale and large-scale images are useful and ought to be used in a complimentary manner for hydrogeological studies. On regional scales (say 1:100000), lineament features may be more than ca. 2 km in length, and represent longer valleys and complex fractured zones; such data can facilitate regional selection. On larger scales (say 1:20000 scale), shorter, local, minor drainage features, individual fracture traces etc. are represented as lineament features; such data are useful in detailed investigations, e.g. for well siting.

Different types of fractures and their characteristics have been described earlier (Chap. 2). The distinction between dilational and shear fractures is very important for groundwater studies as the former are more productive. This distinction can sometimes be made on remote sensing images on the basis of two

Fig. 4.15 Mapping of various types of fractures on IRS-1C LISS-III FCC of Cuddapah region, India. (Courtesy D.P. Rao, A. Bhattacharya and P.R. Reddy)

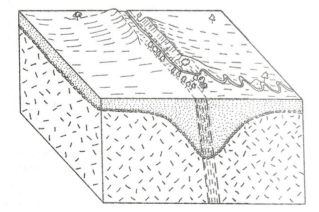

Fig. 4.16 Schematic showing surface manifestation of lineament in terms of topographic/drainage/vegetation alignment. The lineament zone is marked with greater depth of weathering

considerations: (a) relative movement along an individual discontinuity, and (b) orientation and statistical distribution of lineaments. It is always desirable if both these types of evidences are mutually supportive. Sometimes, relative movements along lineaments could be seen to indicate faults and shear fractures (Fig. 4.15); in contrast, lineaments related to dilational fractures would not show any relative displacement. Further, in areas of less complex deformation, statistical analysis of lineament trends could help distinguish between fractures of shear origin and tensile origin. It is always better if such interpretations are supported by evidences of relative movement observed in field or remote sensing data. Hydrotectonic models have also been attempted by some work-

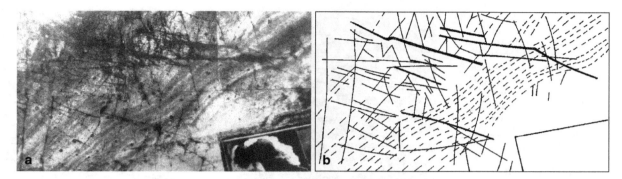

Fig. 4.17 a Thermal IR image showing numerous structural features such as bedding and lineaments, **b** interpretation map (Stilfonstein area, Transvaal, South Africa). (After Warwick et al. 1979)

ers to interpret the erstwhile stress field and thence identify tensile fracture trends as they are believed to be more productive.

On an image, the lineaments can be easily identified by visual interpretation using tone, colour, texture, pattern, association etc. Alternatively, the automatic techniques of digital edge detection can also be applied for lineament detection. However, these techniques have found limited application due to the cropping up many non-meaningful linears. Therefore, the visual interpretation technique is preferred and extensively applied, although it involves some degree of subjectivity, due to human perception (e.g. Sander et al. 1997).

Mapping of lineaments can be done on all types of remote sensing images: stereo panchromatic photographs and various images. The panchromatic, NIR SWIR, and thermal IR images contain information of near-surface, whereas SAR images may provide limited depth penetration (of the order of a few metres at best) in arid conditions. Manifestation of lineaments is related to ground conditions and the sensor spectral band. Figure 4.17 shows an example where aerial thermal IR image exhibits a number of structural features (bedding planes and fractures), which are not seen on the panchromatic band aerial photograph. The structural features appear dark due to evaporative cooling along them. Further, often lineaments are better manifested on SAR images than on panchromatic photographs, owing to the fact that radar response is intensely affected by surface roughness, micro-relief, vegetation and surface moisture. Figure 4.18 shows a comparison of lineament features observed on air photos and SAR images of the same area in Brazil.

Lineaments can be mapped on original data products, as well as on enhanced images. Further, a variety of digital techniques to enhance linear features can also be applied, in the form of isotropic and anisotropic filters (Sect. 4.9). An anisotropic filter can be used to enhance linear information in a certain preferred

Fig. 4.18 a X-band radar image of an area in Bahia, Brazil, and **b** aerial photograph of the same area. Note that the structural features are better manifested on the radar image than on the aerial photograph. (Courtesy A.J. Pedreira)

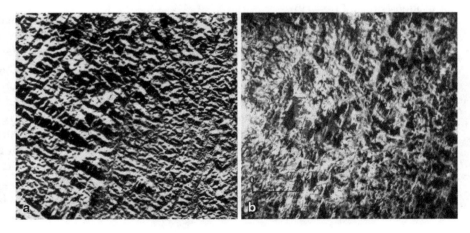

Fig. 4.19 Image processing and dedicated interpretation of a region in the Athur valley (India) using Landsat data. **a**, **c** and **e** are Sobel-filtered, Roberts-filtered and hybrid FCC's respectively, **b**, **d**, and **f** are their respective lineament interpretation maps, **g** is a landform interpretation map, **h** shows selection of groundwater potential targets based on lineament and landform interpretations. ((a), (c) and (e) courtesy of A. Perumal) (for (c) and (e))

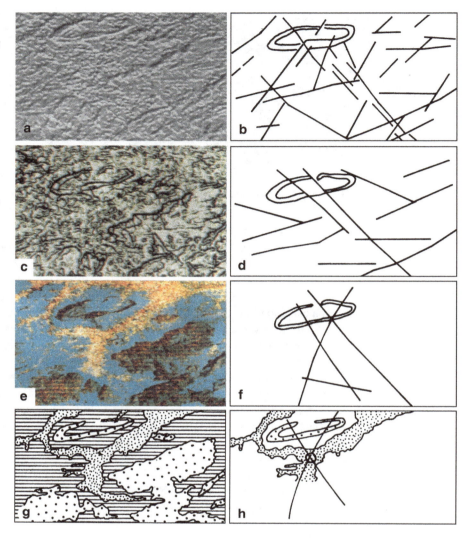

direction; however, due to the presence of artificats, interpretation of such filtered products has to be done cautiously (e.g., Fig. 4.19).

Statistical Treatment of Lineament Data There are a number of ways to assess the statistical distribution of lineaments in an area. One is by considering the number of lineaments per unit area; the second, by measuring the total length of lineaments per unit area; and the third, by counting the number of lineament intersections per unit area. The method of lineament intersection per unit area (density) is generally faster and convenient. The intersections of two (or more) lineaments are plotted as points. The number of points falling within a specified grid area is counted. The data are contoured to give the lineament intersection den-

sity contour map (Fig. 4.20) showing zones of relative intensity of fracturing in the area.

Relation of Lineaments with Other Features As lineaments frequently represent fracture zones in the rock, they are generally associated with zones of greater weathering, thicker soil cover, denser vegetation, higher moisture and valley alignments (Fig. 4.15). They are therefore also, often characterized by lower electrical resistivity values. Correlation between lineament and vegetation anomalies has been shown in several areas (e.g. Gustafsson 1993). Figure 4.21 illustrates that a strong correlation between lineament and vegetation anomalies in fractured rocks may exist even for long linear stretches of 5–10 km.

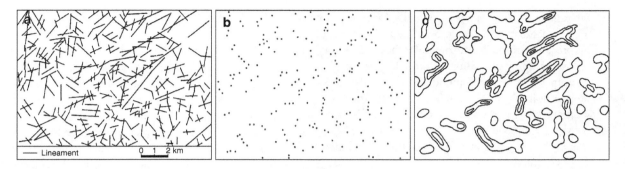

Fig. 4.20 a Lineament interpretation map of the area shown in Fig. 4.10, **b** lineament intersection density point diagram, **c** contour diagram from (b)

Lineament density and drainage density maps can also be combined for regional groundwater exploration; the target zones are usually marked by higher lineament density (representing greater fracturing) and lower drainage density (implying greater infiltration) (see Fig. 13.16).

Lineaments together with other features like landform and drainage have to be considered for delineating groundwater potential areas and for siting of wells. For example, in an area the groundwater potential targets may be marked by the intersection of lineament and valley-fill (Fig. 4.19). Further, it may also be recalled that often lineament traces appear as surface drainage ways but it is advisable to locate boreholes away from drainage lines to avoid chances of flooding.

Lineaments detected on satellite images often correspond to wide zones on the ground, which may be debris/soil covered. Therefore, delineating the main zone of fracturing (lineament) is often necessary and can be best accomplished using geophysical tools. Zeil et al. (1991) describe a case study from Botswana

in which they geometrically coregistered Landsat TM and geophysical (EM) data to locate lineament-fracture zone more accurately in the field. Further, intersection of tensile and shear fractures in association with geophysical anomalies and suitable landforms may produce exceptionally high yields (Boeckh 1992, Fig. 17.10). Integration of remote sensing data and geophysical data (e.g. aeromagnetic) can be highly useful in identifying regional exploration targets (e.g. Ranganai and Ebinger 2008). An additional example of use of lineament conjunctively with other data in a GIS approach for groundwater exploration is discussed in Sect. 6.4.3. Examples of well yield as related to lineaments are given in Sect. 13.5.1 for crystalline rocks and Sect. 15.7 for carbonate rocks.

4.8.11   Faults and Shear Zones

Characteristics of faults and shear zones have been discussed in Chap. 2. Chief criteria for delineating faults

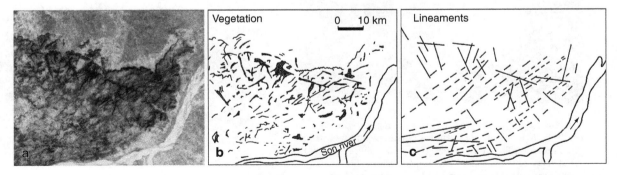

Fig. 4.21 Preferential alignment of vegetation for tens of kilometres along lineaments in the Son river area in Central India. **a** Landsat MSS2 (red band) image, **b** and **c** interpretation maps showing vegetation and lineaments respectively

an remote sensing images are: 1. displacement of beds or key horizons, 2. truncation of beds, 3. drag effects, 4. presence of scarps, 5. triangular facets, 6. alignment of topography including saddles, knobs etc., 7. offsetting of streams, 8. alignment of ponds or closed depressions, 9. spring alignment, 10. alignment of vegetation, 11. straight segment of streams, 12. waterfalls across stream courses, 13. knick points (local steepening of stream gradient), 14. disruption of channels and valleys, etc.

As such, a distinction between faults and lineaments is difficult to make, the main difference being that faults show evidences of movement (Fig. 4.15). On remote sensing images, faults with strike-slip movement are more readily discerned than those with only dip-slip movement. Like lineaments, faults can be delineated on all types of remote sensing images—panchromatic, multispectral, thermal IR, and SAR data. Figure 4.22 (Landsat infrared image) shows the presence of nearly E–W trending prominent extensive Recent fault marked as a lineament. The block on the south of the fault is frequently invaded by coastal waters, playas and salinas and forms the downthrow side.

Fig. 4.22 Landsat infrared image showing the presence of nearly E–W trending prominent extensive Recent fault marked as a lineament. The block on the south of the fault is frequently invaded by coastal waters, playas and salinas and forms the downthrow side. The northern block is covered with dunes. The terrain is dry, bare of vegetation and quite uninhabited. Also note the artificial Indo-Pak boundary running through the dunes

4.8.12 Soil Type

Soil types and humus content can be estimated from remote sensing data, particularly panchromatic photographs. Soil types are intimately related to the bedrock. For example, sandy soils originate from arenaceous and siliceous rocks; clayey soils develop from argillaceous rocks; basalts typically produce black cotton soils. Coarse grained soils and sandy soils are generally light toned. Saline soils are light coloured due to salt encrustation. Fine grained soils and clayey soils are moisture-rich and generally dark toned. Clayey soils also exhibit absorption bands in the SWIR region. Humus content in soils is related to vegetation and moisture which leads to darker tones.

4.8.13 Drainage Characteristics

For hydrogeological investigations, drainage is a very important parameter of study. Drainage channels are characterized by the typical serpentine-sinuous shape and pattern. They may appear differently on different spectral bands depending upon depth of water, turbidity, vegetation cover etc. Condition of the channel including its pattern and density may give a good idea of the hydrogeological characteristics of the terrain. Figure 4.9 shows an IRS-LISSII image of the granitic area, Central India, where drainage is controlled by fractures and the channels are marked by good vegetation (dark tones on the red band image); there is little surface water. Figure 4.23 is a SAR image of basaltic

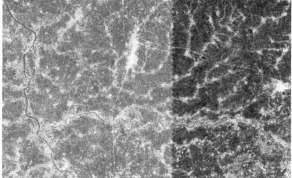

Fig. 4.23 SAR image from Shuttle Imaging Radar-A of a part of Central India. On the SAR image vegetated channels are bright due to higher radar returns whereas water bearing channels are dark due to specular reflection

terrain from Central India showing dendtritic drainage pattern; the drainage-ways are mostly vegetated (bright) with little surface water.

4.8.14 Special Geological Features

Some geological features may be locally important from hydrogeological point of view, such as karst features (in carbonate rocks), alluvial fans and buried channels (in clastic formations), dykes and reefs (in volcanic and crystalline rocks), and unconformities. These are discussed in appropriate lithology chapters.

4.9 Digital Image Processing

4.9.1 Introduction and Methodology

Digital image processing is carried out for image data correction, superimposing digital image data, enhancement and classification. A digital image is an array of numbers, in the form of rows and columns, depicting spatial distribution of a certain parameter, usually intensity of EM radiation in the case of remote sensing images.

Image processing can be carried out for various purposes; for detailed information, the reader is referred to standard texts (e.g. Mather 2004). Digital image processing for geological applications has been discussed by Sabins (1997); Gupta (2003) among others. The general sequence of digital processing of remote sensing data includes: image correction, registration, enhancement, visual interactive interpretation, and finally output.

Image correction is also called image restoration. Several types of radiometric errors and anomalies, including striping, bad data lines and atmospheric scattering effects etc., and geometric distortions such as those related to the Earth's rotation may be present in the image data. These are rectified before further processing. Registration is the process of superimposing images, maps, or data-layers over one-another with geometric congruence.

Image enhancement is the modification of an image to alter its impact on the viewer and render it more interpretable. A wide variety of enhancement techniques are available. Salient features of a few commonly used techniques are summarised in Table 4.8. Contrast enhancement deals with rescaling of gray tones to improve the image contrast and is almost invariably carried out. Edge enhancement is basically a process whereby borders of objects get enhanced. Main applications of edge-enhancement filtering are:

Table 4.8 Common types of digital image enhancements

Name	Description
1. Contrast enhancement	Rescaling of grey levels to improve the image contrast
1.1 Linear stretch	Range of image values is uniformly linearly expanded to fill the range of display device
1.2 Histogram equalized stretch	Assigning new image values on the basis of their frequency of occurrence, resulting in a uniform population density of pixels in the new image, also called ramp stretch; this results in a very high image contrast
1.3 Logarithmic stretch	Image data stretched using a logarithmic function; useful for enhancing features in the lower DN-range
1.4 Exponential stretch	Image data are stretched using an exponential function; useful for enhancing features in the upper DN-range
2. Edge enhancement	Application of a filtering technique to obtain a sharper image; information on local variations, which may be related to local morphology, soil, vegetation, moisture, fractures etc. are enhanced
3. Addition and subtraction	Simplest method to combine multi-image data; pixel-wise addition or subtraction is carried out to generate a new image
4. Ratio images	Generated by dividing the pixel value in one band by the corresponding pixel value in another band for each pixel. On ratio images, effects of illumination/topography are reduced
5. Colour enhancement	Uses colour display; pseudocolour is for a single image at a time such that grey scale is sliced into several ranges and each range is displayed in a separate colour; colour composites are often generated to combine multiple images by concurrently projecting them in different colours (RGB).

(a) to obtain a sharper image with more information on local characters, and (b) to enhance fractures, structural features and lineaments, over-all, or alternatively in a certain preferred direction (e.g. Fig. 4.19).

Addition and subtraction is the simplest method to digitally combine multiple image data sets. After digitally registering image sets, pixel-wise addition of two or more bands produces a new addition image. In a similar manner, pixel-wise subtraction of one image data from another generates a new subtraction image. In general, addition produces a high contrast image and is good for a general study. On the other hand, a difference image has a reduced contrast. A difference image is particularly useful for change detection; for example, change in vegetated area, flood area, or wet ground area etc., between two dates of satellite passes may be detected by this method.

Ratio images are generated by dividing the pixel value in one band by the corresponding pixel value in another band, for each pixel. Ratio images are made to reduce the effects of illumination variation due to topography and enhance spectral information. Ratioing is particularly useful in estimating vegetation density (see Fig. 4.24).

Colour enhancement is a very powerful technique of feature enhancement. Pseudo-colour and RGB (red-green-blue) coding are the two basic techniques of colour enhancement. Pseudo-colour is applied to

enhance differences within a single image. On the other hand, RGB coding is applied on a set of three co-registered images; the three images are co-projected for concurrent display and coded in blue, green and red primary colours. Such colour displays have been found to be extremely useful for geoscientific investigations and terrain evaluation.

Classification is carried out to construct thematic maps, where each pixel is assigned to a particular class by name. The methods of image classification are based on digital image pattern recognition. For subsurface geological exploration, the classification approaches have rather limited applications, and enhancement has been a more rewarding approach for subsurface geological-hydrogeological investigations (c.f. Siegal and Abrams 1976; Gupta 2003).

4.9.2   Digital Image Enhancement of Groundwater Indicators

It may be stated at the outset that there is no universal best digital image processing technique for enhancing all the groundwater indicators, in all conditions. Many combinations have to be attempted by trial and error, and in general, the convergence of evidence principle is very helpful. In these paragraphs some important guidelines for enhancing groundwater indicators are outlined.

1. Structure, faults and lineaments
   For mapping of various structures, viz. faults and lineaments on remote sensing images, the following types of processing are useful:

   (a) NIR band image, suitably, edge-enhanced, contrast-stretched.
   (b) FCCs utilizing edge-enhanced and contrast-manipulated images.
   (c) Difference image of NIR-Red bands, particularly when vegetation is associated with lineaments.
   (d) Hybrid FCCs of SAR image and optical (Red band and NIR band) images.

2. Lithology
   Lithologic discrimination is usually based on indirect evidences such as landform, drainage, soil and vegetation. Therefore, the enhancement of all the geotechnical features is useful.

Fig. 4.24 Ratio NDVI image of a part of the area shown in Fig. 4.9. Brighter tones correspond to greater density of vegetation

3. Vegetation

For vegetation studies, the following procedures of digital image enhancements are generally found to be more useful:

(a) Red and NIR band images, suitably edge-enhanced contrast-manipulated.

(b) CIR composite where vegetation is displayed in shades of red-magenta due to higher reflectance in the NIR band.

(c) The health and vigour of vegetation is indicated by a spectral ratio, involving Red to NIR bands; frequently the ratio is computed as [(NIR–Red)/(NIR+Red)], called normalized difference vegetation index (NDVI). NDVI has higher values for vegetated areas and lower for barren areas (see Fig. 4.24).

4. Landform and Drainage

Landforms and drainage are best studied by streoscopic techniques. Further, the general image quality is vital for such studies. The following techniques of image processing may be particularly helpful:

(a) CIR composites of edge-enhanced contrast-manipulated Green–Red-NIR bands.

(b) Hybrid colour composites involving optical images and SAR images.

4.10 Application of Remote Sensing in Estimating other Hydrological Parameters

The importance of remote sensing in estimating surrogate hydrological data has been emphasized by a few workers and well reviewed by Meijerink et al. (1994). In this way, remote sensing can contribute substantially in groundwater budgeting and modelling (Brunner et al. 2007). Main applications of remote sensing data could be of the following types:

1. *Topography and DTM:* In areas where accurate base maps are not available, topographic data and Digital Terrain Model (DTM) can be generated from stereo photos and images. DTM may be subjected to filtering to produce slope map, flow path map etc., with GIS functions.

2. *Rainfall:* In areas of rainfall <1500 mm per year, regional vegetation patterns are related to spatial rainfall patterns. Therefore, NDVI image can be related to rainfall distribution. As rainfall pattern is affected by topography, local DTM can also provide some information on spatial variation in rainfall. Thus, field data from rain gauge stations, local DTM, and vegetation pattern can be merged in GIS to yield a more realistic isohyetal map.

3. *Snow cover extent:* Owing to synoptic overviews and repetitive coverages, the satellite data are extremely useful for mapping snow cover area and delineating snow cover depletion pattern.

4. *Evaporation and soil moisture:* are important hydrological parameters, governing soil moisture retention and infiltration. Remote sensing data can be used to generate maps showing variation in soil moisture.

5. *Interception and infiltration:* Interception of surface water is affected by vegetation while rooting depth of plants governs infiltration. Remote sensing data can be used to generate maps of various land cover categories for estimating coefficients of infiltration.

6. *Water bodies:* Repetitive remote sensing data is highly useful in mapping temporal variation of surface water bodies, zones of influent/effluent seepage, flooding, etc.

4.11 Emerging Developments and Applications

Future developments and emerging new applications of remote sensing in groundwater may include the following (e.g. Becker 2006):

1. High-resolution gravity survey

High-resolution gravity surveys may be used to estimate groundwater storage (Pool and Eychaner 1995). Gravity surveys measure mass occurring below the satellite that exerts a gravitational pull on the satellite and have no vertical resolving power. Hence, mass of entire vertical column (including that of water at depth in saturated and unsaturated zones, vegetation and atmosphere) affects gravity data and adequate processing is required for accurate/reliable interpretation. The NASA's Gravity Recovery and Climate Experiment (GRACE) satellite is being considered as a possibility. Rodell and Famiglietti (2002) used computer simulations and anticipated GRACE performance parameters to estimate that groundwater level changes as small as about 8–9 mm could be possibly deciphered. However, observational errors and local atmospheric variability limit the applicability of the GRACE data to only very large basins of hundreds of square kilometre area (Swenson et al. 2003). A follow-on mission of GRACE with a higher resolution might provide practical useful data in monitoring water storage changes.

2. Satellite-borne laser altimetry

Laser altimetry involves illumination of the ground by artificially generated laser beams which strike the target and are back-scattered; distance estimation is carried out from the two-way travel time measurements (Fig. 4.25). Aerial laser altimetry has been found to be a powerful tool for various land surface applications where high-resolution terrain mapping/digital elevation models are required, e.g. in detailed town planning.

Measurement of water-level elevation (stage) in rivers and lakes is important from a hydrogeologic point of view as surface water can be considered to represent surface manifestation of groundwater discharge (Winter et al. 1998). River stage measurement in lean season can be used to estimate groundwater discharge. Satellite-borne radar altimetry has been used to measure water-levels in large lakes and rivers. For example, Birkett (1998) (Birkett

et al. 2002) used data from the TOPEX/Poseidon satellite to measure water-levels in large lakes and rivers, with vertical accuracies of 11–60 cm, though only in wide sections (>1 km). Laser altimetry has a much smaller ground footprint and a higher vertical accuracy. For example, ICESAT satellite has a laser sensor with footprint of ~70 m and vertical resolution of 15 cm. Therefore, satellite-borne laser altimetry appears to possess good potential for measuring changes in water-levels in river/lake.

3. SAR interferometry

SAR Interferometry, also called Interferometric SAR (InSAR), is a relatively new technique of remote sensing. It involves imaging the same target from two different angles at different times by Synthetic Aperture Radar (Fig. 4.26). Measurements of phase in SAR return beams yield data on elevation changes to sub-wavelength accuracy. The InSAR data are affected by atmospheric water vapour, and it is generally found that the accuracy of InSAR is ~10 cm in humid regions and ~1 mm in dry regions. In this perspective, InSAR application has a very high potential in arid to semi-arid regions. Applications of InSAR in groundwater studies include the following:

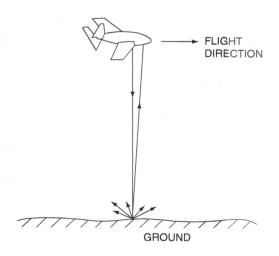

Fig. 4.25 Schematic arrangement of an active laser remote sensing system

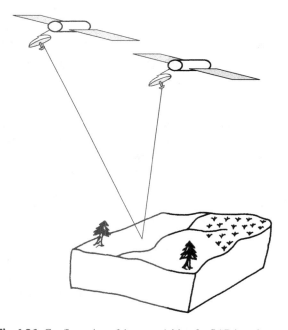

Fig. 4.26 Configuration of data acquisition for SAR interferometry; the same ground/target is sensed from two different angles at different times by the Synthetic Aperture Radar

(a) Monitoring of land subsidence due to groundwater withdrawal: Groundwater withdrawal may lead to compaction of sediments in the aquifer system which may cause elastic or inelastic deformation resulting in surface elevation changes that can be detected by InSAR (e.g. Galloway et al. 1998; Lu and Danskin 2001; Watson et al. 2002; Hoffmann et al. 2003; Schmidt and Burgmann 2003). Further, spatial pattern of subsidence due to groundwater withdrawal may be also influenced by neotectonic Quaternary faults, as revealed by InSAR (Smith 2002).

(b) As InSAR permits monitoring of ground surface elevation changes associated with discharge-recharge of aquifers, using numerical groundwater models it is possible to estimate regional storativity value and its spatial variation using InSAR (e.g. Lu and Danskin 2001; Hoffman et al. 2001; Hoffman et al. 2003).

(c) InSAR has also been proposed as a tool for measuring stream stages (water-level) in rivers (Alsdorf et al. 2003), which in turn is related to groundwater discharge.

and imaging radars. Data from a numerous space sensors are now available that may vary in ground resolution from about 80 m to better than 1 m and could be multispectral and or stereoscopic. The remote sensing data in terms of reflectivity, temperature or radar backscatter can yield valuable information on physical attributes of the ground. Digital image processing techniques can be used to enhance and classify features. In the context of groundwater, the various ground features/parameters of interest are: springs, and discharge zones, areas of recharge, canals, lakes etc., soil moisture, vegetation distribution and density, topography and DEM, landforms, thickness of regolith and weathering, lithology, geological structure, lineaments, faults, joints and fractures, soil types, drainage characteristics, and special geological features such as karst, alluvial fans, buried channels, dykes, reefs, unconformities, ground subsidence etc.

Summary

Remote sensing involves measurements of intensity of electromagnetic radiation reflected or emitted from the ground by sensors flying on aerial or space platforms to infer terrain properties. In groundwater studies, remote sensing is a common first reconnaissance tool. Design of remote sensing systems is constrained by spectral characteristics of the atmosphere, energy available for sensing and sensor technology. Four main types of remote sensors are: photographic systems, line imaging systems, digital cameras,

Further Reading

Becker MW (2006) Potential for satellite remote sensing of ground water. Ground Water, Vol. 44(20), pp. 306–318.

Gupta RP (2003) Remote Sensing Geology. 2nd ed., Springer-Verlag, 655 p.

Lillesand T, Kiefer RW, Chipman J (2007) Remote Sensing and Image Interpretation. 6th ed., John Wiley, 1164 p.

Meijerink AMJ, Bannert D, Batelaan O, Lubcyznski MW, Pointet T (2007) Remote Sensing Applications to Groundwater. IHP-VI, Series on Groundwater No. 16, UNESCO, Paris, 312 p.

Waters P, Greenbaum D, Smart PL, Osmaston H (1990) Applications of remote sensing to groundwater hydrology. Remote Sensing Reviews. Vol. 4(2), pp. 223–264.

Geophysical Exploration

5.1 Introduction

Following the investigations by aerial photographic and satellite remote sensing techniques, geophysical survey is carried out to ascertain the subsurface geological and hydrogeological conditions and aquifer characteristics. Various petro-physical properties utilized in geophysical exploration include electrical resistivity, electrical conductivity, density and elasticity (influencing seismic velocity), electrical permittivity (dielectricity), magnetic susceptibility, and radioactivity. Geophysical methods have the potential to predict distribution and flow of groundwater including sites of hazardous substances in a cost-effective manner. Further, as these methods are non-invasive, as compared to the direct conventional methods (for example, water sampling etc.), they do not disturb the water flow regime and are able to predict parameter distribution more realistically. Depending upon the scale of operations, geophysical surveys can help delineate regional hydrogeological features or even pin-point locations for drilling of water-wells. Geophysical surveys can appreciably reduce much more costly infructuous drilling operations. Details of geophysical methods can be found in several standard texts (e.g. Dobrin and Savit 1988; Parasnis 1997; Telford et al. 1999; Kearey et al. 2002). Geophysical applications specifically for groundwater are reviewed by deStadelhofen (1994), Beeson and Jones (1988), Kirsch (2006) and Ernstson and Kirsch (2006b) among others.

It is essential that geophysical surveys are not undertaken in isolation but are fully integrated with geological, hydrogeological and drilling. A reliable interpretation of geophysical survey data has to take into account prior knowledge of subsurface geology of the area. Therefore, at the outset, these surveys should be carried out at locations where the subsurface geology is better known. This will provide useful controls for interpretation of geophysical data in terms of subsurface geology, which can be later extrapolated to other similar areas.

5.2 Electrical Resistivity Methods

Electrical techniques, especially the resistivity surveys, are the most popular of geophysical methods for groundwater surveys because they often give a strong response to subsurface conditions and are relatively cost-effective (Ernstson and Kirsch 2006a). A combination of techniques can prove particularly useful and many studies are now carried out using a combination of resistivity sounding and electromagnetic traversing.

5.2.1 Basic Concepts and Procedures

Resistivity is defined as the resistance to electric current offered by a unit volume of rock and is a characteristic property of the medium in that state. It is based on the fact that electrical resistivity of a geological formation is dependent upon the material as well as the bulk porosity, degree of saturation and type of fluid. Since electrical resistivity of common minerals is very high, the electrical current flows through the pore fluid (water). The electrical resistivity of water-saturated clay-free material is given by the Archie's Law:

$$\rho_{aquifer} = \rho_{water} \times F \qquad (5.1)$$

where F is the formation factor and $\rho_{aquifer}$ and ρ_{water} are the specific resistivities of aquifer and pore water

B. B. S. Singhal, R. P. Gupta, *Applied Hydrogeology of Fractured Rocks,*
DOI 10.1007/978-90-481-8799-7_5, © Springer Science+Business Media B.V. 2010

Table 5.1 An overview of physical properties of saturated and unsaturated materials likely to be found in geophysical prospecting of groundwater. (Kirsch 2006)

	Seismic	Geoelectric, Electromagnetic		GPR	
	P-wave velocity (m s^{-1})	Resistivity (Ωm)	Conductivity (mS m^{-1})	Permittivity (relative to air)	Wave velocity (cm ns^{-1})
Gravel, sand (dry)	300–800	500–2000	0.5–2	3–5	15
Gravel, sand (saturated)	1500–2000	60–200	5–17	20–30	6
Fractured rock	1500–3000	60–2000	0.5–17	20–30	6
Solid rock	>3000	>2000	<0.5	4–6	13
Till	1500–2200	30–60	17–34	5–40	6
Clay	1500–2500	10–30	34–100	5–40	6

respectively. The formation factor F depends upon porosity, pore shape, cementation etc. The Archie's Law is not valid if grains are conducting (e.g. clay-rich matrix) or if pore water is highly resistive. The electrical resistivity of a dry formation is much higher than that of the same formation when it is saturated with water (Table 5.1).

Resistivity of the ground is measured by injecting current into the ground and measuring resulting potential difference at the surface across selected electrode positions (Fig. 5.1). The data on current flow and potential drop are converted into resistivity values. In case of an inhomogeneous earth, the measured resistivity is influenced in varying proportions by material from a wide depth range in the region covered by the electrodes (Fig. 5.2) and therefore, the field resistivity values are apparent (ρ_a) rather than true.

The arrangement of the four electrodes on the ground (two current and two potential) is referred to as the electrode 'array' or configuration. Most commonly used electrode configurations are **Wenner** and **Schlumberger** types (Fig. 5.3). In Wenner array, the four electrodes are placed collinearly and are equally spaced. In Schlumberger array, the electrodes are collinear but the distance between the two inner potential electrodes is very small in comparison to the distance between the two outer current electrodes. The apparent resistivity (ρ_a) is calculated as:

$$\rho_a = 2\pi\, aR \quad (\textit{Wenner array}) \qquad (5.2)$$

$$\rho_a = \pi\left(L^2/2l\right)R \quad (\textit{Schlumberger array}) \qquad (5.3)$$

where a, L and l are distances as shown in Fig. 5.3 and R is the measured resistance (voltage/current) in each case. Broadly, the depth of investigation of a resistivity survey is directly proportional to the electrode separation, and increases with increasing electrode spacing (Fig. 5.4). There is no single well-defined depth

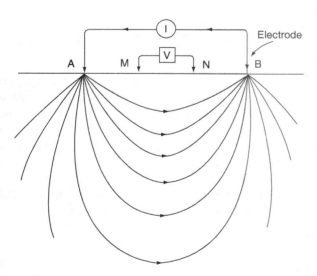

Fig. 5.1 Basic configuration in electrical resistivity surveys. *A* and *B* are current electrodes; *M* and *N* are potential electrodes

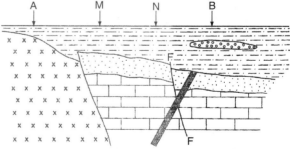

Fig. 5.2 Inhomogeneous geological features at depth often form the target for electrical resistivity surveying

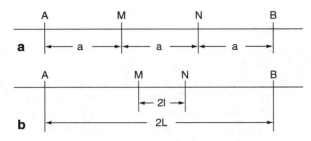

Fig. 5.3 Electrode configurations for collinear resistivity survey. **a** Wenner configuration and **b** Schlumberger configuration

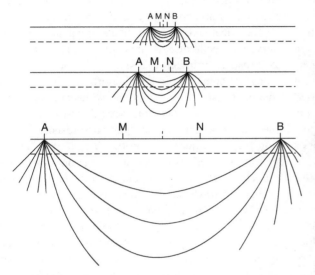

Fig. 5.4 The depth of investigation increases with increasing electrode seperation

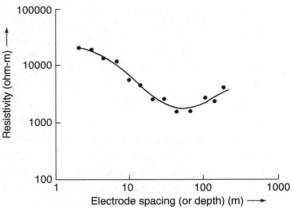

Fig. 5.5 Examples of electrical resistivity curves for depth sounding

to which a resistivity measurement can be assigned. Median depth (0.17–0.19 × distance between the two outer current electrodes) is considered as the most useful concept to describe the depth of penetration.

In geoelectrical methods, a distinction is required to be made whether one has to deal with horizontally layered earth (e.g. sedimentary terrain), or with elongated 2-D bodies like fractured zones and dikes, or with arbitrarily shaped structures (e.g. lenticular bodies or karst caves etc.). Accordingly 1-D (VES), 2-D (electrical imaging) and 3-D geoelectrics is used.

Vertical Electrical Sounding (VES) is applied to near horizontal layered medium, e.g. sedimentary terrain or weathered zones over hard rocks. It is used to determine variations in electrical resistivity with depth. In VES (also loosely called electrical drilling), the distances between electrodes are increased so that the electric current penetrates to deeper and deeper

levels, which allows resistivity measurement of a deeper and larger volume of the earth (Fig. 5.4). The apparent resistivity is plotted against the electrode separation (Fig. 5.5). The task of defining layers in a field survey is quite intricate as several interpretation methods, many involving curve-matching with standard curves, have been developed to provide better delineation of resistivity layers in various conditions (e.g. Koefoed 1979; Bhattacharya and Patra 1968). The interpretation result of a VES survey is the number of layers, their thicknesses and resistivities. In regular sedimentary sequences VES may be more reliable. However, in areas of unconsolidated sediments with rapidly varying thickness, borehole data is valuable in fixing model parameters for obtaining realistic depth estimates. The interpretation can be refined through forward or inverse modelling. Standard interactive computer programs executable on PC's are now available for this purpose. Normally, three to four distinct layers are about the maximum number for a reasonably accurate interpretation of a resistivity sounding curve. The VES remains an extremely powerful technique for delineation of regolith thickness which is vital when the saprolite is potentially thick (>20 m). Table 5.2 gives aquifer prospect as related to resistivity of layered regolith. However, VES is not an appropriate tool for detecting localized fracture systems.

Resistivity mapping is carried out for delineating near surface resistivity anomalies caused by for example, fracture zones, cavities or waste deposits. Any common electrode configuration can be applied.

Table 5.2 Aquifer prospect as related to resistivity (ohm metre) of layered regolith. (After Bernardi et al. in Wright 1992)

0–20	Clays with limited prospect (or saline water)
20–100	Optimum weathering and groundwater prospect
100–150	Medium conditions and prospect
150–200	Little weathering and poor prospect
>250	Negligible prospect

The electrode separation is kept constant and moved along profiles while apparent resistivity is measured. This enables gathering resistivity data over an area for a chosen depth of investigation. Contouring of resistivity data and interpretation provides information on variation in bedrock/soil type, spatial variation in depth of weathering and moisture content etc. Interpretation is generally done qualitatively by locating structures of interest. This method is commonly used for reconnaissance, after which detailed study in the selected target area is made through other geoelectrical methods. Figure 5.6 shows resistivity profiling across a lineament in Zimbabwe.

As mentioned above, formation resistivity is influenced by mainly porosity (primary and secondary), degree of saturation and type of fluid. Therefore, it varies with degree of weathering and seasonal fluctuations in water salinity. In an area, low resistivity values may correspond to clays, highly fractured rocks, or saline sand. On the other hand, high resistivity values may correspond to tight (low porosity) rocks, fresh-water-bearing sands or a relatively clean (clay-free) zone.

Normally, it is difficult to distinguish between permeable and impermeable fractures through electrical methods because their electrical properties are similar, especially when the impermeable fractures are filled with gouge, clay minerals or other alteration products. Such ambiguities which may occur are natural to geophysical methods. Therefore, it is very necessary that geophysical data are interpreted with adequate control on surface and subsurface geology, which may be available from exposures and/or boreholes.

Electrical tomography, or electrical imaging, is a surveying technique for areas of complex geology. In 2-D resistivity imaging, resistivity sounding and profiling are combined in a single process. It is assumed that the resistivity of the ground varies only in the vertical direction and one horizontal direction, i.e., along the profile (assuming that there is no resistivity variation perpendicular to the profile direction). It involves measuring a series of constant separation traverse with the electrode separation being increased with each successive traverse. The measured apparent resistivity values are plotted on a depth section immediately below the centre of the electrode arrangement. The apparent resistivity values are contoured to produce a 'pseudosection', which reflects qualitatively the spatial variation of resistivity in the cross-section. Thus, the method is used to provide detailed information both

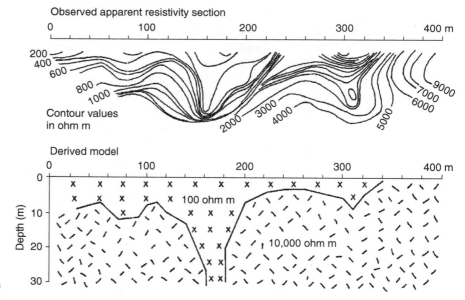

Fig. 5.6 Resistivity traverse data across a lineament in Zimbabwe, with a 2-layer interpretation. (After Griffiths in Carruthers and Smith 1992)

laterally and vertically along the profile so that more complex geological structures can be investigated.

The 3-D resistivity surveying is more complex, time consuming and expensive. A multielectrode (upto 256 and more) resistivity meter with switching is used. Several depth levels may be investigated by increasing the electrode spacing. The observed data of 2-D resistivity survey (also called resistivity imaging) are displayed as a pseudo-section along the profile in which geological–hydrogeological features can be interpreted.

5.2.2 Delineation of Rock Anisotropy

Rock anisotropy due to foliation, bedding, fractured zones etc., invariably leads to electrical anisotropy such that the resistivity in a direction parallel to the strike is generally lower than that in the perpendicular direction. Mapping of resistivity anisotropy is therefore extremely important. There are two methods for delineating rock resistivity anisotropy—square array configuration and azimuthal resistivity survey.

5.2.2.1 Square Array Configuration

This is specially designed for mapping rock resistivity anisotropy and has been widely applied. In this method, the electrodes are arranged to form a square with a pre-selected length side (A) (Fig. 5.7). The

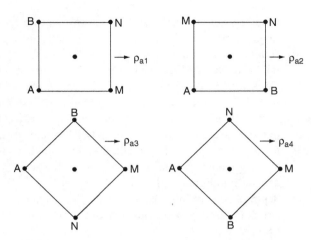

Fig. 5.7 Square array configuration: the apparent resistivity values are measured by rotating the electrodes in four configurations: ρ_{a1}–ρ_{a4}

apparent resistivity is assigned to the mid point of the square and is calculated as:

$$\rho_a = K R \qquad (5.4)$$

where K is called the geometric factor of square ($=2\pi A/2 - \sqrt{2}$), R is the resistance measured. At each location, the square array is rotated by 45° successively, and four apparent resistivity values (ρ_{a1}–ρ_{a4}) are measured. For a single-set of saturated steeply dipping fractures, the square-array method gives apparent resistivity minimum oriented in the same direction as the fracture strike (Fig. 5.8), and the data set can be

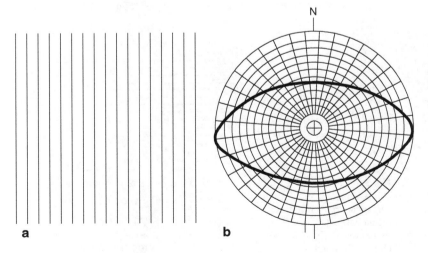

Fig. 5.8 a The case of one set of predominant fractures, and **b** the resulting square array azimuthal resistivity figure

used to compute resistivity anisotropy and approximate strike of the fractured zone. For example, the square array azimuthal resistivity survey in the crystalline rocks at Mirror Lake, New Hampshire, USA, Cook (2003) showed that the minimum resistivity is in N 30° direction, which is also found to be the fracture orientation as revealed from mapping in adjacent outcrops overlain by glacial deposits. The main advantages of the square-array method lie in its higher sensitivity to anisotropy as compared to co-linear arrays, and its requirement of less surface area for a given depth of penetration. This method also does not suffer from the paradox of anisotropy seen in co-linear arrays (see below).

5.2.2.2 Azimuthal Resistivity Survey (ARS)

This is a very powerful method for measuring in-situ characteristics of fractured rocks. It reveals magnitude and azimuthal variation in the permeability of the bedrock. It has many applications, e.g. in a poorly exposed terrain, where field mapping of fracture system may not be possible. Even in areas of good exposures, extensive field data on fracture characteristic is required for evaluating the anisotropic character of the bedrock. The ARS measures bulk properties of fractured rocks. In such rocks, the parallel geometric arrangement of water-bearing fractures makes the resistivity anisotropic. In this regard, fracture connectivity plays a key role in controlling maximum permeability direction.

The technique utilizes conventional resistivity equipment and is performed by rotating a Wenner array (or Schlumberger array) about a fixed centre point. The apparent resistivity is measured as a function of azimuth, say at 10° or 15° angle interval. The electrode spacing of about 5–25 m is used. When the apparent resistivities in different directions are plotted as radii, an anisotropic figure is generated, called apparent resistivity figure (ARF) (Fig. 5.9). In case of one set of parallel fractures, this is an ellipse.

For a single set of steeply dipping saturated fractures, the true resistivity minimum would be oriented parallel to the fracture strike. However, in this type of azimuthal resistivity survey using co-linear array, the apparent resistivity maximum gets oriented parallel to the fracture strike (Fig. 5.9; also Fig. 5.10a). This is known as paradox of anisotropy and owes its origin to the non-uniform distribution of electrical current density in the direction of fracturing (Cook 2003). It appears to be a result of using current magnitude in the calculation of apparent resistivity, whereas the current density determines the actual differences in potential.

The coefficient of anisotropy of apparent resistivity ellipse is λ $\{=\sqrt{(\rho_y/\rho_x)}\}$. It has been shown that joint porosity ϕ can be approximated under non-shale ideal conditions as $\phi = \{\rho_o (\lambda^2 - 1)/\rho_y\}$, where ρ_o is the groundwater resistivity.

Extending the case of single set fractures to multiple set fractures, it is found that the effect of multiple set fractures is additive in nature. If there are two sets of fractures of unequal development, the azimuthal resistivity plot exhibits peaks of unequal magnitude (Fig. 5.10b). Further, if in an area, joint lengths are less than the electrode spacing and joints are poorly developed, the orientation of the ellipse will be intermediate to the trends of joints and will represent the direction of greatest connectivity. It becomes a function of both

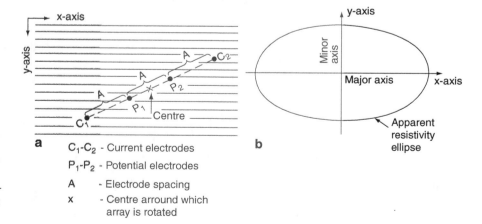

Fig. 5.9 a Scheme of a colinear azimuthal resistivity survey and **b** the resulting azimuthal (apparent) resistivity figure (ARF), which is an ellipse in a simple case

C_1-C_2 - Current electrodes
P_1-P_2 - Potential electrodes
A   - Electrode spacing
x   - Centre arround which array is rotated

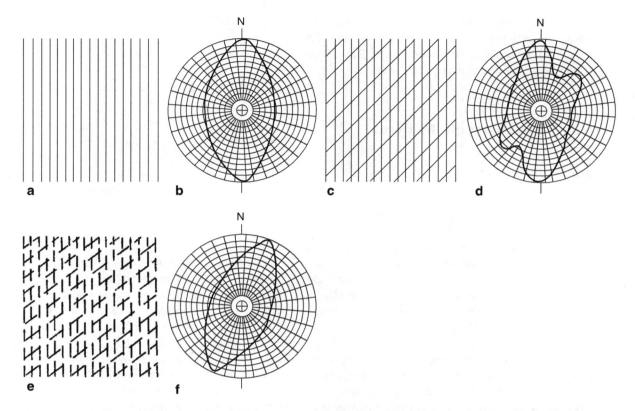

Fig. 5.10 Schematic representation of different configurations of fracture systems and the resulting azimuthal resistivity figures (ARF) using colinear arrays. (Redrawn after Taylor and Fleming 1988), for details see text

average fracture length and fracture frequency of both the joint sets (Fig. 5.10c).

If the spacing of the fractures is large, then different results are obtained at different electrode spacing. Changes in fracture orientation with depth may also lead to different results at different electrode spacing.

Application of azimuthal resistivity survey for detection of fractures has been done by a number of workers, e.g. McDowell (1979), Mallik et al. (1983), Taylor and Fleming (1988), Haeni et al. (1993) and Skjernaa and Jorgensen (1993) in a variety of igneous and metamorphic rocks. Inter-relationship between azimuthal resistivity and anisotropic transmissivity in fractured media has also been shown (Ritzi and Andolsek 1992).

Ideally, the method is valid for homogeneous anisotropic rocks with near-vertical fractures, large fracture length and high fracture frequency, in comparison to the electrode spacing. If the fracture set is not vertical but dips at an angle, the apparent resistivity ellipse has a relatively increased minor axis. For a horizontal set of fractures, it would take the shape of a circle (both axes

equal), as horizontal fractures contribute equally to the horizontal permeability in all azimuthal directions.

A close study of the ARF can immensely help in understanding the natural anisotropy in the bedrock. It provides a good representation of permeability anisotropies, which may be difficult to obtain even from field fracture measurements. A narrow ellipse (large-coefficient of anisotropy or large long to short axes ratio) indicates near-vertical continuous parallel fractures, with large aperture. On the other hand, a broad ellipse suggests dipping or less continuous fractures with low aperture. A single peaked ARF indicates one set of fractures (Fig. 5.10a). A double peaked ARF is formed by two sets of fractures, each peak corresponding to one set of fractures (Fig. 5.10b). In some cases the direction of the long axis of the ARF lies in between the strike directions of two major fracture sets, and probably indicates the most conductive path through the fractured rock (Taylor and Flemming 1988).

Before concluding, it may also be mentioned that the azimuthal resistivity method may not be able to distinguish between clay filled and water filled fractures, which have similar electrical properties but greatly different hydraulic conductivities. Such ambiguities are common in geophysical methods.

5.3 Electromagnetic Methods

5.3.1 Introduction

In areas where surface layers are highly resistive, electromagnetic methods may be used with advantage for groundwater exploration. The inductive coupling avoids the need for direct electrical contact, thus eliminating problems associated with resistive dry or rocky surface conditions. Electromagnetic data can also be collected from aeroplanes or helicopters, allowing survey of large areas at relatively low cost.

In case of deeper regolith (say > 20 m), the variation in thickness of weathered zone in an area is important, which can be estimated from EM profile data. On the other hand, in areas of shallow bedrock (regolith < 10 m), fractures in rocks are the target. Lineaments inferred from aerial photography and remote sensing data need to be surveyed by profiling for precise location and potential. In this context, the EM method has assumed the highest utility as an inventory tool. The recommendations from EM survey can then be checked and confirmed by resistivity techniques in more detail, and interpreted with other exploration data.

5.3.2 EM Method—Frequency Domain

The EM method of frequency domain type (often called Slingram) has been the conventional technique for geophysical exploration. The method utilizes a set of transmitter and receiver coils. A sinusoidal current flowing through the primary transmitter coil at a discrete frequency generates the primary magnetic field which induces eddy current in the sub-surface (Fig. 5.11). This current in turn generates the secondary magnetic field which is dependent on the sub-surface conductivity distribution. The induced magnetic field is picked up by the receiver coil and is interpreted to provide subsurface information. The secondary field is very

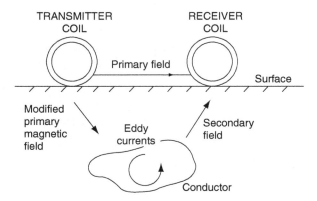

Fig. 5.11 Basic principle of electromagnetic induction prospecting

small in comparison to the primary field and also has a small phase shift with respect to the primary field.

The coil spacing are suitably selected, usually 20/40 m or 50/100 m being more common. The depth of investigation is determined by the intercoil spacing and is conventionally taken to be about one-half of the spacing. A limitation on the depth penetration is provided by the tendency for high frequency energy to propagate close to the surface (skinning effect), so that greater depth penetration is achieved with lower frequencies. The survey is made in a grid such that one axis of the grid is parallel to the main hydraulic conductivity orientation (i.e. rock discontinuity trend). Shorter coil separation and closely spaced EM stations can give better estimates of regolith thickness and dip of the fracture zone.

The EM survey can be carried out in various ways, such as horizontal loop EM (HLEM) (vertical dipole) and vertical loop EM (VLEM) (horizontal dipole). The HLEM is found to be particularly useful for detecting vertical and subvertical fractures (Boeckh 1992). Further, the VLEM system is extremely sensitive to changes in azimuth relative to conductor (fracture) strike. Thus semi-quantitative evaluation of width, depth and strike of fractures zones can be made from the anomaly shapes of both HLEM and VLEM data collectively (Hazell et al. 1992).

A water-bearing fracture acts as a conductor. The vertical dipole EM profile across a vertical conductor is marked by symmetrically placed two apparent conductivity maxima, on either side of a minimum, centred over the conductor (Fig. 5.12a). In case of a dipping conductor, the EM response curve is asymmetrical; the condcutor dips beneath the greater of the two flanking

Fig. 5.12 Schematic representation of horizontal loop EM (HLEM) (vertical dipole) response of a profile across a fracture (conductor body) when **a** the fracture is vertical, and **b** the fracture is inclined

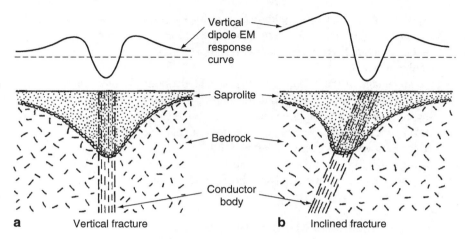

maxima (Fig. 5.12b). This anomaly pattern is extremely useful in deciphering dipping fracture zones (Siemon 2006). It can help avoid faulty borehole siting (Wright 1992; Boeckh 1992). Multi-frequency airborne EM method is suitable for shallow subsurface exploration (less than 100 m) and is widely used for groundwater exploration due to its better resolving capabilities. For deeper targets, ground based or airborne time domain EM method is more suitable.

Another important EM method is the one utilising the very low frequency (VLF) band (15–25 kHz) produced by distant powerful radio stations. The VLF-EM field, gets modified in the presence of an electrically conducting body at depth and this change in the field is measured. From this data, apparent resistivity of buried horizon is computed. Although the interpretation of EM-VLF data based on forward modelling is mostly not unique, still it does provide useful information on subsurface conducting bodies. It is particularly useful in areas covered with resistive layers, e.g. dry desertic sands or resistive hard rocks like basalts overlying fractured-weathered water-bearing horizons. For example, Bromley et al. (1994) successfully used VLF-EM technique for detecting fractured water-bearing horizons covered under basaltic rocks in Botswana.

5.3.3   Transient EM Method

The transient (time domain) EM method is a relatively new development, as compared to the conventional frequency domain EM method and other geoelectrical methods. The method requires very sophisticated electronics for measurement and intensive computer processing for data interpretation. The transient electromagnetic method (TEM) has been specially developed for exploration in areas with extensive and thick cover of relatively low resistivity rocks. The conventional frequency domain methods have difficulties in penetrating the top low resistivity cover, and in such cases the TEM is more suited. It can be employed in ground based as well as airborne mode (Christiansen et al. 2006).

All electromagnetic methods are based upon the fact that the primary magnetic field varying in time induces an electrical current in the surrounding ground conductor and generates an associated secondary magnetic field. The information about subsurface geology, i.e. conductivity of the structures and their distribution is contained in the secondary field. However the secondary field is much smaller in magnitude than the primary field. This means that either the measurement is made very accurately or compensation for the primary field is carried out before the measurements. Normally, the primary and secondary fields are measured collectively without any possibility of differentiating between the two.

In the TEM method, the transmitter transmits a pulse and the current is switched off very quickly; the measurements are then made after the primary field has disappeared, i.e. only on the secondary field. It is necessary to measure the secondary field in a sufficient long interval of time. Thus, the TEM method measures the amplitude of a signal as a function of time, and hence the term time domain.

The TEM requires extremely accurate measurements with high precision, quality and spatial density, as the magnitude of variation involved is very small (variation of only 10–15% with respect to background response may be expected in groundwater exploration) and even a small error can make a significant impact on the interpretation.

Advantage of the TEM method is that the depth of investigation is large compared to the loop size. Though the commercial TEM equipments are very expensive, it can be a cost-effective and powerful tool in geologic conditions with top cover of low resistivity rocks, as the data acquisition is extremely fast and large amount of data are collected over a relatively short period of time.

5.4 Combined EM-Resistivity Surveys

Many of the geophysical surveys are run in combination which helps in resolving ambiguities and confirming interpretation from various angles. For example Randall-Roberts (1993) used EM-VLF, VES and SP techniques for hydrogeological exploration in fractured Precambrian gneiss in Mexico. EM-VLF was used to locate and define fractures in plan. VES soundings brought out horizontal sheeting as zones of low resistivity. SP permitted an interpretation of permeable intersections between vertical and horizontal fracturing. Thus, these data sets enabled a three-dimensional analysis from surface geophysical measurements, which was subsequently confirmed by drilling and pumping tests.

Bromley et al. (1994) describe a combination of aero-magnetic, VLF and coaxial EM surveys for groundwater studies in Botswana. In this region, the main aquifer is of Karoo Formation, broken into a series of grabens and horsts structure by several faults and is completely masked under basalts and Kalahari beds. They used low-altitude (20 m height) airborne geophysical surveys to cover 3300 km^2 area. The magnetic and VLF data were used to penetrate the masking cover. The drilling programme was guided by the geophysical data. Highest yields were obtained from fracture zones associated with VLF anomalies and NW–SE set of lineaments.

5.5 Complex Conductivity Measurements

Spatial distribution of electrical parameters of the subsurface media can yield information that can be used to estimate the characteristics of groundwater and aquifer heterogeneity. It can be used to assess the depth water-table, aquifer vulnerability to pollution, aquifer characteristics such as hydraulic conductivity, sorption capacity, dominant flow regime, water content, water movement and water quality.

Complex electrical measurements involve measurements of real and imaginary part of conductivity (Boner 2006). Models have been developed to relate hydraulic conductivity to electrical parameters. The complex conductivity measurements are sensitive to physiochemical mineral water interaction at the grain surface. In contrast to the conventional geoelectrics, a complex conductivity measurement is influenced by textural and mineralogical properties of the aquifer. Therefore it can yield information on hydraulic conductivity (or the sorption capacity) and distribution of contaminants in the pore space.

The electrical conductivity of water-wet porous rocks is mainly related to the properties of pore fluids, pore geometry, and interaction between mineral matrix and pore water. Migration of substances and electric conduction are governed by the geometry of pore network and microstructure of the mineral grain surfaces. The waste disposals or contamination sources can cause changes in aquifer characteristics, in terms of hydraulic pressure, chemical potential or temperature, which can be reflected in complex conductivity measurements.

5.6 Seismic Methods

5.6.1 Basic Concepts and Procedures

The technique is based on the principle that the elastic properties of materials govern seismic wave velocities. In general, a higher elastic modulus implies higher wave velocity in the material. In seismic surveys, waves are artificially generated by an explosion or impact of a sledge hammer, at the ground surface or at a certain depth. The resulting elastic waves are recorded in order of arrival at a series of vibration detectors (geophones), and the data is interpreted to give wave velocities. Seismic waves follow multiple paths from source to receiver. In the near-surface zone, the waves may take a direct path from source to receiver. Further, the waves moving downward into the earth may be reflected and refracted at velocity interfaces. Figure 5.13 shows

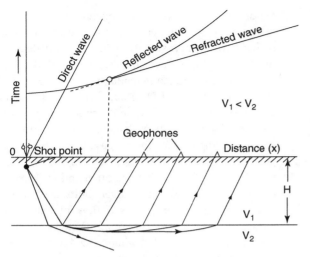

Fig. 5.13 Direct, reflected and refracted waves in a seismic survey

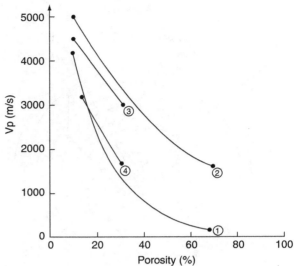

Fig. 5.14 Influence of porosity on P-wave velocity of sandstone. (After Kirsch 2006)

wave travel by direct, refracted and reflected paths. For depth calculations involving two or more layers, various algorithms have been developed. The field procedures for seismic investigations have been made greatly efficient in recent years with the aid of compact, portable computer controlled instruments.

Important rock characters influencing wave velocity are: crystallinity, porosity, cementation, weathering, and discontinuities such as bedding, joints etc. (Rabbel 2006). Massive, compact, crystalline, low-porosity rocks possess higher seismic wave velocities while unconsolidated formations possess lower velocities (Table 5.1). The presence of fractures/porosity in a rock mass causes a reduction in seismic velocity and an increase in attenuation (Fig. 5.14). These effects form the basis for the characterization of fractures by seismic methods. As seismic velocity is influenced by fracturing in rocks, velocities measured in field are much lower than those measured on intact (core) samples in laboratory, for the same rock. Degree of fracturing can be estimated to some extent from a parameter called "velocity ratio", computed as the ratio of the field (in-situ) velocity (V_F) to the laboratory velocity (V_L), in a rock. As the number of fractures decreases, V_F tends to approach V_L. It has been suggested that in general, a velocity ratio (V_F/V_L) of less than 0.5 indicates significantly fractured rock condition. Therefore, the velocity ratio (V_F/V_L) is also sometimes called "fracture index".

Seismic reflection methods are more suited for exploration of deeper structures whereas refraction techniques are more extensively used for investigation of shallower contacts. For groundwater studies, seismic refraction methods are more frequently used, the main application being deciphering the thickness of weathered zone. However, in some cases, optimum use of seismic methods may involve a combination of refraction and reflection principles. P-waves are sensitive to rock porosity and fluid saturation; this makes them a suitable tool for groundwater exploration.

In case the velocity interface is inclined (e.g. dipping strata), it leads to an additional variable. In such cases, recording seismic data in up- and down-dip directions, or reverse profiling is required to obtain true estimates of velocities and depth. The dip of the discontinuity may be calculated by comparing the reverse profile data.

5.6.2 Azimuthal Seismic Refraction Method

Azimuthal seismic refraction method can detect strike direction of major fractures in the bedrock. However, small isolated fractures or fracture zone may not be detected by refraction surveys. A fractured rock mass exhibits anisotropy in wave velocity. For a single set of steeply dipping saturated fractures, a seismic velocity maximum occurs in the direction of the fracture strike and the velocity minimum occurs orthogonal to it.

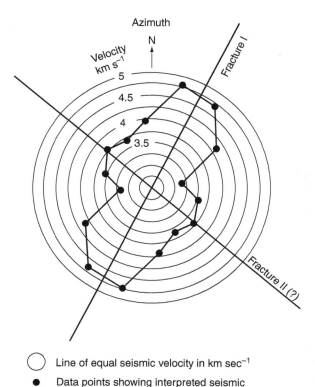

○ Line of equal seismic velocity in km sec⁻¹

● Data points showing interpreted seismic velocity through bedrock in respective azimuth directions

Fig. 5.15 Azimuthal seismic survey indicating the presence of Fracture I and Fracture II (?) at a site in Mirror Lake area, New Hampshire. (After Haeni et al. 1993)

Azimuthal seismic refraction data (using P-wave) can be collected by rotating the survey line at a constant angular increment, about a common centre point. For each survey incidence, a sledge hammer impact may serve as a energy source and geophones are spaced at equal intervals. Data analysis has to be carried out to obtain P-wave velocity for each direction. Figure 5.15 gives an example of the azimuthal plot of seismic velocity data and its interpretation.

5.7 Radon Survey

Radon (^{222}Rn) is an odourless, colourless gas produced by radioactive decay of uranium and thorium in nature. It is the only gas to be radioactive, emitting alpha particles and is therefore, hazardous for health. Its presence and concentration can be detected in water and solid material (soil/rocks) (Ball et al. 1991). Various factors which control radon concentration in groundwater are aquifer mineralogy, fracture characteristics in hard rocks, porosity of sediments and degree of metamorphism (Veeger and Ruderman 1998). A part of the radon generated in nature, may escape in carrier fluids like CO_2 or H_2O through voids and fractures. This property is useful in geothermal and groundwater investigations. Other applications of radon may include delineation of faults, basement structure and possible prediction of earthquake and volcanic activity (Kuo et al. 2006). There is often an increased content of radon in soil-gases over faults and fractured zones, owing to the increased flow of water along these discontinuities. Therefore, Radon survey has been successfully used in some areas for groundwater exploration in basement fractured rocks (Pointet 1989; Wright 1992; Reddy et al. 2006). Further, as the half-life of ^{222}Rn is 3.82 days, it is typically found in higher concentration in groundwater than in surface water. This makes it an ideal tracer for surface water–groundwater interactions such that higher concentrations of ^{222}Rn are expected at places where groundwater is discharging into the stream (Cook et al. 2006).

5.8 Radar Methods

The radar methods can be used from space, aircrafts, ground and boreholes. The radar techniques from aircrafts and space platforms are discussed in Chap. 4. Here we discuss the radar technique from ground (called ground penetrating radar) and boreholes (called borehole radar).

The ground penetrating radar (GPR) is a promising surface geophysical method for hydrogeological studies and has undergone rapid development during the last about two decades (Beres and Haeni 1991; Blindow 2006). It is a method utilizing EM reflection sensing for shallow investigations with high resolution and has found use in groundwater investigations, environmental engineering, and archeological investigations. Its basic principle is quite similar to that of reflection seismic survey with the main difference that in GPR electromagnetic radiation is used where as in seismic surveys elastic waves are used.

The GPR system emits short pulses of radio frequency EM radiation into the sub-surface from a transmitting antenna and the backscattered radiation is

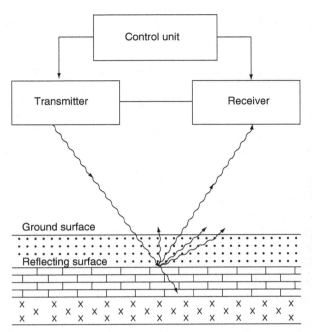

Fig. 5.16 Working principle of ground penetrating radar

sensed by the antenna (Fig. 5.16). The reflected signal is dependent upon the inhomogeneities in electrical properties of the subsurface, such as water content, mineral composition, structure and discontinuities in the rock. The received signal shows the total two-way travel-time for a signal—to pass through the subsurface, get reflected from an inhomogeneity and return to the surface. The GPR system is used in profiles across an area, moving at an average speed of 3–5 km h^{-1}.

Electrical properties mainly electrical permittivity, i.e. dielectric constant and the electrical conductivity are the main physical characteristics of rocks which govern depth of penetration and reflectivity at layer boundaries. Electrical permeability (dielectricity) depends upon the polarization properties of material. It is the dominating factor for the propagation of EM waves in the medium. The speed of EM waves is used for time depth conversion of GPR sections. Typical values of electrical permeability are water = 80, saturated sand = 20–30, air = 1 (Table 5.1). With increasing water saturation, the permittivity systematically increases.

Lower frequencies of EM radiation provide deeper penetration and higher frequencies undergo a skinning effect. However, higher frequencies yield better resolution than lower frequencies. Therefore, a trade-off has to be made with respect to resolution and depth penetration (in terms of frequency). The GPR resolu-

tion depends upon the frequency, polarization of the EM wave and contrast in EM properties of the media. The vertical resolution is better in the wet materials than in dry. GPR and seismic reflection have generally not found application in semi-arid and arid areas.

The GPR profiling enables detection of subsurface conductive zones/layers covered by higher resistivity materials. It can be used for estimating depth to groundwater, detecting lenses of perched groundwater, mapping of clay-rich confining bands and monitoring contaminant transport in the vadose zone. This method is also used for mapping buried objects (drums, pipe lines) and abandoned waste disposal sites including hydrocarbon contaminated sites (Domenico and Schwartz 1998; Blindow 2006).

Borehole radar method is quite similar to the surface radar method, except that the survey is carried out in a borehole, and therefore, the transmitter and receiver both are oriented vertically in a borehole. Radar readings are taken at constant intervals as a function of depth. Data are digitally recorded, processed and displayed. From this data, it is possible to interpret presence of important fractures and master joints in the rock (e.g. Olsson et al. 1992; Haeni et al. 1993). In the Mirror Lake site, New Hampshire, USA, a combination of cross-hole hydraulic test data and GPR was found useful for characterizing the fractured rock aquifer in terms of identifying preferential flow paths (Hao et al. 2008).

5.9 Gravity and Magnetic Methods

Gravity and magnetic methods indirectly yield information about favourable structures for groundwater occurrence. These methods make use of natural fields of gravity and geomagnetism (Ernstson 2006). Changes in gravitational and magnetic fields may be observed on the Earth's surface, which could be related to lateral changes in density and magnetic susceptibility of the material at depth. The variation/contrast in density and susceptibility produce small but measurable changes of corresponding fields. The instruments used for these measurements are the gravimeter and the magnetometer. The practical unit for measuring gravity and magnetic fields are milligal and gamma (nanotesla) respectively. In these surveys, the gravity and magnetic values are measured at the pre-fixed observation points in profiles. Various corrections need to be done on the data (e.g. Dobrin 1976). The corrected values of gravity and magnetic field are plotted at the stations/ profiles and iso-anomaly map is drawn. Quantitative interpretation is done by analysing the nature of contours, highs and lows.

Across a fault plane, a steep gradient of gravity is observed. Gravity high implies denser rocks closer to the ground surface, e.g., basic intrusions. Lower density materials and cavities produce gravity lows. Gravity methods have a rather coarse resolution, the method being suited for detecting large structures, e.g. regional folds, subsurface domes etc. For the specific purpose of aquifer location e.g. solution cavities in karst areas, microgravity measurements have an interesting potential. Both laterally and vertically extended low density features can also be mapped. The technique requires good knowledge of rock density and their variations, derived from measurements on rock samples.

Magnetic surveys (magnetometry) are among the most cost effective of geophysical techniques for geological mapping. These are quite effective in delineating subsurface mafic dykes, as well as quartz and pegmatite veins, the latter due to diamagnetic property

(i.e. negative magnetic susceptibility of quartz). Magnetic surveys can be conducted from space, aerial and ground based platforms. In fact, low-altitude aerial magnetic survey is an extensively used technique in geoexploration.

Magnetic surveys can give an idea of the major geological-structural features. The anticlines would produce positive and synclines negative anomalies. Fault is indicated by a sharp gradient in the magnetic contour map. Since the basement rocks are more magnetic as compared to the overlying sediments, the trend of the magnetic contours is largely related to structural trends in the basement and distribution of basic intrusions. Modelling of magnetic data can bring out highly useful information on structure, dip of faults, contacts etc. This is particularly important as fractured zones along faults and contacts of dykes etc. also form potential aquifer. Orientation of these features can be delineated from magnetic surveys.

Both gravity and magnetic profile data may be portrayed in image mode. This facilitates spatial filtering and image enhancement, which renders image interpretation easy. Figure 5.17 is an example from Botswana, showing the image of aeromagnetic data along with

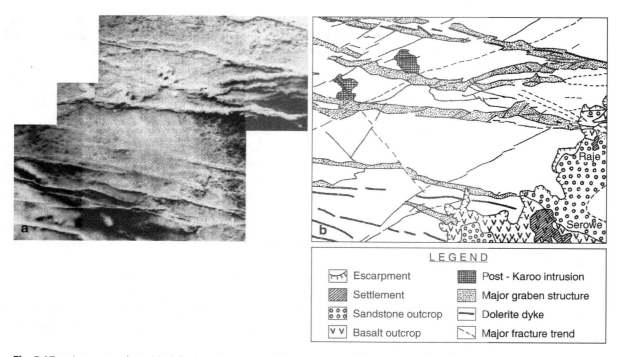

Fig. 5.17 a Aeromagnetic residual field image of an area in Botswana and **b** its structural geological interpretation map. (After Bromley et al. 1994, reprinted by permission of 'Ground Water')

an interpretation geologic map. Further discussion on image manipulation—GIS aspects of various geodata sets, to facilitate a coherent integrated interpretation, is discussed under GIS (Chap. 6).

5.10 Magnetic Resonance Sounding (MRS)

Magnetic resonance sounding is a comparatively new non-invasive geophysical method for groundwater investigations. It was first developed in Russia in 1980s by A.G. Semenov and his co-workers and is a specific application of the well known NMR (nuclear magnetic resonance) to groundwater investigations. The nuclear magnetic resonance (NMR) can be observed only in case of specific isotopes of some elements (e.g. hydrogen, carbon, phosphorous etc.) which possess a net nuclear angular momentum and a magnetic quantum number. In the context of groundwater, hydrogen nucleus, which is made of a single proton, is most important as hydrogen nuclei form a major constituent of water molecules. Such a nucleus behaves like a tiny weak magnet and gets aligned with the local (static) magnetic field. An external excitation magnetic field, usually oriented perpendicular to the ambient magnetic field, is used to momentarily displace the average nuclear magnetic moment from the direction of the ambient magnetic field. This leads to precession around the local magnetic field orientation. After excitation, the precessing nuclei will return to their steady state orientation at a rate determined by various 'relaxation factors' and relaxation time forms an important measurement parameter.

The method measures magnetic resonance signal generated directly from protons of hydrogen molecules in groundwater and therefore it has an advantage of direct detection of subsurface water. The depth and thickness of aquifer is estimated by measurements with varied pulse magnitude. The MRS can also be used for estimating porosity and permeability and in predicting the well yields and correlation-interpolation between boreholes. Broadly, it has been found that there is a good correlation between MRS transmissivity estimates and those obtained from borehole pumping tests (Legchenko et al. 2004) (Fig. 5.18). It is particularly useful in fractured and karstic aquifers where there is spatial variation in hydraulic conductivity (Vouillamoz et al. 2003; Roy and Lubczynski 2003;

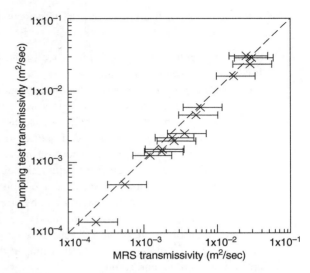

Fig. 5.18 Comparison of transmisivity values derived from MRS and pumping test analysis. (After Legchenko et al. 2004)

Legchenko et al. 2004). MRS technique is also being used in petroleum industry for estimating porosity and permeability of reservoir rocks. The combined use of MRS and electrical/electromagnetic method provides more reliable information about subsurface geology and water quality. However, the available MRS method has limitations for applications in fractured formations with low effective porosity (<0.5%).

5.11 Geophysical Tomography

The advent of Computer Aided Tomography (CAT) has revolutionized medical sciences. Even though similar techniques have been traditionally utilized by seismologists as well as by exploration geophysicists in the field of seismic prospecting for quite sometime, no such special term was coined. But with the emergence of necessary mathematical tools viz., Algebraic Reconstruction Technique (ART), Simultaneous Iterative Reconstruction Technique (SIRT), Back-Projection methods etc, in the field of medical imaging, a new discipline has eventually taken shape in the field of geophysics also, viz. geophysical tomography. Under the umbrella of geophysical tomography, several methods can be included like seismic tomography, electromagnetic tomography, resistivity tomography etc. 'Tomography' simply means a technique used to obtain an image of selected plane of a solid object

(Worthington 1984). It comes from the Greek world '*tome*' meaning a slice.

1. *Seismic travel time tomography*: It basically entails imaging intervening medium between array of receivers and sources. As per geometric configuration of arrangement of sources and receivers, we can have cross–hole, surface–hole and surface–surface travel time tomography. A fracture zone, either air or water filled, in an otherwise massive rock constitutes a low velocity region. Depending upon the orientation of fracture, one among the above modes of seismic imaging can be effective. For example, surface–surface mode is good for horizontal or sub-horizontal fractures whereas near-vertical fractures can be better imaged by cross-hole method, and so on.

   Recently, the cross-hole method of seismic velocity measurement is being more extensively used. It helps in assessing the whole rock mass properties in-situ, between two boreholes. Cross-hole seismic survey is particularly useful in locating cavities or old mine workings in urban areas. Cross-hole seismic measurement underneath the foundations of a building is probably the only effective method to assess the rock mass existing beneath the building. Identification of underground cavities is important to check possible damage to buildings due to subsidence on account of subsurface cavities. In hot dry rock (HDR) systems, cross-hole seismic method is used to delineate cavities created by explosive HDR stimulation. This method is also used for groundwater investigations for mapping fractures (Carruthers et al. 1993). From cross-well seismic investigations in the crystalline rocks near Mirror Lake, Ellefsen et al. (2002) concluded that hydraulic conductivities were higher when P-wave velocity was low ($<5100\,\mathrm{m\,s^{-1}}$), than when it was high ($>5100\,\mathrm{m\,s^{-1}}$), as the fractures increase the hydraulic conductivity and lower the P-wave velocity. This empirical relation helped in preparing a velocity tomogram and thereby creating a map showing zones of high hydraulic conductivity, which was later confirmed from independent hydraulic tests.

2. *EM tomography*: The interest in cross-hole EM tomography is mainly for imaging inter-well electrical conductivity. The sensitivity of electrical conductivity to porosity, fluid type, saturation and temperature has led to the development of cross-well EM systems and imaging algorithms (Rector 1995). Field examples and numerical simulations demonstrate remarkably good resolution of inter-well features when compared to surface EM techniques (Spies and Habashy 1995; Wilt et al. 1995). The cross-well EM tomography can be used to map fractures within highly resistive compact rock (Alumbaugh and Morrison 1993). This method can also track an injected slug of water (Wilt et al. 1995), as conductivity images of data collected before and after injection in a study showed a clear anomaly as a result of salt water plume and indicated the direction of plume migration.

3. *Electrical resistivity tomography* (*ERT*): With the advent of multi-electrode, micro-processor based resistivity measurements (Griffiths and Turnbill 1985), it is now possible to carry out three-dimensional resistivity surveys in a variety of combinations like cross–hole, hole–surface, surface–hole and surface–surface. The methods are being increasingly applied to problems of groundwater flow and pollutant movements, and can as well be used to delineate water filled fractures within moderately resistive host rocks.

   Cross-hole anisotropic electrical and seismic tomograms of fractured metamorphic rock have been obtained at a test site by Herwanger et al. (2004) where they report a strong correlation between electrical resistivity anisotropy and seismic compressional-wave velocity anisotropy apparently related to rock fabric.

5.12 Subsurface Methods

Subsurface methods including exploratory drilling and well logging are essential for confirming results and interpretations made from surface geological and geophysical investigations. Although subsurface investigations are more expensive than surface methods, the precision and reliability of data which they provide more than offset this consideration.

5.12.1 Exploratory Excavation and Drilling

Exploratory excavation may be done by putting pits, trenches, adits and shafts, depending upon the type of problem, topography and needs. The technique provides a means for directly observing and mapping subsurface features, for example, observing how discontinuities continue and behave at depth, and also sampling subsurface rocks. In this stage, various faces and walls need to be very carefully mapped and logged and mutual relationships of various fractures and discontinuities recorded. The excavation methods should be such as to minimally disturb the rock conditions.

However, exploratory excavation being expensive is limited to shallow reaches. For deeper exploration, exploratory drilling is carried out. It helps define the geometry and extent of the aquifer and assessing the groundwater potential. The test holes are preferentially located in such a way that in case of prospects of a good aquifer, the same boreholes can be converted into production wells by redrilling, or reaming to a larger diameter. The test holes also serve as observation wells for monitoring groundwater levels.

The data obtained during drilling is recorded as lithologic log. It is a description of geologic characters of various strata such as lithology, thickness, core recovery etc., encountered during drilling. Drilling time log, consisting of a record of time taken to drill every 2 m of depth is helpful in indicating where there is change of strata or intensity of fracturing and weathering. The rate of penetration of a stratum can be correlated with the formation characteristics and hence with its water-yielding capacity, in a relative manner.

5.12.2 Geophysical Well Logging

Geophysical logs are obtained by lowering a probing tool in a sonde down the borehole. They are used to study the variation of physical properties of subsurface rocks (including fractures etc.) and their fluids. Correlation of logs may reveal the nature of stratification and extension of structures and fractures. The boreholes are logged with a number of geophysical probing tools which provide direct and some quantitative data about the hydraulic characters of subsurface formations.

With the advancement in microelectronics, compact logging units are now available with softwares for interpretation. A number of properties of both formation and interstitial water, such as coefficient of diffusion, formation factor, hydraulic conductivity, specific yield, concentration exponent can be estimated. Fur-

ther, borehole geophysical data can be used to estimate water properties such as salinity, viscosity and density, and formation properties such as porosity and permeability (Jorgensen 1991).

As far as fracture evaluation is concerned, a large number of logs are required to properly detect and interpret fracture characteristics. Fractured zone produces anomaly with respect to normal or constant hole size. Fractures when open, lead to high permeable paths which can be detected by logging in terms of high drilling rate, loss of drilling fluid, poor core recovery and/ or significant increase of borehole size. The treatment here gives a brief resume of the well logging methods with special reference to fractured rocks.

1. *Spontaneous potential log* gives a record of electric potential with depth in the borehole. It is useful for shale vs. sandstone discrimination, but it has limited utility for fracture identification.
2. *Gamma ray log*: Gamma rays are emitted by all natural rock formations as a result of random disintegration of naturally present radioactive elements. The elements producing gamma rays are potassium, uranium and thorium (KUT). The log records total count of gamma rays. The KUT elements naturally concentrate in finer-grained materials (clays, silts) where they are adsorbed in minerals like clays. Therefore, the gamma ray log is regarded as a clay indicator. In fractured igneous and metamorphic rocks, the log response is relatively less consistent. For example, the log may show peak responses due to potassium rich minerals (e.g. feldspars) and/or a clay-rich weathered zone and/or a zone of leaching. Conversely, is some cases a lower activity against the weathered rock may also be observed. Fluid circulation or past fluid circulation in the rock mass can sometimes be also inferred from the gamma ray log. This is due to the presence of uranium oxide which is soluble and highly mobile and can be precipitated in joint and fracture surfaces which form fluid routes within the rock mass. Therefore, gamma logging can identify this local activity where the boreholes intersect such fractures.
3. *Caliper log* measures the diameter of an uncased drill hole as a function of depth. The measurement is obtained with a 3- or 4-arm probe which is electronically opened when the probe is at the bottom of the borehole, and the variation in the borehole diameter is recorded as the probe is winched to the surface. Permeable zones will usually show

a reduction in borehole size due to deposition of thicker mudcake on the borehole. Fractured horizons usually show an increase in the borehole size which may occur due to breaking of the formation wall during drilling. Fracture orientation is likely to affect the borehole ellipticity.

4. *Bore fluid logs* include logging of bore fluid parameters such as electrical conductivity and fluid temperature. Out of these, temperature log has been the most common. Fractured permeable formations are characterised by low temperature anomalies, which occur due to locally increased mud circulation. Logs of fluid temperature and fluid electrical conductivity can be interpreted in terms of groundwater conditions. Fluid logs are run under different hydraulic conditions—usually at rest and also during pumping. A comparison of the two data sets reveals the position where water enters the borehole. These logs are also helpful in investigating inter-aquifer migration of water, adequacy of grouting, quality of groundwater and other related aspects.

5. *Resistivity logs*: An electric log is a record of the apparent resistivity of the subsurface formation with depth. There are numerous variation in the resistivity logs. The electrical logs cannot be run in cased holes and may be operated in dry holes or preferably fluid filled holes. The measured apparent resistivity depends upon the geometric fracture characteristics and the nature of fluid filling the fractures. It is influenced by fracture orientation, size, length, spacing etc.

The range of resistivity in hard rocks is quite large. Fractures filled with water tend to cause decrease in apparent resistivity in hard rocks. A useful method is by using single point resistance (PR) technique. The PR log represents the varying electrical resistance between a single downhole electrode and a fixed surface electrode. It does not measure the true rock resistivity and is strongly influenced by borehole diameter change. However, unlike multi-electrode resistivity logging devices, its response is symmetrical and bed boundaries are recorded in the correct position, and the relative response is useful for recording the junction between rock units and for correlation. Normal resistivity logs in adjacent vertical boreholes can help in mapping the lateral extension of subhorizontal fractures. Vertical or near vertical fractures may not be detected by induction log due to the fact that the induced cur-

rent tends to flow in horizontal loops around the borehole and therefore vertical fracture containing conductive fluid, may go undetected in the log. On the other hand, horizontal fractures filled with conducting fluid appear as conductive anomalies.

Microresistivity logs are likely to miss fractures, as they measure only a small volume of rock around the wellbore. Fractures lead to increased conductivity due to higher local porosity and greater water saturation. Different combinations of laterologs and induction laterologs can be used to decipher presence of fractures close to the wellbore and distant from the wellbore (Van Golf-Racht 1982). However, these advanced techniques are more used in petroleum industry.

6. *Dipmeter log* basically records the dip angle and dip direction of a bedding plane intersecting the borehole. The tool consists of four radial pads positioned at angulary interval of 90°. It is rotated in the borehole at a uniform speed, as it is winched on, yielding four microresistivity curves. The azimuth recording of electrode 1 is continuously made. The dipmeter response may describe all types of discontinuities from horizontal to vertical. Fracture identification log (FIL) is an improved tool for detecting fractures. Higher efficiency in FIL is obtained by superimposing the response of a couple of electrodes, i.e. combination of electrode responses in a specific manner.

7. *Porosity log*: Under porosity logs are included, density, neutron and sonic logs. These logs are capable of detecting fractures and evaluating secondary porosity. In principle, the secondary porosity must be evaluated as the difference between the bulk porosity and the matrix porosity, both of which are measured through logs. A double porosity model is used to link the bulk porosity and the matrix porosity (Sect. 7.2.2). The density log is a gamma–gamma ray log. A gamma ray beam is emitted from an artificial source and a counting system detects the backscattered intensity, which is related to the density of the rock. A higher density causes a relatively lower level of gamma–gamma intensity. Fractures causing higher secondary porosity are indicated by higher gamma–gamma ray count.

Neutron logs respond primarily to the amount of hydrogen present in the formations. In the case of open, water-filled fractures, neutron logs exhibit anomalies indicating higher porosity. The sonic log is very useful in fracture detection, particularly in dense rocks. It uses a transmitter and a wave receiver. As the transmitter makes an energy emission, different types of shear and compressional waves are generated and received at the receiver. The amplitude, velocity and attenuation of different wave types are influenced by the fracture characteristics. A study of amplitude, attenuation and arrival times of shear and compressional waves can provide indication of fracture orientation and lithology.

8. *Borehole Televiewer* (*BHTV*) is used to detect and evaluate fractures and formation boundaries by direct measurements. It may be treated as a partial substitute for continuum well coring. Combined with drill core data, it is a highly valuable tool. BHTV is carried out in boreholes filled with homogeneous, gas-free liquid such as drilling mud, freshwater etc. BHTV includes a source of acoustic energy and a magnetometer mounted in the tool. The tool is rotated at a uniform speed during logging. The changes in the uniformity of the borehole walls such as fractures, rugs, pits, traces etc. are reflected as changes in picture intensity. It produces a two dimensional image of the borehole. The intersection of fractures with the borehole can also be observed in the BHTV image. The fractures perpendicular to the borehole appear as horizontal traces and those parallel to the borehole appear as vertical traces. Fractures intersecting at angles appear as sinusoids. From these data, the dip and strike of fractures can be calculated. Borehole video camera (BVC) which produces either a colour or a black-and-white image of the borehole, is a cost-effective method to locate fractures, changes in lithology and in-flowing zones, especially in shallow boreholes (depth < 100 m) (Delleur 2007).

9. *Hydrophysical logging*: Hydrophysical logging can identify both vertical and horizontal flows in the borehole, since it surveys the entire length of the borehole rather than providing point measurements. Therefore, it is a valuable method for obtaining profiles of flow characteristics and vertical distribution of permeability (Kresic 2007). *Flowmeter log* is a type of borehole fluid log. It is basically a velocity meter and makes a continuous record of flow profile vs. depth. It provides a confirmation of fracture location. Flowmeter logging during pumping can detect increased velocity of water flow moving to the pump, at various inlet points in the borehole, and is therefore highly useful in detecting water-bearing

fracture zones. Packers can be used to isolate portions of the borehole for precise characterisation. Flow-logging tests between boreholes (cross-hole tests) may indicate the degree of connectivity of transmissive zones. Vertical flowmeter logging is practiced to determine vertical flow in a borehole due to differences in hydraulic head between two transmissive units or fractures. Fluid replacement logging is a recently developed technique for identifying the permeable fractures in a borehole. The technique involves replacement of borehole fluid by de-ionized water and subsequent measurement of variations in the electrical conductivity of the fluid in the borehole with time (NRC 1996).

Figure 5.19 shows a suite of well logs from a high-yielding well in Zimbabwe as an example. The

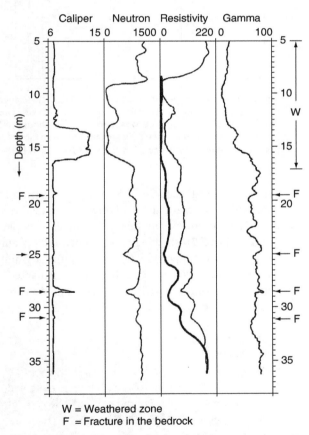

W = Weathered zone
F = Fracture in the bedrock

Fig. 5.19 A suite of well logs including calliper log, neutron log and resistivity log from a site in Zimbabwe. The weathered zone extends up to about 17 m depth. Fractures are indicated at 19.5 m, 25 m, 27.5 m and 31 m levels (scales—calliper: inches, neutron: CPS, resistivity: chm meter, gamma: cps). (After Carruthers and Smith 1992)

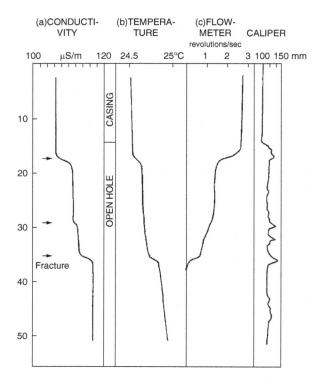

Fig. 5.20 A suite of well logs showing conductivity, temperature, impellor flowmeter and caliper logs in a fractured granite terrain with about 13 m thick weathered rock on the top. The logs were run while the well was being pumped. Note the deflections marking the occurrence of fractures. (After Lloyd 1999)

coincident local excursions in caliper, neutron and resistivity logs mark the individual fractures which provide major inflows to the well. Another example is given by Fig. 5.20 showing conductivity, temperature, flowmeter and caliper logs in a borewell in a granitic terrain. The top 17 m is cased. Fractures are marked by correlated changes in well logs indicating increase in conductivity, temperature and calliper logs.

Summary
Geophysical exploration utilizes the variation in physical properties of rocks such as electrical resistivity, electrical conductivity, density, magnetic susceptibility etc. to differentiate and decipher the various subsurface geologic horizons and their hydrogeological characteristics. Rock anisotropy due to bedding, foliation, fractures leads to anisotropy in geophysical properties. It is essential to integrate geophysical data with field geological and drilling data.

Electrical resistivity surveys have been by far the most popular methods for groundwater studies. Vertical resistivity sounding is used to give depth profile at a place whereas resistivity mapping is carried out for delineating lateral geologic variation. Electromagnetic (EM) methods are particularly useful in areas possessing surface layers of highly resistive nature. In many cases, resistivity and EM surveys are run in an integrated manner. Seismic methods utilize elastic properties of materials. Seismic reflection methods are generally more suited for exploration of deeper horizons whereas seismic refraction techniques are used for shallower depths. In hard rocks, their main application is in delineation of top weathered regoilith zone.

Radon survey is used for detecting fracture and voids etc. and for detecting groundwater discharge into stream. Ground penetrating radar (GPR) has its application in delineating watertable and strata layering/inhomogeneities etc. Magnetic resonance sounding is a relatively new technique to give direct detection of subsurface water. Geophysical tools are also used for logging of drill-wells for subsurface exploration.

Further Reading

Dobrin MB, Savit CH (1988) Introduction to Geophysical Prospecting. McGraw-Hill, New York, NY.

Kearey P, Brooks M, Hill I (2002) An Introduction to Geophysical Exploration. Blackwell Science Ltd., Oxford, UK.

Kirsch R (ed.) (2006) Groundwater Geophysics: A Tool for Hydrogeology. Springer-Verlag, Berlin-Heidelberg.

Parasnis DS (1997) Principles of Applied Geophysics. Chapman and Hall, London.

Telford WM, Geldart LP, Sheriff RE (1999) Applied Geophysics. Cambridge University Press, Cambridge, UK.

Geographical Information System (GIS)

6.1 Introduction

Geographical Information System, also called Geo-based Information System (GIS), is a relatively new technology. It is a very powerful tool for processing, analyzing and integrating spatial data sets (e.g. Star and Estes 1990; Lo and Yeung 2006; Chang 2008; Harvey 2008). A GIS deals with information on locational patterns of features and their attributes (characteristics). It can be considered as a higher-order computer-coded map which permits storage, selective dedicated manipulation, display and output of spatial information. GIS software provides the functions and tools needed to store, analyze, and display information about geographical locations. The key components of GIS software include: (a) a database management system (DBMS); (b) tools that create intelligent digital maps that one can analyze, query for more informa-

tion, or print for presentation; and (c) an easy-to-use graphical user interface (GUI).

Figure 6.1 explains the working concept of a GIS. In a very comprehensive sense, GIS may mean identifying data needs, acquiring data, data management, processing and analysis of data and decision-making. In normal usage, however, GIS means spatial data processing, integration and analysis. GIS has become a standard, rather indispensable, tool for handling spatial information for earth's resources exploration, development and management.

6.1.1 *Why GIS for Groundwater Studies?*

For handling groundwater data, the GIS-technology is aptly suited, for the following main reasons:

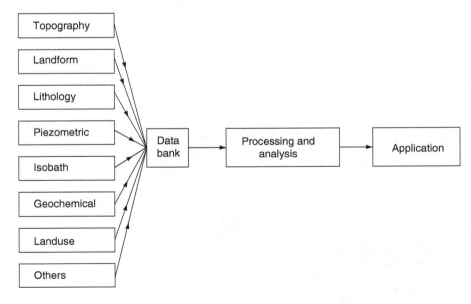

Fig. 6.1 The concept of data integration in a GIS

B. B. S. Singhal, R. P. Gupta, *Applied Hydrogeology of Fractured Rocks,*
DOI 10.1007/978-90-481-8799-7_6, © Springer Science+Business Media B.V. 2010

1. *Concurrent handling of locational and attribute data:* In groundwater studies, one has to deal with information comprising locational data (where it is?) and attribute data (what it is?). GIS packages have the unique capability to handle locational and attribute data; such a capability is not available in other groups of packages (Fig. 6.2).
2. *Variety of data:* Groundwater investigations often comprise diverse forms and types of data, such as: (a) topographic contour maps, (b) landform maps, (c) lithological maps, (d) structural geological maps, (e) isobath map (contour map of equal depth of water-table), (f) isogram (isocone) maps depicting groundwater characteristics by contours of equal concentration of dissolved solids (TDS) or ions, (g) drainage density and other geomorphic maps, (h) tables of various observations and data sets, and (i) point data, say locations and water-levels in observations wells, or spring discharge etc.

   In these, some of the variables are of continuous type, e.g. TDS content, water-level data etc., and some others are of categorical type, such as low/ medium/high drainage density, or gravel/marble/ granite lithology (for continuous and categorical data types, see Sect. 6.2). It is essential to integrate the spatial information for coherent and meaningful interpretation, and to avoid compartmentalization of data. GIS offers technological avenues for integrating the variety of data sets in both qualitative and quantitative terms, hitherto not available through any other route (e.g. Star and Estes 1990; Gupta 2003).

3. *Flexibility of operations and concurrent display:* Modern GIS packages are endowed with numerous functions for computing, searching for and classifying data, which allow processing and analysis of spatial information in a highly flexible manner and concurrent display, interactively.
4. *Speed, time and costs of processing:* Advances in microelectronics and computer technology have made it possible that modern GISes can store, process and analyze large volumes of data which otherwise would be too expensive, tedious and time-consuming to carry out by other methods.

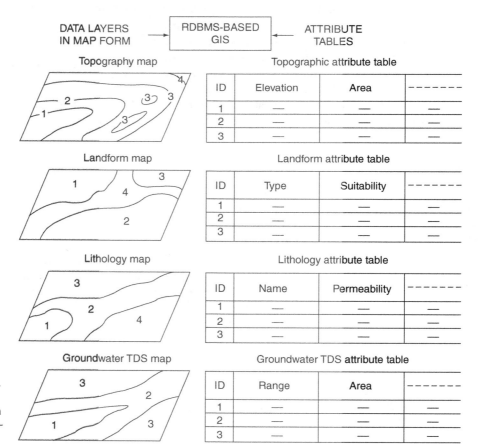

Fig. 6.2 Schematic representation of GIS-working. The GIS maintains a link between the map feature and the corresponding tabular information

Fig. 6.3 Data sources of various types and formats required to be input into a digital data bank

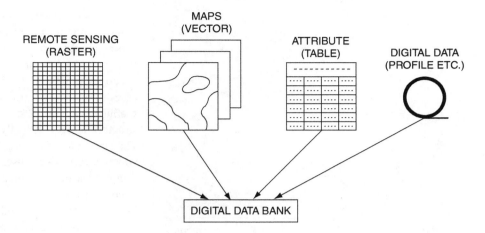

5. *Higher accuracy and repeatability of results:* The technique being digital computer-based, yields higher accuracy, in comparison to manual cartographic products. The results are amenable to rechecking and confirmation.

In a typical hydrogeological investigation, the sources of data could be satellite or aerial sensing, field surveys, geochemical laboratory analyses, geophysical exploration data, etc. and may be available in the form of maps, profiles, point data, tables, lists etc. (Fig. 6.3). If GIS methodology is not used, then integrating such a variety of data sets would involve elaborate manual exercises in order to deduce the relevant information.

6.2 Basics of GIS

We mention here some very basic concepts of GIS (for details, refer e.g. Star and Estes 1990; Chang 2008; Harvey 2008; Lo 2007). The hardware for a GIS comprises basic computer (viz. CPU, storage devices, keyboard and monitor), digitizer and/or scanner for inputting spatial data, high resolution colour monitor for displaying spatial data in image mode, plotter for production of high quality maps, and printer for printing tables, data, raster maps etc.

In GIS, the data-base is created to collate and maintain information. It is a collection of information about things or objects. The geographical or spatial information have two fundamental components (Fig. 6.2):

1. Location (position) of the feature (where it is?), e.g. location of a well, spring, river or city etc.

2. Attribute character of the feature (what it is?), e.g. lithology, or water-table elevation, or TDS in water etc.

Data pertaining to the above two aspects are explicitly or specifically recorded in GIS. There could be other critical characteristics such as time (when did the phenomenon occur?), or spatial-textural relationship (neighbourhood relations with other features), which are not so explicitly recorded in GIS, but could be possibly searched for and deduced.

6.2.1 Location Data

In GIS, the location or spatial position is given in terms of a set of latitude/longitude, or relative coordinates, or reference axes like Easting/Westing. From a geometrical point of view, all features on a map can be resolved into points, arcs and polygons (Fig. 6.4). All features, whether points, lines or polygons can be described in terms of a pair of coordinates, viz. points—as a pair of x–y coordinates; lines—as a set of interconnected points; and polygons—as an area enclosed by a set of lines.

6.2.2 Attribute Data

Attribute data is the information pertaining to what the feature is, i.e. whether the point indicated is a pumping well or an observation well, or information at the specified location pertains to water quality, or depth of

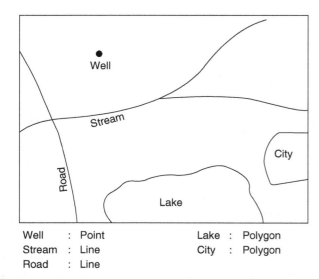

Well	:	Point	Lake	:	Polygon
Stream	:	Line	City	:	Polygon
Road	:	Line			

Fig. 6.4 Types of location data: points, arcs and polygons

comprises of an array of regular cells, in rows and columns. Limitations in raster structure arise from degradation in information due to quantization and cell size. In vector structure, spatial information is digitized and stored as points, arcs (lines) and polygons. In such cases, topology (i.e. mutual relations between various spatial elements) is specifically defined. The vector structure is geometrically more precise and compact. The method of cartographic manual digitization, which is very widely used, utilizes vector mode. However, vector mode is relatively tedious for performing many GIS operations, such as overlay, neighbourhood functions etc. On the other hand, raster structure is simple, easy to handle and understand and suitable for performing most GIS functions, as well as it facilitates all image processing operations. In this work, the examples of groundwater data processing where not otherwise specifically mentioned, utilize raster data structure.

water-table, or topography, or rock type etc. The attribute data is considered to be of two basic types—categorical and continuous. Table 6.1 describes the possible varieties of attribute data with examples drawn from the field of groundwater. A major challenge or problem is to combine all this varied type of information, and for this GIS is a powerful tool.

In GIS, all information is stored in digital form in a computer, in the form of maps (also called coverages) and tables. The link between the tables and the corresponding map feature is maintained. Schematically, the whole system may appear as in Fig. 6.2.

6.3 GIS Methodology

Broadly, a GIS comprises five main segments or stages (Fig. 6.6): 1. data selection and acquisition, 2. data preprocessing, 3. data management, 4. data manipulation and analysis, and 5. data output.

6.2.3 Basic Data Structures in GIS

Two basic types of data structures exist in GIS: vector and raster (Fig. 6.5). Raster is a cellular organization. It

6.3.1 Data Selection and Acquisition

Data acquisition includes identifying data needs and locating and collecting the data sets. For identifying data needs one requires good understanding of the thematic problem to be tackled through GIS. Some data may be available from existing records and some other data may have to be specifically generated.

Table 6.1 Characteristics of continuous and categorical attribute data types

Types of attribute/property	Type of scale	Remarks	Example
Categorical	Nominal	Mutually exclusive categories of equal status	A, B, C; or quartzite, schist, marble etc.
	Ordinal	Hierarchy of states in which all intervening lengths are not equal	Drainage density; low, medium, high
Continuous	Interval	Possess lengths of equal increment, but no absolute zero	Linear contrast stretched image
	Ratio	Possess lengths of equal increment and also absolute zero	Rainfall (mm)

Fig. 6.5 Basic data structures in GIS. **a** A map and the same in **b** vector format and **c** raster format

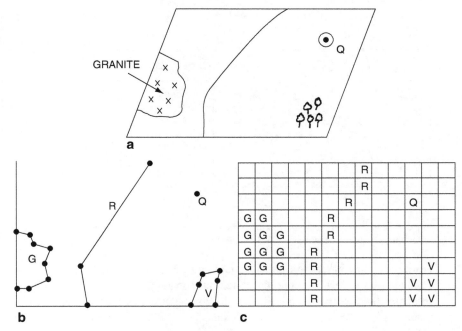

6.3.2 *Data Preprocessing*

Preprocessing is often required to convert the data sets into a form suitable for storage in the GIS data base (or data bank), and make them amenable for integrated analysis. Prepocessing could vary from simple to farily complex operations including format conversion, structure conversion, resampling, interpolation, registration etc. Main preprocessing operations in image-based GIS are: (a) data input, (b) rasterization, (c) interpolation, (d) black-and-white image display, and (e) registration (Fig. 6.7).

Data input means the procedure of encoding data into a computer-readable form, or entering data into the GIS computer. Two types of input methods are used in combination: (a) keyboard entry for recording attribute data, and (b) digitizing for entering locational data. The keyboard entry involves manually entering the data at a computer terminal. All attribute data are entered in this manner into tabular forms, called tables. Feature labels, i.e. numerals to identify points, lines or polygons are also entered through keyboard. *Digitizing* involves entering positional (or locational) data into the computer, i.e. converting maps into digital data forms. This is done by either coordinate digitizing or by scanning, which produce vector and raster data respectively. A vector to raster conversion may be necessary for some applications.

Interpolation is the process of predicting unknown values using the known values in the vicinity. It is a type of neighbourhood operation. Many GIS software carry modules for following types of interpolations. (a) point-based: to estimate values at predetermined locations using points of known position and values (e.g. Thiessen poly-

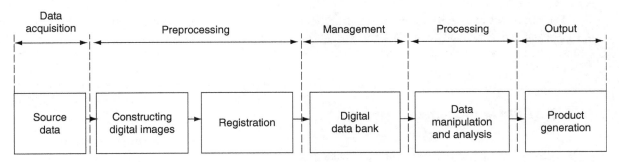

Fig. 6.6 Main segments of a GIS

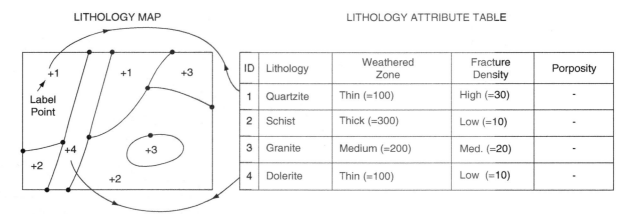

Fig. 6.7 Schematic to show interlinking of map features and their attributes in tables (coded in digital form) during data input (preprocessing) stage for developing a GIS data bank

gon); (b) isoline-based: to estimate values between isolines (e.g. water-level contours); (c) areal interpolation: in this method, polygons, are filled with some selected values (usually in the range 0–255 to suite image display). Often categorical data of nominal and ordinal type are handled by areal interpolation method for a relative representation. Interpolation programmes may employ a wide range of statistical methods for computation.

Image display Employing the standard image processing systems, the rasterized and interpolated spatial data can be displayed as a black-and-white image. Typically, one image displays variation in a single parameter across the scene in shades of gray. The image can be suitably contrast-stretched and enhanced. Examples of groundwater images are given later (Figs. 6.11, 6.12, 6.15).

Registration is the process of superimposing images, maps or datasets, over one-another, with geometric congruence. It means adjusting one data layer so that it can be correctly overlayed on another data layer of the area, cell by cell. Registration is carried out using control points and resampling. The procedure is analogous to rubber-sheet stretching of one data layer to suite the geometry of the other.

The above set of preprocessing procedures yields a co-registered data bank in GIS form, for further image-based data processing, interpretation and analysis (Fig. 6.7).

6.3.3 Data Management

The GIS-software packages are built around Data Base Management Systems (DBMSes) which comprise a set of programs to manipulate and maintain data in a data base. The DBMS approach provides numerous advantages for data processing such as efficient, controlled ordering and organization, sharing of data, data independence, centralized control, data integrity, reduction in data redundancy etc. The DBMS in GISes use relational data models, and hence are also called Relational DBMSes. They provide advantages in operations such as flexible user-preferred views, search functions etc.

6.3.4 Data Manipulation and Analysis

Once the groundwater data such as water-level, aquifer parameters, hydrochemical data etc., i.e. parameters reflecting groundwater characteristics in spatial domain, are co-registered is an image-based GIS, wide possibilities exist for data processing, enhancement and analysis. The whole gamut of standard image processing modules (Table 4.8) and GIS modules (Table 6.2) are available for data processing. Some examples are given in this treatment. Interested readers may find further information elsewhere (e.g. Jensen 1986; Star and Estes 1990; Gupta 2003). Considering the organisation of most commercially available software packages, the discussion is divided into two parts: (a) image processing operations, and (b) GIS analyses functions. However, it may be mentioned that any distinction between raster image processing module and raster GIS module is purely artificial.

6.3.4.1 Image Processing Operations

The objective of image data processing and enhancement is to render the image more interpretable, i.e. to

Table 6.2 Important GIS functions. (Modified after Aronoff 1989)

1. Maintenance and analysis of the spatial data	– Format transformations – Geometric transformations – Editing functions etc.	
2. Maintenance and analysis of the attribute data	– Attribute editing functions – Attribute query functions	
3. Integrated analysis of spatial and attribute data	– Retrieval/classification/measurement	– Retrieval – Classification – Measurement
	– Overlay operations – Neighbouring operations	– Search – Topographic functions – Thiessen polygons – Interpolation – Contour generation
	– Connectivity functions	– Contiguity – Proximity – Network – Spread – Seek – Perspective view
4. Output formating	– Map annotation text, labels, graphic symbols, patterns etc.	

improve the image quality for visual interpretation, and for subsequent computer analysis, if necessary. There is no standard prescription for obtaining best processing and enhancement results, which may be problem- as well as data-dependent. The various techniques of single and multiple image data enhancement available under digital image processing softwares can be applied.

A black-and-white image is typically a single parameter image, i.e. one image displays variation in a single parameter across the scene, in shades of gray. Any continuous parameter (interval or ratio type, Table 6.1) may be used for the purpose, e.g. water-table level, groundwater temperature, TDS content etc. The images can facilitate better interpretation of local and regional features. Also, they can be processed to generate other interesting data images, e.g. a difference image using topographic elevation image minus water-table image, will produce an image showing depth to water-table from the ground surface. Similarly, gradient filtering of water-table image will produce a water-table slope image.

Further, certain parameters (e.g. lithology, land use, soil type) are of categorical type measured on nominal or ordinal scale (Table 6.1). For generating images from such data, the maps are digitized, rasterized and the polygons are filled with certain selected DN-values (usually between 0 and 255). The images may be displayed as black-and-white products or in colour mode. The colour-hue axis may be selected to pertain to any parameter of interest, e.g. stratigraphic age, or perme-

ability, or fracture density of the rocks. These images could be quite informative.

The technique of image colour coding described in Sect. 4.6.3 can be applied for enhancing groundwater data in image form. Single images can be enhanced by density slicing colour coding. Multiple images of groundwater data can also be combined by colour coding to generate false colour composites. On such composites, colour variations would correspond to variations in input component images.

Synthetic stereo is another interesting technique of combining and enhancing image data. It is based on the standard parallax formula used in aerial photogrammetry, the only difference being that parallax is artificially introduced in the image-pair. A set of two images are taken; one acts as the base image and the second corresponds to the heighting image. The pixels in the base image are shifted in such a way that parallax is artificially introduced, the amount of parallax, i.e. 'heights', being introduced correspond to the pixel values in the second image. Thus, a synthetic stereo is generated.

Several examples of image processing operations are given in Chap. 4, and some other ones related to groundwater data are shown in Figs. 6.11, 6.12, 6.15, and 6.16.

6.3.4.2 GIS Analysis Functions

The GIS analysis functions are unique as they can concurrently handle spatial as well as attribute data. Here, the GIS analysis functions are divided into five types: retrieval, measurement, overlay, neighbourhood and connectivity (Table 6.2).

Retrieval operations include selective search on spatial and attribute data in such a way that the geographical locations of features are not changed, i.e. the output shows selectively retrieved data in their original geographic positions. The criteria for selective retrieval could be based on codes of attributes, or Boolean logical conditions (see later) or classification. Common examples could be: select pixels with water-table less than 1 m deep, or select pixels in a specific rock type. As the selective search can be operated on both spatial and attribute data, this becomes a powerful function in handling and processing data in GIS environment.

Measurement operations commonly included in GIS softwares are distances between points, lengths of lines (arcs), perimeters of polygons, areas of polygons, number of points falling in a polygon and number of raster cells in each class. Sample applications in groundwater study could be: find number of wells lying in a district, or find distance between wells, or compute density of wells. These functions are handy while dealing with exploration, planning and management of groundwater resources.

Overlay operations are very frequently required in almost all natural resources investigations, as often one has to deal with several data sets of the same area. Overlaying can be done in both raster and vector structures. However, overlaying in vector structures is rather tedious, for the simple reason that in this case overlaying of polygons leads to intersections and generation of new polygon of various shapes, sizes and attributes which become difficult to handle. In contrast, the overlaying operation in raster proceeds cell by cell, and therefore is relatively simpler.

Overlay operations involving both arithmetic functions as well as logical type can be performed to process the multiple coverages. Overlay arithmetic functions include such computations as addition, subtraction, division and multiplication. This involves arithmetic operation of each value in a data layer by the value at the corresponding location in the second data layer (Fig. 6.8). A logical or Boolean overlay operation involves finding those areas or locations where a specified set of conditions are fulfilled (or not fulfilled). Figure 6.9 displays the Venn's diagrams illustrating the Boolean logic. An example of classification by arithmetic and logical overlay operations is given later (Fig. 6.16).

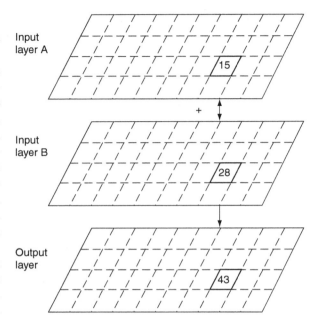

Fig. 6.8 Overlay operation in raster structure

Neighbourhood operations deal with local characteristics, or characteristics surrounding a specific target. These operations are useful in finding local variability and neighbouring/adjoining information. Some

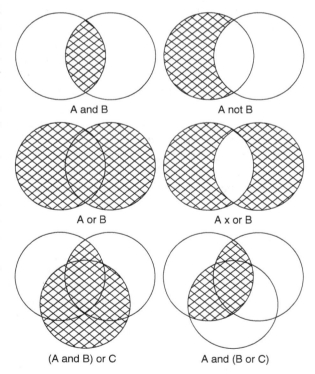

Fig. 6.9 Concept of Boolean conditions

functions such as search, topography and interpolation are usually given in GIS software as neighbourhood operations.

1. *Search function:* It is a very extensively used neighbourhood class of function. In this, on the basis of its neighbourhood characteristics, the target is assigned a certain value; the new value is written to a separate data layer. Sample query could be: "select villages not having water-wells within a search radius of 2 km".
2. *Topographic functions:* A raster data set can be represented in terms of a digital elevation model (DEM). The term topography here refers to the characteristics of such a DEM surface. The topographic functions, namely slope and aspect, are neighbourhood functions. Slope is defined as the rate of change of elevation, and aspect is the direction that a slope faces. In GIS modules, the term gradient is used only for maximum slope. Gradient and slope, both are first-order derivative parameters.
3. *Interpolation:* Interpolation is a typical neighbourhood operation. It involves predicting unknown values at given locations using the known values in the neighbourhood. Interpolation module is commonly provided in GIS packages. Besides the above, a number of image processing subroutines can also be considered as neighbourhood operations, e.g. high-pass filtering, texture transformation, image smoothing, gradient filtering, etc.

Connectivity Operations: Connectivity operations are grouped into contiguity, proximity, network, spread and perspective-view functions.

1. *Contiguity:* Areas possessing unbroken adjacency are classed as contiguous. What constitutes broken/unbroken adjacency in a particular case, can be prescribed depending upon the problem under investigation. For example, impermeable clay layers and dykes etc. are important in the context of groundwater flow, as far as contiguity is concerned. Therefore, this function can facilitate search for contiguous areas in a manner that the notion of contiguity is predefined.
2. *Proximity:* This function is based on the concept of distances between features. The notion of distance can be simple length, or based on any other computed parameter such as velocity, time, hydraulic gradient etc. A very common result of the proximity

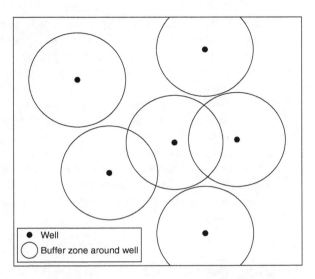

Fig. 6.10 Buffer zone generation around points representing well locations

function is a buffer zone. A buffer zone is defined as an area of a specified width, drawn around the specific location (point, line or polygon) in the map. A sample example is 'buffer the zone within a specified distance from the water-wells' (Fig. 6.10).

3. *Network:* Network functions are commonly used in analysis where resources are to be transported from one location to another. Examples of networks include canals, streams, etc. GIS packages carry modules for network analysis of varying sophistication. Some application problems could be in the field of flow functions and travel route selection. Network functions can be applied in environmental studies and pollutant dispersion investigations.
4. *Spread:* The spread function is endowed with characteristics of both network and proximity functions. In this, a running total of the computed parameter is kept, as the area is traversed by a moving window. The rules for transport and computation of parameter can be outlined beforehand. Therefore, it is a very powerful function particularly for environmental impact assessment and pollution studies.
5. *Perspective view:* The raster image data can be displayed as a three dimensional surface, where the height corresponds to the value at that pixel. This generates a three dimensional model, and forms a valuable and interesting method of enhancement, as the human mind can easily perceive shapes and forms. By defining parameters such as position of illuminating source and viewer's altitude and angle,

shaded relief maps and perspective views can be generated. One more raster data set can be superimposed over the three dimensional model by draping, e.g. a lithological map can be draped over a 3D water-table model.

6. *Classification* is a procedure of subdividing a population into classes and assigning each class a name, using various statistical procedures. In vector data, each polygon may be assigned a name as an attribute. In raster structure, each cell is assigned a new numerical value for class-identification. Classification may be done using a single data set or a multiple data set (Fig. 6.16).

6.3.5 Data Output

The results of GIS analysis may be presented in three types of media: softcopy, hardcopy and electronic. Softcopy output refers to the display on computer monitor. It may be colour, monochrome, or graphic text. Softcopy output is necessary for interactive processing and display. However, it cannot be used for permanent storage, but such screen-outputs can be photographed to provide permanent records. Hardcopy outputs are the permanent means of display. They may use paper, photographic film or other similar materials. Besides, GIS results may be output on electronic media such as disks and cartridges for transfer to other systems.

6.3.6 Sources of Error

As GISes are used to capture, edit and manipulate spatial data sets, errors of several types may creep-in at different stages and affect the data quality. Two basic categories of errors may be present: (a) inherent and (b) operational. Inherent error is present in the source data. Operational error is introduced during the GIS working. An understanding of the types and sources of errors can lead to better job management. Although it is not possible to avoid errors altogether, they can be managed to be kept within permissible limits.

There are a considerable number of powerful GIS packages available on the market, the most commonly used being: ARCGIS/ARCVIEW, Geomedia, Autocad Map, Geomatica, MapInfo, ILWIS and ENVI. These are general purpose software that can be used for any type of geospatial dataset. ArcHydro (Strassberg et al. 2007) is a geographic data model specifically developed for representing spatial and temporal groundwater information within a GIS. It includes two-dimensional and three-dimensional object classes for representing aquifers, wells, bore-hole data and also includes tabular objects for representing temporal information such as water-levels, water quality samples etc., that are related with spatial features.

6.4 Thematic GIS Applications in Groundwater

Many diverse applications of GIS in groundwater studies have been reported ranging from geohydrological exploration, water quality and pollution, to modelling and management. A detailed treatment of each type of application would be beyond the scope of the present work. In the following paragraphs some basic concepts and selected examples are mentioned.

6.4.1 Study of Water-Table Behaviour and Seepage Pattern

The importance of the study of water-table fluctuation and flow/seepage patterns has been discussed in Chap. 3. GIS environment is very aptly suited to process the spatial data of groundwater-table, topography etc., to enable interpretation of water-table behaviour and flow patterns. Often, groundwater data is widespread geographically, and understanding of the regional patterns may be possible only after synoptic viewing, which is easily possible in GIS.

The water-table data can be digitized, rasterized and interpolated to produce an image. Figure 6.11a is a groundwater-table contour map from which an image has been generated (Fig. 6.11b). The gray tones correspond to water-table elevations—darker tones implying lower elevations and lighter tones implying higher elevations. The image provides a prompt synoptic view of the groundwater flow directions. Further, the image indicates that the groundwater flow is largely effluent towards the stream.

In the same manner as above, it is possible to create and display other elevation data images in black-and-

Fig. 6.11 a Water table contour map of Maner river basin (India), **b** corresponding water table image. (a: after Radhakrishna et al. 1976; b: after Srivastava 1993)

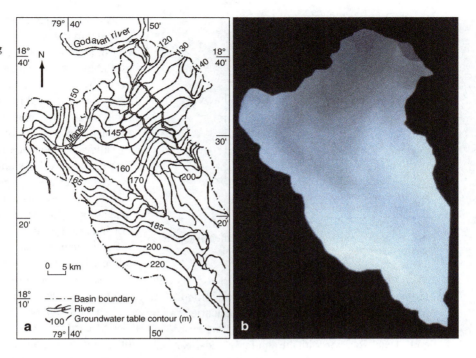

white, such as topographic elevation image (from topographic data), depth to water-table image (from depth to water-table data), depth to bedrock etc.

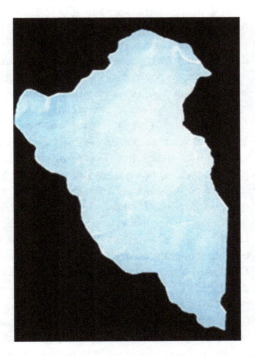

Fig. 6.12 Groundwater table gradient image generated by applying Soble's filter on Fig. 6.11b. (After Srivastava 1993)

Elevation data can be subjected to spatial filtering to generate gradient images, depicting variation in slope, pixel-to-pixel in either x- or y- or both directions (e.g. Jensen 1986). Figure 6.12 is an example of a water-table gradient image generated by applying Sobel's operator on the water-table image (Fig. 6.11b). Note that this image appears artificially illuminated from the northwest, which is due to directional computation in Sobel's filter. Bright pixels pertain to the sites of local higher gradients.

Another application of GIS based groundwater studies could be in extrapolating groundwater patterns to adjacent areas. Sometimes, groundwater data may be sparse in some areas, whereas in adjacent areas with similar hydrogeologic characteristics there could be sufficient available data. In such cases, groundwater patterns can be extrapolated in GIS and useful information can be derived for regions even with insufficient database.

6.4.2   Evaluation of Well Distribution in Different Aquifers

In a multi-aquifer system it is pertinent to have information on the wells tapping various aquifers. GIS can be used to identify and display locations of wells tapping one particular aquifer or a group of aquifers (e.g.

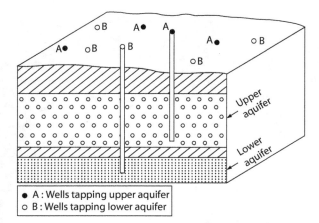

Fig. 6.13 Representation of wells from two different aquifers in GIS (for details see text)

Baker 1991). In such cases, data on location of wells, litholoigic logs and depth of wells are incorporated in GIS. Spatial surfaces of the aquifer boundaries can be generated by interpolation. Then, by relating water-well depth at each location with the aquifer boundaries, wells with respect to source aquifers could be searched for and classified (Fig. 6.13). This may be useful elsewhere, viz. for water management, particularly where different aquifers have water of different chemical quality, and water usage from different aquifers is to be controlled and managed for specific uses (e.g. industrial, agricultural, domestic etc.) according to chemical quality.

6.4.3   Groundwater Exploration and Recharge

Groundwater exploration has to utilize multiple data sets. Often information on spatial pattern and distribution of lithology, soil characteristics, lineament/fracture density, rainfall, topography, landform, landuse, hydraulic parameters etc. are required. GIS tools allow spatial viewing of the relevant factors collectively. The relevant databases in GIS can be created and each data layer can be given a weight relative to its importance. The data set can be integrated to develop recharge and resource models and for assessing the spatial distribution of groundwater resource. Some useful strategies could be, for example: superimposition of geophysical data image, lineament image and landform image, or superimposition of lineament image, lithology image and water-table image etc. Depending upon the problem, different combinations may be selected.

An interesting example of groundwater exploration involving GIS-based approach is furnished by Hansmann et al. (1992). In an area in Sri Lanka, the available data indicates that hydrogeology is influenced by climatic zones. Within a particular climatic zone, the well yield is related to three main hydrogeologic parameters, namely, distance to valley, distance to lineament and lithology. Raster maps of distance to valley and distance to lineament were created by first mapping valleys and lineaments on remote sensing data, and then using proximity functions in GIS (Fig. 6.14). The raster map of lithology was also created as an additional data layer from a geological map. For integrating the above three data layers, they used the statistical approach of 'weights of evidence modelling'. Integration of the data layers yielded a raster map showing posteriori probability of groundwater resource at each cell in the area (Fig. 6.14d).

In a hard rock basaltic terrain, Saraf and Choudhury (1998) used the various thematic input data layers of lithology, landform, lineament distribution, land use and vegetation density. They integrated the datasets through weighting-rating system in a GIS for deriving a relationship with water-level fluctuation in the field, and thus deduced sites suitable for groundwater prospecting and for artificial groundwater recharge.

In a karstic terrain, Shaban et al. (2006) used GIS for a method to assess recharge as a weighted function of the factors: lineament, drainage, lithology, karstic domains and land cover. They used various sensor image data, topographic maps and geologic maps as input data and subsequently related the relative recharge rates to some known recharge rates.

Tweed et al. (2007) provide another interesting example of remote sensing—GIS data application in groundwater studies. In a flood basalt terrain in Australia, they used a time series of Landsat TM/ETM image data to map groundwater discharge areas through temporal variability of NDVI. Topographic depressions were mapped by using a wetness index and DEM. Depth to groundwater image was generated by subtracting groundwater level image from the topographic DEM. Airborne gamma-ray spectrometry was used to distinguish the less-weathered basalt (more recharge potential) from thicker clay/soil profiles. The analysis resulted in maps showing groundwater flow directions, permanent wetlands and lakes, depth to water-table, and groundwater discharge and recharge areas.

GIS was used by Karanga et al. (1990) to generate maps showing actual and near-future water demand in

Fig. 6.14 Raster maps of
a distance to valley, **b** distance to lineament and
c lithology. **d** Superimposition
of the three raster data layers
gives a groundwater potential
map. (After Hansmann et al.
1992)

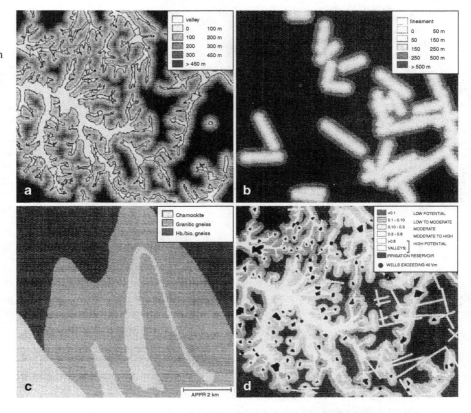

an area in Kenya using census data and drinking water
requirements for human beings and livestock. They
integrated land cover map and water availability map
(based on maps showing landforms and weathered
zones, lineaments, rainfall, and evapotranspiration) to
show spatial pattern in water demand.

6.4.4   Chemical Quality and Water Use

Chemical quality of groundwater is an important attribute which controls water use (Sect. 11.8). This data
being spatially variant, can be processed and analysed
in the GIS environment, in a highly efficient manner.

The chemical quality of groundwater is expressed
in terms of various parameters such as concentration
of various ions or total dissolved solids (TDS). In GIS,
each of these parameters can be treated as a data layer.
The data can be suitably contrast manipulated and
displayed as a black-and-white product. For example,
Fig. 6.15 shows a TDS image generated from the analytical data. Brighter tones indicate higher TDS concentrations in groundwaters. Similar images for other
chemical parameters could be generated and studied to

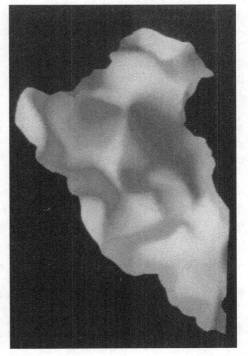

Fig. 6.15 TDS image of the Maner river basin; lighter tones
in the image correspond to higher TDS content. (Courtesy: N.
Srivastava)

Fig. 6.16 Classification using multiple image data sets of TDS, Cl, pH and SO$_4$. The groundwaters have been classified with respect to suitability for drinking purposes, applying WHO norms and Boolean logic. (After Srivastava 1993)

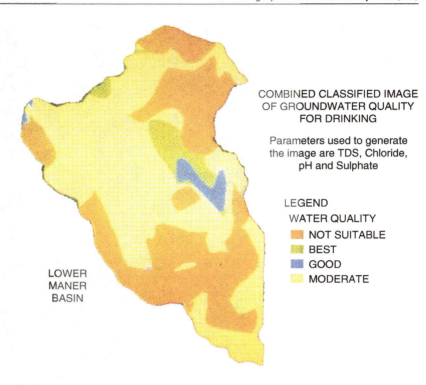

COMBINED CLASSIFIED IMAGE
OF GROUNDWATER QUALITY
FOR DRINKING

Parameters used to generate
the image are TDS, Chloride,
pH and Sulphate

LEGEND
WATER QUALITY
◼ NOT SUITABLE
◼ BEST
◼ GOOD
◼ MODERATE

LOWER
MANER
BASIN

provide a quick pictorial view of the spatial variation in the chemical quality of groundwater. Further the chemical data images can be combined in colour mode to depict spatial variation of chemical parameters.

Overlay operations involving both arithmetic as well as logical functions can be performed to process the multiple coverages, and the output can be used for hydrochemical classification. Figure 6.16 is an example in which four raster data sets of chemical characters, viz. TDS, Cl^{1-}, pH and SO$_4^{2-}$ have been overlayed. Applying WHO norms and Boolean logic, groundwaters of the area have been classified with respect to suitability for drinking purposes.

GIS can also be used to relate hydrochemical quality with lithology. For example, Kalkoff (1993) made one such study in the Roberts Creek watershed, Iowa, relating stream water quality with the geology of the catchment area.

6.4.5 Groundwater Pollution Potential and Hazard Assessment

GIS based spatial model can be used to evaluate groundwater vulnerability to contamination. Most of such studies utilize the raster GIS structure. The proce-dure involves generating relevant hydrogeological data layers, converting the parameters into ordinal numbers (using ratings and weights), and finally integrating the data sets to give a spatial property map.

The DRASTIC model developed by Aller et al. (1987), has been used more frequently for this purpose. It integrates parameters of depth to water-table (D), net recharge (R), aquifer media type (A), soil media character (S), topographic slope (T), impact of vadose zone (I), and hydraulic conductivity of the aquifer (C) for groundwater pollution potential evaluation. It is a model utilizing weights, ranges and ratings on ordinal scale, to provide relative rather than absolute evalua-tions. Each data plane (i.e. layer) is given a (subjec-tive) numerical value or weight, depending upon its importance. Table 6.3a shows that the parameter of depth to water-table (D) is most important and has a weight of 5; in contrast, topography is less important and has a weight of 2. In each of the data layers (i.e. thematic maps), the ranges/attributes are converted into ordinal rating numbers. Table 6.3b gives an example of converting depth to water-table (continuous) data into rating numbers, and Table 6.3c gives an example of converting aquifer media (categorical) data into rat-ing numbers. Thus, various data layers with cell values are generated. Using an additive model the pollution potential can be calculated as follows:

Table 6.3a Weights for DRASTIC features

Features	Weight
Depth to water-table	5
Net recharge	4
Aquifer media	3
Soil media	2
Topography	2
Impact of the vadose zone	5
Hydraulic conductivity of the aquifer	3

Table 6.3b DRASTC ranges and ratings for depth to water-level

Range (ft)	Rating
0.5	10
5–15	9
15–30	7
30–50	5
50–75	3
75–100	2
100+	1

Table 6.3c DRASTIC ranges and ratings for aquifer media

Range	Rating	Typical rating
Massive shale	1–3	2
Crystalline rocks	2–5	3
Weathered zones	3–5	4
Thin bedded rocks	5–9	6
Sandstone massive	4–9	6
Sand and gravel	6–9	6
Basalt	2–10	9
Karst limestone	9–10	10

$$DRASTIC\ index = DrDw + RrRw + ArAw$$
$$+ SrSw + TrTw + IrIw + CrCw$$
$$(6.1)$$

where, r is the cell rating and w is the layer weight. Figure 6.17 shows the working concept. An example of DRASTIC relative hydrogeologic vulnerability map is shown in Fig. 6.18.

Numerous studies using DRASTIC approach have been carried out and reported in the literature, some involving minor adaptations. For example, Guo et al. (2007) investigated the vulnerability of aquifer to arsenic contamination and used (i) Depth to water-table (ii) net Recharge, (iii) Aquifer thickness, (iv) Ratio of cumulative thickness of clay layers to total thickness of vadose zone, (v) Contaminant adsorption coefficient of sediment in vadose zone, and (vi) Hydraulic conductivity of aquifer as the various input data layers, calling

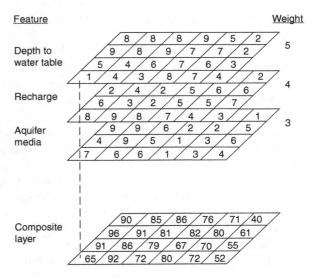

Fig. 6.17 Raster-based scheme for calculating index for groundwater pollution potential; only some of the DRASTIC parameters are shown

the model as DRARCH. In another study, Wang et al. (2007) investigated the vulnerability of groundwater in Quaternary aquifers to organic contaminants. As a variant of DRASTIC model, they ignored topography (for a flat Quaternary terrain), substituted the soil factor (S) by aquifer thickness (M), and hydraulic conductivity (C) by a new factor impact of Contaminant (C), and termed the model as DRAMIC.

DRASTIC-Fm is a modified version of DRASTIC and has been developed for fractured aquifers (Denny et al. 2007). In addition to the seven parameters of D, R, A, S, T, I and C, it uses an additional eighth parameter Fm (fractured media), which takes into account three primary characteristics of fractures—orientation, length and fracture density. Denny et al. (2007) used high resolution remote sensing image data and shaded relief models generated from DEM to map

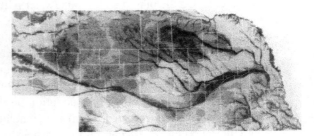

Fig. 6.18 DRASTIC relative hydrogeologic vulnerability map of Nebraska. Increasing shades of darkness indicate greater vulnerability to groundwater contamination. (After Kalinski et al. 1994)

fracture lineament traces. Existing field data on frac-ture characteristics showed that in their area of investigation NE–SW trending fractures were open (wide apertures, extension type) whereas NW–SE fractures were closed (low aperture, compression type). Accordingly, they grouped lineament traces in azimuth ranges and assigned ratings to reflect proximity to zones of extension/compression, giving higher ratings to extension/open type. Fracture length is another measure used by Denny et al. (2007) as it can be used to estimate whether a particular structure is regional or isolated/local. Within the GIS, they calculated lengths of all fracture traces and assigned ratings (higher rating to higher length), as regional structures can often significantly increase the hydraulic conductivity. Further, it is known that fracture intensities increase with proximity to known faults, i.e. frequently there is a zone of fracturing associated with structural faults/contacts. To take this parameter into account, they created buffer distances to all structures—including those mapped in the field, as well to lineament traces derived from DEM and remote sensing image data, and the buffer distance map/dataset was categorized into classes and ratings were assigned. The final Fm-parameter was obtained by integrating all the three characteristics of orientation, length and intensity and computing an average representative value for each raster cell. In this way they used Fm-parameter for incorporating the impact of fractures for the purpose of aquifer vulnerability mapping under DRASTIC approach.

At times a distinction is made between groundwater pollution potential and hazard (e.g. Atkinson and Thomlinson 1994). Pollution potential is derived at any given location based on local groundwater depth, hydraulic conductivity, land surface slope, and soil permeability, from the DRASTIC parameters. It is a natural risk index, not involving the effect of man-made features. On the other hand, groundwater pollution hazard is considered in terms of the probability of groundwater pollution in the area. It is based on pollution potential together with societies' pollution contribution in the area, given in terms of land use/cover and septic system density etc. Halliday and Wolfe (1991) used GIS for assessing groundwater pollution potential from nitrogen fertilizers in Texas. They combined an agricultural pollution susceptibility index map with information on cropped areas, recommended nitrogen fertilizer application rates and aquifer outcrops, and finally generated a nitrogen pollution potential map.

6.4.6   Salinity Hazard Mapping

Dryland salinity occurs in many areas. In northern part of Australia, salinity hazard mapping has been carried out by Tickell (1994) using GIS. The study is a futuristic one, aiming to identify areas which would be vulnerable to dry land salinity, if large-scale clearing of deep rooted vegetation were to take place for dryland agriculture. For this study five indicators viz. groundwater salinity, vegetation, median annual rainfall, aquifer yield and presence/absence of laterite were chosen. It is considered that the dryland salinity hazard in the area is higher with higher groundwater salinity, low open vegetation, low aquifer yield, a certain range of rainfall, and presence of laterite.

Data layers (i.e. thematic maps) for each of the above five indicators were generated from field data. The maps were digitized in vector format. Ratings from each of the data layers were added to produce a salinity hazard index map which was simplified to give a map showing four categories of salinity hazards—high, medium, low, and very low (Fig. 6.19).

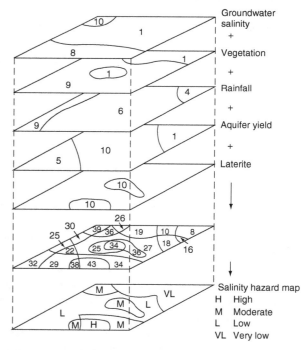

Fig. 6.19 Scheme of salinity hazard mapping. (After Tickell 1994)

6.4.7 GIS and Groundwater Modelling

GIS is endowed with graphical and data handling capabilities. Hence, the GIS environment is conducive to modelling owing to the following main features (El-kadi et al. 1994): (a) information on the database is available on the same screen during data preparation, (b) it provides graphical error checking on a continuous basis, (c) it saves and analyses spatial data and results on any GIS layer, and (d) it has flexibility of overlaying and integrating maps and data.

Some of the main applications of groundwater modelling are in assessing groundwater distribution and resource evaluation, permissible pumping of groundwater and contaminant transport. These applications require large spatial datasets of complex and varied type and therefore the potential of GIS in such studies is obvious. Further, in groundwater modelling, one requires to create a raster/spatial mesh with aquifer parameter values assigned at each cell. With aquifer parameters being known at certain points, interpolation/extrapolation is necessary for unknown points. Solutions may require initial and boundary conditions and results to be spatially displayed for visual interpretation. Therefore, it is appropriate and beneficial to use GIS integratively with numerical groundwater modelling software (e.g. Richards et al. 1993; Roaza et al. 1993; El-Kadi et al. 1994; Pinder 2002).

For GIS based groundwater modelling, there are two options: (a) development and application of models within GIS, and (b) pre/post processing of model data in GIS. Groundwater data modelling may be done directly within GIS, such as by using/integrating overlayed data layers to determine groundwater properties in DRASTIC. Another type of example is given by Lasserre et al. (1999) where they developed a simple model for groundwater nitrate transport within the GIS environment as a subroutine using the Pascal computing language.

Alternatively, a groundwater model may be run as an option-interface in GIS. Nnumerical modeling e.g. MODFLOW through preprocessing tools such as PMWIN (processing MODFLOW) can be interfaced with standard GIS (such as Arc GIS). Various data layers such as topography, groundwater-table or landuse/land cover may be developed in GIS and exported into groundwater model (such as MODFLOW); after running the groundwater model, the result may be imported into GIS for spatial display (Fig. 6.20). Besides, sophisticated modeling packages such as GMS (EMS 2007) equipped with a built-in GIS processor allows automatically conversion of data to facilitate implementation of numerical model.

6.4.7.1 Contaminant Transport Modelling

Pollution of groundwater is a burning problem throughout the world and has been discussed in Chap. 12. The root cause of the problem is the way waste disposal practices have been adopted. GIS can play a very important role in such studies as database on waste disposal sites, contaminant concentrations and groundwater quality can be used as input in groundwater models for predicting contaminant transport and, for example, for planning well sites for potable water. Popular groundwater flow and contaminant transport models like MODFLOW, MODPATH and MT3D can be interfaced with GIS as an option (Fig. 6.20). MODFLOW describes 2D and 3D transient or steady-state flow in anistropic heterogeneous layered aquifer using finite difference

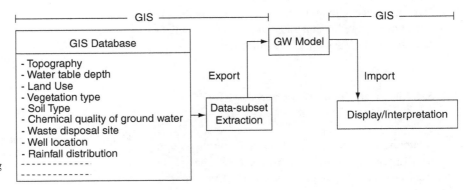

Fig. 6.20 Schematic of interfacing groundwater modelling in GIS

modelling. MODPATH is used to compute path-lines in 3D steady-state groundwater flow system, i.e. it computes position of particles at user-specified time instances. MT3D is a three-dimensional transport model to describe the solute transport. These models can be interfaced with GIS.

6.4.7.2 Salt Water Intrusion

In coastal areas, salt water intrusion is a major problem. Groundwater modelling using GIS can be applied to help predict likely hydraulic gradients under various conditions of pumping, which can be used to help limit pumping thresholds for coastal wells. This can save expensive field work and could be efficient and cost-effective.

6.4.7.3 Recharge/Resource Estimation

Several studies using groundwater modelling for recharge/resource estimation have been reported. For example, in a study in the granitic terrain in Spain, where groundwater availability is highly variable, Lubczynski and Gurwin (2005) used GIS for various geospatial data layers such as topography, groundwater slope, lineament density, apparent resistivity etc., and interfaced the GIS with MODFLOW for modeling spatiotemporally groundwater flow. Similarly, in an area in Washington State, USA, where wastes are stored below the ground, groundwater recharge rate estimation was carried (Fayer et al. 1996) as it was considered that high recharge rate could cause groundwater contamination. The GIS database included landuse, vegetation cover type and soil type. This database was imported into a groundwater recharge model called UNSAT-H. The results of this model were then exported back to GIS, allowing recharge maps to be displayed and interpreted. The above examples illustrate the use of pre- and post-processing of groundwater modelling data in GIS.

6.4.7.4 Well Head Protection Area (WHPA) Modelling

A capture zone represents the zone around a pumping well supplying groundwater to the well. The gradient created around a pumping well may attract contami-

nants if the contamination source is located within the capture zone. The delineation of WHPA utilizes various parameters such as distance, drawdown, travel time, flow boundaries and assimilative capacities. Models for WHPA have been evolved to delineate capture zones. While interfacing with GIS, the procedure includes: (a) extraction of the necessary input data from the GIS data base, (b) running the WHPA Model, and (c) displaying the WHPA map generated from the model in GIS (e.g. Fig. 6.21; Rifai et al. 1993).

6.4.8 Concluding Remarks

Some examples of groundwater data manipulation, processing and analyses in GIS environment have been given above. Availability of groundwater resource and its quality can and ought to play an important role in industrial/township/community planning. GIS based groundwater modelling integrated with other relevant geospatial data sets such as surface topography, forest areas, wetlands, endangered species habitats, landslide areas, etc. can help make informed decisions. For

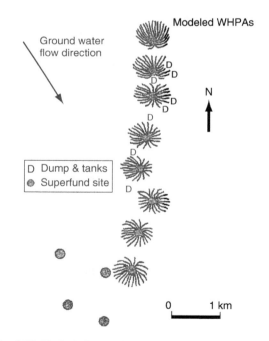

Fig. 6.21 Typical Output from modelling interface in GIS, showing the modelled well head protection areas (WHPAs) and the distribution of dump sites in Harris County, Texas. (Simplified after Rifai et al. 1993)

example, industrial units may be preferably located in sites where groundwater pollution potential is less; housing developments may be permitted in areas with aquifers that yield adequate water in terms of quality and quantity, and the data can also be used to manage/limit well drawdown or aquifer depletion. Thus, integrated community planning in GIS environment considering both resource data and requirements—is likely to play an important role in future.

It is obvious that the GIS-technology offers a very powerful tool for integrating a variety of georeferenced datasets, such as, topographic maps, lithologic maps, borehole data, hydrochemical data, geophysical data, river discharge, geohydrological data etc. This is a rapidly advancing field as new applications are being devised by different workers. In fact, in this methodology one can say that limitations of applications may be imposed only by sophistication in the particular software package and imagination and skill of the investigator.

Summary

Geographic Information System can be considered as computer-coded mappology. It deals with information on locational patterns of features and their attributes. Geometrically, all features can be resolved into points, arcs and polygons, which can be described in terms of coordinates. Attribute data describe the physical characteristics of features. Vector and raster are the two basic data structures in GIS. Typical GIS functions include retrieval operations, measurements, overlay, neighbourhood, search, proximity, contiguity, network etc. In groundwater, the GIS methodology can be used for studying water-table behaviour, seepage patterns, data integration for groundwater exploration, estimating pollution and hazard potential, groundwater modeling etc.

Further Reading

Chang KT (2008) Introduction to Geographic Information Systems. McGraw-Hill, New York, NY.

Lo CP, Yeung AKW (2006) Concepts and Techniques of Geographic Information Systems. Prentice Hall, Englewood Cliffs, NJ.

Meijerink AMJ, de Brouwer HAM, Mannaerts CM, Valenzuela CR (1994) Introduction to the Use of Geographic Information Systems for Practical Hydrology. ITC Publ. No. 23, Enschede.

Star J, Estes J (1990) Geographic Information Systems: An Introduction. Prentice Hall, Englewood Cliffs, NJ.

Principles of Groundwater Flow and Solute Transport

Although mechanism of groundwater flow through porous media is well understood but many uncertainties exist about the flow of water and transport of solutes in fractured rocks. This has acquired greater interest as low permeability formations form potential repositories for high level waste.

7.1 Groundwater Flow

7.1.1 Laminar and Turbulent Flow

Groundwater flow through various rock types may be either laminar or turbulent depending on their coefficients of permeability and the prevailing hydraulic gradient.

In laminar flow, also known as viscous or streamline flow, the flow lines are parallel. The turbulent flow is characterized by high velocities and the formation of eddies. In laminar flow the velocity of flow is proportional to the first power of the hydraulic gradient (Eq. 7.1).

$$V = -KI$$
$$I = \frac{dh}{dl} \tag{7.1}$$

But in turbulent flow, the velocity of flow will be expressed as

$$V = -KI^{\alpha} \tag{7.2}$$

where α is the degree of nonlinearity of the flow ($0.5 \leq \alpha \leq 1$).

Relatively high hydraulic gradient may give rise to turbulent flow through large pores, solution cavities and wide fractures. In laminar flow, the inertial forces are much smaller than viscous forces. The transition from laminar to turbulent flow depends on the Reynolds number (R_e) which is a dimensionless number expressing the ratio of inertial to viscous forces acting on the fluid (Eq. 7.3).

$$R_e = \frac{\rho V d}{\mu} = \frac{V d}{\upsilon} \tag{7.3}$$

where ρ is the fluid density, V is the mean fluid velocity, d is the diameter of the pipe, μ is the dynamic viscosity and υ is the kinematic viscosity. In porous media, d can be replaced by the effective diameter, (d_e) of sand grains which constitute the aquifer (Sect. 9.1.1).

By considering porosity, η the Reynolds number, R_e can be expressed as (UNESCO 1972).

$$R_e = \frac{1}{0.75\eta + 0.23} \frac{V d_e}{\upsilon} \tag{7.4}$$

and by considering the intrinsic permeability (k), instead of effective diameter (d_e), R_e is given by Eq. 7.5 (Van Golf-Racht 1982).

$$R_e = \frac{5 \times 10^{-3}}{\eta^{5.5}} \frac{V \sqrt{k}}{\upsilon} \tag{7.5}$$

Generally, the flow regime is laminar for $R_e < 2000$ and turbulent for $R_e > 2000$ (de Marsily 1986).

It is difficult to determine the Reynolds number in fractured rocks since for a given type of flow it can vary greatly from one point to another along the same fracture.

In fractured rocks, the diameter of the pipe in Eq. 7.3 can be replaced by the hydraulic diameter (D_h) which in a plane fracture is expressed by

B. B. S. Singhal, R. P. Gupta, *Applied Hydrogeology of Fractured Rocks,*
DOI 10.1007/978-90-481-8799-7_7, © Springer Science+Business Media B.V. 2010

$$D_h = 4A/p \qquad (7.6)$$

where A is the cross-sectional area of the fracture through which flow takes place and p is the outside perimeter of the cross-section area of the flow (de Marsily 1986). For a very long fracture, D_h is equal to twice its aperture, i.e. $D_h = 2a$,

For a non-circular conduit such as fracture, the Reynold's number is given as

$$R_e = \frac{\rho V (4R_h)}{\mu} \qquad (7.7)$$

where $R_h = A/P$ is the hydraulic radius which is the ratio of the cross-sectional area, A, to the wetted perimeter, P. For a circular cross section, $D = 4R_h$

The walls of a natural fracture will always have certain degree of roughness causing an additional drop in pressure and thereby influence the flow conditions. The relative roughness, R_r (dimensionless) is defined by

$$R_r = R_f/D_h \qquad (7.8)$$

where R_f is the mean height of the irregularities (asperties) in the fracture. Some measured values of R_r lie in the range of 0.002–0.01. However, as it is often difficult to measure roughness in field, the effect of roughness on groundwater flow is more of academic interest (Van Golf-Racht 1982).

From experimental data, Van Golf-Racht (1982) also showed that the transition from laminar to turbulent flow in fractures takes place at very low values of Reynolds number (600). According to Louis (1974), the transition from laminar ($\alpha = 1$) to completely turbulent ($\alpha = 0.5$) is quite progressive, the exponent slowly changes from 1 to 0.5 as the Reynolds number changes for instance from 100 to 2300. Lee and Farmer (1993) indicated that the critical Reynolds number commonly lies between 100 and 2300, decreasing as the relative roughness of the fracture increases. The flow therefore will change from laminar to turbulent condition with increase in flow velocity, pore size, sinuosity and roughness.

7.1.2 Darcy's Law

In 1856, Henry Darcy, a French hydraulic engineer, investigated the flow of water through filters. His

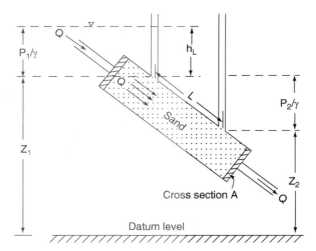

Fig.7.1 Flow through an inclined sand column showing pressure distribution and head loss

experiments demonstrated that the rate of flow i.e. volume of water per unit time, Q is (a) directly proportional to the cross-sectional area, A, and head loss, h_L, and (b) inversely proportional to the length of the flow path, L (Fig. 7.1), i.e.

$$Q \sim A \quad and \quad \sim \frac{h_L}{L} \qquad (7.9)$$

By combining the above relations and introducing proportionality constant (K), the Darcy's law can be written as

$$Q = KA\frac{h_L}{L} \qquad (7.10)$$

or

$$Q = KA\frac{dh}{dl} \qquad (7.11)$$

where K is the hydraulic conductivity of the porous medium and dh/dl is the hydraulic gradient.

Darcy's law can also be written as

$$q = V = \frac{Q}{A} = K\frac{dh}{dl} \qquad (7.12)$$

where V (or q) is the specific discharge or Darcy velocity. Equation 7.12 simply denotes that the specific discharge is the volume of water per unit time through a unit cross-sectional area normal to the direction of flow. Although Darcy velocity, (specific discharge), q implies that flow occurs through the entire cross section of the aquifer, however in nature the actual flow takes place only through the interstitial spaces (pores). Therefore, the average interestitial velocity, $\overline{V}_a$ is

$$\overline{V}_a = \frac{Q}{\eta A} = \frac{q}{\eta} \qquad (7.13)$$

where η is the porosity of the medium.

In order to determine the average (interstitial) velocity (or simply velocity) one should use effective porosity (η_e), instead of total porosity (n).

Therefore,

$$\overline{V}_a = \frac{q}{\eta_e} = \frac{K(dh/dl)}{\eta_e} \qquad (7.14)$$

It could be seen from Fig. 7.1 that

$$Q = KA\{(p_1/\gamma + z_1) - (p_2/\gamma + z_2)\}/L$$

where P is pressure and γ is the specific weight of water. Therefore, the head loss can be defined as energy loss or potential loss which is due to the frictional resistance through the porous medium. As velocity of groundwater movement is low, the kinetic energy of water or the velocity head is neglected. The sum of the pressure head (p/γ) and the elevation head (z) is the hydraulic head.

It may also be seen from Fig. 7.1, that although the pressure (P) at point 1 is less than at point 2 but the piezometric head at 1 ($p_1/\gamma + Z_1$) is more than at 2 ($p_2/\gamma + Z_2$) indicating thereby that the flow takes place from a higher head to a lower one and not necessarily from higher to a lower pressure. It would mean that the flow takes place in the direction of decreasing head and is independent of the inclination of the pipe.

Darcy's law is valid only for laminar flow in porous media and hence the upper limit of its validity is decided by the Reynolds number. Experiments indicate that Darcy's law is valid as far as Reynolds number (R_e) is between 1 and 10 (Todd 1980). As for groundwater flow in porous media, R_e is usually less than 1, therefore, one could conclude that Darcy's law governs the flow of groundwater in porous media except in some situations where either the hydraulic gradients are very steep (as in the neighbourhood of the pumping well or spring outlets) or in rocks with

large cavities, e.g. karstic formations. There may also be some lower limit for the validity of Darcy's law as some threshold hydraulic gradient is required to initiate flow especially in silty and clay formations where the water molecules are adhered to the electrically charged mineral particles. This of course is not of much practical interest (Bear 1979).

7.1.3   General Equation of Flow

The Darcy's law, in a general form can be written as

$$V = -K\frac{dh}{ds} \qquad (7.15)$$

where V, K and h are as defined earlier and s is the distance along the average direction of flow. If an element of saturated material has the dimensions dx, dy and dz (Fig. 7.2) then the velocity components in a rectangular coordinate system can be written as

$$V_x = -K_x\frac{\partial h}{\partial x} \qquad (7.16)$$

$$V_y = -K_y\frac{\partial h}{\partial y} \qquad (7.17)$$

$$V_z = -K_z\frac{\partial h}{\partial z} \qquad (7.18)$$

where h is the total head under steady state conditions, and K_x, K_y and K_z are the coefficients of hydraulic conductivity in the x, y and z directions respectively. x, y, and z directions are the principal permeability directions.

Groundwater satisfies the equation of continuity. It expresses the principle of conservation of mass, i.e. the net inward flux through an elemental volume of an aquifer in the flow field must be equal to the rate at which matter is accumulating within the element (DeWiest 1969). The continuity equation, in its general form can be written as

$$\frac{\partial(\rho V_x)}{\partial x} + \frac{\partial(\rho V_y)}{\partial y} + \frac{\partial(\rho V_z)}{\partial z} + N = \rho S_s\frac{\partial h}{\partial t}$$

$$(7.19)$$

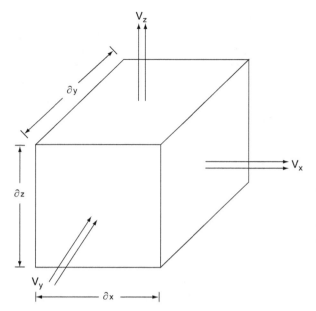

Fig. 7.2 Flow through a saturated soil element

where ρ is the density of the fluid, N is the source/sink term (N is −ve for sink and +ve for source), and S_s is the specific storage defined in Chap. 8.

By eliminating ρ from both sides of Eq. 7.19 for an incompressible fluid and using Darcy's law, we obtain Eq. 7.20 which represents transient flow through a saturated anisotropic medium.

$$\frac{\partial}{\partial x}\left(K_x\frac{\partial h}{\partial x}\right) + \frac{\partial}{\partial y}\left(K_y\frac{\partial h}{\partial y}\right)$$
$$+ \frac{\partial}{\partial z}\left(K_z\frac{\partial h}{\partial z}\right) \pm W = S_s\frac{\partial h}{\partial t} \qquad (7.20)$$

where W is the volume of flux per unit volume of the porous medium (a positive sign for the inflow and negative sign for the outflow).

For a homogeneous and isotropic medium, Eq. 7.20 reduces to

$$\frac{\partial^2 h}{\partial x^2} + \frac{\partial^2 h}{\partial y^2} + \frac{\partial^2 h}{\partial z^2} = \frac{S_s}{K}\frac{\partial h}{\partial t} \qquad (7.21)$$

In a horizontal confined aquifer of thickness, b, the storage coefficient, S will be equal to S_sb and T=Kb. Therefore, Eq. 7.21 reduces to

$$\frac{\partial^2 h}{\partial x^2} + \frac{\partial^2 h}{\partial y^2} + \frac{\partial^2 h}{\partial z^2} = \frac{S}{T}\frac{\partial h}{\partial t} \qquad (7.22)$$

This is the partial differential equation, governing non-steady state groundwater flow in a confined aquifer.

If the flow is steady, ∂h/∂t=0, i.e. velocity and pressure distribution do not change with time, Eq. 7.22 changes to

$$\frac{\partial^2 h}{\partial x^2} + \frac{\partial^2 h}{\partial y^2} + \frac{\partial^2 h}{\partial z^2} = 0 \qquad (7.23)$$

which is the Laplace equation governing for steady state flow of groundwater in a homogeneous and isotropic aquifer.

For groundwater flow to wells, the differential equation governing the unsteady state radial flow in confined aquifer in polar coordinates is expressed as

$$\frac{\partial^2 h}{\partial r^2} + \frac{1}{r}\frac{\partial h}{\partial r} = \frac{S}{T}\frac{\partial h}{\partial t} \qquad (7.24)$$

where r is the radial distance from the well to the point of observation. For steady flow, Eq. 7.24 reduces to

$$\frac{\partial^2 h}{\partial r^2} + \frac{1}{r}\frac{\partial h}{\partial r} = 0 \qquad (7.25)$$

Equations 7.24 and 7.25 are applied to solve problems of well hydraulics including analysis of pumping test data in homogeneous and isotropic aquifers under idealized conditions satisfying the pertinent initial and boundary conditions. In nature, aquifers exhibit complex situations for which necessary modifications are required. The adaptability of these equations to well hydraulics under different geohydrological conditions is discussed in Chap. 9.

7.2 Groundwater Flow in Fractured Rocks

A fractured rock media consists of two different populations i.e. fracture zones and porous media (Fig. 7.3). Fracture zones are characterised by low mean porosity but great lateral continuity. On the other hand, the porous matrix has larger porosity and shorter spatial continuity. In fractured rocks, the groundwater movement mainly takes place along discontinuities, i.e. joints, fractures and shear zones. The interconnec-

Fig. 7.3 Schematic representation of **a** purely fractured medium, **b** double-porosity medium, and **c** heterogeneous medium. (After Streltsova 1977)

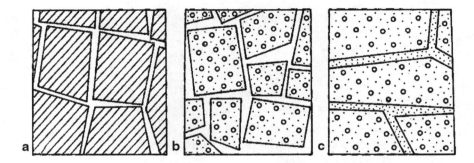

tion between rock discontinuities and their spacing, aperture size and orientation decide the porosity and permeability of such rock masses. Open joints and fractures which are not filled with weathered or broken rock material form potential passage for groundwater movement but their permeability is greatly reduced when filled with clayey material such as smectite or montmorillonite. These filling material form fracture skin which also influence the movement of solutes from the fractures into the porous matrix.

The joint surface roughness is of importance in joints with small apertures. Fractures roughness, which causes point-to-point reduction in fracture causes reduction in hydraulic conductivity (K) (Eq. 7.26), (Domenico and Schwartz 1998).

$$K = \frac{\rho_w g a^2}{12\mu\left[1 + C(x)^n\right]} \qquad (7.26)$$

where a, is the fracture aperture, C is some constant larger than one, x is a group of variables that describe the roughness, and n is some power greater than one. Therefore, roughness causes decrease in hydraulic conductivity.

The effective stress consisting of normal stress, shear stress and fluid pressure components are also important factors governing magnitude of K. Laboratory studies indicate that the permeability decreases significantly with the increase in confining pressure (Indraratna and Ranjith 2001).

Depending on the porosities and permeabilities of the fractures and the matrix blocks, the fractured rock formations can be classified into (a) purely fractured medium, (b) double porosity medium, and (c) heterogeneous medium (Fig. 7.3). In a purely fractured medium, the porosity and permeability is only due to interconnected fractures while blocks are impervious.

In double (dual) porosity medium, both fractures and matrix blocks contribute to groundwater flow but fractures are the main contributors. In a situation, when fractures are filled with clay or silty material, the fracture permeability is considerably reduced and such a medium is termed as heterogeneous.

Some of the models describing flow in a fractured medium are:

1. Parallel plate model
2. Double porosity model
3. Equivalent porous medium model
4. Discrete fracture network model, and
5. Equivalent parallel plate model.

It may be mentioned that various models require knowledge about characteristic features of individual fractures and thereby account for the heterogeneity. Therefore, careful mapping of fractures by surface and subsurface measurements is important (see Sect. 2.4 and 4.8.10). The level of details depends on the purpose for which the model is being developed. A greater accuracy is required for modelling solute transport as the heterogeneity of the fracture system greatly influences the travel time and solute concentration. It is also important to estimate the hydraulic properties of different fracture sets. Studies indicate that fractures parallel to the maximum compressive stress tend to be open, whereas those perpendicular to this direction tend to be closed (NRC 1996).

7.2.1 Parallel Plate Model

The Darcy's law for flow in a single fracture can be written as

$$V = K_f I \qquad (7.27)$$

where K_f is the hydraulic conductivity of the fracture, defined by

$$K_f = \frac{\gamma_w a^2}{12\mu} \tag{7.28}$$

where a is the fracture aperture, γ_w is the unit weight of water, and μ is the viscosity of water.

As the hydraulic conductivity (K) and permeability (k) are related by the expression

$$K_f = \frac{\gamma}{\mu} k \tag{7.29}$$

Therefore, the permeability of the fracture, k_f, can be defined as

$$k_f = \frac{a^2}{12} \tag{7.30}$$

By combining Eqs. 7.27 and 7.28, the average velocity $\overline{V}_a$ in the fracture expressed by a single parallel plate model is given by

$$\overline{V}_a = \frac{\gamma a^2}{12\mu} \frac{dh}{dl} \tag{7.31}$$

Here it is assumed that the fracture walls are impermeable.

Based on Eq. 7.31, the computed groundwater velocities over a range of hydraulic gradients are given in Fig. 7.4. It could be noted that even for small hydraulic gradients, groundwater velocity in discrete fractures is very high compared to velocities typical of porous medium.

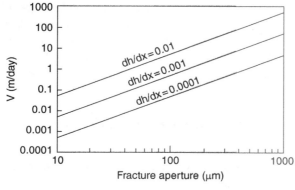

Fig. 7.4 Groundwater velocities in a single fracture as predicted by the cubic law. (After Novakowski et al. 2007)

If the fracture walls are permeable, the average velocity in the fracture will be (Bear 1993).

$$\overline{V}_a = \frac{\gamma a^2}{12\mu + 2a\rho ql} \cdot \frac{dh}{dl} \tag{7.32}$$

where ql denotes the rate of leakage into the wall of the fracture. Equation 7.32 shows that if ql is negligible, or $\mu >> (a\rho ql/6)$, then the conductivity of the fracture can be approximated by Eq. 7.28.

In terms of transmissivity of fracture, T_f, Eq. 7.30 can be written as

$$T_f = \frac{\gamma}{\mu} \frac{a^3}{12} = K_f a \tag{7.33}$$

Many researchers also define T_f as the hydraulic conductivity of fracture (Bear 1993).

The volumetric flow rate (Q_f) per unit plate (fracture) width will be

$$Q_f = \left(\frac{\gamma a^3}{12\mu} \right) I \tag{7.34}$$

Equation 7.34 is referred as the *cubic law* which is valid for laminar flow through parallel wall fractures with smooth surfaces. In natural conditions these assumptions usually do not hold good. Under field conditions it is also difficult to define the representative distance between the fracture walls. The validity of cubic law is discussed by several researchers (Lee and Farmer 1993). At low applied stress, when the fractures are open, the parallel plate approximation for fluid flow through fractures may be valid. However, due to stress the contact area between fractures surfaces will increase and therefore variation in aperture should be considered (Tsang and Witherspoon 1985).

Louis (1984) emperically defined the following five steady state flow regimes, both laminar and turbulent, depending on various degrees of relative roughness, (R_r) fracture aperture, (a) hydraulic head gradient in the fracture plane, (I_f) and the kinematic viscosity (μ/ρ).

Type 1: Smooth laminar:

$$V = \left(\frac{\rho g a^2}{12\mu} \right) I_f \tag{7.35}$$

Type 2: Smooth turbulent

$$V = \left[\frac{g}{0.079} \left(\frac{2\rho a^5}{\mu} \right) I_f \right]^{\frac{4}{7}} \tag{7.36}$$

Type 3: Rough turbulent

$$V = \left(4\sqrt{ag} \; ln \frac{3.7}{R_r} \right) \sqrt{I_f} \tag{7.37}$$

Type 4: Rough laminar

$$V = \left[\frac{\rho g a^2}{12\mu(1 + 8.8R_r)^{1.5}} \right] I_f \tag{7.38}$$

Type 5: Very rough turbulent

$$V = \left(4\sqrt{ag} ln \frac{1.9}{R_r} \right) \sqrt{I_f} \tag{7.39}$$

The domains of validity of the various flow regimes in a fracture are shown in Fig. 7.5.

If the fracture is not completely open i.e. the fracture walls touch at places, the right hand side of Eqs. 7.35–7.39 should be multiplied by the degree of separation, F, which is defined as

$$F = \frac{Open \; fracture \; surface \; area}{Total \; fracture \; surface \; area}$$

Two types of laminar flow are distinguished by Sharp and Mani (Lee and Farmer 1993):

1. Linear laminar: $V = KI$, and
2. Non-linear laminar: $V = K[i - \beta(i - i_{lim})^n]$

where K is the hydraulic conductivity of the fracture, i_{lim} is the limiting gradient for linear laminar flow and β and n are empirically determined constants. They also showed that the flow rates are often not proportional to the cube of the aperture especially in rough fractures with asperities and gave the following exponent values for the relation $Q \propto a^n$

	Rough fractures	Parallel plate
Linear laminar flow	n=2	n=3
Non-linear laminar	1.2<n<2	–
Fully turbulent flow	n=1.2	n=1.5

Several authors have studied the effect of variable aperture on flow through fractures. For example, Bear (1993) quoting the work of Wilson and Witherspoon (1985), expressed the effective aperture, a_{eff} for a series of m discrete segments with different apertures (Fig. 7.6a) as

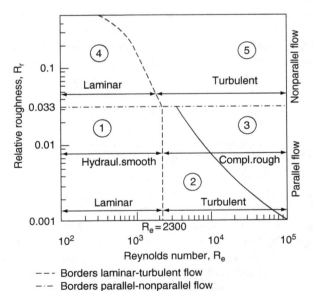

Fig. 7.5 Definitions and range of validity of the various flow regimes in a fracture. (After Louis 1984)

--- Borders laminar-turbulent flow
--·- Borders parallel-nonparallel flow
— Borders hydraulically smooth-completely rough flow

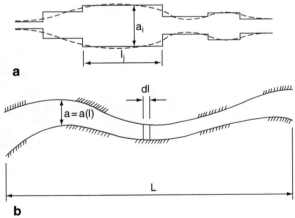

Fig. 7.6 Fractures of variable aperture: **a** discrete aperture variation, and **b** continuous aperture variation. (After Bear 1993)

$$a_{eff}^3 = \frac{\sum\limits_{i=1}^{m} l_i}{\sum\limits_{i=1}^{m} \left(l_i/a_i^3\right)} \qquad (7.40)$$

where l_i is the length of a fracture segment of aperture, a_i.

For an aperture that varies continuously along the fracture (Fig. 7.6b), i.e. $a=a(s)$, Eq. 7.40 is replaced by

$$a_{eff}^3 = \int\limits_0^L \left(\int\limits_0^L \frac{ds}{[a(s)]^3}\right)^{-1} ds \qquad (7.41)$$

where L is the total fracture length.

(a) Flow in Fracture System The volumetric flow rate for a single fracture based on the parallel plate model is given by Eq. 7.34. It can be extended to various types of multiple fracture systems (Bear 1993). In the case of a single family of m fractures of equal aperture oriented parallel to the flow direction (Fig. 7.7), the total discharge, Q is given by

$$Q = m\frac{a^3}{12}\frac{\gamma}{\mu}I \qquad (7.42)$$

and specific discharge, q through a cross-section having a width, L and unit length normal to the flow direction will be

$$q \cong \frac{Q}{L} = \frac{m}{L}\frac{a^3}{12}\frac{\gamma}{\mu}I = K_f I \qquad (7.43)$$

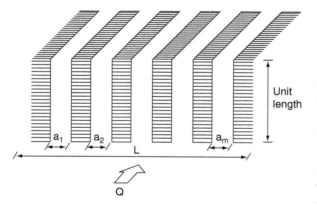

Fig. 7.7 One set of parallel fractures. (After Bear 1993)

with

$$K_f = \frac{ma^3}{12L}\frac{\gamma}{\mu} \qquad (7.44)$$

If the fractures are of varying apertures, $a_i=1, 2, ..., m$, then according to Bear (1993)

$$Q = qL = \frac{\gamma}{12\mu}\left(\sum\limits_{i=1}^{m} a_i^3\right)I = K_f LI \qquad (7.45)$$

(b) Validity of Cubic Law It could be seen from the above discussion that the cubic law is valid for fractures of small aperture and with plane surface. Natural fractures vary widely as far as planerity and surface geometry is concerned. Bedding plane fractures in fine grained sedimentary rocks like shales may be relatively smooth and parallel but in crystalline rocks like granites, fracture surfaces are usually rough and the aperture varies. Further, under greater stress, fracture surfaces are also partially closed and therefore validity of the cubic law is disputed. Witherspoon et al. (1980) concluded that the cubic law is valid for apertures upto 4 μm opening despite local areas of contact at stresses upto 20 MPa. Deviation from cubic law in induced fractures subjected to normal stresses greater than 20 MPa is also demonstrated by Gale (1982a).

An increase in the contact area of the fracture surface will decrease the rate of fluid flow. A reduction of two or more orders of magnitude is reported when the contact area of the surface is greater than 30% as in the case of limestone and sandstone samples where the contact area is estimated to be in the range of 40–70% at an applied stress of 30 MPa (Tsang and Witherspoon 1985).

Channelling, which is a common phenomenon in fractured rocks, refers to preferential flow paths and may apply both to a single fracture and a fracture network. In a single fracture, channelling is related to aperture variations. Channeling effects have been demonstrated by tracer experiments in natural fractures, by laboratory experiments, and computer simulations (Bodin et al. 2003). With increased channelling the first arrival of the solute will be much earlier (Oden et al. 2008), more so at depths greater than 500–1000 m (Tsang and Tsang 1987). These studies have therefore questioned the validity of parallel plate model in nature as the channelling effect leads to preferred paths of flow due to which it may be difficult to predict the flow paths. In practice, however, the channelling concept is rarely taken into account for interpreting hydraulic or tracer tests as the fracture media is usually considered a parallel plate system with constant aperture.

7.2.2 Double Porosity Model

This is also known as *dual porosity model*. The behaviour of a fractured aquifer for regional groundwater investigations can best be represented by a double-porosity model. The concept of double-porosity model was first developed by Barenblatt et al. (1960) and has been explained in detail by Streltsova-Adams (1978) and Gringarten (1982). The model assumes two regions-the porous block and the fracture, which have different hydraulic and hydromechanic characteristics. Three types of distribution of matrix block are considered-horizontal slabs (strata-type), spherical blocks, and cubes (Fig. 7.8). The fissured medium with horizontal fractures has been considered to be analogous to alternate aquifer-aquitard system (Boulton and Streltsova 1977). The block consists of fine pores which are separated from fractures. The blocks supply fluid to the fractures and act as uniformly distributed source. Such an approach with some modifications has been also used by several researchers in analysing

behaviour of fractured oil and geothermal reservoirs (Earlougher 1977; Horne 1990; NRC 1996).

In the double-porosity model, the porous blocks have high primary porosity but low hydraulic conductivity while the adjacent fractures have although low storativity but high conductivity. The differences in pressures between the porous blocks and the fractures lead to flow of fluid from porous blocks to adjacent fractures.

The characteristics of a double (dual) porosity aquifer system are described in terms of K_f, K_m, S_f, S_m, where subscripts f and m are for fractures and marix respectively. The other two parameters are storativity ratio (ω) and the transmissivity ratio (λ), Storativity ratio (ω), is the ratio of fissure to total system (blocks plus fissures) storativities and can be expressed as

$$\omega = \frac{S_f}{S_f + S_m \beta} \tag{7.46}$$

where S_f and S_m are the storativities of fracture and blocks respectively, and β is a factor depending on matrix block geometry which is taken to be equal to 1 for strata type.

The transmissivity ratio, or interporosity flow coefficient (λ) (dimensionless) is given by (Gringarten 1982).

$$\lambda = \alpha r_w^2 \frac{K_m}{K_f} \tag{7.47}$$

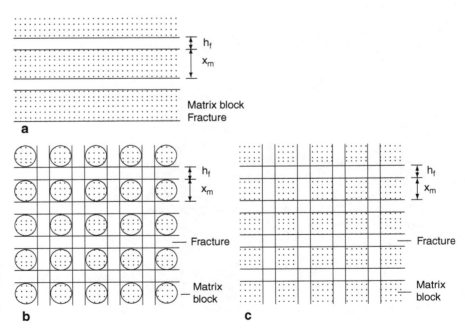

Fig. 7.8 Types of double porosity aquifers: **a** horizontal fractures and matrix blocks, **b** spherical matrix blocks, and **c** cubical matrix blocks

where α is the shape factor parameter depending on the geometry of the aquifer matrix and fracture (dimension reciprocal area), r_w is the radius of production well and α is defined as

$$\alpha = \frac{A}{l\upsilon} \qquad (7.48)$$

where A is the surface area of the matrix block, υ is the matrix volume, and l is the characteristic length.

If the matrix blocks are in the form of horizontal slab blocks (Fig. 7.8a), as may also be the case in thinly layered formations, $\alpha = 12/x_m^2$, where x_m is the thickness of a matrix block (Kruseman and de Ridder 1990), and x_f is the thickness of the high permeability layer, i.e. fracture. When matrix blocks are in the form of cubes or spheres (Fig. 7.8b and c), $\alpha = 60/x_m^2$, where x_m is the length of a side of the cubic block, or the diameter of the spherical block.

Values of ω are generally in the range of 10^{-1}–10^{-4} (Kruseman and de Ridder 1990). However, in a special case when $\omega=1$, the matrix porosity will be zero and the reservoir will be characerized by single porosity instead of double porosity (Horne 1990). Values of λ are usually very small (10^{-3}–10^{-10}). A larger value of λ indicates lesser heterogeneity and thereby points towards a reservoir of single porosity (Horne 1990).

Based on the continuity equation and Darcy's law, the differential equations describing the one dimensional confined flow within the fractures and matrix block can be obtained as

$$T_f \frac{\partial^2 h_f}{\partial x^2} = S_f \frac{\partial h_f}{\partial t} - V_d \qquad (7.49)$$

$$T_m \frac{\partial^2 h_m}{\partial x^2} = S_m \frac{\partial h_m}{\partial t} + V_d \qquad (7.50)$$

where V_d is the water transfer per unit time from blocks to fractures in a column of aquifer having unit horizontal area; h_f and h_m are the mean piezometric heads in the fracture and matrix block respectively.

According to Streltsova (1976a), the quasi-steady state transfer of water between porous blocks and fractures can be expressed as

$$V_d = \frac{K_m}{l}(h_f - h_m) \qquad (7.51)$$

where, l is the characteristic dimension of block.

As the hydraulic conductivity of porous blocks is much less than their storativity, the term $T_m(\partial^2 h_m/\partial x^2)$ may be neglected as compared to $S_m(\partial h_m/\partial t)$ in Eq. 7.50 (Barenblatt et al. 1960; Streltsova-Adams 1978). Therefore, Eqs. 7.49 and 7.50 can be expressed as

$$T_f \frac{\partial^2 h_f}{\partial x^2} = S_f \frac{\partial h_f}{\partial t} + S_m \frac{\partial h_m}{\partial t} \qquad (7.52)$$

$$S_m \frac{\partial h_m}{\partial t} = \frac{K_m}{l}(h_f - h_m) \qquad (7.53)$$

The flow from block to fissure will be time dependent, which can be expressed as

$$V_d = S_m \frac{\partial h_m}{\partial t} = K_m \frac{h_f - h_m}{l} \qquad (7.54)$$

or

$$\frac{\partial h_m}{\partial t} = \frac{K_m}{S_m} \frac{h_f - h_m}{l} = \psi(h_f - h_m) \qquad (7.55)$$

where

$$\psi = \frac{K_m}{S_m l}$$

Finally, according to Streltsova-Adams (1978), the general differential equation for the confined flow in fissure to a well in a double porosity formation will be

$$T_f \left(\frac{\partial^2 h_f}{\partial r^2} + \frac{1}{r} \frac{\partial h_f}{\partial t} \right) = S_f \frac{\partial h_f}{\partial t} + \psi S_m \int_0^\infty \frac{\partial h_f}{\partial t_{t=\tau}}$$
$$exp\left[-\psi(t - \tau)\right] dt \qquad (7.56)$$

An identical solution was obtained by Boulton and Streltsova considering block-and-fissure units of a regular pattern (Streltsova-Adams 1978).

The fissure-porous block flow is a time-dependent process caused by the readjustment of the block and fissure pressures, the duration of which would depend on the elastic properties of blocks and fissures, their permeability and dimensions (Streltsova 1976b). It is therefore necessary to define, in a representative elementary volume, the different pressures, one in the fractures and the other in the matrix (blocks), as well as exchange of mass between the intergranular porosity and the porosity of the fracture (de Marsily 1986).

Due to differences in the permeabilities of fractures and blocks, flow mechanism is different during early, intermediate and long times of pumping. At early times, flow from the matrix block is essentially zero as initially fluid is removed primarily from the fractures which have high permeability. At long times, the fractured reservoir behaviour is equivalent to that of a homogeneous porous medium with a permeability equal to the fissure permeability. At intermediate times, there is a transition from fracture flow to flow from fractures and matrix blocks, during which the drawdown remains more or less constant. As the flow response from fractures will be recorded first in the well and the response from blocks later, the time-drawdown data will be plotted on two separate semi-log straight lines. The characteristic of time-drawdown curves are discussed in Sect. 9.2.3. The separation between the two straight lines will depend on storativity ratio (ω), and transmissivity ratio (λ) (Gringarten 1982; Horne 1990). Therefore, the response of a dual porosity system is similar to that of delayed yield in unconfined aquifer (Boulton and Streltosva 1977). It could, therefore, be concluded that the flow in a porous fractured rock differs from an ordinary porous medium only during the initial stages of transient flow and only in the vicinity of pumping well. Beyond a relatively short period, the flow regime will be identical to that of a single continuum (Bear 1993). Methods of analysing pumping test data from double porosity aquifers are discussed in Sect. 9.2.4.

7.2.3 Equivalent Porous Medium Model (EPM)

It is also known as equivalent continuum model. This is a conceptually simple and commonly used approach in estimating flow and transport in fractured media as it would avoid characterization of fractures. Many researches have shown that flow in a large enough volume of fractured medium can be reasonably represented by flow through a porous medium i.e. by an equivalent continuum model. This would be true when (a) fracture density is high, (b) apertures are constant rather than varying, (c) orientations are distributed rather than constant, and (d) larger sample sizes are tested (Long et al. 1982), and (e) the interest is mainly on volumetric flow, such as for groundwater supplies. The equivalent continuum approximation is also dependent on the relative orientations of the joint set. In case where the rock has two sets of joints, chance for equivalent continuum behaviour is found to increase as the angle between joint sets approaches 90° (Panda and Kulatilake 1995). In such cases conventional forms of the groundwater flow equations developed for granular porous media can be adopted.

7.2.4 Discrete Fracture Network Models

Network models use fracture characteristics and heterogeneity of rock mass based on field data. The hydraulic behaviour of a discrete fracture network is a function of fracture aperture, length, density, orientation, connectivity of fractures and fracture filling material as discussed earlier. Numerical tools have been developed to distinguish between fractured systems that can be treated as EPM and those that should be treated as discrete fracture networks (Long et al. 1982).

Two-dimensional and three-dimensional fracture network models have been suggested by Long et al. (1982), and Lee and Farmer (1993). These models evaluate flow in fractures or fracture sets with synthetic distributions of apertures, orientations, spacings, and dimensions and take into account various surface roughnesses, flow channelling, and mixing phenomena at fracture intersections. The application of these theoretical models to natural systems has been limited. These models are useful where the area of interest is small as in the study of the effect of radionuclide transport through fractures as a result of nuclear waste storage in geological formations over a long period of time. The disadvantages of the discrete modeling are that statistical information about fracture characteristics may be difficult to obtain. These models are also complex and there is no guarantee that a model reproducing the apparent geometric properties of a fracture network will capture its essential flow or transport features (NRC 1996).

7.2.5 Equivalent Parallel Plate Model

Sharp (1993) proposed the equivalent parallel plate model which takes into consideration to a greater degree geological controls than other models. The approach is based on the assumption that the area of interest in a fractured rock mass can be divided into smaller near-homogeneous domains, each having flow characteristics different from the other. In this way, fracture characterization can be based on a direction dependent parameter, which has been called here as "integrated fracture density of apertures" (IFDA) function. The various fractures are grouped according to their strike trends. For each strike trend, field data on aperture (a) and density (N_i) for all fracture sets are collected, and their IFDA is calculated from the relation:

$$IFDA = \sum N_i a_i^2 \qquad (7.57)$$

IFDA values are plotted graphically for each direction (Fig. 7.9). In this figure the trend of each line, with double arrows, represents the strike of the dominant fracture sets and the length corresponds to the respective IFDA value. Thus, based on this characterization, the area of interest can be divided into near-homogeneous smaller domains. Each domain would have its own characteristic distribution of hydraulic conductivity which will provide an input into two-dimensional finite-element models.

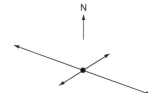

Fig. 7.9 Graphical representation of IFDA (for details see text)

Based on the above concept, an IFDA map of a part of Bundelkhand granitic terrain in India was prepared (Fig. 7.10). The different domains are demarcated based on field measurements of fracture characteristics (orientation, aperture and spacing) at 64 sites, except domains VII b, c and d, which were mapped based on the trends of lineaments on remote sensing (aerial photographs and IRS-LISS-II) data as these areas are covered with weathered rock and soil. The trend of the IFDA tensor in the adjacent area (domain VIII a) also helped in deciding the IFDA tensor in the domains VII b, c and d. Figure 7.10 indicates that in this area, hydraulic tensor of the type I is by far the most extensive one, characterized by a strong NW–SE trend, though minor trends exist in other directions also.

7.3 Flow in the Unsaturated Zone

The zone between the water-table and the ground surface is the zone of aeration or the unsaturated (vadose) zone (Sect. 1.3). The mechanism of downward move-

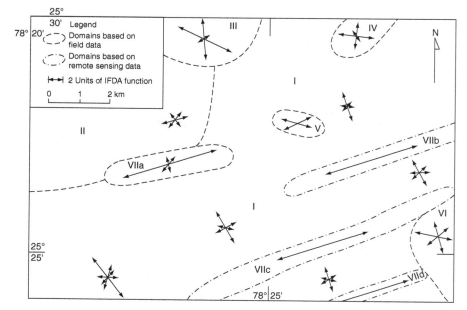

Fig. 7.10 IFDA map of a part of Bundelkhand area, India. (Modified after Gupta 1996) (Note: In this map, one unit of IFDA function = 1 μm)

ment of water through this zone is of importance both from the point of view of natural and artificial recharge to phreatic aquifer as well as downward movement of contaminants from the ground surface. There is also upward vertical movement of water in the unsaturated zone by evaporation and transpiration. This has an important bearing on crop water requirements, drainage and soil salinization processes. The hydraulic properties of the unsaturated zone also influence the rise or fall of water-table during recharge or abstraction of water due to pumping or drainage. Unsaturated flow is actually a two phase flow involving air and water. Air can flow through very fine fractures whereas the water phase preferably flows through larger fractures which offer much lesser resistance.

7.3.1 Flow Through Unsaturated Porous Media

The *volumetric water content* (θ) in unsaturated material is defined as $\theta = V_w/V$ where V_w is the volume of water and V is the total volume of a soil or rock. It is usually reported as a decimal fraction or a percentage. For saturated flow, $\theta = \eta$ and for unsaturated flow, $\theta \leq \eta$, where η is the total porosity.

In the unsaturated zone, although there will be localised pockets of water (perched water) having positive water potential but the potential (head), h in general is negative due to the surface tension of water. The water present is subjected to capillary and adsorptive forces and energy is required to remove it. With decrease in water content, the remaining water will have more negative water potential. As the water content decreases, the remaining water is left adhered to the surfaces of mineral grains or walls of fractures as thin film. The flow characteristics of this thin film is also important, which depends on the thickness of the film. It is often difficult to estimate the thickness of the film and thereby the conditions under which the films will coalesce to cause saturation of the pores or the fractures.

Unsaturated flow in the vadose zone (zone of aeration) can be analysed on the basis of Darcy's law by considering the unsaturated hydraulic conductivity, (K_u), instead of normal hydraulic conductivity (K). Unsaturated hydraulic conductivity, K_u is dependent on the water content θ which is related to the (negative) pressure head, h'. Therefore, K_u can be expressed as $Ku(\theta)$ or as $Ku(h')$ depending on whether it is con-

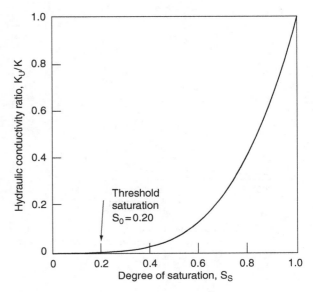

Fig. 7.11 Ratio of unsaturated (K_u) to saturated (K) hydraulic conductivity as a function of degree of saturation. (After Irmay 1954)

sidered in relation to θ or h'. In unsaturated material, the presence of air in the pore spaces obstructs water movement. Therefore, the lower the value of the water content, the lower will be the unsaturated hydraulic conductivity. Based on the experimental data, the relation between K_u, K, and degree of saturation is given as (Todd 1980).

$$\frac{K_u}{K} = \left(\frac{S_s - S_0}{1 - S_0}\right)^3 \qquad (7.58)$$

where S_s, is the degree of saturation and S_0 is the threshold saturation (nonmoving water held by capillary forces). These data when plotted in Fig. 7.11, show that K_u varies from zero at $S_s = S_0$ to $K_u = K$ at $S_s = 1$, i.e. at full saturation.

The relation between Ku/K and negative pressure, h' are S-shaped as shown in Fig. 7.12. The approximate equation for this type of relation is (Todd 1980).

$$\frac{K_u}{K} = \frac{a}{\frac{a}{b}(-h')^n + a} \qquad (7.59)$$

where a, b and n are constants that decrease with decreasing grain size of the material. It can be seen that at atmospheric pressure when h'=0, then $K_u = K$.

Under unsteady unsaturated flow condition, K_u and Q continuously increase during infiltration of rain

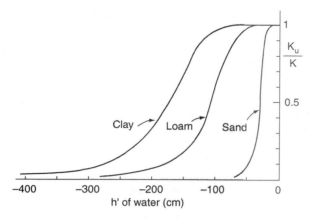

Fig. 7.12 Relationship between hydraulic conductivity ratio (K_u/K) and soil-water pressure (h') for sand, loam and clay. (After Bouwer 1978)

and will decrease after the cessation of recharge in the upper part of the wetted front. Evaporation and transpiration will also reduce K_u and Q in the upper part of the profile.

Darcy's law, for unsaturated flow, can be written as

$$q = -K(\theta)\nabla H \qquad (7.60)$$

where q = volumetric flux vector, $K(\theta)$ is hydraulic conductivity tensor as a function of water content, θ, and ∇H = total hydraulic gradient in three dimensional space.

The hydraulic head is the sum of elevation and pressure head. However, for unsaturated conditions, the pressure head will be negative, as mentioned earlier. The flow will also be influenced by osmotic potential and temperature gradients. In addition to water content, $K(\theta)$ depends upon the total porosity, pore size distribution as well as the viscosity and density of the fluid. It is therefore difficult to measure $K(\theta)$ under the field conditions.

The general differential flow equation for a three dimensional unsaturated flow (Eq. 7.61) was derived by Richards (Bouwer 1978) using the analogy to heat flow in porous media, in the same way as for saturated flow.

$$\frac{\partial}{\partial x}\left(K(h)\frac{\partial h}{\partial x}\right) + \frac{\partial}{\partial y}\left(K(h)\frac{\partial h}{\partial y}\right)$$
$$+ \frac{\partial}{\partial z}\left(K(h)\frac{\partial h}{\partial z}\right) + \frac{\partial K(h)}{\partial z} = \frac{\partial \theta}{\partial t} \qquad (7.61)$$

The term $\partial K(h)/\partial z$ is included to account for gravity flow. For two-dimensional flow in a vertical plane (stream seepage to a deep watertable, infiltration from irrigation channel etc.), the second term in Eq. 7.61 will be zero. For one-dimensional vertical flow (infiltration of rainfall, evaporation from water-table), the first two terms are zero. For steady flow systems $\partial \theta/\partial t$ is zero (Bouwer 1978).

Richard's equation is based on the assumptions that the fluid is incompressible and the flow takes place under isothermal conditions. However, the movement of water and contaminants is influenced by temperature gradients in the unsaturated zone. As the ground surface temperatures are usually higher than at depth, there will be a tendency of water vapour to move upward. This will be opposite to the movement of water which is downward due to gravity. The natural temperature gradients will also be disturbed by the disposal of radioactive waste in underground repositories. Under certain conditions, the circulation of air brought about by barometric changes will also affect the moisture content in the unsaturated zone as discussed in Sect. 7.3.2 Therefore, although theoretical reasoning suggests that the Richard's equation does not fully represent the flow in the unsaturated zone and may even be inaccurate in some circumstances, but the errors resulting from ignoring multi-phase nature of the flow process in natural conditions in most of the cases will be negligible.

As most of the soils are anisotropic and heterogeneous, transport of a part of infiltrating water will be faster than that of the average wetting front. This *preferential flow* of water in the unsaturated zone is of importance for simulating the field water balance and therefore for the calculation of crop water use, solute transport and pollution of groundwater (Feddes et al. 1988). Preferential flow, also known as bypass flow or short circuiting, in unsaturated zone, may take place through macropores, caused by cracking and shrinkage of clayey soils, by plant roots or by tillage operations. Various models have been developed to simulate preferential flow by modifying the basic partial differential equation (Feddes et al. 1988).

7.3.2 Flow Through Unsaturated Fractured Rock

The flow mechanism in the unsaturated unconsolidated material, which has been studied in detail in the past is also applicable to weathered zone in hard rocks. However, the processes affecting flow in unsaturated fractured rock are not well understood so far. It has gained importance in the recent years only mainly for the study of migration of contaminants from the high-level repositories of radioactive waste and sanitary landfills in the unsaturated (vadose) zone. This has led to theoretical, laboratory and field investigations regarding flow mechanism in the unsaturated fractured rocks (Evans and Nicholson 1987). One such site for radioactive waste disposal in a desert environment which has been studied in detail is the Yucca Mountain in Nevada, USA (Wang and Narasimhan 1993; NRC 1996). The potential repository is located in the volcanic tuff at a depth of approximately 350 m below the ground surface and 225 m above the water-table. These studies have provided useful information about the fluid flow in partially saturated fractured porous medium for predicting the migration of nuclear waste from the repository to the available environment over thousands of years.

The movement of water in unsaturated fractured rocks is essentially a problem of multiphase fluid flow (water and gas phases) as in the unsaturated porous granular medium. However, in a fractured rock there is likely to be a time lag between the movement of fluids through the fractures and the matrix blocks. Under fully saturated conditions, the fractures will control the fluid flow while under unsaturated or partially saturated condition, the role of fracture flow is minimal and the flow is mainly governed by matrix-block characteristics. Under unsaturated conditions, due to capillary mechanism, the fractures will drain easily as fracture apertures are usually larger than matrix pore sizes. Therefore, fractures will remain dry under partially saturated conditions while the water will be held by capillarity in the finer pores of the matrix. In such a situation where porous blocks are separated by desaturated fractures, there may not be any flow of water from one porous block to the other.

However, at greater depths due to compressional stress, the walls of the fractures are pressed together forming asperities. Further, the aperture of a fracture will be smaller near the asperities and larger at other places within the fracture plane. At these places, the fractures will be partly saturated forming bridges for liquid water to move across the fractures from one porous block to the adjacent one (Fig. 7.13). Borehole data show that about 12% of the fracture surfaces in Yucca Mountain have coatings of secondary minerals,

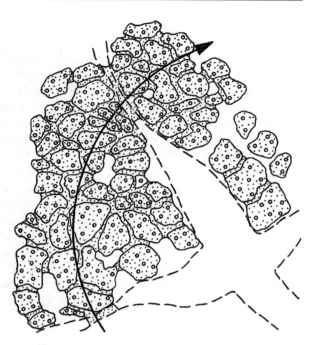

Fig. 7.13 Schematic representation of water flow in a fractured porous medium. (After Wang and Narasimhan 1993)

e.g. zeolite, clay and calcite, indicating that the coating area of 12% corresponds to contact area of fractures through which inter-block flow would take place across the fractures under partially saturated condition. (Wang and Narasimhan 1993).

In the radioactive waste repositories, the natural temperature gradients will also be disturbed due to the heat and large temperature gradients caused by the radioactive waste decay thereby influencing the movement of water as well as contaminants both in liquid and the vapour phases. The heating part will loose moisture while the condensation of water vapour will take place away from the heat source. Readers may refer to the work of Evans and Nicholson (1987) and Pruess and Wang (1987) for experiments to study the effect of temperature gradients on movement of water vapour and numerical modelling of isothermal and non-isothermal flow in unsaturated fractured rock.

The convective transport of vapours in the unsaturated fractured rock is also dependent on air pressure differences which are controlled by changes in barometric pressure, thermal gradients and changes in rock saturation due to addition or removal of water. In areas of steep relief, as in the Yucca Mountain, it is observed that barometric effects will influence the air circulation and thereby drying of the rock in the upper parts of the unsaturated zone (Weeks 1987). Therefore, the hydrological properties and vapour transport in the unsaturated zone will be affected by topographically enhanced air circulation under natural conditions. The drying of rocks would reduce the potential for deep percolation through the proposed repository site (Weeks 1987).

Single phase approach to unsaturated flow behaviour holds good for all practical purposes except where the effect of entrapped air is significant. Two-phase (water-gas, water-oil) flow through fractured rocks has gained greater importance in petroleum engineering and nuclear waste disposal. As multiphase flow in fractured rocks involves a set of complex phenomena, no comprehensive model has been developed to date, which includes factors such as the interaction between each phase, change of fluid properties, and associated joint deformation (Indraratna and Rajnith 2001).

7.4   Principles of Solute Transport in Fractured Media

The transport of dissolved substances (solutes) in groundwater is important to study the movement of natural constituents due to rock water interaction and also of contaminants introduced by man's activities as a result of disposal of domestic, industrial or agricultural wastes. The solutes could either be non-reactive (conservative), viz. chloride, or reactive substances, viz. sodium. The non-reactive substances are not subjected to any exchange or reaction while moving through the matrix, thereby excluding radioactive decay and adsorption.

The assessment of contaminant transport through fractured media has gained greater importance in recent years in view of the programmes of permanent storage of high level nuclear fuel wastes in crystalline rocks in several countries, especially in North America and Europe (Sect. 12.3.3).

7.4.1   Mechanism of Solute Transport

Similar to transport of contaminants in ordinary porous media, the main mechanisms of transport of solutes through fractured porous rocks are by advection, hydrodynamic dispersion, molecular diffusion, radioactive or biological decay, rock water interaction and retardation including adsorption and desorption (Fig. 7.14).

The governing equation for solute transport in a fracture (Fig. 7.14) which incorporates various mechanisms of solute transport is written as (Novakowski et al. 2007).

$$a \left[R_a \frac{\partial c}{\partial t} + \bar{v} \frac{\partial c}{\partial x} - D_L \frac{\partial^2 c}{\partial x^2} + \lambda R_a c \right]$$
$$+ 2q = 0, \quad 0 \le x \le \infty \tag{7.62}$$

where the x-coordinate is in the direction of the fracture axis, a is the fracture aperture, λ is a decay constant, q is the diffusive flux perpendicular to the fracture axis, R_a is the retardation factor is pore velocity, D_L is the

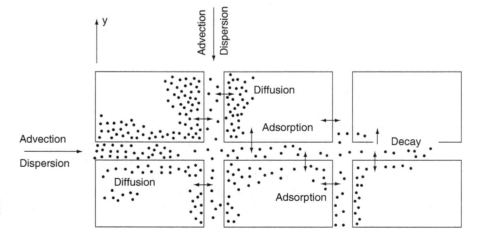

Fig. 7.14 Schematic representation of various mechanisms of solute transport in an idealized fractured porous medium. (After Germain and Frind 1989)

hydrodynamic dispersion coefficient and $\overline{V}$ is the fluid velocity.

As most of the fractured rock aquifers show double porosity character, like fluid flow, the mechanism of contaminant transport depends on the relative porosities and permeabilities of the fractures and the matrix blocks. If the rock matrix is impermeable, has negligible porosity and is compact, the advective transport through the fractures will be prominent. If the matrix blocks are porous but have negligible permeability, the main mechanism in the matrix will be by molecular diffusion and transport through the rock matrix by advection will be generally insignificant because of its low permeability. In such a case, as a part of clean-up operation, the contaminants from the fractures may be flushed out relatively quickly but the porous blocks will only slowly release the contaminants into the fractures. In a third situation, when porous block has the same permeability as the fracture, the transport of solute will take place simultaneously in the two media (fracture and matrix) by advection, dispersion and diffusion depending on differences in the head and the concentration of the solutes in the matrix blocks and the fractures.

7.4.1.1   Advection

The main mechanism of transport of non-reactive solutes is by advection (convection) due to which the solutes are transported at an average rate equal to the average linear velocity, $\overline{V}_a$ of the water in the porous medium is given by

$$\overline{V}_a = \frac{q}{\eta_e} = \frac{K_x \frac{dh}{dx}}{\eta_e} \qquad (7.63)$$

In fractured media, the average velocity of flow parallel to a planar fracture $(\overline{V}_a)$ can be described by

$$\overline{V}_a = \frac{\gamma a^2 (dh/dl)}{12\mu} \qquad (7.64)$$

The fractured rocks have much lower effective porosity $(10^{-2}-10^{-5})$ as compared with porous granular material (0.5–0.03). Therefore, groundwater velocities in fractured rocks is several order of magnitude higher than in the granular rocks. This can be illustrated by

computing groundwater velocities by using the modified Darcy's Eq. 7.65.

$$\overline{V}_a = \frac{K}{\eta_f} \frac{dh}{dl} \qquad (7.65)$$

where η_f is the bulk fracture porosity.

If we assume that a fractured medium (granite) has a fracture porosity (η_f) of 10^{-4}, and hydraulic conductivity of $10^{-5}\,m\,s^{-1}$ and if the hydraulic gradient is 10^{-2}, as is generally the case in the field, the groundwater velocity using Eq. 7.65 will estimated to be equal to $86.4\,m\,d^{-1}$. This compared to a porous medium (say silty sand) with the same values of hydraulic conductivity and hydraulic gradient and effective porosity of 0.26 will have a velocity of about $0.03\,m\,d^{-1}$ indicating thereby that the groundwater velocity in a fractured medium can be several times higher than in the porous rocks. Therefore, advection transport is much faster in fractured media than in porous rocks provided the matrix is impermeable and has negligible porosity. However, numerical studies show that tracer velocities may be less or greater than the average fluid velocity depending on the fracture roughness and contact area. The tracer velocity is found to be less than the average fluid velocity if flow is transverse to the roughness and greater if flow is parallel to the roughness (NRC 1996).

Zuber and Motyka (1994) emphasised the importance of matrix porosity to determine contaminant velocity at large scales in a fractured rock (Eq. 7.66)

$$v_1 \eta_p \approx (\Delta H / \Delta H)\,K \qquad (7.66)$$

where v_1 is the mean velocity of a conservative tracer or pollutant, η_p is the matrix porosity, $(\Delta H/\Delta X)$ is the hydraulic gradient and K is the hydraulic conductivity.

7.4.1.2   Dispersion

In addition to advection, the solute will have a tendency to spread by hydrodynamic dispersion which causes the dilution of the solute. During flow through a cylindrical pipe, the velocities are distributed in a parabolic form causing a faster movement of solute along the axis and a comparatively slower movement near the surface due to drag effect (Fig. 7.15a). The same

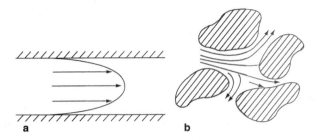

Fig. 7.15 Distribution of velocities due to hydrodynamic dispersion through: **a** a pipe, and **b** porous medium

phenomena would occur in porous media or fractures with more complexities due to variation in velocities as the pores and fracture openings may be of different sizes, shapes and orientations (Fig. 7.15b). Dispersion will also be affected by the tortuosity and connectivity of multiple flow paths. Dispersion is a nonsteady irreversible process.

The dispersion phenomena can be demonstrated by simple laboratory experiments using cylindrical sand column (Fig. 7.16a). The concentration of solute will decrease with time in the direction of flow which can be demonstrated by breakthrough curves (Fig. 7.16b).

Due to dispersion, the solute will spread both in the longitudinal and transverse directions. Dispersion in the direction of flow (longitudinal dispersion) is much more stronger than in any direction normal to the flow (transverse dispersion) (Fig. 7.17).

In fractured media, dispersion is due to heterogeneities of the velocities in a fracture (parabolic profile) and also due to differences in velocities from one fracture to another (due to differences in aperture sizes and tortuosity) as well as due to intersection and channelling of fractures which cause transverse mixing and dispersion. Reader may refer to Leap and Mai

(1992) for the influence of pore pressure and Smith and Schwartz (1993) for a detailed discussion on the effect of fracture intersection on dispersion. The effect of channelling on solute transport in fractured rocks is discussed by Oden et al. (2008).

In homogeneous and isotropic media for the two-dimensional case, the equation for dispersion for a conservative tracer has the form

$$\frac{\partial C}{\partial t} = \frac{\partial}{\partial x}\left(D_L\frac{\partial C}{\partial x}\right) + \frac{\partial}{\partial y}\left(D_T\frac{\partial C}{\partial y}\right)$$
$$- V_x\frac{\partial C}{\partial x} - V_y\frac{\partial C}{\partial y} \qquad (7.67)$$

where C is the relative tracer concentration ($0 < C < 1$), D_L and D_T are longitudinal and transverse coefficients of hydrodynamic dispersion respectively, V_x and V_y are the components of fluid velocities (X is the coordinate in the direction of flow and y is the coordinate normal to flow), and t is the time. Dimensions of the dispersion coefficients are L^2T^{-1}.

The longitudinal and transverse coefficients of hydrodynamic dispersion can be expressed as

$$D_L = \alpha_L V + D$$

$$D_T = \alpha_T V + D$$

where α_L and α_T are the characteristic properties of the porous medium known as *intrinsic dispersion coefficient or dispersivity* with the dimension of L, and D is the coefficient of molecular diffusion for the solute in the porous medium (L^2T^{-1}).

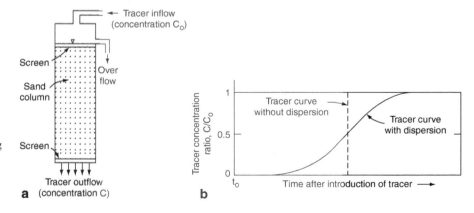

Fig. 7.16 Longitudinal dispersion of a solute passing through a sand column: **a** experimental setup, and **b** break through (dispersion) curve

Fig. 7.17 Longitudinal and transverse dispersion of a tracer introduced into a porous medium: **a** one time source, and **b** continuous source

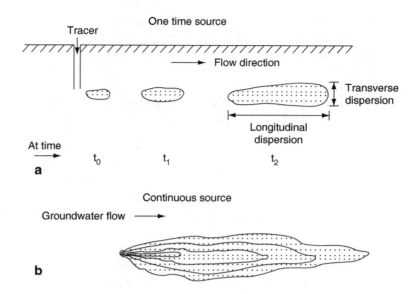

At low velocity, as dispersion is mainly on account of diffusion; the coefficient of hydrodynamic dispersion, D_L equals the diffusion coefficient, D. At high velocity, mechanical mixing will be the dominant mechanism of solute transport, therefore $D_L = \alpha_L V$. Larger dispersivity value of the medium produces greater mixing of the solute front as it advances. Laboratory studies on homogeneous sandy material have given values of longitudinal dispersivity, α_L in the range of 0.1–10 m while the transverse dispersivity α_T values are lower by a factor of 5–20. Laboratory values are significantly smaller than those obtained from field studies (Freeze and Cherry 1979) which is attributed to the effect of heterogeneities on field scale (Fetter 1993). Neuman (2005) has also shown that longitudinal dispersivities observed worldwide from laboratory and field tracer tests increase with the scale of observation. A similar trend is seen in porous media indicating the effect of travel distance or travel time on the spread of the solute plume. Therefore, the values of dispersivities obtained from laboratory study will not be of much help in predicting dispersion in the field. Methods of determining dispersivity in field are given by Freeze and Cherry (1979). Field studies indicate that longitudinal dispersivity, α_L typically ranges from 10 to 100 m and transverse dispersivity, is 10–30% of α_L (Table 7.1). The ratio of longitudinal to transverse dispersivity (α_L/α_T) of an aquifer is an important criteria which influences the shape of the contaminant plume.

Dispersion will be greater in consolidated (fractured) media than in unconsolidated media as the pore-size distribution is wider in the consolidated rocks and thus the distribution of velocities is also wider. A greater compactness also leads to greater branching of flow lines, hence greater lateral dispersion (Rasmuson and Neretnieks 1981). It can be also noted from Table 7.1 that the fractured basalt of Idaho, shows greater value of α_T as compared with α_L which is due to fracturing.

7.4.1.3 Molecular Diffusion

Diffusion mainly causes a slowdown in the migration of the solute and a decrease in the concentration peak. It is of importance in assessing the risk of pollution in an aquifer. This physical phenomenon causes movement of particles (ions or molecules) due to concentration gradient between two neighbouring points from zones of high concentration to those of low concentration. Diffusion which is also a dispersion phenomena is of

Table 7.1 Values of dispersivity (α_L and α_T) in some rock types estimated from field studies. (After Fetter 1988; de Marsily 1986)

Rock	Location	Dispersivity
Basalt	Hanford, Washington, USA	$\alpha_L = 30$ $\alpha_T = 20$
Basalt	Snake River, Idaho, USA	$\alpha_L = 91$ $\alpha_T = 137$
Limestone	–	$\alpha_L = 152$
Alluvium	–	$\alpha_L = 10$–60 $\alpha_T = 5$–50

importance only at lower velocities. Diffusion can take place even if the fluid is stationary as the movement is induced primarily under the influence of concentration gradient. The molecular diffusion coefficient decreases with decrease in temperature.

The mass flux of diffusing substance is proportional to the concentration gradient and is given by Fick's first law (Eq. 7.68)

$$F = -D_0 \frac{\partial C}{\partial x} \qquad (7.68)$$

where, F=mass flux i.e. the mass of solute per unit area per unit time, D_0=molecular diffusion coefficient $[L^2 T^{-1}]$, C=solute concentration and $-\partial C/\partial x$=concentration gradient. The negative sign would indicate movement of solute from areas of greater concentration to those of lower concentration.

In case where concentration changes with time, Fick's second law will apply (Eq. 7.69)

$$\frac{\partial C}{\partial t} = \frac{D_0 \partial^2 C}{\partial x^2} \qquad (7.69)$$

where $\partial C/\partial t$=change in concentration with time.

Matrix diffusion may vary from one tracer to the other due to molecule size.

With the exception of H^+ and OH^-, values of D_0 range from 5×10^{-6} to $20 \times 10^{-6}\,cm^2 s^{-1}$; ions with greater charge have smaller values (Schwartz and Zhang 2003).

The molecular diffusion coefficient in porous media, D_0 will be small as compared with that in water due to the resistance in the porous medium by the length and tortuosity of the flow paths. The ratio of D/D_0 varies from 0.1 in clayey sands to 0.7 in sands; in crystalline rocks, D/D_0 is <0.02 (UNESCO 1980).

Molecular diffusion is a slow process. In formations of high permeability, the effect of molecular diffusion are masked by higher velocities but in low-permeability material, such as clay or silt, where groundwater velocities are low, molecular diffusion can be a significant process influencing the distribution of various ions. Matrix diffusion also has a greater control on solute transport in fractured rocks when the matrix porosity is high (e.g. chalk). The geometry of the fracture network and hydrodynamic dispersion will have a secondary role (Bodin et al. 2003). This also depends on the relative diffusion coefficients of

ions and whether they are reactive to the matrix or not. Tritium (^3H) profiles in fractured clayey till show that tritium has moved by active groundwater circulation in fractures upto about 7–10 m and has travelled 1 or 2 m beyond that depth by molecular diffusion in the matrix (Ruland et al. 1991). Diffusive properties of a rock also depend on its petrography and mineral alteration. alteration may increase the diffusion properties by a factor of 20–200 (Bodin et al. 2003).

The diffusion properties of solute are often determined from laboratory experiments on rock samples (Bodin et al. 2003). As the diffusion process is slow, the tests are to be run for longer duration. It is observed that the in-situ values of diffusion are lower than those estimated from laboratory experiments by a factor of 2–2.5 due to changes in the physical properties e.g. stress conditions and porosity. Extensive studies on matrix diffusion of radionuclides in fractured rocks has been carried out at the Aspo Hard Rock Laboratory, Sweden for site characterisation and performance assessment of nuclear waste (Hodgkinson, et al. 2009).

Channelling also influences the diffusion mechanism as it reduces the effective contact surface between the solute and the matrix. As diffusion and channelling have similar effect on tracer movement in fractured rocks, multi-tracer studies are suggested to distinguish between the two processes as diffusion effect will vary from one tracer to another whereas channelling remains unchanged (Bodin et al. 2003).

As mentioned earlier, transport through the rock matrix by advection in fractured rocks is generally insignificant because of its low permeability. However, molecular diffusion from the fractures into the porous blocks will cause attenuation by removing solute (contaminant) mass from the fracture flow thereby retarding the advance of contaminants along the fracture. Figure 7.18 presents the effect of matrix diffusion on concentration distribution of a nonadsorbing and adsorbing nondecaying solute migrating through a fracture in a nonporous, and a porous matrix. In Fig. 7.18a, as the matrix is non-porous and dispersion in the fracture is neglected, the concentration profile has a rectangular shape but in a porous medium diffusion will be prominent which will cause greater retardation of the solute (Fig. 7.18b). In tight fractured granite, diffusion would be minimal but in clays where the matrix is porous and has low permeability, solute will diffuse into the rock matrix resulting in a decrease in the concentration of the contaminant

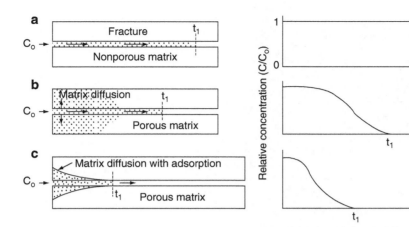

Fig. 7.18 Effect of advection, matrix diffusion and adsorption on contaminant distribution in fractured medium: **a** advective transport along a fracture in a nonporous matrix **b** advective transport with matrix diffusion into porous matrix due to molec-ular diffusion **c** advective transport with molecular diffusion and adsorption. Profiles of relative concentration (C/Co) of the con-taminant along the fracture at time, t_1 in these cases, are also shown. (Modified after Vogel and Giesel 1989)

with increasing distance along the fracture. Diffusion is therefore an important process causing retardation in the movement of soutes in fractured porous rocks. Tracer study using bromide in fractured clay demon-strated the strong retardation effect due to molecular diffusion (Cherry 1989). A retardation of about three orders of magnitude is reported from fractured clays (Germain and Frind 1989). Figure 7.18c shows the added effect of adsorption in the surface of the fracture which will cause further retardation of the migration of contaminant in the fracture. We return to this sub-ject in the latter part of this chapter. The presence of fracture coatings may also create a zone of diffusion having different diffusion coefficients from that in the unaltered matrix.

Migration of contaminants has been studied for various fracture openings, spacing and other fracture geometry by number of workers (Sudicky and Frind 1982; Vogel and Giesel 1989; Rowe et al. 1989). Large fracture openings or high velocities will reduce matrix diffusion due to increase in the solute-transported to solute stored ratio. Studies by Grisak and Pickens (1980) and Sudicky and Frind (1982) show that the fracture spacing will have a significant effect on the advance rate of a contaminant. A small fracture spac-ing will result in a greater penetration distance along the fractures because of the limited capability of the porous matrix to store solute resulting in insignificant molecular diffusion in the matrix.

A parallel fracture system was considered by Vogel and Giesel (1989). Rowe et al. (1989) gave a semi-

analytical solution for 1D, 2D and 3D contaminant diffusion into the matrix in a orthogonally fractured system. Their study shows that in fractured shales, three dimensional effects are minimal for very highly fractured systems (i.e. fracture spacing of 0.01 m or less) but the three-dimensional diffusion can appre-ciably alter the shape and extent of the contaminant plume for wider fracture spacings (e.g. 1 m).

The transport of solutes in the fractures is also affected by channelling because effective porosity and fracture surface area for diffusive flux is reduced caus-ing a reduction in the attenuation of solute. Therefore, small-scale laboratory experiments may significantly overestimate the effective porosity and specific surface area unless network effects are taken into account. This will lead to overestimation of the capacity of retarda-tion by means of sorption and matrix diffusion (Dver-storp, et al. 1992).

The effect of roughness of fracture walls on the transport of solutes is demonstrated by Raven et al. (1988). They have shown that wall-irregularities in fractures promote formation of immobile liquid zones (Fig. 7.19). In these parts the solutes get stored in the early stages and are released later when the solute con-centration in the mobile liquid decreases.

7.4.1.4 Retardation

Solutes dissolved in groundwater may undergo chemi-cal changes due to reaction with water and aquifer

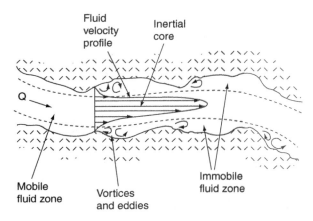

Fig. 7.19 Zones of mobile and immobile water in a fracture. (After Raven et al. 1988)

material by ion exchange, radioactive decay, biodegradation and other processes.

Ion exchange or adsorption reactions are of importance in the movement of contaminants in groundwater systems. The mechanism of ion exchange (adsorption) is further elaborated in Sect. 11.6.1. Absorption reactions are very rapid relative to the flow of groundwater. In a purely fractured rock, adsorption will take place only on the surface area of the fractures which is relatively small. However, in porous fractured rock, adsorption will also take place on the very large solid surface area within the porous blocks. Adsorption will cause retardation of contaminant front as is evidenced from column studies (Fig. 7.18c).

The retardation of the solute or the tracer due to adsorption will be much greater by the matrix material due to its greater surface area as compared with the adsorption on the fracture surface. The extent of adsorption depends on type of ions and composition of matrix material.

Most radionuclides that are stored in underground repositories are subject to sorption in crystalline rocks. Fracture coatings of clay minerals, calcite or Fe-oxyhydroxides greatly influence the retardation properties of fractured rocks. Sorption properties of a rock also varies with the specific mineral groups and the nature and degree of mineral alteration. Sorption reactions are also affected by fluid properties, such as pH and Eh (Bodin et al. 2003).

In case when two tracers are used, one reactive and the other non-reactive, the reactive tracer front will lag behind the nonadsorbed tracer which will spread

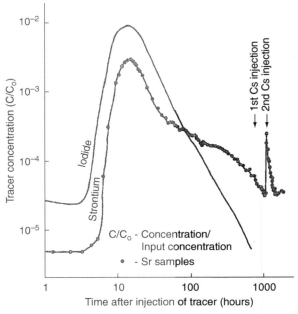

Fig. 7.20 Breakthrough curves for iodide and strontium following simultaneous injection of a pulse of each solute (initial iodide background elevated by previous experiment). (After Heath 1985)

out as a result of dispersion. Breakthrough curves for simultaneous injection of non-sorbing iodide and sorbing strontium followed by highly sorbing cesium in fractured granite are given in Fig. 7.20. The strontium peak is lower than that of iodide which is explained due to the adsorption of strontium. The secondary peak of strontium after about 1000 h is related to the injection of cesium which caused elimination of strontium. These investigations therefore demonstrate that migration of a single solute may differ from those involving simultaneous injection of several solutes having different sorbing properties. Similar trends are also observed by using conservative tracer (uranin) and reactive tracers (^{22}Na and ^{85}Sr) in the crystalline (granodionte) rock mass at the Grimsel Test Site, Switzerland (Keppler et al. 1996).

Equation 7.70 known as the retardation equation, gives a simplified expression for the retardation of a contaminant (A) in an aquifer (UNESCO 1980).

$$\overline{V}_A = \frac{\overline{V}_a}{1 + \dfrac{\rho_b}{\eta} K_d^A} \qquad (7.70)$$

Table 7.2 Measured field K_d values (in ml g^{-1}) for caesium and strontium. (After UNESCO 1980)

Material	Element	
	Cs	Sr
Granite	34.0	1.7
Limestone	13.5	0.2
Basalt	792–9520	16–135
Sandstone	102.0	1–4
Sand	53–523	4–19
Alluvium	450–950	40.0

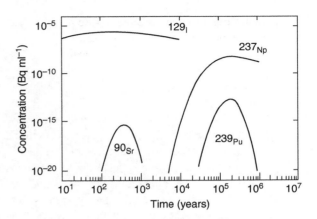

Fig. 7.21 Concentration profiles of some typical radionuclides. Q = 1 Bq, Z = 10 m, Va = 10 cm^3 yr^1, η = 0.005, a = 0.02 cm. (After Krishnamoorthy et al. 1992)

where $\overline{V}_A$ is the average velocity of a contaminant (A), $\overline{V}_A$ is average groundwater velocity, ρ_b is the bulk density, η is the porosity and K_d^A is the distribution (retardation) coefficient which expresses the ratio of the amount of contaminant (A) adsorbed or precipitated per gram of soil or aquifer material to the amount of the same contaminant remaining in solution, i.e. groundwater (units ml g^{-1}). Consequently, higher value of K_d^A will indicate greater adsorption or precipitation of a contaminant.

Equation 7.70 illustrates that if the value of K_d for a particular contaminant increases, there must be a corresponding decrease in the average velocity of that contaminant $\overline{V}_A$ meaning thereby its retardation relative to the average groundwater velocity. K_d values in different rock type for two important radioactive elements (caesium and strontium) which have long-lived radionuclides of great toxicity, are given in Table 7.2.

For a fractured media, the retardation Eq. 7.70 becomes (Freeze and Cherry 1979)

$$\frac{\overline{V}_a}{\overline{V}_A} = \frac{1 + 2K_d^A}{a} \qquad (7.71)$$

Where $\overline{V}_a/\overline{V}_A$ is known as the relative velocity and α is the fracture aperture. Equation 7.71 indicates that retardation decreases with increase in fracture aperture. This equation will be valid only when rock matrix is nonporous. In case of porous rock matrix adjacent to fractures, a part of contaminant will migrate by molecular diffusion from the fracture into the rock matrix thereby causing attenuation of the contamination in the fracture as discussed earlier.

Further, Eq. 7.71 assumes an uniform value of retardation coefficient which will not be true in the case of heterogeneous fractured media due to variation in fracture aperture and orientation, which will not only

influence the advective flow but also the retardation factor (Wels and Smith 1994).

Krishnamoorthy et al. (1992) computed the concentration and flux of radionuclides commonly encountered in high level waste (HLW) assuming certain paramteric values in fractured granite (Fig. 7.21). It was noted that various nuclides show varying trends depending on their K_d values. For example, ^{129}I reaches a peak value of 3×10^{-6} Bq ml^{-1} after about 300 years but does not show any marked decrease due to lack of adsorption. ^{239}Pu ($K_d = 500$ ml g^{-1}) and ^{237}Np ($K_d = 100$ ml g^{-1}) reach peak concentration around 2×10^5 year; ^{239}Pu decays faster but ^{237}Np continues to have significant concentration even after 1 million years due to relatively low K_d (1.7 ml g^{-1}). The variation in their peak concentrations is also explained due to differences in K_d values. For examples ^{90}S$_r$ shows a peak much earlier (around 400 years) due to its low K_d (1.7 ml g^{-1}).

The above studies indicate that most of the radionuclides will decay to insignificant level at distances of over 50 m or so. However, even this small fraction may lead to high concentration of those radionuclides which have low K_d and long half-life due to lower flux of water into the fractures.

Although matrix diffusion and adsorption greatly retard the transport of contaminants including radionuclides but these may create problems in undertaking the clean-up of the aquifer because the contaminant above a permissible limit may persist for many years following the cut off of the source. At higher matrix porosity this problem is more severe as the contaminant may persist for several decades.

Summary

Groundwater flow through porous media is governed by the Darcy's law. However, in fractured rocks, as fracture aperture and their connectivity influence the groundwater flow, cubic law is advocated. Depending on the porosities and permeabilities of the fractures and the intervening matrix blocks, various types of groundwater flow models have been suggested, the most popular being the double-porosity model. Variations in fracture aperture largely influence the flow of water and solutes.

Flow through the unsaturated zone is essentially a multi-phase flow involving water and gases. It is of importance in the study of drainage and soil salinization. Unsaturated flow in fractured rocks has gained greater importance from the point of view of migration of contaminants. The main mechanisms of solute transport through porous and fractured media are by advection, dispersion, diffusion, radioactive decay and retardation including adsorption and desorption. In double-porosity aquifers, which is commonly the case in fractured rocks, the contaminant transport depends on the relative porosities and permeabilities of both the fractures and the matrix.

Further Reading

Bear J, Tsang CF, Marsily G de (eds) (1993) Flow and Contaminant Transport in Fractured Rock. Academic Press, Inc., San Diego.

Delleur JW (2007) Elementary groundwater flow and transport processes, in The Handbook of Groundwater Engineering (ed. Delleur JW), 2nd edn, CRC Press, Boca Raton.

de Marsily G (1986) Quantitative Hydrogeology. Academic Press, Inc., New York.

Fetter, CW (2001) Applied Hydrogeology. 4th edn, Prentice Hall, New Jersey.

Lee CH, Farmer I (1993) Fluid Flow in Discontinuous Rocks. Chapman and Hall, London.

NRC (1996) Rock Fractures and Fluid Flow- Contemporary Understanding and Applications. National Academic Press Washington, DC.

Hydraulic Properties of Rocks

8

8.1 Basic Concepts and Terminology

Hydraulic properties of water bearing formations are important as they govern their groundwater storage and transmitting characteristics. These are described below.

Porosity (η) Porosity, η is a measure of the interstices or voids present in a rock formation. It is defined as the ratio of volume of voids, V_v to the total volume, V of the rock mass (Eq. 8.1)

$$\eta = \frac{V_V}{V} \qquad (8.1)$$

Porosity, (η) is given as a percentage or as decimal fraction. Porosity is of two types, primary and secondary. Primary porosity is the inherent character of a rock which is developed during its formation. Secondary porosity is developed subsequently due to various geological processes, viz. fracturing, weathering and solution activity. In unconsolidated rocks, primary porosity is of importance but in hard rocks secondary porosity is of greater significance.

Porosity is controlled by: (a) shape and arrangement of constituent grains, (b) degree of sorting, (c) compaction and cementation, (d) fracturing, and (e) solution. The geometrical arrangement of constituent grains (packing) and sorting have important influence on porosity. Well sorted clastic material has high porosity irrespective of grain size. In poorly sorted material, porosity is less as small-size grains occupy pore spaces between bigger grains. Compaction and cementation reduces porosity. In unconsolidated formations, porosity at deeper levels will be less due to compaction, e.g. shales have lower porosity than clays.

The fractured rock formations are made up of two porosity systems, (a) the intergranular porosity or matrix porosity (η_m), formed by intergranular void spaces, and (b) the secondary porosity developed due to fractures and solution cavities, termed as fracture porosity, (η_f). Therefore, in fractured rocks, the total porosity is the sum of matrix and fracture porosities, i.e.

$$\eta = \eta_m + \eta_f \qquad (8.2)$$

The two porosities can be expressed in the conventional manner as

$$\eta_m = \frac{Matrix\ void\ volume}{Total\ bulk\ volume}$$
$$\eta_f = \frac{Fracture\ void\ volume}{Total\ bulk\ volume}$$

In laboratory, porosity of rock samples can be estimated by immersion in liquids, density test and by gas-porosity meters (UNESCO 1984b). In field, geophysical logging methods, viz. resistivity, neutron and gamma methods can be used for determining porosity.

In the field, fracture porosity can be estimated from scan line method by the relation $\eta_f = Fa$, where F is the number of joints per unit distance intersecting a straight scan line across the outcrop and a is the mean aperture of fractures. The porosity of natural materials may range from almost zero in hard massive rocks to as much as 60% in clays (Table 8.1).

Laboratory measurements on a variety of fractured rocks show that fracture porosity, η_f is considerably less than the matrix porosity. η_m is reported to generally vary from 0.1% to 8% and n_f from 0.001% to 0.01% (Lee and Farmer 1993).

B. B. S. Singhal, R. P. Gupta, *Applied Hydrogeology of Fractured Rocks,*
DOI 10.1007/978-90-481-8799-7_8, © Springer Science+Business Media B.V. 2010

Table 8.1 Representative values of porosity (η), specific yield (S_y) and specific retention (S_r) of geological materials. (After Morris and Johnson 1967; Hamill and Bell 1986)

Geological formation	η(%)	S_y(%)	S_r(%)
Unconsolidated deposits			
Gravel	28–34	15–30	3–12
Sand	35–50	10–30	5–15
Silt	40–50	5–20	15–40
Clay	40–60	1–5	25–45
Dune Sand	40–45	25–35	1–5
Loess	45–50	15–20	20–30
Rocks			
Sandstone	15–30	5–25	5–20
Limestone, dolomite	10–25	0.5–10	5–25
Shale	0–10	0.5–5	0–5
Siltstone	5–20	1–8	5–45
Till	30–35	4–18	15–30
Dense crystalline rock	0–5	0–3	–
Fractured crystalline rock	5–10	2–5	–
Weathered crystalline rock	20–40	10–20	–
Basalt	5–30	2–10	–

Void Ratio (e) This is generally used in soil mechanics and is expressed as

$$e = \frac{V_v}{V_s} \qquad (8.3)$$

where V_S is volume of mineral grains and V_v is as defined earlier. Void ratio has large numerical variation. In natural soils, where total porosity ranges from 0.3 to 0.6, the corresponding void ratio range is 0.45–1.5. The relation between total porosity, η and void ratio e, can be expressed as

$$\eta = \frac{e}{1+e}, \quad \text{or} \quad e = \frac{\eta}{1-\eta} \qquad (8.4)$$

Specific Retention (S_r) This is a measure of the volume of water retained by the rock material against gravity on account of cohesive and intergranular forces. It can be expressed as

$$S_r = \frac{V_r}{V} \qquad (8.5)$$

where V_r is the volume of water retained. Specific retention depends on the *specific surface* of constituent mineral grains which in turn is influenced by the grain size, shape and type of clay minerals present. Specific surface is defined as the area per unit weight of the material and is expressed in $m^2 g^{-1}$. The specific surface values of coarse grained material, such as gravel and sand, is small compared with silt and clay size fractions. Among clay minerals, non-swelling clays have specific surfaces in the range of $10–30\,m^2 g^{-1}$, but swelling clays such as montmorillonite have large values of about $800\,m^2 g^{-1}$. A similar property is the specific surface area, (S_{Sp}), defined by

$$S_{SP} = \frac{\text{Total surface area of the interstitial voids}}{\text{Total volume of the medium}}$$

S_{sp} has the dimensions of L^{-1}. In fine-grained material, S_{sp} will be more, viz. in sands, S_{sp} will be of the order of $1.5 \times 10^4\,m^{-1}$ but in montmorillonite, it is about $1.5 \times 10^9\,m^{-1}$ (de Marsily 1986). These properties are of importance in the adsorption of water molecules and ions on the surfaces of mineral grains, especially on clay minerals.

Specific Yield (S_y) This is defined as the ratio of the volume of water that an unconfined aquifer will release from storage by gravity, to the total volume of fully saturated aquifer material. It is expressed either as a decimal fraction or as a percentage. Specific yield depends on the duration of drainage, temperature, mineral composition of water, grain size and other textural characteristics of aquifer material. Values of specific yield (S_y) of some common rock materials are given in Table 8.1.

Effective Porosity (η_e) Effective porosity or *kinematic porosity* is the same as specific yield (de Marsily 1986). The concept of effective porosity indicates that all the pores do not participate in the flow of water. Fine grained and poorly sorted materials have low effective porosity as compared with coarse grained and well sorted material, due to the greater retention of water on account of intergranular forces. Instead of total porosity, the effective porosity (η_e) is more important for estimating the average velocity of groundwater and transport of contaminants as discussed in Sects. 7.1.2 and 7.4.1.

In crystalline and other hard rocks, the size and interconnection of fractures are mainly responsible for imparting effective porosity to the rock mass. In such rocks, although total porosity may be high but due to unconnected fractures, the effective or kinematic porosity will be less, viz. in granites and other crystalline rocks, although total porosity may be 1–10% but the effective porosity is very small $(5 \times 10^{-5} - 1 \times 10^{-2})$. Similarly, in dolomites, which are formed as a result of diagenesis, although the rock may acquire high secondary porosity due to reduction in volume of mineral grains, but effective porosity will be less.

Hydraulic Conductivity (K) This is a measure of the ability of a formation to transmit water. It has the dimensions of $L\,T^{-1}$ and is usually expressed in $m\,s^{-1}$.

In terms of Darcy's law (Eq. 7.12), K can be expressed as

$$K = \frac{V}{dh/dl} \qquad (8.6)$$

where K is hydraulic conductivity, V is groundwater velocity, and dh/dl is hydraulic gradient.

In the USA, K was earlier expressed in two forms– as the Meinzer's (Laboratory) coefficient of permeability, K_m and also Field coefficient of permeability (K_f), both using units of $gal\,d^{-1}\,ft^{-2}$. The main difference between the two being that K_m is expressed at a constant temperature of 60°F (15.6°C) while K_f is measured at the actual field temperature. As field conditions do not influence the groundwater temperatures to any significant extent, the distinction between K_m and K_f has now been discarded.

The hydraulic conductivity depends both on the properties of the medium (rock material) as well as of the fluid. In sedimentary formations, grain size characteristics are most important as coarse grained and well sorted material will have high hydraulic conductivity as compared with fine grained sediments like silt and clay. Increase in degree of compaction and cementation reduces hydraulic conductivity. In hard (fractured) rocks, K depends on density, size and inter-connection of fractures (Sect. 8.2).

Permeability (k) It is a more rational concept than hydraulic conductivity (K) as it is independent of fluid properties and depends only on the properties of the medium. Fluid properties which influence hydraulic conductivity are viscosity (μ), expressing the shear resistance, and specific weight, (γ), expressing the driving force of the fluid.

The relation between hydraulic conductivity, (K) and properties of the medium and the fluid can be expressed as

$$K = \frac{cd_e^2\,\gamma}{\mu} = \frac{k\gamma}{\mu} \qquad (8.7)$$

where c is a dimensionless constant also known as shape factor and d_e is effective grain size; c depends on porosity and packing etc. Equation 8.7 when substituted in Darcy's equation gives

$$k = \frac{\mu\,Q/A}{\gamma\,(dh/dl)} \qquad (8.8)$$

The value of k can be given in darcy units which in terms of Eq. 8.8 is expressed as

$$1\,darcy = \frac{\left[1\,(cm^3/sec)/cm^2\right]\,1\,cP}{1\,atm/cm} \qquad (8.9)$$

Thus, a porous saturated medium will have a permeability of one darcy if a fluid of 1 centipoise (1cP) viscosity will flow through it at a rate of $1\,cm^3\,s^{-1}$ per cm^2 cross-sectional area under a pressure or equivalent hydraulic gradient of $1\,atm\,cm^{-1}$.

According to Eq. 8.9, k has the units of area. As the value of k is very small, it is also expressed in square micrometres $(\mu m)^2$.

$$1(\mu m)^2 = 10^{-12}m^2$$

By substitution of appropriate units in Eq. 8.9, it can be shown that,

$$1\,darcy = 0.987(\mu m)^2 = 10^{-8}cm^2$$
$$= 10^{-5}ms^{-1}(approx.)$$

The range of values of hydraulic conductivity, K and permeability, k are given in Table 8.2. Conversion factors for the various common units of K and k are given in Appendix. It could be noted from Table 8.2 that the permeability of natural materials has wide variation. As the permeability of crystalline rocks and other tight formations is usually very small, k in such cases is usually expressed in millidarcy (md) which is approximately equal to $10^{-8}\,m\,s^{-1}$. The permeability of dense unfractured rocks is usually very low being up to $1\,md^{-1}$, and usually below 0.001–$0.5\,md^{-1}$. Fractures increase the permeability by several orders of magnitude above the

Table 8.2 Range of values of hydraulic conductivity and permeability for various types of geological materials

Hydraulic conductivity, K (ms⁻¹)	·1	10^{-1}	10^{-2}	10^{-3}	10^{-4}	10^{-5}	10^{-6}	10^{-7}	10^{-8}	10^{-9}	10^{-10}	10^{-11}	10^{-12}	10^{-13}
Permeability, k (darcy)	10^5	10^4	10^3	10^2	10	1	10^{-1}	10^{-2}	10^{-3}	10^{-4}	10^{-5}	10^{-6}	10^{-7}	10^{-8}
Relative values	Very high		High		Moderate			Low				Very low		

Representative materials
Unconsolidated deposits
Gravel
Clean sand
Silty sand
Clay till (often fractured)
Rocks
Shale & siltstone (unfractured)
Shale & siltstone (fractured)
Sandstone
Sandstone (fractured)
Limestone & dolomite
Karst limestone & dolomite
Massive basalt
Vesicular & fractured basalt
Fractured & weathered crystalline rock
Massive crystalline rock

solid rock mass. The fracture permeability can be 100 and even 1000 md⁻¹.

Transmissivity (T) This is defined as the rate of flow of water at the prevailing field temperature under a unit hydraulic gradient through a vertical strip of aquifer of unit width and extending through the entire saturated thickness of the aquifer (Fig. 8.1). Transmissivity (T) has dimensions of L^2T^{-1} and is expressed in m^2d^{-1} or m^2s^{-1}. Darcy's law, in terms of T, can be written as

$$Q = TIL \qquad (8.10)$$

where Q=rate of flow, I=hydraulic gradient, L=width of the flow section, measured at right angles to the direction of flow.

In confined aquifer, T=Kb, where b is the saturated thickness of the aquifer. In unconfined aquifer, the saturated thickness will be less than the true thickness. Here it is assumed that K is isotropic and constant across the thickness of the aquifer which may be horizontal or dipping. Transmissivities greater than $1000\,m^2d^{-1}$ represent good aquifers for groundwater exploitation. In geothermal reservoir, transmissivity is usually expressed in terms of permeability-thickness (kb) in units of d-m ($1\,m^3 = 10^2\,d$-m). We return

to this subject with respect to geothermal reservoirs in Chap. 18.

Storativity (S) Storativity of an aquifer is defined as the volume of water which a vertical column of the

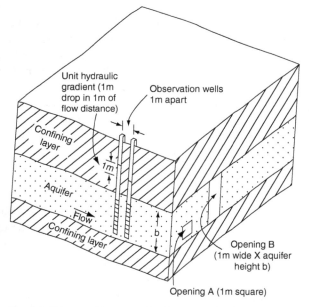

Fig. 8.1 Diagram illustrating coefficients of hydraulic conductivity (K) and transmissivity (T). Flow of water through opening A will be equal to K and that through opening B equal to T

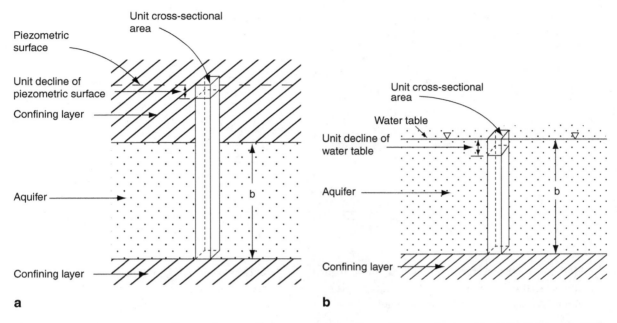

Fig. 8.2 Diagrams illustrating coefficient of storage (S) for: **a** confined aquifer, and **b** unconfined aquifer. (After Ferris et al. 1962)

aquifer of unit cross sectional area releases from storage as the average head within this column declines by a unit distance (Fig. 8.2). It is therefore dimensionless. In a confined aquifer, where water released from or taken into storage is entirely due to compressibility of aquifer and water, the storage coefficient is given by $S = bS_s$, where S_s is the specific storage, defined later. Value of storativity in confined aquifer is of the order of 10^{-3}–10^{-6}. In an unconfined aquifer, storativity S, is given by $S = S_y + bS_s$. Usually $S_y \gg bS_s$, thus storativity of unconfined aquifer for all practical purposes is regarded equal to its specific yield or effective porosity, (η_e) (Hantush 1964). Storativity in unconfined aquifers ranges from 0.05 to 0.30.

The relation between storativity and the compressibility of the aquifer material and of water can be expressed as

$$S = \gamma \eta b \left(\beta + \frac{\alpha}{\eta} \right) \qquad (8.11)$$

where, β is the compressibility of water ($4.7 \times 10^{-10}\,\mathrm{Pa^{-1}}$), α is the compressibility of solid skeleton of the aquifer and S, η and b are defined earlier.

Specific Storage (S_S) This is the volume of water which a unit volume of the confined aquifer releases from storage because of expansion of water and com-

pression of the aquifer under a unit decline in the average hydraulic head. It has the dimension of L^{-1}. S_s is used exclusively in confined aquifer analysis, as in unconfined aquifer the water released from storage is mainly due to gravity drainage and not due to the compressibility of aquifer material or of water. S_s is a more fundamental parameter as compared to S as the latter depends upon both the specific storage and the aquifer geometry. In terms of Eq. 8.11

$$S_s = \gamma \, (\alpha + \eta\beta) \qquad (8.12)$$

Therefore, S_S depends both on coefficient of compressibility (α) and porosity (η) of the rock. As hard dense rocks have low porosity and low coefficient of compressibility, S_S is also less as compared with sands and clay formations. Some representative values of α and S_S, are given in Table 8.3.

Table 8.3 Range in values of the coefficient of compressibility of solid material (α) and specific storage (S_s). (After Domenico 1972; Streltsova 1977)

Rock type	α (Pa^{-1})	S_s (m^{-1})
Dense rock	10^{-12}–10^{-10}	10^{-7}–10^{-5}
Fissured and jointed rock	10^{-10}–10^{-9}	10^{-5}–10^{-4}
Sand	10^{-9}–10^{-8}	10^{-4}–10^{-3}
Clay	10^{-8}–10^{-7}	10^{-4}–10^{-2}

Fig. 8.3 Measured values of hydraulic diffusivity (κ) for a variety of low permeability rocks plotted against effective stress and the equivalent depth. (After Neuzil 1986)

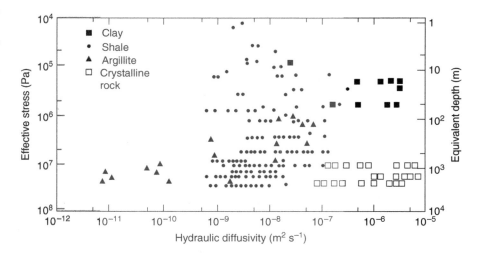

Hydraulic Diffusivity (κ) This is a single formation characteristic that couples the transmission properties, K and storage property, S_s or alternately T and S.

$$\kappa = T/S = K/S_s \qquad (8.13)$$

k has dimensions of L^2T^{-1}. It is a significant property of the medium for transient flow and has a major influence on the drawdown response around a pumped well. A comparison of measured κ in rocks of low permeability under varying stress is illustrated in Fig. 8.3. The equivalent depth is also given. It could be noted that in argillaceous materials, the hydraulic diffusivity, κ is generally less $(10^{-9}–10^{-7} m^2 s^{-1})$ than in crystalline rocks, viz. gabbros and granites $(10^{-7}–10^{-5} m^2 s^{-1})$, which appears to be due to the low S_s of crystalline rocks.

Leakage Coefficient or Leakance (Dimensions T^{-1}) This is the property of the semi-confining (aquitard) layer. It is equal to K^1/b^1, where K^1 and b^1 are the vertical hydraulic conductivity and thickness of the aquitard respectively.

Hydraulic Resistance (C) It is the reciprocal of leakage coefficient. It indicates resistance against vertical flow in an aquitard. It is equal to b^1/K^1 and has the dimensions of time. If hydraulic resistance, C=∞, the aquifer is confined.

Leakage Factor (B) This determines the distribution of leakage through an aquitard into a leaky aquifer. It is defined as

$$B = \sqrt{TC} \qquad (8.14)$$

Leakage factor has dimensions of length and is expressed in metres. High values of B indicate greater resistance of the semi-previous strata to leakage.

Boulton's Delay Index (1/α) This is a measure of the delayed drainage of an unconfined aquifer. It has the dimensions of time. The value of 1/α may vary from about 50 min in coarse sand to 4000 min in silt and clay.

Drainage Factor (D) Drainage factor ($D=\sqrt{T/\alpha S_y}$) is a property of unconfined aquifer. It has the dimensions of length and is usually expressed in metres. Large values of D indicate fast drainage. If D=∞, the yield is instantaneous with the lowering of the water-table, i.e. the aquifer is unconfined without delayed yield.

Storativity Ratio (ω) This and the interporosity coefficient (λ) are the properties of fractured aquifers described in Sect. 7.2.2. Methods of estimating the hydraulic properties are described in Chap. 9.

8.2   Hydraulic Conductivity of Fractured Media

Fractures control the hydraulic characteristics of low permeability rocks, viz. crystalline, volcanic and carbonate rocks. Also in some clastic sedimentary formations, viz. sandstones, shales, glacial tills and clays, fractures form the main pathways for movement of fluids and contaminants.

In fractured rocks, a distinction can be made between hydraulic conductivity of fracture, K_f and of intergranular (matrix) material, K_m. As fractures form the main passage for the flow of water, the hydraulic conductivity of fractured rocks mainly depends on the fracture characteristics described in Chap. 2. The matrix permeability (k_m) in granite is estimated in the order of $10^{-19}\,m^2$ whereas fracture permeability (k_f) can vary from 10^{-12} to $10^{-15}\,m^2$ depending on the fracture aperture and interconnectivity. The role of some important parameters, e.g. aperture, spacing, stress, infilling (skin), connectivity etc. is discussed below (also see Sect. 19.6.3.2).

8.2.1 Relationship of Hydraulic Conductivity with Fracture Aperture and Spacing

The relationship between hydraulic conductivity (K_f) of a single plane fracture with aperture (a) is given by Eq. 8.15.

$$K_f = \frac{\gamma\, a^2}{12\mu} \qquad (8.15)$$

The equivalent hydraulic conductivity of a rock mass, (K_s) with one parallel set of fractures is expressed by

$$K_s = \frac{a}{s} K_f + K_m = \frac{\gamma\, a^3}{12\, s\mu} + K_m \qquad (8.16)$$

where s is fracture spacing. Usually K_m is very low except when rock matrix is porous and/or fractures are filled with impervious material. Therefore,

$$K_s = \frac{\gamma\, a^3}{12\, s\mu} = \frac{g\, a^3}{12\, vs} \qquad (8.17)$$

where g is gravitational acceleration ($981\,cm\,sec^{-2}$) and v is the coefficient of kinematic viscosity which is $1.0 \times 10^{-6}\,m^2\,s^{-1}$ for pure waer at 20°C.

In fractures with infillings, the hydraulic conductivity of fracture will depend on the permeability of the filling material, assuming that this permeability is still significantly greater than that of the rock matrix.

The hydraulic conductivity of a rock mass with three orthogonal joint sets with the similar spacing and constant aperture in all directions, in the three dimensional space, is given by Eq. 8.18 (Lee and Farmer 1993).

$$K_s = \frac{2\,\gamma\, a^3}{12\, s\mu} + K_m \qquad (8.18)$$

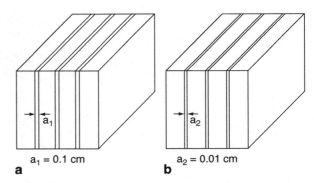

a $\quad a_1 = 0.1\ cm \qquad$ b $\quad a_2 = 0.01\ cm$

Fig. 8.4 Two sets of fractures with same spacing but different apertures

Figure 8.4 shows fractures with a common spacing of 1 joint per metre but with different apertures. In Fig. (8.4a) fracture apertures, $a_1 = 0.1\,cm$, and in Fig. (8.4b) $a_2 = 0.01\,cm$. Substituting these values in Eq. 8.17 will give equivalent hydraulic conductivity of K_1 to be $8.1 \times 10^{-4}\,m\,s^{-1}$ and K_2 for the second type will be $8.1 \times 10^{-7}\,m\,s^{-1}$ i.e. K_2 will be about three orders of magnitude less than that of set 1.

Figure 8.5 gives the hydraulic conductivity values of fractures with different apertures and frequencies. It could be seen that one fracture per metre with an aperture of 0.1 mm gives rock hydraulic conductivity of about $10^{-6}\,m\,s^{-1}$, which is comparable to that of porous sandstone. With a 1 mm aperture and the same spacing, the hydraulic conductivity will be $10^{-3}\,m\,s^{-1}$, similar to that of loose clean sand.

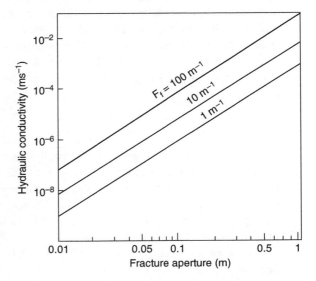

Fig. 8.5 Variation in hydraulic conductivity with fracture frequency and conducting aperture. (After Lee and Farmer 1993)

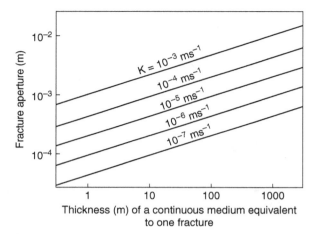

Fig. 8.6 Comparison between the hydraulic conductivity of the porous medium and that of the fractured medium as a function of aperture. (After Maini and Hocking 1977, reproduced with permission of the Geological Society of America, Boulder, Colorado, USA)

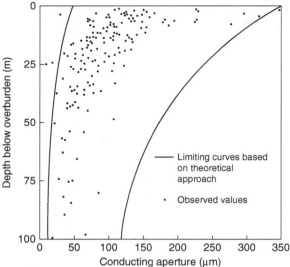

Fig. 8.7 Relationship between conducting aperture and depth based on field data. The right hand side envelope is based on initial conducting aperture of 350 μm and left hand side of 50 μm. JCS = 95.8 MN m^{-2}, JRC = 7.4. (After Lee and Farmer 1993)

The equivalence between hydraulic conductivity in a fractured rock and that of porous material is depicted in Fig. 8.6. As an example, the flow from a 10 m thick cross-section of a porous medium with a hydraulic conductivity of 10^{-4} m s^{-1} could be the same as from one single fracture with an aperture of about 1 mm. This demonstrates the large amount of flow which can be expected from fractures of even small openings.

A distinction is made between real or mechanical aperture (a_r) and conducting or hydraulic aperture (a_c). The real aperture is usually larger than conducting aperture if the fracture surfaces are rough. In smooth and wide fractures, the mechanical aperture and the conducting aperture will be equal. The empirical relation between a_r and a_c can be expressed as (Lee et al. 1996)

$$a_c = \frac{a_r^2}{JRC^{2.5}} \qquad (8.19)$$

where a_r and a_c are in μm and JRC is Joint Roughness Coefficient, having a range from 20 (roughest surface) to 0 (smoothest surface). In natural fractures which are highly irregular, JRC may vary from 3 to 12 from one part of the fracture to another. Increase in JRC results in an exponential decrease in flow rate.

Assuming a maximum initial conducting aperture (a_c) equal to 350 μm and minimum initial conducting aperture to be 50 μm and by considering other mechanical properties of similar rock types, Lee and Farmer

(1993) obtained two curves showing variation in conducting aperture with depth (Fig. 8.7). These curves indicate a good agreement between the range of calculated and observed values. A similar trend of variation in fracture aperture was obtained by Oda et al. (1989).

8.2.2 Effect of Stress on Permeability

The mechanical behaviour and fluid flow in fractures is greatly influenced by the effective stress which is taken to be the normal stress on fracture minus the fluid pressure. Effective stress values are usually positive but in some cases as in hydrofracturing where fluid pressure exceeds the normal stress, effective stress values will be negative.

The combined effect of normal and shear stress on permeability and void structures is of importance in geotechnical and nuclear waste disposal studies and recovery of oil, gas and geothermal fluids from reservoirs (Rutqvist and Stephannson 2003). It is shown, both by theory and experimental studies, that stress reduces fracture aperture and thereby permeability of fractured rocks. Stress being a directional phenomenon, its state determines the relative permeability of different fracture sets in a rock mass. Fractures parallel to the maximum stress tend to be open, whereas those perpendicular to it tend to be closed.

Several experiments have been designed to estimate permeability variation in a variety of rocks under varying stress and thermo-mechanical conditions. (Brace 1978; Gale 1982a, 1982b; Oda et al. 1989; Read et al. 1989; Jouanna 1993; Azeemuddin et al. 1995; Indraratna and Ranjith 2001; Rutqvist and Stephannson 2003). Snow (Gale 1982a) proposed an empirical model of the form of

$$k = ko + \left(K_n a^2 / s\right) (P - Po) \quad (8.20)$$

where k is the permeability of horizontal fractures after loading, ko is the permeability at an initial pressure Po, K_n is the normal stiffness of the fracture; a and s are defined earlier.

Lee and Farmer (1993) quoting the work of Brace et al. showed a decrease in permeability of Westerly Granite from 350 nd at 10 MPa pressure to 4 nd at 400 MPa pressure (1 nd = 10^{-18} m²). In Berea sandstone, a decrease in permeability from 10^{-10} m² to 10^{-11} m² due to increase in hydrostatic pressure from ambient to 30 MPa is reported by Read et al. (1989). The permeability reduction from an uniaxial strain test on Berea sandstone was estimated to be 20% but it was drastic (75%) in Indiana limestone which is attributed to pore collapse (Azeemuddin et al. 1995). Laboratory tests on shale, granite and sandstone show that shale has most stress-sensitive permeability while the sandstone is very sensitive at low stress but appears to attain a residual permeability at higher stress. The differences in the stress-permeability relationship in different rock types are explained by differences in pore shapes (Rutqvist and Stephansson 2003).

The effect of stress on permeability of jointed rocks also depends on the direction of stress in relation to joint orientation. Laboratory studies on jointed granite showed that an uniaxial stress of 12 MPa, parallel with the joint, raised fracture permeability (k_f), whereas 3 MPa, normal to the joint, decreased k_f to half of the initial value. However, even at the highest pressures (100 MPa), the permeability of the rock containing joints was at least a factor of 10^3 higher than the matrix permeability (k_m) (Brace 1978).

Studies indicate that when normal stress is applied to a natural fracture in the laboratory, there occurs a reduction in permeability indicating fracture deforma-

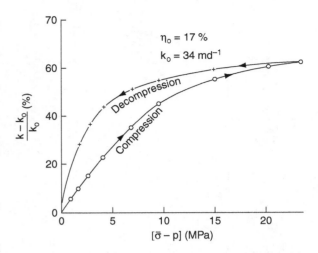

Fig. 8.8 Change in percent permeability with stress during compression and decompression in a fractured limestone. (After Van Golf-Racht 1982, with kind permission from Elsevier Science-NL, Amsterdam, the Netherlands)

tion. The change in fracture permeability is higher in the initial stages (Fig. 8.8). Further investigations indicate that under confining stress, both in air and water, the average permeability decreases by almost 90% above 8 MPa compared to permeability values at zero confining pressure (Indraratna and Ranjith 2001). This is attributed to considerable reduction in joint aperture upto a certain value of confining stress. It is also noted that the reduction in permeability also depends on the roughness of fracture – the greater the roughness the lower the rate of reduction of permeability (Indraratna and Ranjith 2001). Varied responses of permeability during compression and decompression are also indicated (Van Golf-Racht 1982; NRC 1996). After a cycle of compression and decompression, the rock permeability may either return to the original permeability (Fig. 8.8), or may get reduced due to permanent deformation (Fig. 8.9). Laboratory experiments also show that in a mica schist, on application of normal stress (applied perpendicular to the foliation plane), the decrease in rate of flow is greater when normal stress increases than when the stress decreases (Fig. 8.10).

Although effect of changes in normal stress on fracture permeability has been studied to a large extent, there have been very few controlled studies on the effect of shear stress on fracture permeability. The combined effect of normal and shear stresses on flow and void structures are not very well known so far (NRC. 1996; Rutqvist and Stephansson 2003). Such conditions are

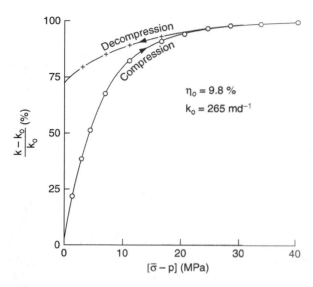

Fig. 8.9 Change in percent permeability with stress. Note that in this case the original permeability is not achieved after decompression. (After Van Golf-Racht 1982, with kind permission from Elsevier Science-NL, Amsterdam, The Netherlands)

likely to occur in civil and mining works such as underground excavations, dams and rock slopes.

It is expected that shear-stress will cause fracture dilation, especially in rough fractures due to displacement under low to moderate normal stress, resulting in significant changes in fracture permeability (Gale 1982a). In rocks like granites and quartzites, with porosity less than about 5%, the dilatancy effects have been found to be quite conspicuous. In granite, the permeability increased nearly fourfold, while in sandstone the increase was about 10–20%, but permeability of sand decreased considerably. The different behaviour of these materials indicates their varied response to stress (Brace 1978). Increase in permeability (k) and specific storage (S_S) due to the growth of dilatant cracks is also demonstrated by triaxial test (Read et al. 1989). On the other hand, in soft rocks, like mica schist, increase in shear stress showed a conspicuous decrease in rate of flow along the schistosity planes (Jouanna 1993) (Fig. 8.11).

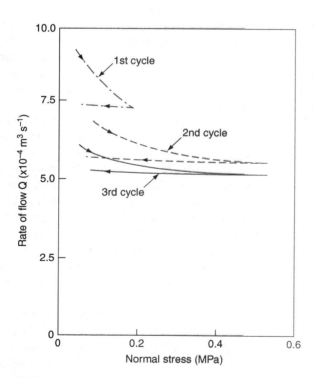

Fig. 8.10 Effect of normal stress on rate of flow in a mica schist. (After Jouanna 1993)

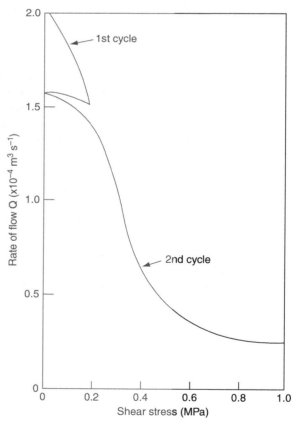

Fig. 8.11 Influence of shear stress on rate of flow in a mica schist. (After Jouanna 1993)

8.2.3 Relationship of Permeability with Depth

The depth dependence of permeability can be expressed by Eq. 8.21 given by Black (1987).

$$k = az^{-b} \qquad (8.21)$$

where a and b are constants and z is the vertical depth below ground surface. Based on the data given by Snow (1968b) about the variation in the permeability of fractured crystalline rocks with depth from Rocky Mountain, USA, Carlsson and Olsson (1977) gave the Eq. 8.22.

$$K = 10^{-(1.6\,log\,z + 4)} \qquad (8.22)$$

where K is the hydraulic conductivity in $m\,s^{-1}$ and z is depth in metres. Similar empirical relations have been given by Louis (1974), and others. For example, based on tests in a number of wells in crystalline rocks in Sweden, Burgess (in Lee and Farmer 1993) gave Eq. 8.23.

$$log\,K = 5.57 + 0.352\,log\,Z - 0.978\,(log\,Z)^2 + 0.167\,(log\,Z)^3 \qquad (8.23)$$

where K and Z have the same units as in Eq. 8.22.

Equation 8.23 can be transformed into Eq. 8.24 by relating stress and hydraulic conductivity on the basis of $\sigma = \gamma Z$

$$log\,K = 5.57 + 0.352\,(log\,\sigma/\gamma) - 0.978\,(log\,\sigma/\gamma)^2 + 0.167\,(log\,\sigma/\gamma)^3 \qquad (8.24)$$

The decaease in permeability with depth in fractured rocks is usually attributed to reduction in fracture aperture and fracture spacing (Fig. 8.12). A least square fit to the packer-test data from boreholes in the granites of Stripa mine in Sweden, also indicated a general decreasing trend of permeability with increasing depth (Fig. 8.13a). The relationship between fracture frequeny and permeability from the same area is illustrated in Fig. 8.13b. A decrease in fracture aperture with depth is also reported from several other studies,

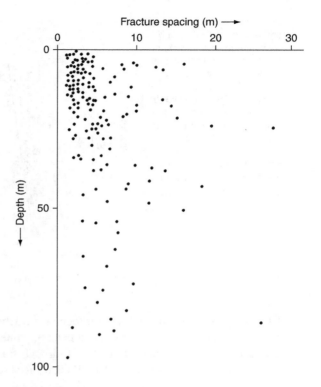

Fig. 8.12 Variation of fracture spacing with depth. (After Snow 1968a)

e.g. from a radioactive waste depository in andesite rocks in Taiwan (Lee et al. 1995).

Although, a decrease in permeability with increasing depth is demonstrated from several other places also, but this decrease may not be systematic, especially at greater depths (>50 m). The permeability can also vary by several orders of magnitude at the same depth (Fig. 8.14). Higher permeabilities at shallow depths (<50 m) can be attributed to greater influence of surficial phenomena like weathering etc. and development of sheeting joints due to unloading. Further, fractures at the same depth below the ground surface but with different orientations may be subjected to different stresses and therefore may have different permeabilities.

Even at depths of more than 1000 m, appreciable permeabilities are reported in fractured rocks. For examples, Fetter (1988) reported higher permeabilities from fractures at depths of 664–1669 m in granitic rocks of Illinois, USA. Recent studies under the Continental Deep Drilling Project in Germany (Kessels and Kuck 1995) and HDR experiments in the Rhine

Fig. 8.13 Variation of permeability with: **a** depth, and **b** fracture frequency based on borehole packer tests in granites at Stripa, Sweden. (After Gale et al. 1982). The *central full line* in figures (a) and (b) is the least square regression line; on each side of this line the 95% confidence limits are shown for individual predicted values (*full lines*) and for mean predicted values (*dashed lines*)

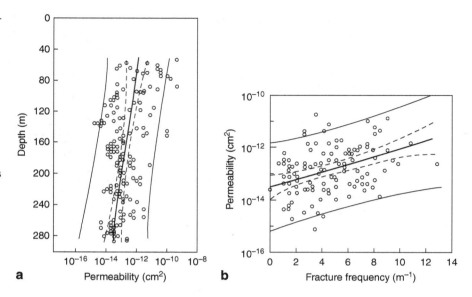

Graben, France (Gerard et al. 1996; Stober and Bucher 2005) also show the existence of good hydraulic interconnection between adjacent boreholes through fractures even at a depth of more than 3000 m (also see Sect. 13.7.2).

On the basis of above discussion, it may be summarized that although, generally in fractured rocks a decrease in permeability with depth is observed at several places but there is not much justification of such an universal rule. Therefore, site specific studies are necessary.

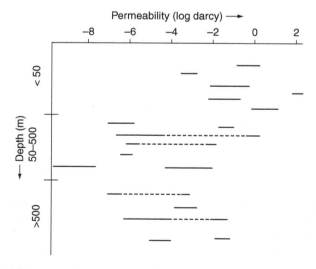

Fig. 8.14 Ranges of permeability with depth in crystalline rocks. (After Brace 1980)

8.2.4 Influence of Temperature on Permeability

Significant changes in rock temperature can occur due to natural weather conditions viz. alternate freezing and thawing and due to man induced changes. Formation of ice in fractures will cause extension of fractures and can also block the movement of water producing pressure build-ups. The influence of temperature on rock permeabilities is important for disposal of radioactive waste and in harnessing geothermal energy. An increase in temperature will cause a volumetric expansion of the rock material leading to reduction in fracture aperture and an overall decrease in rock permeability. Studies at Stripa mine in Sweden demonstrated a reduction in permeability of granites by a factor of three when temperature was increased by 25 °C by circulating warm water. Similarly, in another experiment, a tenfold reduction in permeability was observed in a fractured gneiss when the temperature was increased by 74 °C (Lee and Farmer 1993). On the other hand, thermal contraction causes development of new fractures during hydraulic fracturing in HDR experiments. However, doubts are created that these may not have significant positive long-term influence on development of geothermal energy from HDR systems (Zhao and Brown 1992) (see also Sect. 18.3). Changes in temperature may also change the effective stress in the rock mass. These stress changes may cause deformation of fractures as described earlier. Temperature changes will also cause precipitation and dissolution of minerals thereby affecting rock permeabilities.

8.2.5 Effect of Fracture Skin on Permeability

Fracture surfaces are usually covered with altered or detrital material, such as clay, iron or manganese oxides. These filling materials form fracture skin, which reduce the permeability of fractures and also influence the movement of solutes from the fractures into the matrix blocks. The presence of clay in the fractures may also increase the mechanical deformability of fractures.

8.2.6 Interconnectivity of Fractures

The interconnectivity of the different fracture sets is important for deciding the hydraulic continuity, which depends on fracture orientation, density, spacing and fracture size. Inter-connectivity can be expressed in terms of the ratio of average fracture spacing to fracture trace length. A ratio in the range of 1/20 to 1/50 has been suggested to indicate continuum (Lee and Farmer 1993). Long and Witherspoon (1985) studied numerically the effect of both the magnitude and nature of the fracture interconnection on permeability. The results showed that for a given fracture frequency, as fracture length increases, the degree of interconnection increases and thereby permeability also increases. Rouleau and Gale (in Lee and Farmer 1993) suggested an empirical interconnectivity index, I_{ij} between two fracture sets given by Eq. 8.25.

$$I_{ij} = \frac{l_i}{s_i} \sin \theta_{ij} \ (i \neq j) \qquad (8.25)$$

where l_i is the mean trace length for set i, s_i is the mean spacing of set i and θ_{ij} is the average angle between fractures of set i and j.

8.3 Anisotropy and Heterogeneity

Geological formations usually do not exhibit uniformity in their texture and structure either spatially or in different directions. Accordingly, their hydraulic characteristics such as hydraulic conductivity, and storativity also vary.

Anisotropy is usually a result of the rock fabric. In a porous rock consisting of spherical grains, the hydraulic conductivity will be the same in all directions and therefore it is said to be isotropic (Fig. 8.15a). On the other hand in a stratified formation, the constituent

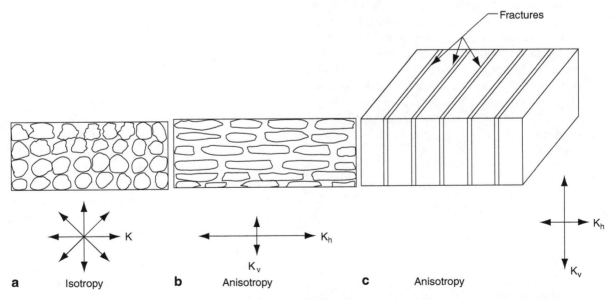

Fig. 8.15 Isotropic and anistropic aquifers: **a** isotropic sedimentary aquifer, **b** anisotropic sedimentary aquifer, and **c** anisotropic fractured aquifer

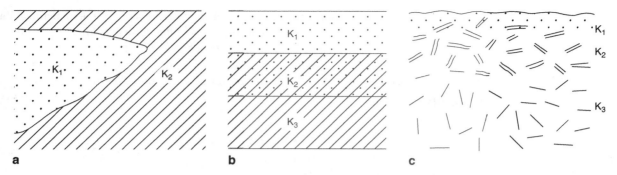

Fig. 8.16 Heterogeneous aquifers: **a** wedge-shaped aquifer **b** layered aquifer, and **c** fractured rock aquifer ($K_1 > K_2 > K_3$)

mineral grains of tabular nature, are usually laid with their longer axes parallel to each other. In such a case the hydraulic conductivity parallel to the bedding plane K_h, is higher than vetical conductivity, K_v (Fig. 8.15b). In fractured rocks, anisotropy is due to the presence of fractures which may have different orientations. The conductivity along the fracture, will be significantly higher than that normal to the fractures (Fig. 8.15c). In foliated rocks, like schists and phyllites also, conductivity parallel to foliation plane is greater than that normal to the foliation. Therefore, hydraulic conductivity is a tensorial property. The ratio of K_h to K_v may vary from 10^{-2} in fractured rocks to 10^3 in stratified sedimentary rocks as in the former permeability is mainly due to vertical fractures while in the latter the bedding planes which form easy passage for water are mostly horizontal. For analytical purpose, values of $K_h/K_v = 1$ to 10 are commonly used (Lee and Farmer 1993).

In a sedimentary rock sequence, it may be necessary to determine the average value of hydraulic conductivity normal to bedding (K_v) and parallel to bedding (K_h). If the total thickness of the sequence is H and the thickness of the individual layers are h_1, h_2, h_3, ..., h_n, with corresponding values of the hydraulic conductivity K_1, K_2, K_3, ..., K_n, then K_v and K_h can be obtained by

$$K_v = \frac{[H]}{h_1/K_1 + h_2/K_2 + h_3/K_3 + \cdots + h_n/K_n}$$

(8.26)

and

$$K_h = \frac{h_1 K_1 + h_2 K_2 + h_3 K_3 + \cdots + h_n K_n}{[H]}$$

(8.27)

The heterogeneity in sedimentary formations could be due to lateral variation in aquifer thickness even if the hydraulic properties such as hydraulic conductivity and specific storage remain constant. Such situations are observed in wedge shaped aquifers (Fig. 8.16a), showing lateral variation in thickness, or where alternate beds of sediments with different textures are formed under varying depositional conditions. Figure 8.16b shows a layered sedimentary formation where individual bed has a homogeneous hydraulic conductivity K_1, K_2, ... but the entire sequence shows a heterogeneous character due to vertical variation in hydraulic conductivity. In carbonate rocks, hetrogeneity can develop due to variation in degree of solution and in crystalline rocks it is a result of varying density of aperture sizes of fractures (Fig. 8.16c).

8.4 Representative Elementary Volume (REV)

REV is the smallest sample volume which is representative of the rock mass. The concept of REV is necessary to define the distribution of aquifer characteristics, such as hydraulic conductivity. In aquifer modelling also, it is necessary to have an idea of the minimum volume of rock which should be sampled having representative value of rock mass properties. In unfractured homogeneous rocks, the hydraulic conductivity tends to become constant beyond a particular rock volume. This least volume is known as REV (Fig. 8.17a). However, in fractured rock mass, hydraulic conductivity value may not become exactly constant with increase in sample volume but its variation may become rather insignificant (Fig. 8.17b). REV will also depend on

Fig. 8.17 Representative elementary volume (REV) in: **a** unfractured rock, and **b** fractured rock

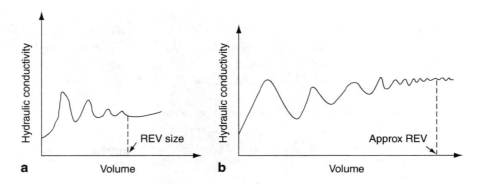

the flow system and the geometrical parameters of the fractured medium.

In fractured rocks, the concept of REV will be applicable when there is a constant hydraulic gradient and linear flow lines as in a truly homogeneous anisotropic medium. Further, the following criteria must be met in order to replace a heterogeneous system of given dimensions with an equivalent homogeneous system for the purposes of analysis (Long et al. 1982).

(a) There is an insignificant change in the value of the equivalent permeability with small addition or substraction of the test volume.
(b) An equivalent permeability tensor exists which predicts the correct flux when the direction of gradient in a REV is changed.

Point (a) indicates that the REV size is a good representative of the sample considering rock mass heterogeneities and, point (b) implies that the boundary condition will produce a constant gradient throughout a truly homogeneous anisotropic sample.

REV increases in size with increase in discontinuity spacing. Therefore, in order to define REV, one should have sufficient knowledge of rock discontinuities. The influence of fracture geometry on REV is shown in Fig. 8.18. In granular rocks without discontinuities, small REV can be representative of the rock mass (Fig. 8.18a), but in fractured rocks, REV should be large enough to include sufficient fracture intersections to represent the flow domain (Fig. 8.18b). The size of the REV will be large compared to the size of the fractures lengths in order to provide a good statistical sample of the fracture population. However, in case of large scale features, such as faults and dykes, REV may not be feasible as it will be too large an area (Fig. 8.18c). This implies that the concept of REV may not be true and practical in every rock mass.

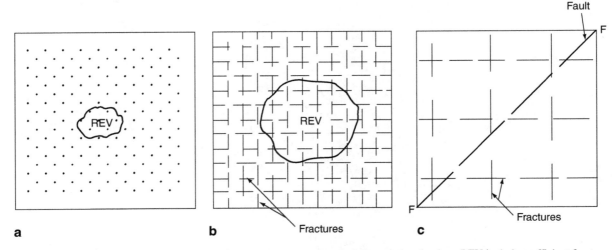

Fig. 8.18 REV in different rock conditions: **a** homogeneous porous rock, **b** fractured rock where REV includes sufficient fracture intersections, and **c** rocks with large scale discontinuities where REV is either very large or nonexistent

Summary

Important aquifer properties which control the occurrence and movement of groundwater are porosity (η), hydraulic conductivity (K), permeability (k), transmissivity (T) and storativity (S). In fractured rocks, a distinction is made between hydraulic properties of fractures and that of matrix, viz. K_f, K_m, S_f and S_m etc (the subscript f denotes properties of fractures and m that of matrix material).

The hydraulic conductivity of fractures depends on fracture characteristics, viz. aperture, spacing and interconnectivity. The fluid flow in fractures is also influenced by stress which in turn depends on the fracture orientation in relation to direction of stress. In general, permeability decreases with depth due to reduction in fracture aperture, though under certain specific tectonic conditions, high permeabilities are reported from very deep levels (>2 km).

Further Reading

Domenico PA, Schwartz FW (1998) Physical and Chemical Hydrogeology. 2nd ed., John Wiley & Sons, New York.

Indraratna B, Ranjith P (2001) Hydromechanical Aspects and Unsaturated Flow in Jointed Rock. A.A. Balkema Publ., Tokyo.

Kresic N (2007) Hydrogeology and Groundwater Modeling. 2nd ed., CRC Press, Boca Raton, FL.

Schwartz FW, Zhang H (2003) Fundamentals of Ground Water. John Wiley & Sons Inc., New York.

Hydraulic properties of rock materials can be estimated by several techniques in the laboratory and in the field. The values obtained in the laboratory are not truly representative of the formation. However, the advantage of laboratory methods is that they are much less expensive and less time consuming.

9.1 Laboratory Methods

The laboratory techniques are based on (a) indirect, and (b) direct methods.

9.1.1 Indirect Methods

In unconsolidated material, hydraulic conductivity can be determined from grain size analysis. The hydraulic conductivity of unconsolidated material is found to be related empirically to grain-size distribution by a number of investigators (e.g. Hazen 1893; Krumbein and Monk 1942; and Uma et al. 1989, among others). Hazen, as far back as 1893 developed the empirical relation (Eq. 9.1) between hydraulic conductivity (K) and effective diameter (d_e)

$$K = Cd_e^2 \qquad (9.1)$$

where C is a coefficient based on degree of sorting (uniformity coefficient) and packing. If K is in centimetre per second and d_e is in cm, the value of C in Eq. 9.1 ranges between 45 in very fine poorly sorted sand to 150 in coarse well sorted sand; a value of C = 100 is used as an average.

Effective diameter, (d_e), is the diameter of the sand grain (d_{10}) such that 10% of the material is of smaller size and 90% is of larger size. It can be estimated by plotting the grain size distribution curve. Uniformity coefficient being the ratio of d_{60} to d_{10} is a measure of the degree of sorting (Fig. 9.1).

On comparison of values of K obtained from pumping tests and Hazen's formula, Uma et al. (1989) showed that Hazen's formula gave consistently higher values of K which is perhaps due to the reason that degree of compactness is not considered in the Hazen's formula. They suggested that for unconsolidated and poorly cemented sandy material, value of C in Hazen's formula is 6.0, for moderately cemented sandstones 3.8, and for well compacted and cemented sandstones its value is 2.0.

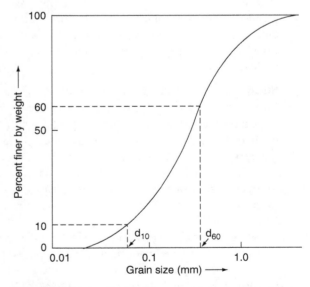

Fig. 9.1 Grain size distribution curve of unconsolidated material based on sieve analysis data

B. B. S. Singhal, R. P. Gupta, *Applied Hydrogeology of Fractured Rocks,*
DOI 10.1007/978-90-481-8799-7_9, © Springer Science+Business Media B.V. 2010

One of the well known equations for determining hydraulic conductivity by indirect method is Kozeny–Carman equation (Lee and Farmer 1993), which has the form

$$K = \frac{1}{C_T C_0 S_{SP}^2} \frac{e^3}{1+e} \frac{\gamma}{\mu} \qquad (9.2)$$

where C_T is the pore tortuosity factor, and C_0 is a pore shape factor; S_{sp}, specific surface area, and e, void ratio, were defined earlier in Chap. 8.

Values of hydraulic conductivity of unconsolidated sands obtained from grain size analysis, slug test, pumping test and numerical modelling show variations of one to two orders of magnitude (Cheong et al. 2008).

9.1.2 Direct Method

In the laboratory, hydraulic conductivity can be measured directly in three ways: (a) steady or quasi-steady flow techniques, (b) hydraulic transient flow tests, and (c) mechanical transient flow tests. The last two methods which analyse the time dependent behaviour can be used for the estimation of both hydraulic conductivity and specific storage. Steady state flow techniques are applicable in rocks of high conductivity, viz. sand, sandstone etc. while transient flow tests are recommended in low permeability tight formations like crystalline rocks, carbonate rocks and shales. Mechanical flow tests are generally used in compressible media such as clays.

Steady or Quasi-steady Flow Techniques These involve use of various types of permeameters—constant and variable head types. In these methods the rate of fluid flow through the specimen and the hydraulic gradient across the specimen are measured. Darcy's Law can be used to calculate hydraulic conductivity. The description of these methods is given in many texts on groundwater hydrology, e.g. Todd (1980), and Fetter (1988).

When the conductivities are very small (of the order of 10^{-10} m s^{-1}), steady state flow can be achieved by taking small lengths of samples in the flow direction and using large hydraulic gradients (Neuzil 1986). Falling (variable) head permeameters are also used for

the estimation of hydraulic conductivity of both coarse grained and tight formations. Some researchers have used a closed reservoir instead of an open stand pipe which is advantageous for clayey soil (Neuzil 1986).

In low permeability formations, e.g. hard rocks, use of gas permeants can be advantageous. Gas permeameters are more commonly used in oil industry. Compressed air is a satisfactory fluid in most cases. Other gases such as nitrogen, oxygen, hydrogen, helium and carbon dioxide have also been used in special cases. The advantage of using gas is that they have low viscosity and they do not react significantly with the rock material in the dry state. Therefore, problems of swelling of clay minerals, bacterial growth and other chemical changes which can considerably affect permeability, are avoided.

Hydraulic Transient Flow Test The transient groundwater flow equation in one dimension can be written as

$$K = \frac{\partial^2 h}{\partial z^2} = S_S \frac{\partial h}{\partial t} \qquad (9.3)$$

Transient tests are used for estimating very low permeabilities (10^{-10} to 10^{-17} m s^{-1}). Neuzil (1986) performed these experiments on shale specimens for estimating permeability, k and specific storage, S_s. In most of the cases a constant lateral and longitudinal load on the specimen was applied. In order to study the effect of rock deformation on permeability and specific storage, experiments based on transient pressure pulse method were designed by Read et al. (1989). Trimmer et al. (1980) used transient technique for determining permeabilities of intact and fractured granites and gabbros in the laboratory under high confining and pore-water pressures to have an understanding of fluid flow behaviour in igneous rocks at large depths from the point of view of their suitability as host rock for radioactive waste disposal.

Mechanical Transient Flow Test Permeability and specific storage can also be determined from transient mechanical behaviour of the specimen due to drainage of pore water as a result of loading. In soil mechanics, such a type of test (consolidation test) is used for low permeability compressible media. The values of permeability obtained by consolidation test and hydraulic test vary considerably in the case of highly deformable media such as clays.

Scale Effects The values of permeability determined in the laboratory from core samples especially of fractured rocks, are usually several orders of magnitude lower than those existing under natural conditions due to the smaller size of the sample and in-situ formation heterogeneity (Clauser 1992; Rovey and Cherkauer 1995, Sanchez-Vila et al. 1996, Ilman and Neuman 2003; Shapiro 2003). In this context a reference may be made to Fig. 13.12. The main factors which influence permeability values determined in the laboratory on core samples are:

1. Core (specimen) length: If the fracture spacing is more than the core length, the measured permeability will be representative of the matrix only.
2. Fracture orientation and connectivity: The permeability estimates are considerably influenced by the orientation of fractures in relation to flow direction. The radial flow to a well during field test will be quite different from the linear flow through the sample examined in the laboratory.
3. Aperture size: As permeability is dependent on aperture size, an estimate from core samples obtained from deeply buried rocks will not be representative of in-situ condition.
4. Duration of testing: The values of permeability are found to decrease with time during extended testing in the laboratory. This is attributed to clogging of pore spaces by finer particles, swelling of clay minerals and other chemical reactions between permeant and pore fluids.

Therefore, as compared to laboratory methods, the field or in-situ methods provide a better estimate of hydraulic characteristics of rock formations as a larger volume of the material is tested.

In a recent study, Shapiro (2003) has shown that although the hydraulic conductivity measured from borehole tests in individual fractures varies over more than six orders of magnitude (10^{-10}–10^{-4} m s^{-1}), the magnitude of the bulk hydraulic conductivity of the rock mass was the same from aquifer tests over 10s of meters and kilometrer-scale estimates inferred from groundwater modelling. In contrast, the magnitude of the formation properties controlling chemical migration viz., dispersivity and matrix diffusion increases from laboratory size tests to field tests on kilometre scale. A reference may be made to the concept of REV (representative elementary volume) which has been discussed in Sect. 8.4.

9.2 Field Methods

Table 9.1 gives a list of commonly used field methods. The choice of a particular method depends on the purpose of study and scale of investigations. For small scale problems, as in geotechnical investigations, seepage of water to mines and contaminant transport problems, especially in fractured rocks, packer tests, slug tests, cross-hole tests and tracer tests are preferable. In case of groundwater development and management on a regional scale, pumping test methods should be preferred. The choice is also governed by practical limitations or expediencies. The applicability of these methods in geothermal reservoirs has been discussed in Chap. 18.

9.2.1 Packer Tests

A packer test, also known as injection test, is used in uncased borehole to determine the hydraulic

Table 9.1 Field rest methods for the estimation of hydraulic characteristics of aquifers. (Modified after UNESCO 1984a)

Purpose of investigation	Size of area under investigation	Distribution of fractures	Test method
Geotechnical investigations, mine drainage, waste disposal, etc.	A few square kilometres	Random	Packer (Lugeon) test; slug test, tracer injection test
		Systematic fractures of 1, 2 or 3 sets	Modified packer test; crosshole hydraulic test; tracer injection test
Groundwater development; water resources investigation	>100 km^2	Random and closely interconnected	Pumping test
Geothermal and petroleum reservoirs	A few square kilometres	Random	Well interference test; tracer injection test

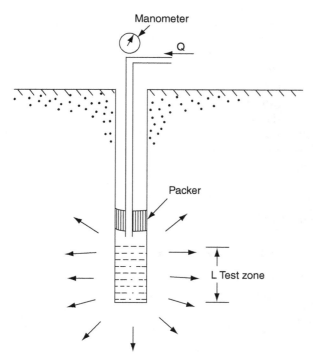

Fig. 9.2 Standard Lugeon test with single packer

conductivity of individual horizon by isolating it with the help of pakcers (Fig. 9.2). This method is widely used for estimating the hydraulic characteristics of fractured rocks in various geotechnical and waste disposal investigations (Louis 1974; Black 1987; Levens et al. 1994, Kresic 2007; Novakowski et al. 2007, among others). In a packer test water is injected under pressure at a constant hydraulic head in an isolated portion of the borehole measuring the flow rate at a steady-state condition. Generally a two-packer system having a single isolated zone is adequate for testing moderately fractured rocks (Fig. 9.3). The borehole should be flushed before hand to remove coatings from the wall of the borehole in order to get reliable results. The following three types of packer test could be used in fractured rocks, depending on details of the information required:

1. Standard Lugeon test, which gives average hydraulic conductivity.
2. Modified Lugeon test, which gives directional hydraulic conductivity on the basis of relative orientation of the test hole to the system of fractures, and
3. Cross-hole hydraulic test, described under Sect. 9.2.5.

Standard Lugeon Test The Lugeon method of testing was introduced by Maurice Lugeon, a French engineer, mainly for rock grouting in geotechnical works. It is relatively a low cost method especially for determining variations in hydraulic conductivity with depth and also in different strata. The test is made either in a completed borehole, or as the hole advances during drilling.

Lugeon test can be carried out by using either one or two packers (Figs. 9.2, 9.3). In the single packer method, the packer is placed at some selected distance above the bottom of the hole. After the test is over, drilling can further be resumed and the test can be repeated in deeper horizons. The single packer method is recommended when the rock mass is weak and intensely jointed and there are chances of the hole to collapse. In a completed borehole, two packers can be used to isolate the required section (3–6 m long) of the hole from the rest of it (Fig. 9.3) Tests can be carried out to depths up to 300 m.

Water is injected under pressure into the test section with increasing pressure from 0 to 10 bars (0–1 MPa) and then it is decreased from 1 to 0 MPa, in fixed steps at prescribed time intervals. However, testing at pressures as high as 1 MPa is questionable in estimating permeability, as such high pressures are likely to increase the permeability locally by inducing new fractures and widening the existing ones. It also enhances the possibilities of turbulence. The time interval is commonly 15 min. The flow rate of water in the borehole is measured under a range of constant pressures.

The flow pattern around the test zone depends on the orientation of the fractures in relation to the axis of the borehole (Fig. 9.3). The flow rates will be very high when the intercepted fracture is parallel to the borehole.

In a test where the flow is cylindrical and an observation well (piezometer) is used (Fig. 9.4), hydraulic conductivity (K) can be determined from Eq. 9.4

$$K = \frac{Q/L}{2\pi(h_0 - h)} \, ln\frac{r}{r_w} \qquad (9.4)$$

where, K is hydraulic conductivity perpendicular to the axis of the bohrehole, in (m s^{-1}), Q is rate of inflow, in (m^3 s^{-1}), L is thickness of the test zone, in (m), h_0 and h are the piezometric heads in (m) measured in the test well and at distance (r) in the observation well, and r_w is radius of the test borehole, in (m).

Fig. 9.3 Influence of direction of fractures on the flow during Lugeon tests **a** fractures parallel to the borehole, **b** fractures not parallel to the borehole

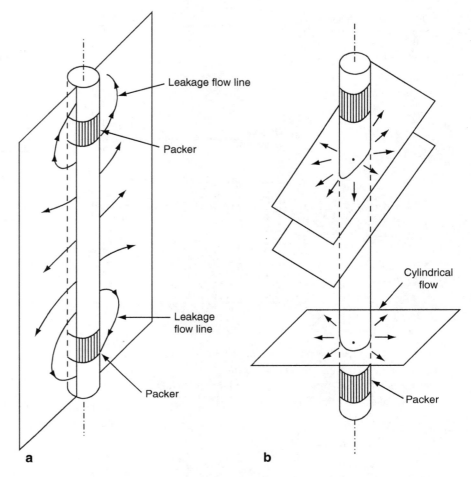

Specific permeability (q) which is also a measure of rock permeability can be obtained from Lugeon test. It is defined as the volume of water injected into the borehole per 1 m length during 1 min under a pressure of 1 m of water (0.1 atm)

$$q = \frac{Q}{Hl} (L\,min^{-1}\,m^{-1}\ at\ 0.1\ atm)$$

where Q is the injection rate, H is the effective injection pressure, and l is the length of the test interval.

In order to reduce the cost of the test, very often piezometers are not installed. For such conditions, Eq. 9.4 becomes

$$K = \frac{Q/L}{2\pi\Delta h}\,ln\frac{R}{r_w} \qquad (9.5)$$

where, Δh is the hydraulic head, in (m) in the test well causing flow and R is the radius of influence i.e. dis-

tance (m) at which initial water-level conditions do not change due to injection. Errors in the evaluation of R do not affect test results very much as $R \gg r_w$. Therefore,

$$\frac{ln(R/r_w)}{2\pi} = constant \qquad (9.6)$$

As, lnR/r_w does not vary much, it can be assumed that (UNESCO 1984b)

$$ln(R/r_w) \cong 7 \qquad (9.7)$$

Therefore, for numerical computations, Eq. 9.5 can be rewritten as

$$K = 1.12 \times \frac{Q/L}{\Delta h} \qquad (9.8)$$

where units are the same as in Eq. 9.4.

Fig. 9.4 Standard Lugeon test—experimental setup and definitions necessary to interpret test results

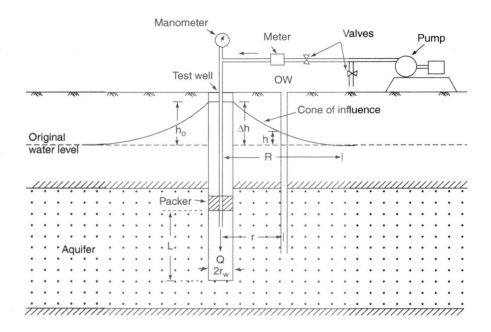

If it is assumed that only one fracture intersects the test section, an equivalent single fracture aperture, a_{eq}, can be determined from the test results using Cubic law (Novakowski et al. 2007).

$$a_{eq} = \left[T \cdot \frac{12\mu}{\rho g} \right]^{1/3} \qquad (9.9)$$

The vertical distribution of T and a_{eq} can be determined by using packers at the desired levels.

Some of the important rules which should be followed in conducting a packer test are:

1. The test should be carried out in a saturated zone.
2. Each test at each pressure is continued until steady state conditions are reached.
3. The test may be repeated with two or three increasing steps of pressure but care should be taken that it may not cause fracture-dilation or hydro-fracturing.
4. Pressure is measured in piezometres around the test hole.

Hydraulic conductivity, can be expressed in Lugeon unit (L_u). Lugeon is defined as the rate of flow of water per minute under a pressure of water injection of $1\,\mathrm{MPa\,m^{-1}}$ length of the tested material, i.e.

$$1\,L_u = 1\,l\,\mathrm{min^{-1}\,m^{-1}} \quad \text{under} \quad 1\,\mathrm{MP_a}$$

or

$$1\,L_u = 1 \times 10^{-7}\,\mathrm{m\,s^{-1}} \text{ (approx.)}.$$

It is generally assumed that the test section intercepts a number of fractures. Therefore, the estimated value of K represents an average value of hydraulic conductivity of the rock mass in the plane perpendicular to the borehole. The conductivity of individual fracture, K_f can be expressed by

$$K_f = \frac{K}{N} \qquad (9.10)$$

where N is the total number of fractures in the test section as determined from borehole core logs. It is assumed that fracture apertures are constant.

A comparison of transmissivity values determined from packer test and those obtained from pumping test indicates that although the two values are correlatable, the values obtained from pumping test tend to be higher (Fig. 9.5). This appears to be due to the reason that in a packer test only the properties in the immediate vicinity of the borehole are reflected but pumping test covers a larger volume of rock mass around the borehole. Therefore, pumping tests are preferred to packer tests in oder to obtain representative values of hydraulic parameters of rocks for the purpose of groundwater development.

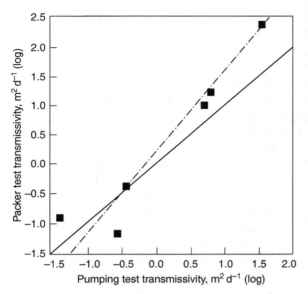

Fig. 9.5 Comparison of transmissivity values obtained from Lugeon tests and pumping tests. (After Karundu 1993)

In a standard packer test, the flow rate (Q/L) is plotted against applied pressure (P) as a flow–pressure curve. The flow–pressure curve will show different characteristics, depending on the permeability of the formation and changes brought about by water injection during the test (Fig. 9.6). A typical flow–pressure curve shows a lower flow rate during the increase in pressure as compared with values during the fall in pressure (Fig. 9.6a). A similar pattern (Fig. 9.6b) is also expected from the cleaning of the existing fractures or development of new fractures due to hydraulic fracturing during the test. A reverse situation might also be observed due to the clogging of the fractures during the test (Fig. 9.6c).

Modified Lugeon Test In the standard Lugeon test, the borehole is vertical irrespective of the position and orientation of fractures in the rock mass. This is the only alternative when fractures are randomly disttributed and are very large. Under such conditions the rock mass can be considered continuous. When there are well defined fracture sets as in Fig. 9.7 (F_1, F_2 and F_3), it is necessary to determine the hydraulic conductivities of each fracture set separately, i.e. K_1, K_2 and K_3 especially in geotechnical and contaminant transport problems. This can be achieved by having separate test holes for separate fracture sets; keeping the orientation of borehole perpendicular to the considered fracture set.

A triple hydraulic probe to measure the directional permeabilities was proposed by Louis (1974) In this method records are taken in three appropriate directions and measurements are made at different pressures. In order to measure the hydraulic heads during the test in the vicinity of the testing area, piezometers are installed. Four piezometric measurements enable to determine the three dimensional distribution of the anisotropic permeabilities.

Some of the problems in conventional packer test are (Louis 1974).

1. Turbulence effect
2. Deformation of the medium due to high injection pressure
3. Influence of other fracture sets, i.e. effect of F_2 and F_3 when test hole is perpendicular to F_1
4. Entrance losses.

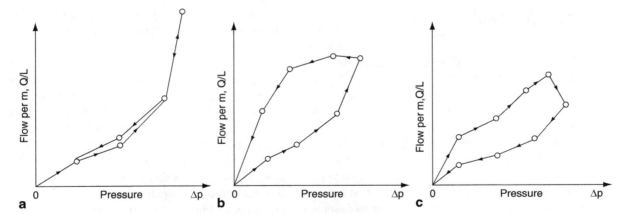

Fig. 9.6 Characteristic flow–pressure curves for packer (Lugeon) test: **a** normal type with reversible cycle, **b** cleaning of existing fractures or development of new fractures, **c** reverse type due to the clogging of fractures

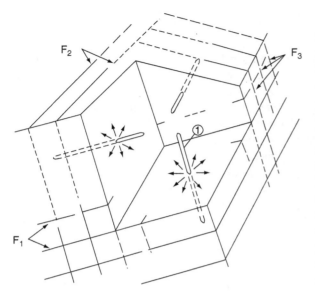

Fig. 9.7 a Modified Lugeon test in rock mass with three sets of fractures. Borehole (*1*) is parallel to F_2 and F_3 for testing the fracture set F_1

A Multi-function Bedrock Aquifer Transportable Testing Tool (BAT[3]) has been developed at the USGS for carrying out a single-hole hydraulic test in fractured rocks by withdrawing/injecting water. It is also used to identify hydraulic head as a function of depth and to collect discrete-interval groundwater samples for chemical analysis. The water should be withdrawn in case of highly transmissive fractures and injected when fractures are less transmissive. BAT[3] can estimate transmissivity ranging over approximately eight order of magnitude (Kresic 2007).

9.2.2 Slug Test

A slug test involves sudden injection of a known volume (or slug) of water into a well and measurement of fall of water-level with time. Alternatively a known volume of water can be withdrawn and rise of water-level is noted.

The advantage of slug test is that it is cheap, as it requires less equipment and manpower and the duration of the test is also relatively short. Observation wells are also not required. It can be carried out even if nearby wells are being used. Contaminated aquifers can also be tested as there will be no problem of extraction and disposal of contaminated water.

Slug tests can also provide reliable values of T in fractured rocks in spite of rock heterogeneties. Discrete fractures, if suitably located by geophysical or other methods can be tested by using packers (Butler 1998). Slug tests have been also performed for site selection of nuclear waste repositories suspected to behave as a double-porosity medium (Butler 1998). Values of hydraulic conductivity determined from slug tests, fluid injection tests and pumping tests in fractured crystalline rocks are quite comparable (Shapiro and Hsieh 1998; Nativ et al. 2003). However, as slug tests stress only a small volume of the aquifer, this method may not be used to interpret formation heterogeneity or large-scale formation properties. Although slug tests provide reliable values of T but estimates of S may have dubious values (Beckie and Harvey 2002). Therefore, slug test cannot replace the conventional pumping test as the latter have several additional advantages.

The duration of the slug test depends on the permeability of rocks—it has to be larger in low permeability media. The well radius also influences the duration of the slug test. In low permeability formations, it is necessary to have boreholes or standpipes of small diameter in order to reduce the duration of the test.

The major factors affecting test duration for a partially penetrating well is given as (Butler 1998)

$$\beta = \frac{K_f b t}{r_c^2} \qquad (9.11)$$

where β is dimensionless time factor, K_f is radial component of the hydraulic conductivity, b is effective screen length, t is the total time since the start of the test, and r_c is effective casing radius. Equation 9.11 indicates that the test duration (t) is inversely proportional to the effective screen length (b) and directly proportional to the square of the effective casing radius, r_c.

Since the hydraulic conductivity calculated from a slug test is an estimate for a small part of the aquifer around the well, skin effect (especially skin with low permeability) will greatly influence the computed hydraulic conductivity. Both field slug test results and numerical modelling demonstrate that well bore skin significantly affects water-level recovery and drawdown in low permeability geologic formations. Therefore, only properly developed wells will provide reliable estimates of aquifer characteristics from slug tests. Wellbore skin effect can be minimized by using the proper segment of the time-drawdown curve. However,

Fig. 9.8 Slug test in a fully penetrating well of finite diameter in a confined aquifer

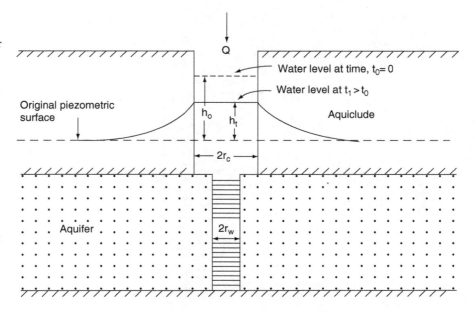

there is no general agreement on which segment will have no influence from the wellbore skin effect. Values of permeability estimated from slug tests are generally low as compared with those obtained from pumping tests which is attributed to incomplete well development and skin effect (Butler and Healey 1998).

The commonly used methods of analysing slug test data for confined aquifer is given by Cooper et al. (1967) described below. Bredehoeft and Papadopulos method which is especially suited to low permeability rocks is also given in this text. Readers may refer to Kruseman and de Ridder (1990) and Butler (1998) for other methods.

Cooper's Method Cooper et al. (1967) developed a curve-matching method for slug test in fully penetrating wells in confined aquifers. This method does not require any pumping or observation well. The sudden injection of a known volume of water, V in a well of diameter $2r_w$ will cause an immediate rise of water-level, h_0 in the well (Fig. 9.8) which can be expressed as

$$h_0 = \frac{V}{\pi r_c^2} \qquad (9.12)$$

The change in water-level with time is given by Eq. 9.13.

$$\frac{h_t}{h_0} = F(\alpha, \beta) \qquad (9.13)$$

where,

$$\alpha = \frac{r_w^2 S}{r_c^2} \qquad (9.14)$$

$$\beta = \frac{Tt}{r_c^2} \qquad (9.15)$$

h_0, h_t, r_c and r_w are as shown in Fig. 9.8.

Values of function $F(\alpha, \beta)$ against α for different values of α and β are plotted on semi-log paper as family of type curves (Fig. 9.9). The value of h_0 is calculated from Eq. 9.12 based on the volume of water injected into or withdrawn from the well. Ratio of h_t/h_0 is calculated for different values of t.

A field data curve (log h_t/h_0 against corresponding values of t) is superimposed over the family of Cooper's type curves (Fig. 9.9) to find the best match of the field data plot with one of the type curves. The value of β for the corresponding type curve is noted. The value of t for corresponding value of $\beta = 1$ is read from the data curve. T is calculated by substituting values of t, r_c and $\beta = 1$ in Eq. 9.15. The value of S can be determined from the value of α of the type curve with which data curve is matched using Eq. 9.14.

The type curves especially for small values of α are quite similar in shape and therefore it is difficult to select the type curve for a unique match. This may result in large errors in calculating S. However, T can

Fig. 9.9 Family of Cooper's type curves. (After Papadopulos et al. 1973)

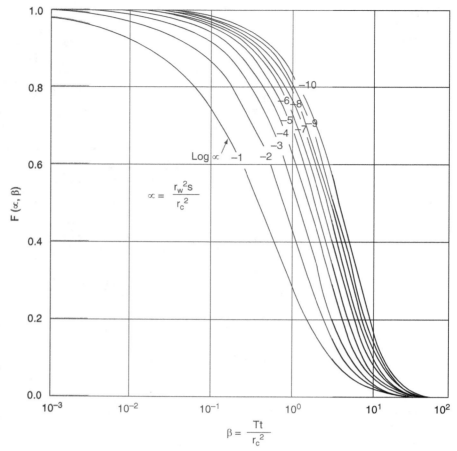

$$\propto = \frac{r_w^2 S}{r_c^2}$$

$$\beta = \frac{Tt}{r_c^2}$$

be determined with greater accuracy but it will be representative of the formation in the immediate vicinity of the well.

Bredehoeft and Papadopulos Method A modified slug test, i.e. pressure pulse method was suggested by Bredehoeft and Papadopulos (1980) in which the test well is filled with water to the surface, and suddenly pressurized with an additional amount of water. The water is then shut-in and the fall in pressure is noted with time. Packers are used to isolate the intake part of the open hole. This method is advantageous in reducing the time of test which is the main problem in tight formations. Bredehoeft and Papadopulos (1980) showed that a formation of $T = 10^{-11} \, m^2 s^{-1}$ and $S = 4 \times 10^{-4}$, can be tested adequately in a few hours using the pressure pulse method, whereas years may be required by the conventional slug—testing methodology. Therefore Bredehoeft and Papadopulos

method is feasible for in-situ estimation of hydraulic properties of tight formations, viz. shales and massive crystalline rocks having low hydraulic conductivities of the order of $10^{-13} - 10^{-16} \, m s^{-1}$. Bredehoeft and Papadopulos also argued that for $\alpha > 0.1$, type curve of $F(\alpha, \beta)$ vs. the product, $\alpha\beta$ are more suitable for analysing test data than the Cooper's type curves (Fig. 9.9) where $F(\alpha, \beta)$ is plotted against β only. The same procedure of matching the field data curve (h_t/h_0 vs. t) plotted on semi-logarithmic paper with the type curve $F(\alpha, \beta)$ against $\alpha\beta$ can be adopted for estimating either T and S separately or only the product TS.

This method is based on the assumptions, among others, that (1) the flow in the tested interval is radial, and (2) the hydraulic properties of the tested interval remain constant throughout the test. In many cases, the first assumption holds good as the ratio of K_h/K_v is quite high. In order to satisfy the second assumption, especially in the fractured rocks, the fluid pressure in the tested interval should be kept low to avoid the formation of new fractures. This can be taken care of by initial pulses of 1–10 m (10–100 kPa).

Neuzil (1982) argued that the Bredehoeft–Papadopulos method does not assure the approximate equilibrium condition i.e. equal hydraulic head in the well and the formation at the beginning of the test. Therefore, the test data will not give correct values of K. To avoid this problem, Neuzil (1982) suggested a modified test procedure to ensure near-equilibrium condition at the start of the test and also the use of two packers and two transducers to monitor pressures both below the lower packer and between the packers to detect leakage.

Air-pressurized slug tests are carried out by pressurizing the air in the casing above the water-level in a well and the declining water-level in the well is noted. Afterwards the air pressure is suddenly released to monitor the rise in water-level. T and S can be estimated from the rising water-level data using type curves (Kresic 2007). Such tests are known as 'prematurely terminated air pressurized slug tests'.

The use of nitrogen gas, instead of water, as permeant is suggested by Kloska et al. (1989) for determining permeability of unsaturated tight formations. This is advantageous in unsaturated or less saturated low permeability rocks as nitrogen will not change the rock properties.

9.2.3   Pumping Tests

9.2.3.1   Introduction

In a pumping test, a well is pumped at a known constant or variable rate. As a result of pumping, water-level is lowered and a cone of water-table depression in an unconfined aquifer and cone of pressure relief in a confined aquifer is formed. The difference between the static (non-pumping) and pumping water-level is known as drawdown. As pumping advances, cone of depression expands until equilibrium conditions are established when the rate of inflow of water from the aquifer into the well equals the rate of pumping. The distance from the centre of the pumped well to zero drawdown point is termed as *radius of cone of depression* (*R*). In an ideal uniform, isotropic and homogeneous aquifer, the cone of depression will be symmetrical and the contours of equal drawdown will be circular or near-circular. In contrast, in a fractured aquifer, due to anisotropy, the plotted drawdown cone, based on data from a number of observation wells, will be linear, highly elongated or irregular; the longer axis will be parallel to the strike of water conducting fractures. The slope of the cone of depression and its radius (R) depend on the type of aquifer and its hydraulic characteristics. In aquifers of high transmissivity, the gradient of cone of depression is less and its radius is large, as compared with aquifers of low transmissivity (Fig. 9.10).

The speed of propagation of the cone is inversely proportional to storativity, S. In an unconfined aquifer

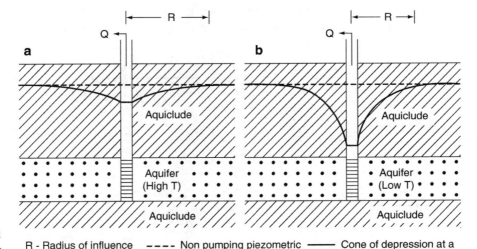

Fig. 9.10 Comparison of cones of depression formed due to pumping in a confined aquifer of **a** high transmissivity, and **b** low transmissivity

Table 9.2 Spread of the cone of depression in metres. (After Schoeller 1959) ($T = 1.25 \times 10^{-3} \text{m}^2 \text{s}^{-1}$)

	1 min	1 h	1 day	10 days	100 days
Unconfined Aquifer ($S = 0.2$)	0.91	7.11	34.8	110	348
Confined aquifer ($S = 1 \times 10^{-4}$)	41	318	1558	4930	15580

as the storativity is high, being of the order of 10^{-1}–10^{-2}, corresponding to the specific yield, the spread of cone is very slow and therefore it would take a long time to reach the edges of the aquifer. In contrast to this, in confined aquifers the storativity is very small, of the order of 10^{-3}–10^{-6}. Hence the cone extends very rapidly in confined aquifer than in unconfined aquifer (Table 9.2).

In case of heterogeneous aquifers, multiple and sequential pumping tests are useful to account for spatial variation in T and S (see Sect. 9.2.6).

9.2.3.2 Planning a Pumping Test

Selection of Site

Pumping tests are expensive and therefore should be carefully planned. The cost depends on number of observation wells and duration of the test. Before conducting a pumping test, it is necessary to know the geological and hydrological conditions at the test site. Subsurface lithology, and aquifer geometry are of help in the proper interpretation of pumping test data. Existing wells in the area can provide important information about the sub-surface lithology. A preliminary estimate of the transmissivity of the aquifer can be made from

subsurface lithology and aquifer thickness. At places, where subsurface geological information is not directly available, geophysical methods can be used to ascertain the lithology (Chap. 5). The presence of hydrological boundaries in the form of rivers, canals, lakes or rock discontinuities (faults and dykes etc.) should also be noted. In the absence of this information, the test data are liable to be interpreted in different ways, viz. the effect of a recharge boundary on the time-drawdown curve may almost be the same as for a leaky aquifer or an unconfined aquifer. This aspect is further elaborated in the latter part of this section.

The test site for a confined aquifer should not be close to railway track or highway to avoid effect of loading which produces water-level fluctuations. The site should be also away from existing discharging wells to avoid well interference.

Design of Pumping and Observation Wells

The well field for a pumping test consists of one pumping well and one or more observation wells. The pumping well should be at least of about 150 mm diameter so that a submersible electric pump can be installed. The diameter of the observation wells should be able to accommodate an electric depth sounder or some other water-level measuring device, viz. automatic water-level recorder.

In order to avoid entry of pumped water into the aquifer, it should be conveyed by pipes or lined drains to a distance of 200–300 m away from the discharging site.

The pumping well should tap the complete thickness of the aquifer to avoid effect of partial penetra-

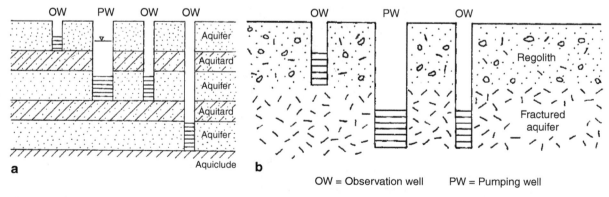

OW = Observation well   PW = Pumping well

Fig. 9.11 Layout of pumping well (*PW*) and observation wells (*OW*) in: **a** multilayered aquifer system, and **b** fractured aquifer

tion. In thick aquifers where partial penetration cannot be avoided necessary correction in the drawdown data should be made. Observation well(s) should have the same depth and tap the same aquifer(s) as the pumping well. However, in some cases, where the interconnection between aquifers due to leakage is suspected some observation wells should also tap the overlying and underlying aquifers (Fig. 9.11a). This is also required in fractured rock aquifers to know the vertical component of flow (Fig. 9.11b).

The length and position of the screen in the pumping well is important as this will affect the amount of drawdown. Partial penetration causes lengthening of flow lines and thereby excessive drawdown (Fig. 9.12). In unconfined aquifers, as pumping causes

dewatering of the aquifer, the well screen should be put in the lower one-half or one-third of the aquifer where the flow (stream) lines will be mainly horizontal, otherwise the observation well may be located at distances one and a half times the saturated thickness of the aquifer (Fig. 9.12a). The desirable position of screen in a confined aquifer is shown in Figs. 9.12c, d and 9.13. The number of observation wells depends on the purpose of study and financial considerations. It is desirable to have at least one observation well but three observation wells at different distances from the pumping well are better. This will help in analysing the test data by both the time-drawdown and distance-drawdown methods and identify the boundary conditions. In order to assess the anisotropy and heteogeneity of

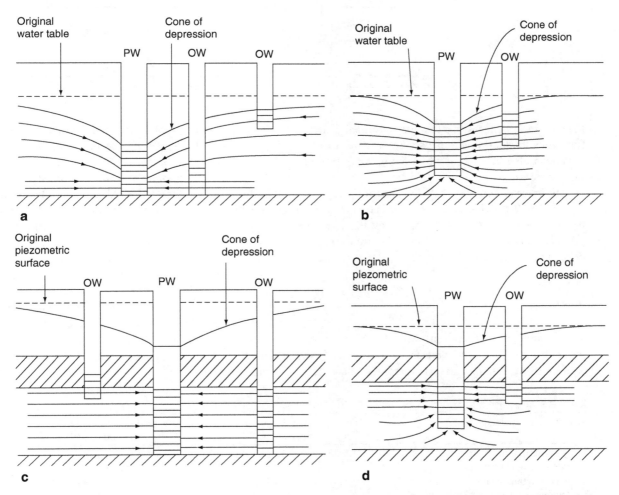

Fig. 9.12 Positioning of screens in observation wells, **a** fully penetrating pumping well in an unconfined aquifer, **b** partially penetrating pumping well in an unconfined aquifer, **c** fully penetrating pumping well in a confined aquifer, and **d** partially penetrating pumping well in a confined aquifer

Fig. 9.13 Schematic section of a cluster of piezometers in a multilayered aquifer system

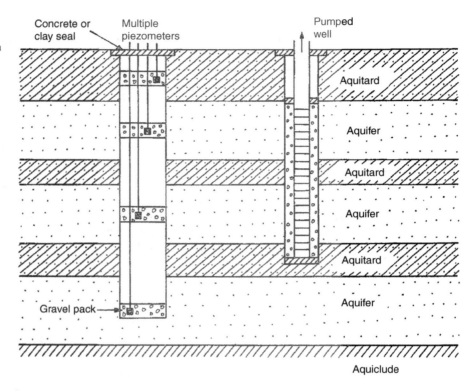

the aquifer, it is necessary to have additional observation wells in other direction also. Some of the existing wells can also be used as observation wells to minimize the cost.

An observation well located upstream from pumping well will provide information of storativity both for the upstream and downstream side but it cannot provide much information about the transmissivity downstream (Jiao and Zheng 1995).

Well Spacing

The spacing of observation wells from the pumping well depends on the well penetration, type of the aquifer, its hydraulic characteristics and rate and dura-

tion of pumping (Table 9.3). In unconfined aquifer, as the radius of cone of depression is comparatively small, observation well should be within a distance of 100 m from the pumping well. In a confined aquifer, this distance could be upto 200 or 250 m. For semiconfined aquifers, intermediate values may be taken. Transmissivity also influences the shape and size of the cone of depression as described earlier. In aquifers with high transmissivity, observation wells can be put farther away from the pumping well than in aquifers with low transmissivity. One observation well should be selected which is outside the cone of depression of the well field. This helps to assess the effect of atmospheric or other natural causes of water-level fluctuations so that necessary correction in the drawdown data may be made.

Table 9.3 Suggested spacing of the observation wells. (After Hamill and Bell 1986)

Aquifer and well condition	Minimum distance from pumped well to nearest observation well	General distance within which observation wells should be located
Fully penetrating well in unstratified confined or unconfined aquifer	1–1.5 times aquifer thickness	20–200 m in confined aquifer, 20–100 m in unconfined aquifer
Fully penetrating well in a thick or stratified confined or unconfined aquifer	3–5 times aquifer thickness	100–300 m in confined aquifer, 50–100 m in unconfined aquifer
Partially penetrating well (<85% open hole) in a confined or unconfined aquifer	1.5–2 times aquifer thickness	35–200 m in confined aquifer, 35–100 m in unconfined aquifer

In fractured rocks, study of structural data viz., orientation, spacing and interconnectivity of fractures is important in planning a pumping test (observation well network and pumping test duration etc.).

Installation of observation wells in fractured rocks depends on whether the fractures are discrete or interconnected and also the permeabilities of rocks on the two sides of the fractures (matrix blocks) and whether fractures are open or filled with impervious material. Multiple observation wells are preferred tapping both the pumped intervals comprised of highly transmissive fractures and the intervening less transmissive horizons. In heterogeneous aquifers, a piezometer nest is recommended for recording of water-levels at different depth horizons.

Duration of Test

The duration of pumping test depends on the type of the aquifer and the degree of accuracy required in estimating the aquifer properties. Usually the test is continued till the water-level is stabilised so that both the non-steady and steady-state methods of analysis could be used for computing aquifer parameters. The duration of constant rate pumping test depends on the type of aquifer, viz. in confined and leaky aquifers it could be about 24 h and 20 h respectively but in unconfined aquifers, as it takes longer time to reach steady state flow condition, the duration of test may be about 72 h (Hamill and Bell 1986; Kruseman and De Ridder 1990). In fractured rocks of low hydraulic conductivity, pumping test can be only of shorter duration (10–15 h).

9.2.3.3 Measurements

Two sets of measurements are taken during a pumping test: (1) groundwater levels (both during drawdown and recovery phases) in pumping and observation wells, and (2) well discharge. Water-level measurements can be made by using steel tape or electrical depth sounder. The automatic water-level recorders are better for obtaining continuous water-level records in observation wells. Fully automatic microcomputer— controlled systems are also now available for accurate recordings of water-levels.

As drawdown is faster in the observation wells close to the pumped well during early part of the test, water-level observations should be made at an interval of 1/2–1 min for the first 10 min of pumping. The time interval of measurements can later be increased to 2, 5 and 10 min. In tests of longer duration extending for a period of 1 day or more, water-level measurements for later part of the test could be taken at 1 h or even longer time intervals. After the pumping is stopped, rate of recovery in pumped well and also in observation wells is noted. As the rate of recovery is faster in the early part of recuperation, water-level measurements during this period should be taken at shorter time intervals; later the time interval can be increased. The water-levels measured during the test are likely to be also influenced by other extraneous reasons such as barometric pressure changes etc. Therefore, necessary corrections should be made to isolate the effect of such changes.

Well discharge can be measured by using different types of weirs; circular orifice weir being the most common (Driscoll 1986).

9.2.3.4 Types of Drawdown Curves

The drawdown data are plotted as time-drawdown and/ or distance drawdown curves. A time-drawdown graph on a semi-logarithmic scale may show different slopes (Fig. 9.14). The earliest time response is usually due to

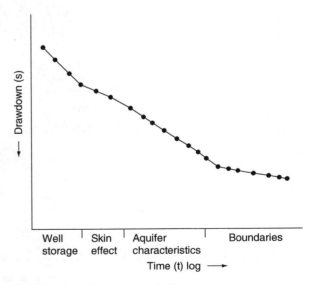

Fig. 9.14 Drawdown response in time sequence

wellbore storage as most of the water at this stage comes from the well itself. Next part of the time-drawdown curve is influenced by the contact between the well and the aquifer due to 'skin effect'. At a later stage, the type of aquifer and its hydraulic characteristics (T and S) influence the rate of drawdown. Finally, boundary conditions affect the trend of the time-drawdown curve. There could be various possible combinations of these effects. A proper interpretation of pumping test data would therefore depend on the recognition of these effects. The duration of the test is therefore important to recognise the various components (segments) of a time-drawdown plot. The influence of various factors on time-drawdown curve is discussed below:

Aquifer Types

A log–log time drawdown plot for an ideal unconsolidated confined aquifer tapped by a fully penetrating well will match with the Theis type curve (Fig. 9.15a) and a semi-log time-drawdown plot will be characterized by an uniform slope (Fig. 9.15a'). On the other hand, unconfined aquifers which are characterized by delayed yield, show typical S-shape on a log–log plot with three time-segments (Fig. 9.15b); the early part (first segment) conforms to the Theis type curve for confined aquifer due to instantaneous release of water from aquifer storage while the middle part (sec-

ond segment) shows flattening due to vertical drainage. The last or the third segment also conforms to the Theis type curve as by that time the effect of vertical drainage becomes negligible. The semi-log plot for an unconfined aquifer is characterized by two parallel straight line slopes of segments 1 and 3 (Fig. 9.15b').

The leaky aquifer shows a behaviour similar to segments 1 and 2 of an unconfined aquifer as in this case additional water is transferred to aquifer by leakage through the aquitard. However, segment 3 of the time-drawdown curve for unconfined aquifer is not observed because the process of recharge by leakage continues and drawdown stabilises after sometime as is indicated by both log–log and semi-log graphs (Fig. 9.15c, c').

Fractured Rock Aquifers

Fractured rocks show various types of drawdown and pressure build-up curves, depending on the nature of fractures and matrix blocks, boundary conditions and wellbore storage (Davis and DeWiest 1966; Streltsova 1976b; UNESCO 1979; Gringarten 1982; Kruseman and de Ridder 1990). Similar trends are also observed in geothermal and petroleum reservoirs (Grant et al. 1982; Horne 1990). In case of dewatering of fractures, due to decrease in the effective T-value, an increase in the drawdown slope with time is observed.

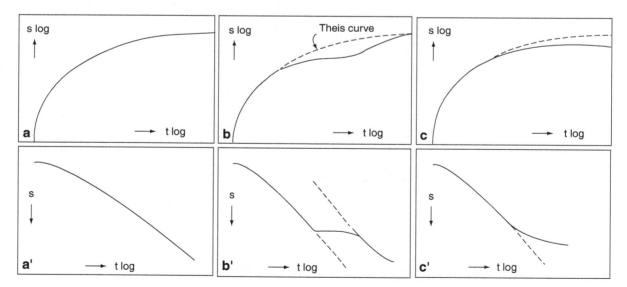

Fig. 9.15 Comparison of log–log and semi-log plots of the theoretical time-drawdown relationship in unconsolidated aquifers, **a** and **a'** confined aquifer, **b** and **b'** unconfined aquifer, and **c** and **c'** leaky aquifer

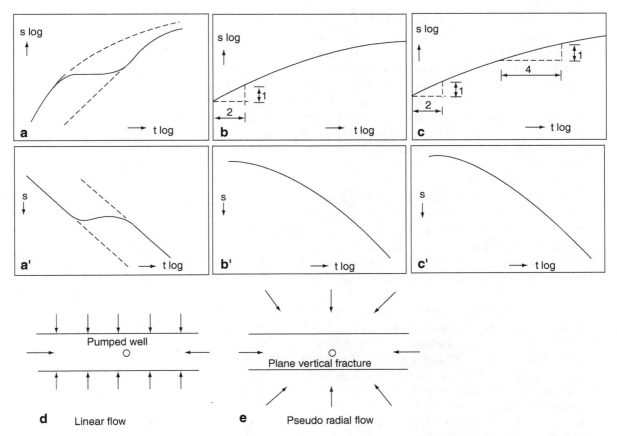

Fig. 9.16 Comparison of log–log and semi-log plots of the theoretical time-drawdown relationship of fractured aquifers **a** and **a'** double porosity confined aquifer, **b** and **b'** single plane vertical fracture, **c** and **c'** fractured permeable dyke in a massive country rock of low hydraulic conductivity. **d** and **e** Show linear flow and pseudo-radial flow respectively. (Modified after Kruseman and deRidder 1990)

The response to pumping in a fractured dual-*porosity aquifer*, having the matrix blocks divided by fractures is similar to that of delayed yield in unconfined aquifer (Fig. 9.16a, a'). A similar trend is shown by karst aquifers (Chap. 15) consisting of dissolutional conduits intersecting a network of diffuse fractures in which the groundwater flow is slow. The first and third segments of the time-drawdown curve on a semi-log plot may have parallel slopes (Fig. 9.16a'). The first segment, with a uniform slope, is due to removal of water from storage in the fracture or in large conduits. This is also characteristic of confined aquifers. Unconfined aquifers with storage coefficient significantly less than 1×10^{-3} will also show a similar response (Kresic 2007). The second segment with a flattening character indicates additional contribution from matrix blocks or smaller fractures or fissures towards the larger fractures or dissolutional conduits in karst aquifers.

A similar trend is shown by unconfined intergranular aquifers due to delayed gravity drainage (Fig. 9.16c, c'). If the pumping is carried out for longer duration, the time-drawdown curve shows a third segment with a relatively uniform slope which is characteristic of both the unconfined and dual porosity aquifers (Fig. 9.15b' and Fig. 9.16a, a').

The effect of pumping from a well in a single vertical fracture of infinite hydraulic conductivity in a confined homogeneous and isotropic aquifer of low permeability is also shown in (Fig. 9.16b, b'). The response at early times of pumping is characterized by a 1/2 slope straight line on the log–log plot of drawdown against time which is explained due to linear flow (Fig. 9.16d). At later time as the flow regime changes to pseudo-radial (Fig. 9.16e), the shape of the curves resembles those of (a) and (a'). Parts (c) and (c') of Fig. 9.16 present the response of a pumping well in a fractured dyke of high permeability traversing a confined aquifer of low permeability and high storativity. The log–log time-drawdown plot in such a case is characterized by two straight line segments. The first segment has 1/2 slope like that of the well in a single vertical fracture of infinite permeability. At intermediate times, the log–log plot shows a 1/4 slope due to contribution from the host rock. At late times, the flow in the host aquifer is pseudo-radial which is reflected by a straight-line segment in the semi-log plot (Kruseman and de Ridder 1990).

In view of the above discussion, it could be concluded that results of short duration pumping tests may not be able to reflect the true aquifer characteristics and therefore the test duration should be long enough, say of few days, as the late time-drawdown data are of importance to distinguish between different aquifer types.

Some anomalous trends showing greater drawdown in wells away from the pumping well than those in the vicinity have also been reported from fractured rocks (Streltsova-Adams 1978; Gringarten 1982). Also there are field evidences of rise of water-level near the pumping well. These anomalous behaviours in literature are referred as 'reverse water-level fluctuation' or *Noorbergum effect* which can be explained by the following phenomena (Streltsova 1976b).

1. Delayed release of water from storage in the porous matrix of a double porosity medium in response to pumping resulting in local increase of water-level in the fracture system.
2. Decrease of the storage capacity of fractures due to their deformation as a result of pumping.
3. Additional recharge of fractures due to the recycling of pumped water discharged on the ground surface.

Skin Effect

The concept of skin effect or 'skin factor' on head loss in a well was first introduced in oil well industry, where due to mud filtrate at the well face, greater well loss was observed (Horne 1990). The idea was extended later to water-wells also. The 'skin effect' will be positive when the effective well bore radius and permeability is reduced due to mud filtrate. Conversely, when the effective well bore radius is increased due

to acidization or other methods of well stimulation (Sect. 17.4), the skin effect will be negative. In fractured rocks, the skin effect will be negative due to good contact between the well and the water bearing fractures but when there is a clay coating on the walls of the fractures, the skin effect will be positive resulting in greater drawdown.

9.2.3.5 Method of Analyzing Test Data

Several analytical and digital techniques are available for analysing pumping test data under different geohydrological conditions. The advantage of numerical methods is that they can take care of complex geohydrological conditions. The graphical procedures are subjective and the number of type curves also become prohibitively large when the number of parameters exceeds three.

Graphical methods, although cumbersome, have an advantage as one can visualize the extent to which the field data match with the assumed conditions and the investigator can be selective in choosing the best part of the data which would give reliable results under given hydrogeological conditions. In this text only graphical methods of pumping test data analysis are described. For computer assisted methods, the reader is referred to other texts (e.g. Rushton and Redshaw 1979; Boonstra and Boehmer 1989; Boonstra and Soppe 2007).

The methods described in this section were originally developed for unconsolidated aquifers but can also be applied to fractured rocks if the fractures are interconnected.

The various commonly used methods are:

Thiem's equilibrium method
Theis non-equilibrium type curve method
Jacob's non-equilibrium straight line method
Theis' recovery method
Walton's type curve method for leaky confined aquifers
Boulton' type curve method for unconfined aquifers
Neuman's type curve method for unconfined aquifers
Papadopulos-Cooper type curve method for large diameter wells.

An excellent account of various methods with examples is given in Kruseman and de Ridder 1990; and Boonstra and Soppe 2007) among others. Although the above methods were primarily developed for homogeneous granular formations, but have been also used successfully for estimating hydraulic parameters in fractured basement and carbonate rocks with solution cavities (for examples see Kresic 2007).

In case of hetrogeneous aquifers, multiple and sequential pumping tests are useful to account for spatial variation in T and S (Li et al. 2007; Straface et al. 2007). In a sequential test, each well from a well field is pumped in sequence while the other wells are used as observation wells (see Sect. 9.2.6 in this chapter).

Constant Discharge Tests

Confined Aquifers

The pumping test data from a confined aquifer can be analysed both by steady state and unsteady state methods. The steady-state Eq. 7.25 for groundwater flow in a confined homogeneous aquifer is given in Sect. 7.1.3.

Thiem's Equilibrium Method

Thiem (in Wenzel 1942) was one of the first workers to use drawdown data for estimating T from unconfined and confined aquifers. Thiem's equation based on steady-state flow for a confined aquifer can be written as

$$Q = \frac{2\pi T(h_2 - h_1)}{ln(r_2/r_1)} \qquad (9.16)$$

or

$$Q = \frac{2\pi T(s_1 - s_2)}{2.30\,log(r_2/r_1)}$$

where Q is well discharge in $m^3 d^{-1}$, T is transmissivity in $m^2 d^{-1}$, s_1 and s_2 are drawdowns in metres in the two observation wells located at distances r_1 and r_2 in metres respectively from pumping well (Fig. 9.17).

After the system has reached steady-state (equilibrium) condition, the drawdown data from different observation wells are plotted on a semi-logarithmic paper. (drawdown, s_1 and s_2 on arithmetic scale and distances, r_1, r_2... on logarithmic scale). The slope Δs of this line is determined and is substituted in Eq. 9.17 for estimating T.

$$T = \frac{2.30\,Q}{2\pi\Delta s} \qquad (9.17)$$

where Δs is the difference of drawdown per log cycle of r.

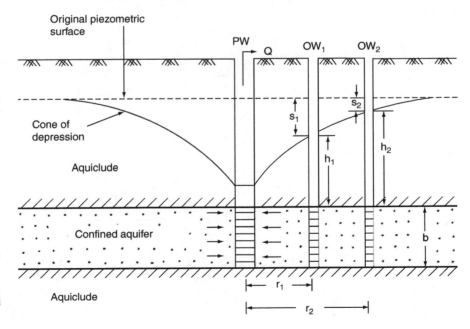

Fig. 9.17 Schematic cross section of a pumped confined aquifer

Theis Non-equilibrium Type Curve Method

The differential Eq. 7.24 for unsteady flow in plane polar coordinates is given in Sect. 7.1.3. Based on the analogy between flow of water in an aquifer and flow of heat in an equivalent thermal system, Theis (1935) obtained Eq. 9.18

$$s = \frac{Q}{4\pi T} \int_u^\infty \frac{e^{-y}}{y}\, dy \qquad (9.18)$$

where,

$$u = \frac{r^2 S}{4Tt} \qquad (9.19)$$

s=drawdown in metres in a piezometer or observation well at distance, r in metres from the pumping well, S is storativity of the aquifer (dimensionless); t is time since pumping started, in days; Q and T are as defined earlier.

Although the Theis equation is based on several assumptions, all of which usually do not hold good under natural field conditions, it has been very useful in solving many groundwater flow problems. The exponential integral in Eq. 9.18 is written symbolically as W(u) which is generally read as well function of u or Theis well function.

Equation 9.18 can therefore be written as

$$s = \frac{Q}{4\pi T}\, W(u) \qquad (9.20)$$

where

$$W(u) = -\,0.5772 - lnu + u$$
$$-\frac{u^2}{2.2!} + \frac{u^3}{3.3!} - \frac{u^4}{4.4!} \qquad (9.21)$$

and $u = \frac{r^2 S}{4Tt}$ as already indicated.

Values of W(u) for various values of u are given in several standard text books on Groundwater Hydrology, e.g. Todd (1980).

From Eqs. 9.18 and 9.19 it can be shown that if s is known for one value of r and several values of t or for one value of t and several values of r, and if Q is known, then T and S can be determined.

The first step in this method is to plot on a double-log paper, the Theis type curves of W(u) vs. u, W(u) vs. $\sqrt{u}$, and W(u) vs. 1/u (Fig. 9.18). The field data curve can be matched with one of the Theis type curves as given in Table 9.4. A reverse type curve W(u) vs. 1/u can be matched with a data curve of s vs. t/r^2 or if only one observation well is available then s is plotted against t. This will save computation time. If drawdown data is available for one value of t and several values of r then a distance-drawdown curve (s vs. r) is plotted which is matched with a type curve of W(u) vs. $\sqrt{u}$. As an example, the matching of the Theis type curve, W(u) vs. u with the field data curve, r^2/t vs. s is shown in Fig. 9.19. An arbitrary match point is selected anywhere on the overlapping portion of the sheets and the coordinates of this point are noted. These data on

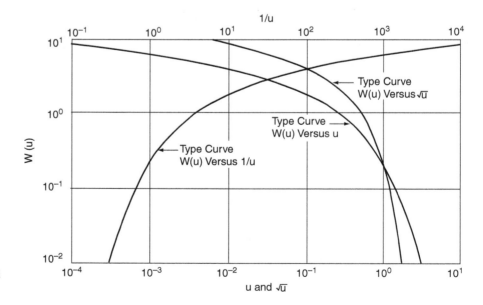

Fig. 9.18 Theis type curves W(u) vs. u, W(u) vs. $\sqrt{u}$, and W(u) vs. 1/u

Table 9.4 Various combinations of type curves and data curves in Theis method of analysis

Type curve	Data curve
W(u) vs. u	s vs. r^2/t or s vs. 1/t
W(u) vs. 1/u	s vs. t/r^2 or s vs. t
W(u) vs. √u	s vs. r

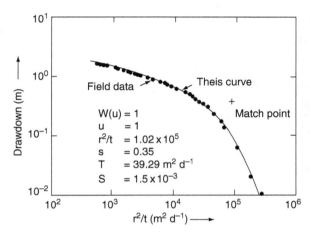

Fig. 9.19 Analysis of pumping test data by Theis type curve method for a confined sandstone aquifer. (After Radhakrishna et al. 1979)

substitution in Eqs. 9.20 and 9.19 will give values of T and S respectively.

Jacob's Method

For values of u < 0.01, Eq. 9.20 reduces to Eq. 9.22.

$$s = \frac{Q}{4\pi T}\left(ln\frac{4Tt}{r^2 S} - 0.5772\right) \quad (9.22)$$

In decimal logarithms, Eq. 9.22 reduces to

$$s = \frac{2.30\,Q}{4\pi\,T}\,log\frac{2.25\,Tt}{r^2 S} \quad (9.23)$$

which is known as the Jacob or Theis-Jacob equation.

According to Eq. 9.23, the graph of drawdown, s vs. log t will be in the form of a straight line. Equation 9.23 can be rewritten as

$$T = \frac{2.30Q\,(log\,t_2/t_1)}{4\pi\,(s_2 - s_1)} \quad (9.24)$$

or

$$T = \frac{2.30\,Q}{4\pi\,\Delta s} \quad (9.25)$$

where Δs is the change in drawdown over one log cycle of time.

The storativity, S can also be determined from the same semi-log plot by extending the time-drawdown curve to zero drawdown axis. Solving for storativity S, the equation in its final form becomes

$$S = \frac{2.25T t_0}{r^2} \quad (9.26)$$

where t_0 is the time intercept in days for s = 0.

The reliability of Theis type curve and Jacob's straight line methods for estimating T and S in heterogeneous aquifers has been demonstrated by several workers viz. Sanchez-villa et al. 1999; Li et al. 2007; Straface 2007). However, the hydraulic connectivity of individual fractures which controls the movement of fluids and contaminants cannot be well established.

Theis Recovery Method

After pumping is stopped, the water-levels in pumping and observation wells will start rising. This is known as the recovery or recuperation phase. Theis equation for recovery phase can be written as

$$s' = \frac{Q}{4\pi\,T}\left[W(u) - W(u')\right] \quad (9.27)$$

where

$$u = \frac{r^2 S}{4Tt'} \quad u' = \frac{r^2 S}{4Tt'} \quad (9.28)$$

Q, T, S and r are defined earlier, t is the time in days since pumping started, t' is the time in days since pumping stopped and s' is the residual drawdown. As in Jacob's equation, for small values of r and large values t', Eq. 9.27 can be written as

$$s' = \frac{2.30Q}{4\pi\,T}\,log\frac{t}{t'} \quad (9.29)$$

Therefore a plot of residual drawdown, s' on arithmetic scale vs. t/t' on logarithmic scale should form a straight line. Equation 9.29 can be rewritten as

$$T = \frac{2.30Q}{4\pi\,\Delta s'} \quad (9.30)$$

where $\Delta s'$ is the change in residual drawdown per log cycle of t/t'.

Leaky (Semi-confined) Aquifers

In nature, truly confined aquifers are rare as the confining layers are not completely impermeable. Therefore, certain amount of water is contributed to the pumped aquifer due to leakage through the aquitard. Such aquifers are known as leaky or semi-confined aquifers.

Pumping from a fully penetrating well in a leaky aquifer draws water from storage in the leaky aquifer as well as from storage in the aquitard and the water-table aquifer. The flow in the leaky aquifer is radial (horizontal) towards the well but the flow through aquitard is vertical due to head difference between the two aquifers (Fig. 9.20). As pumping advances, the cone of depression expands and the head differences in between the two aquifers increases leading to greater amount of leakage. The amount of leakage also depends on the vertical hydraulic conductivity of the aquitard (K'). A steady state situation may arise when the rate of withdrawal becomes equal to the amount of water contributed from the pumped aquifer plus that contributed by leakage from the adjacent aquifer(s).

It was earlier assumed that during pumping the hydraulic head in the unpumped aquifer remains constant and that the rate of leakage into the pumped aquifer is proportional to the hydraulic gradient across the aquitard (Hantush and Jacob 1955). The first assumption will only hold good if there is a constant source of recharge to the unpumped aquifer which may not be possible. The second assumption which ignores the effects of storage capacity of the aquitard is justified when the flow has reached steady-state condition. Under unsteady-state condition, the effect of aquitard storage cannot be neglected. Therefore, effects of aqui-

tard storage as well as decline of hydraulic head in the unpumped aquifer were also considered by later workers (Freeze and Cherry 1979). A review of evaluation of aquifer test data in leaky aquifers is given by Walton (1979) and Kruseman and de Ridder (1990).

Walton's Method

The unsteady state flow in a leaky aquifer without water released from storage can be given in an abbreviated form by Eq. 9.31

$$h_0 - h = s = \frac{Q}{4\pi T} \int_u^\infty \frac{1}{y} exp\left(-y - \frac{r^2}{4B^2 y}\right) dy$$

(9.31)

or

$$s = \frac{Q}{4\pi T} W(u, r/B)$$

(9.32)

where

$$u = \frac{r^2 S}{4Tt}$$

(9.33)

$$r/B = \frac{r}{\sqrt{T/(K'/b')}}$$

(9.34)

or

$$B = \left(Tb'/K'\right)^{\frac{1}{2}}$$

(9.35)

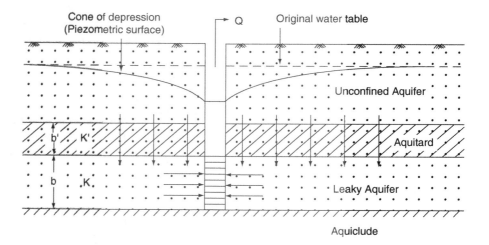

Fig. 9.20 Schematic cross section of a pumped leaky aquifer

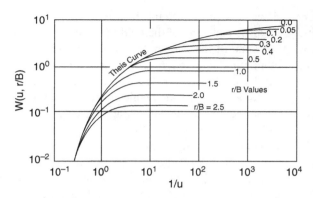

Fig. 9.21 Family of Walton's type curves of W(u,r/B) vs. 1/u for different values of r/B for a leaky aquifer. (After Walton 1962)

where, B is the leakage factor, K' is the vertical hydraulic conductivity of the aquitard, and b' is the thickness of the aquitard (Fig. 9.20).

It can be seen that for large values of B, r/B would tend to be zero and therefore Eq. 9.32 will approach the Theis Eq. 9.20.

W(u.r/B) is read as the well function for leaky aquifers. Values of W(u.r/B) in terms of the practical range of u and r/B are plotted on a double-log paper to provide a family of type curves for leaky aquifers (Fig. 9.21). Pumping test data (s vs. t) is plotted on another log–log paper which is matched with one of the leaky aquifer type curves for a particular value of r/B and a match point is selected. The coordinates of the match point W(u.r/B) and l/u are read from the type curve sheet and s and t from the data curve sheet. The values of Q and match point coordinates, i.e. W(u.r/B) and s are substituted in Eq. 9.32 to calculate T and S is obtained from Eq. 9.33. The value of r/B of the type curve with which data curve was matched is substituted in Eq. 9.34 to determine K' and also leakage,

K'/b' or its reciprocal, hydraulic resistance, C from the relation B = √TC. A straight line method for estimating hydraulic properties of leaky aquifers is given by Hantush (1956, 1964).

Unconfined Aquifers

Pumping from an unconfined aquifer causes dewatering of the aquifer resulting in vertical component of flow in addition to the radial flow (Fig. 9.22). This is in contrast to a confined aquifer where there is no dewatering of the aquifer and the water is obtained due to compaction of the aquifer and expansion of water.

The time-drawdown curve in response to pumping from an unconfined aquifer is characteristically S-shaped due to delayed yield as described earlier. In unconfined aquifer, if drawdown, s is small as compared to its saturated thickness, b, the vertical component of flow can be neglected and the Theis equation can be used to determine aquifer characteristics. However, when drawdowns are significant, the vertical component of flow cannot be ignored and therefore the use of Theis equation is not justified. In such conditions where gravity drainage is consisderable, methods for analysing pumping test data based on the concept of delayed yield are given by Boulton (1963) and Neuman (1975).

Boulton's Method

Boulton (1963) gave Eq. 9.36 for drawdown in an unconfined isotropic aquifer with delayed drainage,

$$s = \frac{Q}{4\pi T} W\left(U_{AY} \frac{r}{D}\right) \qquad (9.36)$$

where W(U_{ay}, r/D) is called the 'well function of Boulton'. Under early-time conditions, describing the first

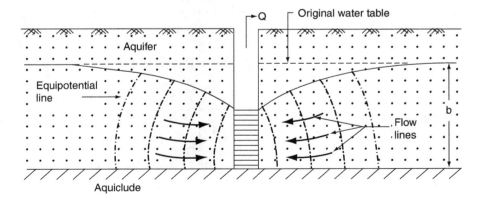

Fig. 9.22 Schematic cross-section of a pumped unconfined aquifer

segment of the time-drawdown curve, Eq. 9.36 can be written as

$$s = \frac{Q}{4\pi T} W\left(U_A \frac{r}{D}\right) \quad (9.37)$$

where

$$U_A = \frac{r^2 S_A}{4Tt} \quad (9.38)$$

and S_A = early time storage coefficient.

Under later-time conditions for the third segment of the time-drawdown curve, Eq. 9.36 reduces to

$$s = \frac{Q}{4\pi T} W\left(U_y \frac{r}{D}\right) \quad (9.39)$$

where

$$U_y = \frac{r^2 S_y}{4Tt} \quad (9.40)$$

and, S_y = specific yield. D is called the drainage factor, which is defined as

$$D = \sqrt{\frac{T}{\alpha S_y}} \quad (9.41)$$

D has the dimensions of L, $1/\alpha$ is the 'Boulton delay index' which is an empirical constant having the dimensions of time. It is used to determine the time, t at which the delayed yield ceases to affect the draw-down. The procedure for the use of Boulton's method involves matching of the field data curve in two steps, with one of the Boulton's type curves (Fig. 9.23) for computing aquifer properties (Kruseman and de Ridder 1970). Prickett (1965) was among the first workers to demonstrate the applicability of Boulton's type curve method to unconfined aquifers. An example of field data analysis using Boulton's type curve method in a granitic aquifer is given in Fig. 9.24.

Neuman (1972) showed that Boulton's method can be used only for large values of pumping time. In general, the limitation of Boulton's method becomes more severe as the ratio of horizontal permeability to vertical permeability increases, the thickness of the aquifer increases and the distance from the pumping well decreases. Neuman (1975) suggested an alternative method as described below.

Neuman's Method

Neuman's method for determining aquifer characteristics of anisotropic unconfined aquifer does not involve such semi-empirical quantities as Boulton's delay index, $1/\alpha$. It also takes into account aquifer anisotropy. Neuman's drawdown equation can be written as

$$s = \frac{Q}{4\pi T} W(U_{A'} U_{B'} \beta) \quad (9.42)$$

where, $W(U_A, U_B, \beta)$ = well functions for water-table aquifer with fully penetrating wells having no storage capacity (dimensionless).

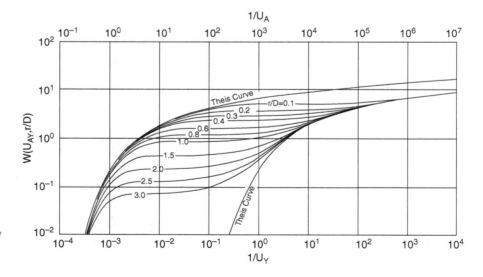

Fig. 9.23 Family of Boulton's type curves for an unconfined aquifer. (After Prickett 1965, reproduced by permission of the journal of Ground Water)

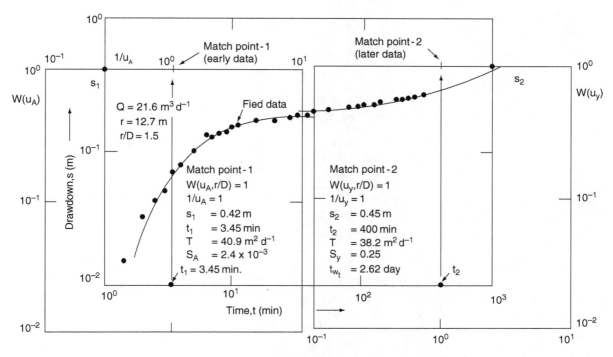

Fig. 9.24 Analysis of pumping test data by Boulton method from a granitic aquifer in Lower Maner Basin, Andhra Pradesh. (After Radhakrishna et al. 1979)

Under early time conditions Eq. 9.42 reduces to

$$s = \frac{Q}{4\pi T} W(U_{A'} \beta) \qquad (9.43)$$

where

$$U_A = \frac{r^2 S_A}{4Tt} \qquad (9.44)$$

S_A = Elastic early-time storativity.

Under late-time conditions the third segment of the time-drawdown curve, Eq. 9.42 reduces to

$$U_B = \frac{r^2 S_y}{4Tt} \qquad (9.45)$$

Neuman's parameter β is defined as

$$\beta = \frac{r^2 K_v}{b^2 K_h} \qquad (9.46)$$

where K_v and K_h are the vertical and horizontal hydraulic conductivities of the aquifer respectively and Q, T, R, t, S_y, and b are as defined earlier.

For an isotropic aquifer $K_v = K_h$. Therefore,

$$\beta = \frac{r^2}{b^2}$$

Figure 9.25 presents the family of Neuman's type curves of $W(U_A, U_B, \beta)$ vs. $1/U_A$ and $1/U_B$ for various values of β.

The aquifer characteristics can be determined by type-curve and straight-line methods (Kruseman and de Ridder 1990). Neuman (1975) showed that contrary to the assumption of Boulton, α is not a characteristic constant of the aquifer but decreases linearly with the logarithm of r.

The difference between Boulton's theory and that of Neuman (as far as fully penetrating wells are concerned) is that the former only enables one to calculate α, whereas the latter enables one to determine the degree of anisotropy, as well as the horizontal and vertical hydraulic conductivities.

Bounded Aquifers

In the various aquifer test methods mentioned above, it was assumed that the aquifer is of infinite areal

Fig. 9.25 Neumann's type curves for unconfined aquifer

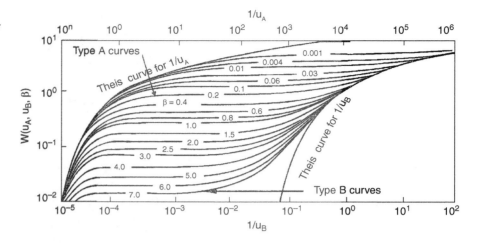

extent. However, in nature such an assumption will not always hold good as the aquifers are limited by some hydrogeological boundary. Two types of boundaries are easily identified, i.e. a recharge boundary like a river or a canal which causes replenishment of the aquifer and a barrier or impermeable boundary which is a result of either thinning or termination of an aquifer against a low-permeability formation or dyke or a fault which limits movement of water (Fig. 9.26).

The method of images, commonly used in the study of heat conduction in solids, has been used for the solution of boundary problems in groundwater flow. The boundary condition is simulated by an image well or imaginary well located at the same (equivalent) distance from the boundary but on the opposite side. Depending upon the nature of the boundary, i.e. recharging or barrier type, the flow is characterized by assuming a recharging or discharging image well. Thus the real bounded system is replaced by an equivalent hydraulic

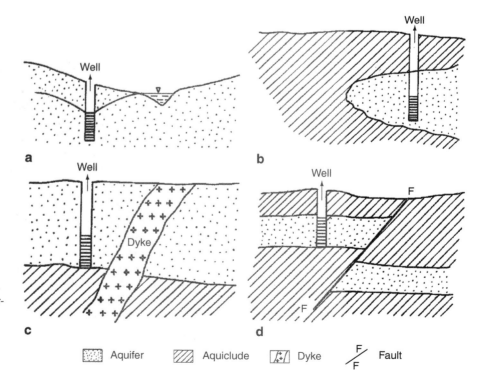

Fig. 9.26 Diagram illustrating various types of boundaries. **a** Recharging boundary due to stream. Impermeable boundaries due to: **b** lateral termination, **c** dyke, and **d** fault

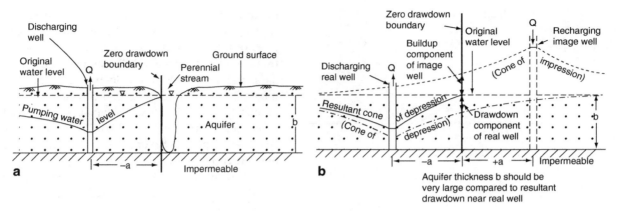

Fig. 9.27 Schematic section of: **a** a recharging well in a semi-infinite aquifer bounded by a perennial stream, and **b** of the equivalent hydraulic system in an infinite aquifer. (After Ferris et al. 1962)

system, i.e. an imaginary system of infinite areal extent. Both the recharge and discharge image wells are located perpendicular to the boundary, but on the opposite sides at equal distances from the boundary (Figs. 9.27, 9.28). In a recharging system (Fig. 9.27), the recharge image well recharges the aquifer at the same rate as the rate of withdrawal from the real well. As a result of this the build-up due to cone of impression (recharge) and the drawdown caused due to pumping from the real well exactly cancel each other at the recharging boundary resulting in a constant head along the line source. In cases where there is more than one boundary, more image wells are to be considered depending upon the configuration and nature of the boundaries. Several such geometric configurations are considered by Ferris et al. (1962) and Bear (1979), among others.

The effect of boundary conditions on the time-drawdown trend in an observation well will depend on the nature of the boundary, the time when the cone of depression intercepts the boundary and the distance of the observation well from the pumping well. On a log–log plot, the boundary effects are observed by the deviation of the field data from the Theis curve. When time-drawdown data are plotted on semi-log paper, the recharging boundary will cause a flattening of the time-drawdown curve and a barrier (impermeable) boundary will cause further steepening of the curve (Fig. 9.29). The time of inflection of the time-drawdown curve from the normal trend (when the aquifer is of infinite areal extent) depends on the distance of the observation well from the pumping well. Note that the effect of recharging boundary on time-drawdown curve is the same as that for leaky aquifers or partially penetrating wells. Therefore, for a proper interpretation of pumping test data, necessary information about the hydrogeological system is necessary.

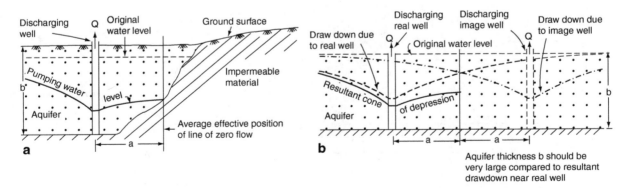

Fig. 9.28 Schematic section of: **a** a discharging well in a semi-infinite aquifer bounded by an impermeable formation and **b** the equivalent hydraulic system in an infinite aquifer. (After Ferris et al. 1962)

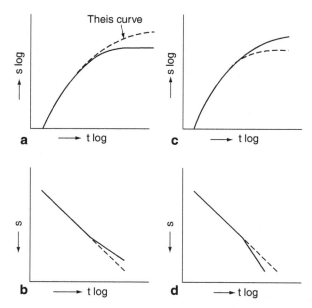

Fig 9.29 Effect of a recharge boundary (a and b) and impermeable boundary (c and d) on the time-drawdown curves in a confined aquifer

The recharging boundary, viz. rivers, canals, or lakes can be easily observed in the field but barrier boundaries are at times hidden as a subsurface dyke or a fault. Such boundaries can be identified and mapped by the analysis of observed time-drawdown data provided that pumping is carried out for sufficiently long time so that the cone of depression intercepts the boundary. As mentioned above, the nature of the boundary (recharging or barrier type) can first be ascertained from the trend of the time-drawdown data. The next step is to determine

the distance of the image well from the observation well (r_i) for which time-drawdown data from at least three observation wells is necessary. The values of r_i from various observation wells can be estimated from the semi-log time drawdown plot by choosing any arbitrary value of drawdown, s_w and corresponding value of time, t_r, before the effect of boundary influences the time-drawdown curve (Fig. 9.30a). The time-drawdown curve is extrapolated beyond the time of inflection. The time intercept, t_i of an aqual amount s_w of divergence caused by the image well is read from the data curve. The distance r_i, can be determined from Eq. 9.47.

$$r_i = r_r \sqrt{t_i / t_r} \qquad (9.47)$$

where r_r is the distance of the observation well from the pumping well and r_i, t_i and t_r are as defined earlier.

After the values of r_i are computed for the individual observation wells, circles are drawn with their centres at the respective observation wells and their radii equal to the respective estimated values of r_i. The intersection of the arcs will give the location of image well. The perpendicular bisector of the line joining the image well and the real (pumped) well will mark the strike of the boundary (Fig. 9.30b).

Partially Penetrating Wells

One of the assumptions in the Theis and other methods as discussed earlier, is that the pumped well penetrates the entire thickness of the aquifer so that the flow

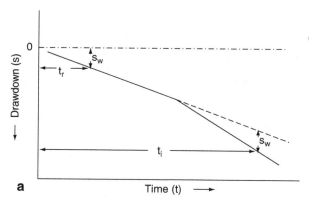

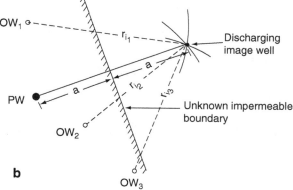

Fig. 9.30 Diagrams illustrating application of image well theory to locate an unknown impermeable boundary; **a** time-drawdown plot showing effect of impermeable boundary, **b** location of the unknown impermeable boundary. *OW* Observation well, *PW* pumping well

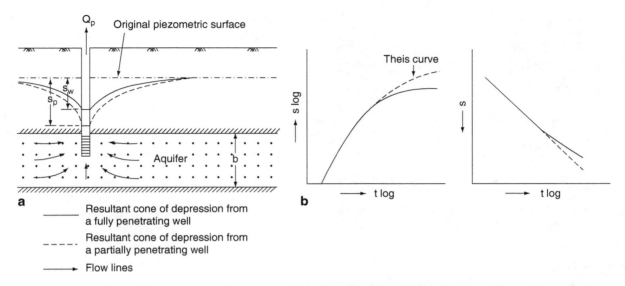

Fig. 9.31 a Effect of partially penetrating well on drawdown in a confined aquifer, **b** Effect of wells partial penetration on the time-drawdown relationship in a confined aquifer

towards the well is horizontal. However, under field conditions especially when aquifers are very thick, it is usually uneconomical to provide screen along the entire thickness of the aquifer. Such a well which taps only a part of the aquifer thickness is known as partially penetrating well.

In a partially penetrating well, the flow pattern differs from the radial horizontal flow which pressumably exists around a fully penetrating well. Therefore, flow towards a partially penetrating well will be three-dimensional due to vertical flow component. Accordingly, the average length of a flow line will be greater in partially penetrating well as compared with the fully penetrating well resulting in additional drawdown (Fig. 9.31b). The discharge from a partially penetrating well therefore will be less than a fully penetrating well for the same drawdown. It would also mean that if they are pumped at the same rate, the drawdown in a partially penetrating well (s_p) will be greater than that in a fully penetrating well (s_w). The effect of partial penetration is more near the well face and will decrease with increasing distance from the pumping well. Anisotropy causes greater effect of partial penetration on drawdown as compared to isotropic aquifers as in the former the flow is three dimensional. In an anisotropic aquifer, the effect of partial penetration will be negligible at a distance of $r \geq 1.5\,b\,\sqrt{K_h/K_v}$ and at $t > Sb^2/2T$ (Kruseman and de Ridder 1990). Methods of analyzing test data from partially penetrating wells for steady and unsteady conditions in confined, uncon-

fined, leaky and anisotropic aquifers are described by Hantush (1964) and Kruseman and de Ridder (1990).

Large Diameter Wells

The various methods described above assume that pumping well has an infinitesimal diameter which will not be valid in large diameter (dug) wells. Therefore, these methods are not applicable to large diameter wells due to the significant effect of well storage. Dugwells are common in low permeability hard rocks of India and other developing countries. The earliest method to estimate aquifer properties from large diameter fully penetrating well in a confined aquifer was given by Papadopulos and Cooper (1967). Later, several other analytical and numerical solutions of unsteady flow to large-diameter wells were developed (Boulton and Streltsova 1976; Rushton and Holt 1981; Rushton and Singh 1983; Herbert and Kitching 1981; Barker 1991; Chachadi et al. 1991; Rushton 2003).

Papadopulos–Cooper Method

Papadopulos and Cooper (1967), gave Eq. 9.48 to describe drawdown in a fully penetrating large-diameter well (with storage) in a confined aquifer (Fig. 9.32).

$$s_w = \frac{Q}{4\pi T} F(U_w, \beta) \qquad (9.48)$$

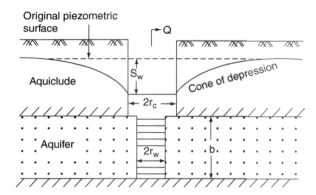

Fig. 9.32 Schematic cross-section of a confined aquifer pumped by a large diameter well

where,

$$U_w = \frac{r_w^2 S}{4Tt} \qquad (9.49)$$

$$\beta = \frac{r_w^2 S}{r_c^2} \qquad (9.50)$$

r_w = radius of the well screen or open well

r_c = radius of the well casing over which the water-level is changing.

As in other type curve matching methods, the time-drawdown curve is plotted on a double-log paper which is matched with one of the type curves for large diameter wells (Fig. 9.33). A match point is selected and the values of $F(U_w, \beta)$, $1/u_w$ and t are noted. The value of β

of the type curve with which the observed data curve is matched is also noted. T is estimated from Eq. 9.48. Values of S can be computed by two methods (1) by substituting the values of $1/u_w$, t, r_w and T into Eq. 9.49; and (2) by substituting the values of β, r_w and r_c into Eq. 9.50.

The main problem in curve matching is that the early part of type curves (Fig. 9.33) differs only slightly in shape from each other and therefore uncertain value of β will make significant difference in the computed value of S. Therefore, unless time of pumping is large, reliable value of S cannot be obtained but transmissivity can be computed without such a problem.

The well storage dominates the time-drawdown curve upto a time, t, given by t = (25 r_c^2/T), after which the effect of well storage will become negligible and hence reliable values of aquifer characteristics T and S can be obtained. Therefore, for obtaining a representative value of S, the well should be pumped beyond this time, t. In hard rocks, such as granites and basalts where T is about $10\,m^2 d^{-1}$ and r_c is 2 m, t will be 10 days which is impractical. Therefore, Papadopulos–Cooper type curve method may not give a correct value of S but T values can be reliable.

The Papadopulos–Cooper method although takes into account the effect of well storage but it is applicable only to fully penetrating abstraction well in a confined aquifer. A method for analysing test data from a partially penetrating well in an anisotropic unconfined aquifer is given by Boulton and Streltsova (1976). The Papadopulos–Cooper method also assumes constant

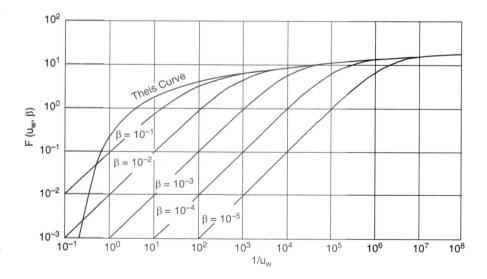

Fig. 9.33 Family of a Papadopulos type curves for large-diameter wells

rate of abstraction which is difficult to maintain when a large diameter well is discharged by a centrifugal pump. A method to maintain constant discharge during the pumping test is suggested by Athavale et al. (1983). A curve matching technique considering the falling rate of abstraction during the test is given by Rushton and Singh (1983).

The effect of seepage face which is developed in the unconfined aquifer due to differences in the water-level outside and inside a large diameter well was considered by Rushton and Holt (1981), Rushton and Singh (1987) and Sakthivadivel and Rushton (1989). The advantage of using recovery data over drawdown data is advocated by Herbert and Kitching (1981), Singh and Gupta (1986) and others. This is because during recovery: (a) well storage does not play any important role as all water is derived from the aquifer, and (b) well losses are small especially during the later stages of recovery.

The use of numerical methods for computing aquifer properties under complex geohydrological conditions are advantageous. The discrete Kernel approach suggested by Patel and Mishra (1983), Rushton and Singh (1987), and Barker (1991) is advantageous for estimating aquifer parameters from both drawdown and recovery data under varying rates of pumping. Chachadi et al. (1991) also considered storage effect both in the production and observation wells.

9.2.4 Pumping Tests in Fractured Rock Aquifers

In areas where the fractures are closely spaced and are inter-connected, conventional pumping test methods assuming confined and leaky confined models can be used. However, the double porosity model (Sect. 7.2.2) is more representative of uniformly fractured aquifers including petroleum and geothermal reservoirs (Earlougher 1977; Horne 1990; Grant et al. 1982).

9.2.4.1 Double Porosity Model

The flow characteristics in a double (dual porosity) aquifer were discussed earlier in Sect. 7.2.2. In a double porosity aquifer the rock mass is assumed to consist of a number of porous blocks as well as large number of randomly distributed, sized and oriented fractures. A method of estimating hydraulic properties of such a double porosity aquifer is described below.

Streltsova–Adam's Method for Confined Fractured Aquifer

Streltsova–Adams (1978) assumed that a confined fractured aquifer representing a double porosity model, consists of matrix blocks and fracture units in the form of alternate horizontal slabs (Fig. 7.8a). The thickness of the matrix units is greater than that of fracture units.

The drawdown distribution in the fracture (s_f) is given by Eq. 9.51

$$S_f = \frac{Q}{4\pi T_f} W\left(u_f, r/B_f, n\right) \tag{9.51}$$

where

$$u_f = \frac{r^2 S_f}{4 T_f t} \tag{9.52}$$

$$B_f^2 = \frac{T_f}{\alpha_f S_m} \tag{9.53}$$

and

$$n = 1 + \frac{S_m}{S_f} \tag{9.54}$$

Subscripts f and m represent properties of fractures and matrix blocks respectively, B_f is the drainage factor. Accordingly, the equation for drawdown in matrix block, s_m is also developed.

Values of the drawdown function $W(u_f, r/B_f, n)$ computed for $n = 10$, 100, and 1000 and various assumed values of parameters r/B_f and $1/u_f$ are given by Streltsova–Adams (1978).

Streltsova–Adam's method is based on the assumptions that: (a) fractures and blocks are compressible, (b) the abstraction well is fully penetrating the fractured aquifer and receives water from it, (c) pumping is at constant rate, (d) the radius of the pumped well is vanishingly small, (e) the flow in the block is vertical, (f) flow in the fissure is horizontal, and (g) flow in both blocks and fissures obey Darcy's law.

Fig. 9.34 Streltsova-Adam's type curves for double porosity confined aquifer (n=10)

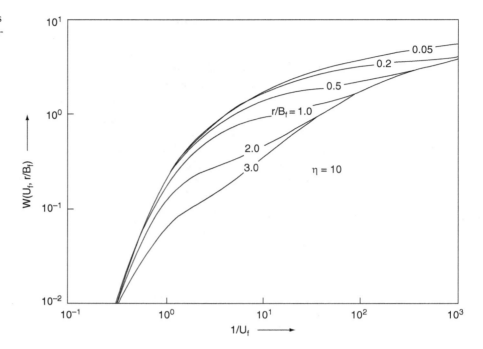

Based on the values of well function, and values of r/B_f and $1/u_f$, type curves can be plotted on double-log paper. One such set of type curves for n=10 is given in (Fig. 9.34). Time-drawdown data is plotted on another double log paper and matched with one of the type curves. A match point is selected for which coordinate values are noted and substituted in Eqs. 9.51–9.53 to obain values of T_f, S_f, T_m, S_m and B_f. Other characteristics which can be estimated are α_f, and C_f. α_f, a characteristic of the fissure flow (days^{-1}) is given by

$$\alpha_f = \frac{T_f}{B_f^2\, S_m} \qquad (9.55)$$

and C_f (hydraulic diffusivity of the fissure)

$$= \frac{T_f}{S_f}\; \left(m^2\, d^{-1}\right) \qquad (9.56)$$

As an example, the time-drawdown data curve from an observation well in fractured basalt is given in Fig. 9.35. The field data curve was matched with the Theis type curve, Hantush leaky aquifer type curve, and Boulton's type curves for unconfined aquifer. However, keeping in view the hydrogeological situation and because the best match was obtained with Streltsova–Adam's type curve (n=10), hydraulic pa-

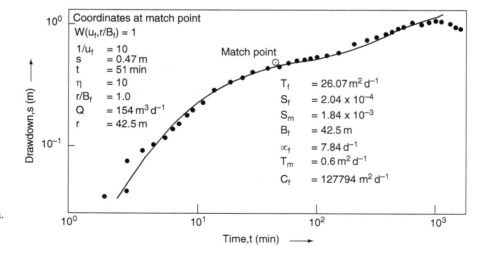

Fig. 9.35 Time-drawdown plot of pumping test data from an observation well in fractured basalt, South India. (After Singhal and Singhal 1990)

rameters were computed assuming double porosity model (Fig. 9.35).

The equations given by Streltsova–Adams for double-porosity aquifer, involve delayed yield from blocks to fissures. Hence Streltsova–Adam's type curves are identical to Boulton's type curves for unconfined aquifer involving delayed yield in unconsolidated formations. Therefore, for proper interpretation of pumping test data, hydrogeological framework of the aquifer system should be known.

Bourdet–Gringarten's Method

The hydraulic characteristics of the fractures and the matrix block in a double porosity aquifer can also be estimated using Bourdet–Gringraten's type curve method (Bourdet and Gringarten 1980; Gringarten 1982). The difference between Streltsova–Adam's approach and that of Bourdet–Gringarten is that the former considered blocks and fissures of regular shape and orientation in the form of slabs, while the latter model is applicable to different geometrical shapes of blocks. Readers may refer to Kruseman and de Ridder (1990) for further details of this method.

The effect of fracture skin in a fissured double porosity aquifer is discussed by Moench (1984) and its application to estimate hydraulic parameters of fissured rock system is given by Levens et al. (1994).

9.2.4.2 Single Vertical Fracture Models

The majority of oil and gas reservoirs are hydraulically fractured to produce a single vertical fracture to augment well production. In Hot Dry Rock (HDR) experiments also, such fractures are induced for circulating water to deeper parts in the crust for tapping geothermal energy. Naturally occurring vertical fractures are also intersected during drilling for water and geothermal wells. Therefore, drawdown and pressure buildup in wells penetrating a single vertical fracture is of importance to estimate its hydraulic characteristics. The time-drawdown characteristics of a vertical fracture of high permeability was discussed earlier.

Gringarten–Witherspoon's Method

The drawdown in an observation well as a result of pumping from a single plane, vertical fracture in a homogeneous, isotropic confined aquifer is given by Eq. 9.57.

$$s = \frac{Q}{4\pi T} F\left(u_{vf}, r'\right) \qquad (9.57)$$

where

$$U_{vf} = \frac{Tt}{Sx_f^2} \qquad (9.58)$$

$$r' = \frac{\sqrt{\left(x^2 + y^2\right)}}{x_f} \qquad (9.59)$$

x_f = half length of the vertical fracture (m)

x, y = distance between observation well and pumped well, measured along the x and y axes, respectively (m).

Equations 9.57 and 9.58 indicate that the drawdown in the observation well depends on its location with respect to orientation of the fracture. The observation well could be located on the x axis or y axis or along a line at an angle of 45° to the strike of the fracture (Fig. 9.36).

Type curves of drawdown function F(u_vf, r') for different values of U_{vf} and r' and location of observation wells are given in Kruseman and de Ridder (1990). If one knows about the location of observation well with respect to the fracture, a particular set of type curves can be selected for matching the test data plot. By matching the data curve with the type curve, coordinate values of the match point are noted and T and S are calculated from Eqs. 9.57 and 9.58.

The various methods of analysing test data from fractures are primarily developed for hydraulically fractured petroleum reservoirs with the assumption

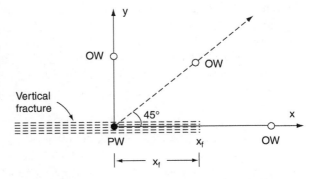

Fig. 9.36 Plan view of a vertical fracture with a pumped well (*PW*) and observation wells (*OW*) at three different locations

that fractures have large hydraulic conductivity. There-fore, their application to natural fractures is limited as natural fractures will not have infinite hydraulic conductivity due to infilling. Moreover, the assumption of negligible storage may also not hold good as natural fracture zones may be quite wide.

Kresic (2007) compared the values of T and S obtained from karst aquifer in South Dakota, USA, estimated by Jacob's straight line method, Neuman's unconfined aquifer type curve method and dual porosity methods. It was noted that the values of T and S, estimated by Jacob's solution for both the early and late time-drawdown data, are similar to those obtained using double porosity model. Further, the storativity estimated by Jacob's method from early data is almost identical with the fracture and conduit storativity obtained from dual porosity model.

9.2.4.3 Intrusive Dykes

Dykes are intrusive bodies of igneous rocks, commonly of dolerite composition which may extend to long distances of the order of several kilometres but are of limited width. They cut across different type of rocks and at places cause fracturing of the country rocks at the contact due to baking effect. The dykes are themselves fractured into a system of joints usually of columnar type which impart secondary permeability. At other places, dykes may be comparatively massive and impermeable forming barrier boundary.

Methods of analysis of pumping test data from fractured dykes are given by Boonstra and Boehmer (1986, 1989) and Boehmer (1993). It is assumed by these researchers that the dyke is of infinite length and has finite width and a finite hydraulic conductivity. The upper part of the dyke is less permeable due to weathering of the dyke rock into clayey material, the middle part is fractured and is permeable while the deeper part is massive and impervious as the fractures tend to die out with depth. Therefore, middle fractured part of the dyke forms a confined horizon which is continuous with the adjacent confined part of the aquifer in the country rock (Fig. 9.37).

The time-drawdown data obtained from a pumping well in a dyke is characterized by three segments as mentioned earlier (Fig. 9.16). The hydraulic characteristics of the dyke and the country rock can be determined by analysing time-drawdown data from

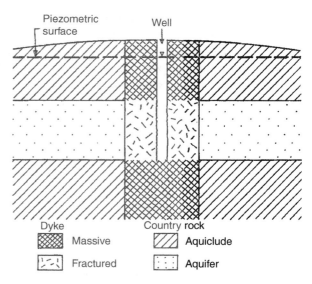

Fig. 9.37 Composite dyke-aquifer system. (After Kruseman and deRidder 1990)

observation wells in a dyke using a type-curve method developed by Boonstra and Boehmer (1986). A distance-drawdown type curve method is also suggested. They have demonstrated the applicability of their method for estimating hydraulic characteristics of a dolerite dyke of the Karoo system in the Republic of South Africa (Boehmer and Boonstra 1987; Boonstra and Boehmer 1989).

9.2.5   Cross-hole Tests

As Lugeon tests, slug tests and conventional pumping tests provide information of only a part of the aquifer in the close vicinity of the well, cross–hole tests are preferred to determine three dimensional properties in both saturated and unsaturated porous and fractured rocks from an array of randomly oriented boreholes. Cross-hole tests are usually carried out for estimating relative permeabilities and interconnection between various fracture sets which are of importance in the assessment of potential repositories for the disposal of radioactive waste and solute movement.

A cross-hole project was implemented at Stripa mine, Sweden, for characterization of high level waste repository where a fan shaped array of boreholes, 200–250 m long, were drilled below the ground emanating from the end of a drift, for geophysical and hydrogeo-

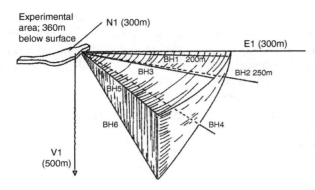

Fig. 9.38 General layout of the crosshole programme borehole array at Stripa, Sweden. The fan shaped array of six boreholes (*BH1–BH6*) lies within an orthogonal array of three existing boreholes, *E1* (east), *N1* (North) and *V1* (vertical). (After Black and Holmes 1985)

logical testing (Fig. 9.38). A polar plot of directional permeabilities would indicate whether the rock mass behaves as an anisotropic continuum. An ellipsoidal configuration of directional permeabilities would support continuum assumption. In Stripa mine, Sweden, the plotted values of hydraulic diffusivity (ratio of permeability to specific storage) were not strictly ellipsoidal indicating a limited interconnection of discrete fractures (Black 1987).

In unsaturated fractured rocks also, single-hole and cross-hole pneumatic injection tests using air rather than water in vertical and inclined boreholes are carried out to characterise the bulk pneumatic properties and connectivity of fractures. The test holes can be either vertical or inclined depending on the geometry of fractures. A single-hole pneumatic injection test provides information for only a small volume of rock close to the borehole and therefore may fail to provide information for the rock heterogeneity. Therefore, cross-hole injection tests are preferred in which air is injected into an isolated interval within the injection well and also the other adjacent boreholes especially designed for this purpose. This facilitates the estimation of bulk pneumatic properties of larger rock volumes between observation boreholes and also the degree of interconnectivity on the scales ranging from meters to several tens of meters (Ilman and Neuman 2001).

The design and conduct of such tests and interpretation of test data by steady-state, transient type curve and asymptotic analysis from unsaturated fractured tuff at the Apache Leap Reserch Site (ALRS) near Superior, Arizona, USA are given by Ilman and Neuman

(2001, 2003) and Ilman and Tartakovsky (2005). Later, Ilman and Tartakovsky (2006) suggested an asymptotic approach to analyze test data from cross-hole hydraulic tests in saturated fractured granite at the Grimsel Test Site in Switzerland.

9.2.6 Variable-Discharge (Step-drawdown) Test

In a variable-discharge (step-drawdown) test, the well is pumped in three or more steps with increase in rate of discharge, each step being of about 1 h duration. Drawdown measurements in the pumped well are taken at frequent intervals.

Step-drawdown tests are useful for the estimation of (1) aquifer transmissivity, and (2) hydraulic characteristics of wells, described in Sect. 17.5.

The drawdown in a pumping well at a given time, s_w includes two components, the aquifer loss BQ and the well loss (CQ^n) (Fig. 9.39). Aquifer or formation loss (BQ) is due to Darcy type linear flow of water in the aquifer. Well losses (CQ^n) are both linear and nonlinear. As the well loss is mainly due to turbulent flow

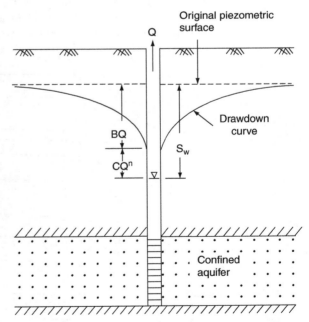

Fig. 9.39 Schematic cross section of a fully penetrating well in a confined aquifer showing relation of well loss (CQ^n) and formation loss, BQ to total drawdown, S_w

it is expressed as CQ^n, where the value of n may vary from 1.5 to 3.5, depending on the extent of turbulence or rate of discharge. The value of n is usually taken to be 2, i.e.

$$s_w = BQ + CQ^2 \qquad (9.60)$$

where, s_w = drawdown in pumping well; B = formation loss coefficient; BQ = formation loss; C = well loss coefficient; CQ^2 = well loss.

9.2.6.1 Estimation of Transmissivity from Specific Capacity Data

Aquifer parameters viz. transmissivity and hydraulic conductivity can be estimated from specific capacity values (see Sect. 17.5) to avoid the cost of long duration pumping test (Walton 1962, Brown 1963, Mace 1997). Several graphs relating theoretical specific capacity with transmissivity for various values of S, T, and r_w are given by Walton (1962). One such plot relating specific capacity to T and S for a given value of t and r_w is illustrated in Fig. 9.40. If specific capacity has been determined from field (pumping) test, S is known *a priori* from either well log data or water-level fluctuations, and r_w and t are known, transmissivity T can be computed by using such a graph. These graphs are based on the assumption that the well is 100% efficient

and that the effective diameter is the same as the diameter of the well screen. The second assumption may hold good in consolidated rocks but in unconsolidated sediments, effective diameter is usually more than the diameter of the screen.

The other approach to estimate T from specific capacity values is based on the Thiem's equilibrium formula

$$T = \frac{2.3\, Q \, log\, R/r_w}{2\pi\, s_w} = A(Q/s_w) \qquad (9.61)$$

where A is the dimensionless constant depending on the radius of influence, R and well radius, r_w; Q/s_w is the specific capacity of the well. Values of A usually range from 0.9 to 1.53 with a mean of 1.18 (Rotzoll and El-Kadi 2008). Values of A would also depend on the hydraulic characteristics of the aquifer and degree of confinement. It can also be estimated from values of r_w and R. In cases where exact values of R are not available, approximate values of R can be used as given in Table 9.5 as value of A in Eq. 9.67 will not be much affected by poor estimates of R.

Huntley et al. (1992) have given an empirical relation Eq. 9.62 between T and specific capacity (Q/s_w) for fractured rock aquifer

$$T = A\left(Q/s_w\right)^{0.12} \qquad (9.62)$$

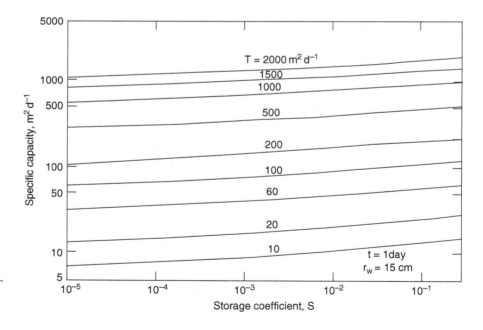

Fig. 9.40 Relationship between specific capacity, transmissivity and storativity based on Jacob's non-equilibrium equation. (After Meyer 1963; Todd 1980)

Table 9.5 Approximate values of radius of influence, R. (After UNESCO 1972)

Type of formation	Type of aquifer	R (m)
Fine and medium-grained sands	Confined	250–500
	Unconfined	100–200
Coarse-grained sands and gravel-pebble beds	Confined	750–1500
	Unconfined	300–500
Fissured rocks	Confined	1000–1500
	Unconfined	500–1000

In Eq. 9.62, both T and Q/s_w are expressed in $m^2 d^{-1}$. Here it is assumed that well loss is negligible. This will result in the under-estimation of transmissivity in alluvial aquifers due to significant well loss.

Table 9.6 gives the summary of results obtained by several studies from fractured aquifers including methods of estimation of S_c and T. The table indicates that the values of A are site specific depending on hydrogeological characteristics of aquifer and method of determining S_c and T. Based on studies in the basalts of Hawaii islands, Rotzoll and El-Kadi (2008) concluded that correction for well loss significantly improves the correlation between T and S_c. In fractured rocks, due

to larger effective radius of well and shorter duration of pumping, one will expect overestimation of transmissivity based on specific capacity values (Razack and Huntley 1991). Further, the value of T estimated from specific capacity data applies only to the aquifer adjacent to the well which is affected by drilling and well development. Therefore, values of T estimated from specific capacity data are greater than those obtained from aquifer tests.

9.2.7 Hydraulic Tomography (HT)

HT is a sequential cross-hole hydraulic test followed by inversion of all the data to map the spatial distribution of aquifer hydraulic properties depending on the distribution and connectivity of fractures, which is important for water resources management and also for groundwater contamination, prevention and remediation (Hao et al. 2008). It provides more information than a classical pumping test by providing reasonable estimates from the same well field. HT is

Table 9.6 Summary of results giving empirical relationships between transmissivity and specific capacity. (Modified after Rotzoll and El-Kadi (2008))

Study (see Rotzoll and El-Kadi 2008)	Aquifer	Location	Methods to determine hydraulic parameters		Regression coefficients A
			Specific capacity	Transmissivity	
Razack and Huntley (1991)	Heterogeneous alluvium	Haouz plain, Morocco	Constant-rate test (uncorrected)	Cooper-Jacob	15.30
Huntley et al. (1992)	Fractured rock	San Diego, California	Constant-rate test (uncorrected)	Cooper-Jacob (Neuman), Gringarten	0.12
Jalludin and Razack (2004)*	Sediment, fractured basalt	Djibuti, Horn of Africa	Step-drawdown test (corrected)	Cooper-Jacob, Boulton, Theis Recovery	3.64
Razack and Lasm (2006)*	Fractured rock	Man Danane, Ivory Coast	Step-drawdown test (corrected)	Theis Recovery	0.33
Eagon and Johe (1972)	Karst	NW-Ohio	Constant-rate test (corrected)	Cooper-Jacob	3.24
Mace (1997)	Karst	Edwards, Texas	All steps from step test (uncorrected)	Theis, Cooper-Jacob, Theis Recovery	0.76
Choi (1999)*	Volcanic island	Jeju, Korea	Constant-rate test (uncorrected)	Cooper-Jacob	0.45
Hamm et al. (2005)*	Volcanic island	Jeju, Korea	Constant-rate test (uncorrected)	Moench (leaky)	0.99
Adyalkar and Mani (1972)	Volcanic (Deccan trap)	India	Thiem method		0.37–0.63
Fernandopulle et al. (1974)	Volcanic rock	Gran Canaria Spain			0.39

Regression coefficients are for S_C in m^2/d

* For these references, see Rotzoll and El-Kadi (2008)

an application of the concept of Computerized Axial Tomography (CAT) in medical sciences and tomographic surveys in geophysics for imaging subsurface hydraulic heterogeneity.

HT involves installation of multiple wells in an aquifer which are partitioned into several intervals along the depth using packers. A sequential aquifer test at selected intervals is conducted by injecting or pumping at selected intervals and its response is noted in this well and the other observation wells. The test is repeated by pumping from another interval and also from intervals in other wells (Yeh and Liu 2000; Yeh and Lee 2007).

The results of a sequential pumping test in the alluvial formation near Naples, Italy are discussed by Straface et al. (2007). The test data were analysed by both the conventional type curve and straight line methods using distance-drawdown and time-drawdown data. They also used HT technique for characterizing the aquifer and concluded that the HT technique provides useful information about the heterogeneity pattern giving spatial distribution of hydraulic properties over a large volume of geologic material without resorting to a large number of wells.

test are preferred. Field tests are expensive and therefore need proper planning about the design and location of observation wells, rate of pumping and duration of test etc. A prior knowledge of the hydrogeology of the area is necessary in planning these tests. The choice of a particular method also depends on the purpose of study. For hydrogeologic characterisation of fractured rocks, double-porosity model is generally considered more realistic. Cross-hole tests are preferred in both saturated and unsaturated fractured rocks for ascertaining the spatial distribution of hydraulic properties. Lately, sequential pumping test (hydraulic tomography) involving pumping/injection of water at different depths and measurement of corresponding responses at various intervals is found more useful.

Summary

Estimation of aquifer parameters (T and S) is one of the most difficult tasks. Laboratory methods are simple but they do not provide realistic values due to the limited sample size and formation heterogeneities in the field. Therefore field methods which include pumping test and tracer

Further Reading

Boonstra J, Soppe R (2007) Well hydraulics and aquifer Tests, in Handbook of Groundwater Engineering (Dlleur JW ed.). 2nd ed., CRC Press, Boca Raton.

Kresic N (2007) Hydrogeology and Groundwater Modeling. 2nd ed., CRC Press, Boca Raton, FL.

Kruseman GP, de Ridder NA (1990) Analysis and Evaluation of Pumping Test Data. 2nd ed., Intl. Inst. for Land Reclamation and Improvement, Publ. No. 47, Wageningen.

Rushton KR (2003) *Groundwater Hydrology: Conceptual and Computational Models*. John Wiley & Sons, Chichester, UK.

Schwartz FW, Zhang H (2003) Fundamentals of Ground Water. John Wiley & Sons Inc., New York.

Tracer and Isotope Techniques

10.1 Introduction

Tracers are defined as chemical substances (inorganic or organic molecules, including isotopes), present naturally or introduced in the environment. A variety of tracers are used in geohydrological investigations for estimating the rate and direction of groundwater movement, groundwater recharge and its residence time. Tracers are also used for the study of origin of groundwaters including saline and geothermal waters, contaminant transport including site characterisation of repositories for nuclear waste, interconnection between surface water and groundwater, stream discharge measurement and sediment transport. There are reports of the use of tracers viz. chloride, fluorescein and bacteria in karst aquifers in Europe even in the late 1800s and early 1900s.

An ideal tracer should have the following characteristics (though no single tracer may meet all these requirements) (Todd 1980):

1. Quantitatively detectable in very small concentrations,
2. More or less absent from the natural water to be traced,
3. Neither adsorbed by the aquifer material nor chemically react with natural water causing its precipitation or biological degradation,
4. Non-toxic to the people and the ecosystem,
5. Inexpensive and readily available, and
6. In case of chemical tracers, the resulting solution should have approximately the same density as water.

The selection of tracer is site-specific, depending on both the formation fluid and rock characteristics and temperature.

The study of groundwater movement using conventional approach of analysis of water-level gradients and transmissivities generally envisages a continuous through flow. However in actual groundwater systems because of certain geological controls, especially in the fractured rocks, there are hydrogeological discontinuities. Thus in such rocks, groundwater flow—direction and velocity—can be best studied using tracers because the interpretation of these tests does not require the assumptions of porous medium i.e. continuum approach (see Chap. 7).

10.2 Types of Tracers

A tracer could be natural or artificial, inert (stable) or radioactive (Table 10.1). Natural tracers are those which are present in the natural waters. For example, surface water with comparatively high concentration of nitrate can be used as a natural inert tracer. In other situations, injection of fresh water into a deep seated formation having highly saline water may act as tracer and can be useful in studying the mixing phenomenon and thereby estimating hydraulic conductivity of formations.

Silica is one of the good examples of natural reactive tracers to trace the movement of water in hot fractured formations because its solubility increases with rise in temperature. Increase of silica caused by rock dissolution during an injection backflow experiment in Hot Dry Rock (HDR) experiment in the Rhine Graben gave an approximate idea of the quantity of rock dissolved during the interaction (Jouanna 1993).

Sometimes, it is also necessary to use multiple tracers to have a better understanding of the aquifer

B. B. S. Singhal, R. P. Gupta, *Applied Hydrogeology of Fractured Rocks*,
DOI 10.1007/978-90-481-8799-7_10, © Springer Science+Business Media B.V. 2010

Table 10.1 List of commonly used tracers in groundwater studies

Dyes	Sodium fluorescein, methylene blue, congo red
Salts	Sodium chloride, calcium chloride, ammonium chloride, lithium chloride
Organic compound	Chlorofluorocarbons
Radioactive isotopes	^{3}H, ^{14}C, ^{32}P, ^{36}Cl, ^{51}Cr-EDTA, ^{60}Co, ^{82}Br, ^{86}Rb, ^{129}I
Stable isotopes	^{2}H, ^{3}He, ^{15}N, ^{18}O

Table 10.3 Recommended doses of salt tracers for injection into wells. (After UNESCO 1972)

Salt	Doses (kg)	Distance between injection and observation wells (m)
Sodium chloride	10–15	5–7
Calcium chloride	5–10	3–5
Ammonium chloride	3–5	2–5
Lithium chloride	0.010–0.015	2–5

characteristics. One of the limitations of the artificially injected tracers is that they can, at best, give the hydraulic characteristics of only a small segment of the aquifer investigated. Secondly, they provide the information about the system valid only for the duration of the experiment. A good knowledge of the mineralogical characteristics of the rock helps to obtain information about the adsorption behaviour of various tracers. Matrix diffusion, especially in fractured rocks, is also important as it considerably influences the flow pattern of tracers (Sect. 7.4.1).

10.2.1   Dyes and Salts

Fluorescent dyes viz. fluorescein and rhodomine are commonly used as groundwater tracers. Automatic samplers or in situ spectrofluoro-photometers are of much help to determine very low concentration of dye. Fluorescein has a lesser tendency to get adsorbed. It is a conservative tracer and is also safe to use. It has also been used in brackish water as a groundwater tracer. Other dyes, like congo red and methylene blue are also used commonly as groundwater tracers. The amount of dye required for determining the groundwater velocity depends on the permeability and thickness of the aquifer as well as the distance between the injection and observation wells (Table 10.2).

Salt solutions (Table 10.3) are also used as tracers to explore the role of fractures on groundwater recharge

Table 10.2 Dye quantities used in measuring groundwater velocity (in grammes per 10 m of flow path). (After UNESCO 1972)

Dye	Type of water-bearing formation			
	Clayey	Sandy	Fractured	Karstic
Fluorescein	5–20	2–10	2–20	2–10
Congo red	20–80	20–60	20–80	20–80
Methylene Blue	20–80	20–60	20–80	20–80

and flow (Rugh and Burbey 2008). Water soluble salts like chlorides, bromides and sulphates, can be detected chemically or by measuring electrical conductivity of groundwater using especially designed electrodes. Approximate doses of the salt tracers used for injection and recommended distances between injection and observation wells are given in Table 10.3.

10.2.2   Organic Compounds

Chlorofluorocarbons (CFCs), also known as freons, are very stable synthetic organic compounds containing chlorine and fluorine. They are used in refrigeration, air conditioning and as cleaning agents and solvents. Their concentration in air since 1950 is well known.

Although, CFCs are not environmental isotopes, they are of great value in groundwater studies as tracers and for estimating rates of recharge to shallow aquifers. They are also useful for dating groundwater of recent origin (younger than 45 years) except in urban and industrialized areas, due to the release of CFCs from man-made sources locally (Busenberg and Plummer 1992). Proper care should be taken in the collection of water samples and analysis to assure that the samples are completely isolated from the air. Further, CFC concentrations in anaerobic environments may be reduced by microbial degradation. Sorption will also effect CFC concentrations (Domenico and Schwartz 1998). It is better to use CFCs with other environmental isotopes like ^{3}H and ^{15}N to study a particular hydrogeological problem (Long 1995).

Since 1990, the use of CFCs has been restricted as they deplete the ozone layer and contribute significantly to the greenhouse warming. As a result of such restrictions, the concentration of CFCs has been declining in the air. Therefore, CFC dating of modern groundwater will become less precise in future.

Sulfur hexafluoride (SF_6) is another organic compound which is used as a conservative, artificial, nontoxic gaseous tracer in many groundwater studies. Environmental SF_6 from atmospheric sources is also used as a dating tool for young groundwaters. The dating range of SF_6 is for water recharged in 1970 to present. However, non-atmospheric (terrigenic) SF_6 derived from rock weathering can cause significant young bias to the SF_6 groundwater age.

10.3 Tracer Injection Tests

Tracer tests are a valuable tool for determining flow characteristics viz. groundwater flow direction, flow rates, solute transport properties, and basin boundaries, especially in heterogeneous aquifer rocks because the interpretations of these tests does not assume the continuum approach. These tests are therefore also helpful in establishing interconnection of fractures and flow continuity between both the injection and observation wells. A variety of tracers, non-radioactive and radioactive can be used (Sect. 10.2). Non-sorbing tracers viz. ^3H, ^{131}I, and uranine are often used as they follow the flow of water and do not interact with the rock material. Moderately sorbing radionuclides viz. Cs, Sr, and Rb are used to study the transport properties viz. dispersivity and adsorption studies which are essential for the site characterisation (SC) and performance assessment (PA) of radioactive waste disposal sites (Karlsson 1989; Hodgkinson et al. 2009) (see Sect. 12.3.3). In addition to selecting a suitable tracer, the tracer injection procedure is also important. Injection can be at the well head (open hole testing) or in selected zones isolated by packers by down-hole injectors. In low permeability fractured rocks, there are problems in collecting water samples representative of in situ conditions. In such cases it is necessary to purge the borewell water by pumping for a period of time before collecting the sample (NRC 1996).

10.3.1 Groundwater Velocity Method

This procedure, also known as natural gradient test, involves injection of a tracer in a well and its time of arrival is noted in a grid of wells located down gradient from the injection point maintaining natural gradient. The average interstitial velocity, $\overline{V}$ is given by:

$$\overline{V} = \frac{K}{\eta_e}\frac{h}{L} \tag{10.1}$$

where K is the hydraulic conductivity, h is the difference in height between water-levels in the injection and observation wells, η_e is effective porosity and L is the distance between the observation wells. $\overline{V}$ can also be obtained from

$$\overline{V} = \frac{L}{t} \tag{10.2}$$

where t is the time of travel of tracer over the distance involved, L. Hence

$$K = \frac{\eta_e L^2}{ht} \tag{10.3}$$

These tests have an advantage as they do not disturb the natural flow system and give realistic values of aquifer characteristics. However, the disadvantage is that such tests require a large number of observation wells to monitor the tracer plume, due to uncertainties about the flow direction. Further, water samples must be collected from different depths in each monitoring well to define the concentration distribution in three dimensions (NRC 1996). Hence these tests are more expensive and require more time.

10.3.2 Point Dilution Method

Point dilution method is used to determine horizontal average linear velocity of groundwater in the vicinity of a borehole. The method is based on the assumption that the aquifer is homogeneous and isotropic and the flow is horizontal. Packers are used in the well screen to isolate the portion of the aquifer to be tested. In fractured media also, packers can be used to isolate individual fracture zones. A tracer, viz. ^3H, ^{131}I or common salt (NaCl) is introduced into the isolated part of the well segment which gets diluted with time due to mixing with groundwater. Measurements of tracer concentration with time are made. Analysis of the resulting dilution curve gives the groundwater velocity.

The average bulk velocity across the centre of the well, ($\overline{V}^*$) is computed by using Eq. 10.4.

$$\overline{V}^* = I - \frac{W}{At}ln\left(\frac{C}{C_0}\right) \tag{10.4}$$

where W is the volume of the well segment where tracer is released, A is the vertical cross sectional area through the centre of the isolated segment of the well screen, C_0 is the initial concentration of the tracer at t=0, and C is the concentration of tracer in water at time, t.

The average linear velocity of groundwater in the aquifer $\overline{V}$ is obtained by the relation.

$$\overline{V} = \frac{\overline{V}^*}{\eta_e \alpha} \qquad (10.5)$$

where α is a factor that depends on the well and aquifer characteristics. The usual range of α for tests in sand or gravel formations is from 0.5 to 4 (Freeze and Cherry 1979). After $\overline{V}$ is known and hydraulic gradient (I) is determined from field observations; K can be calculated from Eq. 10.6 using Darcy's law:

$$K = \frac{\eta_e \overline{V}}{I} \qquad (10.6)$$

In fractured media, the average bulk velocity, $\overline{V}$, is given by Eq. 10.7 (Louis 1977).

$$\overline{V} = \frac{\pi r_0 L}{4\pi Na} ln \frac{C/C_0}{t} \qquad (10.7)$$

where r_0=radius of the test section, L=length of the test section, N=number of conducting fractures, a=average aperture of conducting fractures; C, C_0 and t are as defined earlier.

The average hydraulic conductivity of the fractures ($\overline{K}_f$) intersecting the tested section is obtained from Eq. 10.8.

$$\overline{K}_f = \frac{\overline{V}}{I} \qquad (10.8)$$

Radon (^{222}Rn) is also used as a natural version of a point dilution method for estimating the groundwater flow rates (Cook 2003) by using the ratio of radon concentrations in an undisturbed well to that in the groundwater after purging. As the half-life of Rn is small (3.8 days), if the water in the well is stagnant all the radon might decay. On the other hand, significant amount of Rn in well-water will indicate greater flow into the well. For example, this method has been used to estimate the vertical variation in groundwater flow rates and thereby identifying preferential flow zones in a fractured rock aquifer in Australia (Cook 2003).

10.3.3   Well Dilution Method

Unlike point dilution tests, well dilution tests do not require use of packers to isolate parts of a well. In this method, measurements of tracer concentration are made over the length of a well with a view to know the vertical variation in flow rates. In a well dilution test, a tracer is added to the well and changes in its concentration with depth are measured which indicate the variation in flow inputs from the aquifer into the well. The well is also not pumped so that natural groundwater flow rates are maintained. Figure 10.1 gives the results of a well dilution test using saline water as a tracer in a borehole drilled through basalts underlain by metamorphic rocks in Atherton Tablelands, Queensland, Australia. The figure shows that within basalt (up to a depth of 18 m), the specific conductivity decreases with time but in the underlying metamorphics no such change is manifested. Groundwater flow rates were estimated from the rate of decrease of tracer concentration at different depths.

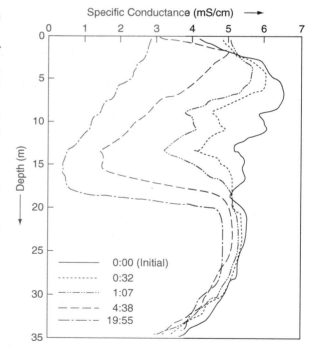

Fig. 10.1 Results of well dilution test on a borehole from the Atherton Tablelands, Australia. Depths are shown as metres below the water table. Times are in hours and minutes since the first profiling after injection. (After Cook 2003)

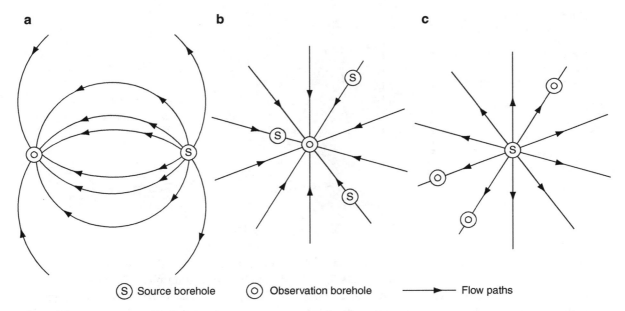

Fig. 10.2 Flow field geometries in **a** injection withdrawal method, **b** radial convergent method and **c** radial divergent method

10.3.4 Forced Gradient Tracer Tests (FGTTs)

In a forced gradient (induced hydraulic) test, the tracer is added to the recharge well as a pulse or step increase in concentration. Three flow-field geometries are commonly used (1) injection–withdrawal, (2) radial convergent, and (3) radial divergent (Fig. 10.2). However, the flow paths are likely to be highly complex in fractured rocks due to their heterogeneous character. In an injection–withdrawal system, a tracer is added into the injection well and its concentration is monitored at the pumping well. In a radial-convergent flow tracer test, water is pumped from a well. Later, a tracer is added into adjacent injection wells and the arrival of tracer is monitored in the pumping well. In a radial-divergent format, a tracer is injected into a single well either as a pulse or by a step increase and tracer concentration is passively monitored at one or more observation wells down gradient of the injection well (Novakowski et al. 2007).

The plot of a tracer concentration ratio vs. time (breakthrough curve, BTC) is used to estimate the fracture and matrix properties which control various flow mechanisms (viz. advection, dispersion and diffusion) of solute transport (see also Sect. 7.4). Breakthrough curves for some tracer tests in fractured rocks show multiple peaks which is attributed to flow through individual channels (NRC 1996). It is advantageous to use combination of two or more tracers with different diffusion coefficients. For example, in a radially convergent flow tracer test in a fractured chalk formation, four tracers (deuterium, uranine, iodine, and ^{13}C) were simultaneously injected in a well about 10 m away from the pumping well. The observed breakthrough curves at the pumping well are shown in Fig. 10.3. Each tracer curve is normalized by dividing the concentration by the injected mass so that the comparison becomes more objective. It may be noted that uranine, which has the largest molecular weight and thereby lowest coefficient of molecular diffusion, has a high peak while deuterium and iodine with higher coefficients of molecular diffusion show lower peaks due to their greater diffusion into the chalk matrix. The very low peak of ^{13}C suggests that it was lost by adsorption in the matrix (Maloszewski and Zuber 1992). From a similar study in highly fractured metasedimentary rocks, Sanford et al. (2002) observed that the breakthrough curve of helium was retarded relative to bromide due to greater aqueous diffusion coefficient of helium. By fitting the characteristics of the tracer BTCs, effective fracture aperture, fracture interconnectivity, and the travel time distances were determined (Weatherill et al. 2006).

There are several problems when using results obtained under a forced gradient tracer test to predict transport under natural conditions due to heterogeneous

Fig. 10.3 Breakthrough curves of four tracers during a radially converging flow tracer test performed by Garnier et al. (1985). (After Maloszewski and Zuber 1992)

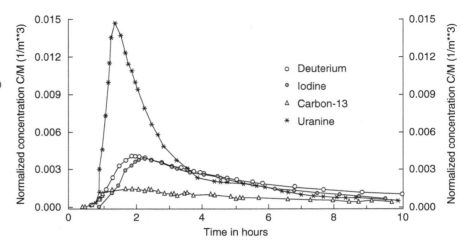

character of fractured rocks (Hsieh and Tiedeman 2001; Tiedeman and Hsieh 2004; Weatherill et al. 2006). The interpretation of forced gradient tracer can be made with a greater confidence if the fracture intersections are ascertained from cross-hole hydraulic tests (Novakowski et al. 2007).

A comparison of values of longitudinal dispersivity (α_L) obtained from radial-flow (natural gradient) tracer tests and two-well (forced-gradient) tracer test indicates that the values of α_L obtained from radial flow test are lower as compared with those obtained from forced gradient tracer tests (Hsieh and Tiedeman 2001). Simulation studies by Tiedeman and Hsieh (2004) further demonstrate that the longitudinal dispersivity (α_L) estimates from forced gradient tests depend on the distance between the tracer source well and pumping well and suggested that the distance between the two wells should be large enough to minimize the effect of high velocity pathways, otherwise such tests will underestimate dispersivity than those which exist under natural flow gradient on a regional scale. Inspite of these limitations, forced-gradient tracer tests are useful hydrogeological tools to study the fracture geometry, and interconnectivity and also solute transport characteristics in fractured rocks (Cook 2003; Novakowski et al. 2007).

10.4   Isotopes

Isotopes of a particular element have the same atomic number but different atomic weights, e.g. ^1H, ^2H, ^3H. The isotopes may be radioactive or stable. A host of radioisotopes produced naturally by the cosmic ray interaction with atmosphere get incorporated into the hydrological cycle. Added to the cosmogenically produced isotopes, there are isotopes introduced into the environment through testing of nuclear weapons and atomic installations. Together, these are called environmental isotopes. The isotopes of hydrogen and oxygen, viz. ^2H, ^3H and ^{18}O are natural tracers that tag water directly and therefore are of particular interest in hydrological investigations. Isotopes are used for a variety of purposes in hydrogeological investigations, viz. as tracers, for determination of age and origin of groundwater, estimation of recharge, hydraulic characteristics, and transport processes in aquifers (IAEA 1983, 1989; Clark and Fritz 1997; Kehew 2001; Aggarwal et al. 2005). A list of isotopes commonly used in hydrogeological studies is given in Table 10.4.

10.4.1   Stable Isotopes

Among stable isotopes, ^2H, and ^{18}O are used for determining the source of water and its genesis (viz. mixing of waters of different origin, salinisation processes etc.), whereas ^{13}C, ^{15}N, and ^{34}S are used in pollution studies.

A convenient way of expressing the ratio of two isotopes of an element is by the delta relation. For example, $\delta^{18}O$ is defined as

$$\delta^{18}O\%_{00} = \frac{(^{18}O/^{16}O)_{standard} - (^{18}O/^{16}O)_{standard}}{(^{18}O/^{16}O)_{standard}} \times 1000$$

$$(10.9)$$

Table 10.4 Isotopes of poetential use in hydrogeological studies. (Modified after Davis 1989)

Isotope	Half-life (years)	Examples of possible hydrogeological applications
^{2}H	*	Interpretation of palaeoclimates, recharge elevation, evaporation prior to recharge
^{3}H	12.3	Presence of water less than 40 years old, travel time
^{3}He	*	Presence of water of crustal origin
^{13}C	*	Source of carbon in groundwater
^{14}C	5730	Residence time of water in the 500–50 000 year range
^{15}N	*	Origin of nitrate, pollution studies
^{18}O	*	Interpretation of palaeoclimatic conditions during recharge, geothermal activity
^{32}Si	100	Possible use for residence time in the 50–500 year range
^{34}S	*	Natural tracer for SO_4 in water, index of biological activity, identifies sources of pollution
^{36}Cl	3.01×10^5	Residence time of water in the 5×10^4–2×10^6 year range
^{37}Cl	*	Possible natural tracer, origin of brine
^{39}Ar	269	Residence time of water in the 100–1 000 year range
^{40}Ar	*	Buildup of ^{40}Ar suggests either geothermal activity or very old water
^{81}Kr	2×10^5	Possible use for residence time in the 5×10^4–2×10^6 year range
^{85}Kr	10.4	Residence time of groundwater in the 1.0–20 year range
^{87}Sr	*	Tracer in water, residence time of marine brine in halite
^{129}I	1.7×10^7	Origin of water, residence time in the 5×10^6–4×10^7 year range
^{222}Rn	3.8 days	Presence of groundwater in surface water, groundwater tracer

* Stable isotopes

For ^{2}H and ^{18}O, the normal standard reference is SMOW, an acronym for Standard Mean Ocean Water.

The ratio of stable isotopes δ^2H and δ^{18}O is used for a variety of geohydrological problems based on fractionation depending on altitude, distance from the ocean and evaporation. It is therefore possible to identify the source of groundwater through stable isotope composition. For example groundwater which has the source of recharge at high altitudes, will be richer in lighter isotopes and may be distinguished from the groundwater recharged at lower altitudes. On the other hand, the contribution from surface water bodies like lakes, ponds and reservoirs, subjected to evaporation process, can be identified by enrichment in heavier stable isotopes.

Deuterium (^{2}H) and ^{18}O data from hot and cold springs help in determining the source of recharge, altritude of recharge, mixing of cold water, thermal diffusion and flow pattern. A similarity of δ^2H values between spring water and precipitation would indicate local recharge. An increase in δ^{18}O values in hot spring waters is attributed to water rock interaction (see Sect. 18.7). Some case histories from India are given by Sukhija et al. (2002).

Stable isotope ratio of carbon (^{13}C/^{12}C) is used to study the source of CO_2 which plays an important role in the dissolution of carbonate minerals. δ^{13}C is also helpful to determine the source of methane in groundwater (Kehew 2001).

^{15}N, particularly the ^{15}N/^{14}N ratio has been used for identifying sources of nitrate pollution in groundwater from agricultural fields and household septic systems. Its use so far has been limited because of analytical problems and poor understanding of ways to distinguish various sources of nitrates in natural waters. Preferably, in addition to ^{15}N, other tracers may also be used to give more dependable results. As an example, Barraclough et al. (1994) used NO_3 and ^{2}H as tracers for studying fracture flow mechanism in the British Chalk which helped in obtaining a better understanding of pollutants.

δ^{34}S is useful in determining the source of sulphur which could be from marine evaporite deposits, seawater or pyrite.

10.5 Radioactive Isotopes

The advantage of radioactive tracers is that they can be detected even in low concentrations in water. Radioactive isotopes viz. ^{3}H, ^{14}C and ^{36}Cl are commonly used for dating groundwater, and are also of importance for recharge estimation and groundwater management. The data on half-lives of some radioactive isotopes used in groundwater studies is given in Fig. 10.4. Repetitive sampling for radioactive isotope analysis can be used to more accurately estimate the groundwater ages and transit times (Bruce et al. 2007).

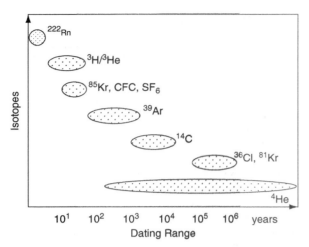

Fig. 10.4 Dating ranges of some environmental tracers. (after Loosli and Purtschert 2005)

10.5.1 Recharge Estimation

Tritium (^{3}H), produced in the upper atmosphere, enters the hydrological cycle in precipitation. The concentration of ^{3}H has varied from time to time, depending upon the detonation of the nuclear devices since 1952 by different countries. Tritium concentration is reported in terms of tritium units (TU). 1 TU corresponds to one atom of ^{3}H in 10^{18} atoms of ^{1}H. In precipitation, the tritium concentration is generally less than 20 TU. The tritium concentration reached peak values of around 1000 TU in 1963 in Canada due to nuclear testing (Schwartz and Zhang 2003).

^{3}H isotope infiltrates through the soil in the same manner as water does and therefore, its concentration in soil profile provides useful data regarding groundwater recharge. There are two methods of computing groundwater recharge by using tritium: (a) environmental tritium method, (peak and integral methods), and (b) injected tritium method. In the environmental tritium peak method, recharge is estimated from the equation

$$R_p = \frac{100 S_m}{p} \qquad (10.10)$$

where R_p is recharge from precipitations, S_m is the total soil moisture in the soil column, from the surface to the depth where the 1963 peak is located, and P is the total rainfall since 1963 to the time of investigation. In the integral method, the total amount of tritium in rain since 1952 to the time of investigation is determined and also the amount of tritium and moisture content present in the soil profile. Recharge as percentage of rainfall R_p, is calculated from the Eq. 10.11.

$$R_p = \frac{\sum a_x m_x}{\sum A_i P_i} \times 100 \qquad (10.11)$$

where a_x is tritium concentration (TU) of soil at depth x, m_x is moisture content of soil at depth x (in cm), A_i is tritium concentration (TU) in precipitation and P_i is precipitation (in cm) in different years since 1952.

Groundwater recharge can also be estimated by monitoring the vertical movement of injected tritium. This is known as Injected Tritium or Tritium Tagging method. It has been used quite widely in India for estimating groundwater recharge in sedimentary as well as hard (fractured) rock formations (see Sect. 20.2.3 and Table 20.1). This method is based on the assumption that the soil water in the unsaturated zone moves downward in discrete layers. Any fresh water added near the soil surface due to precipitation or irrigation will move downward by pushing the older water beneath and this in turn will push still older water further below, thereby ultimately the water from the unsaturated zone is added to groundwater reservoir. This flow mechanism is known as piston flow model. Therefore, if tritium (^{3}H) is injected at any particular level, the vertical movement of this tagged layer can be monitored by measuring the concentration of ^{3}H at different depths. The position of the tracer will be indicated by a peak or maximum in the tritium activity vs. depth. Molecular diffusion, dispersion, and other aquifer heterogeneities may cause broadening of the peak.

The average infiltration flux (q_i) can be estimated from Eq. 10.12.

$$q_i = \frac{\Delta z}{\Delta t} \theta_v \qquad (10.12)$$

where Δz is the depth to maximum tritium activity, Δt is the elapsed time between sampling and maximum historic atmospheric tritium activity, and θ_v is the volumetric soil–water content.

For example, in a part of the western Gangetic Plains, several paleochannels of the river Ganga are present (Fig. 10.5a). As the paleochannels form potential sites of artificial groundwater recharge, it is very

Fig. 10.5 Estimation of groundwater recharge by monitoring vertical movement of injected tritium. (**a**) Is an IRS-LISS-III image of a part of the western Gangetic plains showing several paleochannels amidst the alluvial plains. Injection of tritium was carried out at site *S1* in the paleochannel and at *S2* in the adjacent alluvial plain. Injection was uniformly made at 70 cm depth below ground level at pre-monsoon time. After monsoon, the tritium peak was found to have shifted by 160 cm in the paleochannel area (**b**), and by 40 cm in the alluvial plains (**c**). Salient hydrogeological characteristics and computed recharge rates of the paleochannel area and the alluvial plains are given in (**d**). (Courtesy: R.K. Samadder)

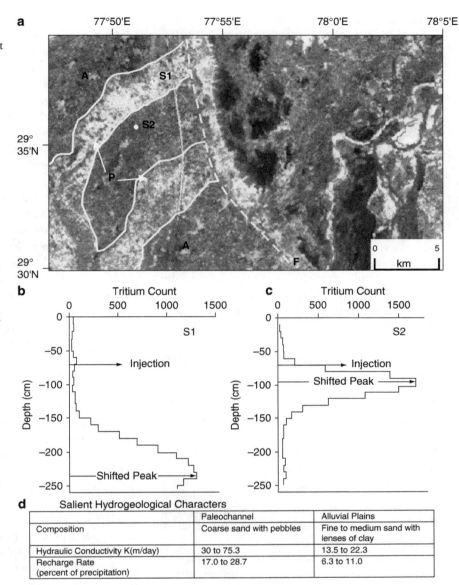

d Salient Hydrogeological Characters

	Paleochannel	Alluvial Plains
Composition	Coarse sand with pebbles	Fine to medium sand with lenses of clay
Hydraulic Conductivity K(m/day)	30 to 75.3	13.5 to 22.3
Recharge Rate (percent of precipitation)	17.0 to 28.7	6.3 to 11.0

important to estimate natural rate of groundwater recharge. The major source of recharge to groundwater in the area is precipitation, more than 85% of which occurs during monsoon period (June–September) only. Conventional methods for estimating groundwater recharge require large volume of hydrometeorological and hydrogeological data accumulated over a considerable time span, which is normally inadequately available, lacking or even unreliable in many cases (Jacobus et al. 2002; Scanlon et al. 2002; Mondal and Singh 2004; Chand et al. 2005). Therefore, tritium tagging method was used for soil moisture movement

analysis and estimating groundwater recharge for both the areas of paleochannels and the adjacent alluvial plains. Tritium injection was made immediately before the monsoon period (year 2006). The soil samples collected with depth before and after the monsoon yielded the positions of the original and shifted peaks of tritium (Fig. 10.5b, c). This has indicated a higher rate of recharge in the paleochannels (17.0–28.7%) as compared to the adjacent alluvial plains (6.3–11.0%).

Tritium studies are also useful for determining preferential flow of water through fractures and thereby to predict the movement of contaminants and evaluating potential waste disposal sites (Ahn 1988; Clark and Fritz 1997; Kehew 2001; Offerdinger et al. 2004). An excellent example of the importance of ^{3}H and stable isotopes (^{2}H and ^{18}O) along withdissolved ions (Cl^{-1}, HCO_3^{-1}) for studying flow mechanism in the crystalline rocks of the Mont Blanc tunnel in Central Austrian Alps is given by Fontes et al. (in Mazor 1991). This study demonstrated the preferential flow of groundwater through vertical fractures. A similar study in the crystalline rocks of Seoul, Korea, indicated that the main recharge to crystalline rock aquifers is by downward percolation from the overlying alluvial and unconsolidated deposits (Ahn 1988).

Tritium is also used to determine the suitability of sites for underground disposal of radioactive waste as presence of tritium above 1 TU in deep brines would signify hydraulic connection with the surface indicating thereby non suitability of the repository.

^{14}C has also been used for the estimation of groundwater recharge from equation

$$Q = \eta_e H/t \; ln \; H/h \qquad (10.13)$$

where

Q = recharge rate(mm year^{-1})
η_e = effective porosity of the saturated zone in the aquifer
H = total thickness of the aquifer (m)
h = saturated thickness of the aquifer (m)
t = ^{14}C age of water (years)

This method has been used successfully in several groundwater recharge studies in Southern Africa (viz. Sibanda et al. 2009).

As compared to ^{14}C, ^{3}H has the advantage that chemical processes in groundwater do not affect it. However, a relative disadvantage with ^{3}H is comparatively short half-life which allows dating up to about 50 years only.

10.5.2 Age Dating

Isotopes such as ^{3}H, ^{3}He, and ^{222}Rn, having shorter half-life are used for dating groundwaters of up to about 50 years (Fig. 10.4). As ^{3}He is a decay product of tritium and can also be generated from other sources viz. mantle derived rocks such as basalts, ^{3}He ages should be interpreted with care (Heilweil et al. 2009).

^{222}Rn is a naturally occurring radioactive decay product of ^{226}Ra which is present in rocks and soils. ^{222}Rn is a colourless, odourless radioactive gas that is soluble in water. It has a half-life of 3.8 days and is therefore typically found in higher concentrations in groundwater than in surface water. Therefore higher concentrations of ^{222}Rn are expected at places where groundwater discharges into the stream (Cook et al. 2006). As radon concentration in soil gas is related to CO_2, it is also used for preliminary assessment of the permeability of geothermal fields. ^{222}Rn has also been considered as a precursor for earthquake prediction (e.g. Kuo et al. 2006).

^{14}C is used for the age determination of comparatively older (upto 30000 years) waters. Lloyd and Heathcote (1985) have given several interesting case histories of the use of ^{14}C for dating of groundwaters. Karlsson (1989) has shown the usefulness of ^{14}C and ^{3}H in the study of residence time of groundwater at greater depths in the crystalline rocks of Sweden. As the origin of groundwater is also related to climatic cycles, the presence of isotopically different water can be attributed to recharge in pluvial intervals during glacial epochs (Payne 1967). For example, ^{14}C dating of deep well-waters from some arid zones in United Arab Republic and Saudi Arabia, gave ages of 20000–30000 years which shows that the waters were recharged during the Wisconsin ice age when the area had wet climate (Todd 1980). Studies in the arid tracts of western Rajasthan, India also show that the shallow groundwater is recharged from river channels and flash floods but the groundwater in deeper aquifers has negligible tritium and depleted ^{18}O indicating that it was recharged during pluvial times in the past (Navada et al. 1993). Radioisotope (^{3}H and ^{14}C) measurements in groundwaters from the fractured crystalline rocks of Vedavati basin in south India indicate that shallow groundwater is quite young but the deeper groundwater is old (200–4000 years) indicating lack of interconnection between fractures (Sukhija and Rao 1983). Stable and radioactive isotope (^{18}O, ^{14}C and ^{36}Cl) studies in the groundwater of fractured granite in central India also indicate groundwater ages ranging from about 50 to 5000 years (Sukhija et al. 2002). Similar results have been obtained from studies in the crystalline rocks of the Bohemian Massif, Czech Republic (Silar 1996).

Michael and Voss (2009) have quoted ^{14}C ages in the range of 5500–21000 years estimated by IAEA from the deep alluvial aquifers (100–300 m) in Bangladesh while the shallow aquifers have groundwater of less than 100 years as determined from He/^{3}H ratios.

^{36}Cl (half-life, 301 000 years) is suitable for dating much older groundwaters within a range of 5×10^4–1×10^6 years, which is beyond the normal range of dating. It is chemically stable and is not adsorbed on soil particle surfaces. In addition to dating of groundwater, ^{36}Cl has also been used for recharge studies in the vadose zone especially in arid and semi-arid regions (Long 1995). However, as in deep groundwaters, there may also be other sources of Cl, due to either dissolution of evaporite deposits or membrane filtration; therefore, necessary correction in the observed concentration of ^{36}Cl is required for dating (Long 1995). If Cl-bearing minerals are present, apparently too low ^{36}Cl groundwater ages may be obtained (Kresic 2007).

The first use of ^{36}Cl dating was made in the Jurassic aquifer of the Australian Great Artesian Basin where satisfactory agreement was obtained between ages computed from hydrodynamic simulation and ^{36}Cl ages. In the Great Artesian Basin, ^{36}Cl groundwater ages ranging from less than 100 000 to over 1 000 000 years have been obtained (Bentley et al. 1986).

^{36}Cl along with other isotope ratios (^{15}N, ^{87}Sr, ^{11}B) and hydrochemical data is also being increasingly for knowing the origin of salinity in crystalline basement rocks in shield areas (Edmunds 2005) (also see Sect. 13.7.2).

^{81}Kr, which has a similar half-life as ^{36}Cl (Table 10.4), gave an age of one million years for the groundwater in Sahara (Glynn and Plummer 2005). The age estimation of old waters is also helpful in finding suitable sites for high level radioactive waste disposal (see Sect. 13.7.2).

introduced artificially into the aquifer for a variety of hydrogeological investigations such as rate and direction of groundwater movement, groundwater recharge, its residence time, interaction between surface water and groundwater, and solute transport etc. Isotopes are found to be particularly useful in such studies especially in fractured rocks where conventional methods applied to porous media may not provide actual flow and transport characteristics due to formation heterogeneities. The forced gradient tracer tests (FGTTs) which include injection–withdrawal, radial convergent, and radial divergent methods yield data on hydraulic characteristics of fractures and matrix blocks. It is advantageous to use a combination of two or more tracers with different diffusion characteristics for solute transport studies.

Isotopes may be radioactive or stable. They are increasingly being used for a variety of hydrogeological studies, viz. as tracers, estimation of recharge, age and origin of groundwater, including flow and transport characteristics of aquifers. The commonly used isotopes for these purposes are ^{2}H, ^{18}O, ^{15}N, ^{3}H, ^{222}Rn, ^{14}C, ^{36}Cl and ^{131}I.

Summary

Tracers are chemical substances (inorganic viz. common salt and organic viz. CFCs and SF$_6$) including isotopes present either naturally or

Further Reading

Aggarwal PK et al. (eds) (2005) Isotopes in the Water Cycle. Past, Present and Future of a Developing Science. Springer, Dordrecht.

Cook PG (2003) A Guide to Regional Groundwater Flow in Fractured Rock Aquifers. CSIRO, Australia.

Domenico PA, Schwartz FW (1998) Physical and Chemical Hydrogeology. 2nd ed., John Wiley & Sons Inc., New York.

Kehew AE (2001) Applied Chemical Hydrogeology. Prentice Hall, New Jersey.

Mazor E (2003) Chemical and Isotopic Groundwater Hydrology. 3rd ed., Mercel Dekkar, Inc., NY.

11.1 Introduction

The quality of water is as important as its available quantity. Rain and snow are the purest form of water which undergo many complex chemical changes after coming in contact with soil and other rock materials. Man's activities also have a considerable influence on water quality.

The problem of groundwater contamination is quite widespread especially in developing countries due to lack of proper sanitary conditions and piped water supply. In these countries, a large number of people still use water from shallow dugwells or ponds and rivers. The WHO estimated that in 1980, about 1320 million (57%) of the developing World (excluding China) were without a clean water supply, while 1730 million (75%) were without adequate sanitation. Keeping this in view, the United Nations declared the decade 1981–1990 as the International Decade of Drinking Water and Sanitation. In India also, the Federal Government has launched a programme of Drinking Water Mission to provide safe drinking water.

11.2 Expressing Water Analysis Data

The concentration of the dissolved inorganic constituents in water is expressed in ionic form, while those salts which occur in undissociated or colloidal form, are reported as oxides or as an uncombined element. For example, the sodium concentration is expressed as cation Na^+, iron is reported as the element Fe and, silicon as oxide SiO_2.

The concentrations of dissolved salts or ions in groundwater are usually expressed in parts per million (ppm) by weight. Parts per thousand unit is commonly used in reporting the composition of sea-water. Parts per billion (ppb) or parts per trillion (ppt) are used in reporting concentration of trace elements. In weight per volume units, the concentration of ions is expressed in milligrams per litre ($mg\,l^{-1}$). The ppm and $mg\,l^{-1}$ units are numerically almost the same, if the density of water is nearly 1.0, and the concentration of dissolved solids is less than 7000 $mg\,l^{-1}$. In highly mineralized water, the conversion can be made by Eq. 11.1.

$$Parts\ per\ million\ (ppm) = \frac{Milligram\ per\ litre}{Specific\ gravity\ of\ the\ water} \qquad (11.1)$$

A unit which is more convenient for geochemical studies, is equivalent per million (epm) or milligram equivalents per litre ($meq\,l^{-1}$). The concentration of an ion in $meq\,l^{-1}$ can be determined by multiplying $mg\,l^{-1}$ value with the reciprocal of its combining (equivalent) weight. The concentration of an ion in $meq\,l^{-1}$ is designated by putting symbol gamma (γ) before the chemical symbol, viz. γCa. This unit gives a better idea of the chemical character of water and is also of help in ascertaining the accuracy and completeness of chemical analysis.

The accuracy of chemical analysis of water samples can be checked by calculating the cation–anion balance, as the sum of the major cations should be equal to the sum of major anions expressed in $meq\,l^{-1}$. If it is not so, the analysis is either erroneous or incomplete. The percentage error i.e. 'ion-balance error' (e), can be determined by Eq. 11.2 (Matthess 1982).

$$e = \frac{\Sigma yc - \Sigma ya}{\Sigma yc + \Sigma ya} \qquad (11.2)$$

B. B. S. Singhal, R. P. Gupta, *Applied Hydrogeology of Fractured Rocks*,
DOI 10.1007/978-90-481-8799-7_11, © Springer Science+Business Media B.V. 2010

where γc represents the cation sum and γa the anion sum in meq l^{-1}. In general the value of e should be less than 5% and certainly less than 10%. In this procedure, colloidal or suspended matter such as Al_2O_3 and SiO_2 is not included.

11.3 Isotopic Composition

Natural water contains mainly hydrogen (H) of mass 1(^{1}H) and oxygen (O) of mass 16 (^{16}O). In addition, it also contains small amounts of deuterium (^{2}H), tritium (^{3}H), and isotopes of oxygen (^{17}O and ^{18}O). The relative abundance of these isotopes is given in Table 11.1. Of the six isotopes of hydrogen and oxygen (^{1}H, ^{2}H, ^{3}H, ^{16}O, ^{17}O and ^{18}O), five are stable, while tritium (^{3}H) is radioactive with half-life of 12.3 years. The var-

ious isotopes of hydrogen and oxygen combine to form 18 types of water molecules. The common water molecules are $^1H_2{}^{16}O$, $^1H_2{}^{17}O$, $^3H_2{}^{16}O$, $^1H_2{}^{17}O$ and $^1H_2{}^{18}O$. Water with isotopic composition $^1H_2{}^{16}O$ and molecular weight of 18 forms 99.8% of the total water on the earth while the proportion of heavy water $^2H_2{}^{16}O$ is about only 0.2% (Matthess 1982). Isotopic analysis of water is of importance in hydrology as it can provide data on the history and origin of water (see Chap. 10).

11.4 Dissolved Constituents

The relative abundance of various elements in groundwater mainly depends upon their chemical mobility. The mobility of an element in the hydrosphere is determined by the solubility of its various compounds, the tendency of the ion towards adsorption and base-exchange, and the degree to which it is bound in biosphere. Behaviour of various elements from the point of view of their hydrochemical mobility is given in Table 11.2 where a comparison of the distribution of common elements in igneous and sedimentary rocks is made with their distribution in fresh groundwater and sea-water.

Silicon, aluminum and iron which are most abundant in igneous rocks have low mobility in the hydrosphere. Chlorine which is relatively scarce in the

Table 11.1 Relative abundance of hydrogen and oxygen isotopes in natural waters. (After UNESCO 1984b)

Isotope	Relative abundance	Half-life
^{1}H	99.985	Stable
^{2}H	0.015	Stable
^{3}H	Trace	12.3 year
^{16}O	99.76	Stable
^{17}O	0.04	Stable
^{18}O	0.20	Stable

Table 11.2 Average chemical composition (mg l^{-1}) of three major rock types, sea water and groundwater. Relative mobility of common elements is also indicated. (After Chebotarev 1950; Hem 1989)

Constituent	Igneous rocks	Sandstone	Carbonates	Average sea water	Groundwater	Relative mobility
SiO_2	285 000	359 000	34	6.4	1–30	0.20
Al	79 500	32 000	8970	<2.0	1–2	–
Fe	42 200	18 600	8190	0.01	0–5	–
Ca	36 200	22 400	272 000	400	10–200	3.00
Mg	17 600	8100	45 300	1350	1–100	1.30
Na	28 100	3870	393	10 500	1–300	2.40
K	25 700	13 200	2390	380	1–20	1.25
Sr	368	28	617	8	<10	–
HCO_3	–	–	–	142	80–400	–
SO_4	–	–	–	2700	10–100	57.00
Cl	200	150	–	19 000	1–150	100.0
F	715	220	112	1.3	0.1–2	–
Br	2.4	1.0	6.6	65	<0.5	–
B	7.5	90	16	4.6	<2	–
TDS	–	–	–	35 000	100–1000	–

Table 11.3 Sources of ions in water

	Source
Major ions ($>1\,mg\,l^{-1}$)	
Calcium Ca^{2+}	Carbonates, gypsum
Magnesium Mg^{2+}	Olivine, pyroxene, amphiboles
Sodium Na^+	Clays, feldspars, evaporites, industrial waste
Potassium K^+	Feldspar, fertilizers, K-evaporites
Bicarbonate HCO_3^-	Soil and atmospheric CO_2, carbonates
Chloride Cl^-	Windborne, rain water, sea water and natural brines, evaporite deposits; pollution
Sulphate SO_4^{2-}	Gypsum and anhydrite, sea water, windborne, oxidation of pyrite
Nitrate NO_3^-	Windborne, oxidation of ammonia or organic nitrogen, contamination
Silica SiO_2	Hydrolysis of silicates
Minor ions (1–$0.1\,mg\,l^{-1}$)	
Iron Fe^{2+}	Oxides and sulphides, e.g. hematite and pyrite; corrosion of iron pipes
Manganese Mn^{2+}	Oxides and hydroxides
Boron B	Tourmaline, evaporites, sewage, sea water
Fluoride F^-	Fluorine-bearing minerals, viz. fluorite, biotite
Trace elements ($<0.1\,mg\,l^{-1}$)	
As	Arsenic minerals, e.g. arsenopyrite, arsenic insecticides
I	Marine vegetation, evaporites
Zn	Sphalerite, industrial waste
Heavy metals (Hg, Pb, Cd, Cr)	Industrial waste and igneous rock weathering, under mild reducing conditions
Radioactive elements e.g. U, Ra, etc.	Uraniferous minerals, nuclear tests and nuclear power plants

Earth's crust is very mobile and widespread in the hydrosphere. Sodium is more mobile than potassium although both occur in almost equal amounts in the igneous rocks (Table 11.2). The various chemical constituents, based on their concentration in water, can be classified as: (1) Major constituents ($>1\,mg\,l^{-1}$) viz. Ca, Mg, Na, K, HCO_3, SO_4, Cl, NO_3 and SiO_2; (2) minor constituents (1–$0.1\,mg\,l^{-1}$) viz., Fe, Mn, F, B; and (3) trace elements ($<0.1\,mg\,l^{-1}$), viz. Al, As, Hg, Pb, Cr, Zn and other heavy metals. The other constituents are pesticides, and radioactive materials. The sources of various ions are also given in Table 11.3.

11.5 Graphical Presentation of Chemical Data

The water analysis data, for their proper evaluation can be plotted in: (a) Hydrochemical maps, (b) Hydrochemical diagrams, and (c) Hydrochemical sections (Zaporozec 1972; Freeze and Cherry 1979; Matthess 1982; Lloyd and Heathcote 1985; and Kehew 2001. A brief description of these is given below.

11.5.1 Hydrochemical (Isocone) Maps

In hydrochemical maps, the spatial variation in the hydrochemical characteristics of groundwater can be depicted by plotting contours (isocones) of either equal concentration of each ion or of equal ionic ratios in $mg\,l^{-1}$ or $meq\,l^{-1}$. Information about the lithology and structural features of rocks as well as the groundwater regime is necessary for proper interpretation of hydrochemical maps. Isochlor maps, depicting distribution of chloride ion, are useful to determine the direction of groundwater flow as normally the chloride concentration increases in the flow direction. Isochlor and isoconductivity maps are also useful in the study of sea-water intrusion in coastal aquifers (see Sect. 20.7.2).

Equal ratio maps provide useful information about the relative dissolution of aquifer material, ion exchange and the mixing of waters of different sources. Schoeller (1959) suggested equal ratio contour maps for $\gamma SO_4/\gamma Cl$, $\gamma Mg/\gamma Ca$ and other ionic combinations, where concentration of individual ions is expressed in $meq\,l^{-1}$ (Fig. 11.1). The ratio of $\gamma Mg/\gamma Ca$ in water is particularly useful in the study of

Fig. 11.1 Equal ratio isocone maps of the Mornag alluvial aquifer, Tunisia, **a** map of $\gamma SO_4^{2-}/\gamma Cl^{-1}$ ratios, **b** map of $\gamma Mg^{2+}/\gamma Ca^{2+}$ ratios. Piezometric contours are also shown. (After Schoeller 1959)

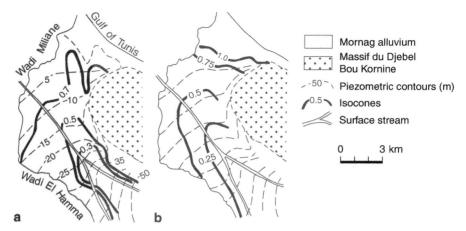

water chemistry from carbonate rocks (see Sect. 15.6). In limestone and chalk aquifers $\gamma Mg/\gamma Ca$ ratio is usually in 0.5–0.7 range and in dolomite aquifers it is in 0.7–1.1 range. In the olivine-bearing (Mg-rich) silicate rocks, such as basalts, $\gamma Mg/\gamma Ca$ ratio in groundwater is grater than 1. The $\gamma Cl/\gamma HCO_3$ and $\gamma SO_4/\gamma Cl$ ratios have been used as indicators of sea-water intrusion in coastal aquifers. The $\gamma HCO_3/\gamma Cl$ ratio is useful in determining the direction of groundwater movement and distance of a given water sample from its initial recharge area as this ratio will decrease in the direction of flow due to greater dissolution of Cl. Figure 11.2 illustrates an example of increase in TDS and SO_4 and decrease in HCO_3/Cl ratio in the direction of groundwater flow. The ratio of sodium to total cations is used in studying the effect of ion exchange in sediments

having cation exchange properties. The ratio of Br/Cl, I/Cl and B/Cl are recommended for determining the origin of deep oil field waters (Zaporozec 1972). Br/Cl ratio is also a reliable parameter for identification of salinization sources (Fontes et al. 1989; Vengosh and Rosenthal 1994). GIS methods using colour coding can be used to show the distribution of various ions (see Sect. 6.3.4).

11.5.2 Hydrochemical Pattern Diagrams

The purpose of hydrochemical pattern diagrams is to depict the absolute or relative concentration of different cations and anions in terms of either $mg\,l^{-1}$ or

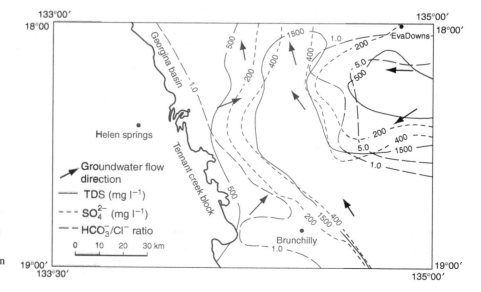

Fig. 11.2 Isocones of TDS, SO_4^{2-} and HCO_3^-/Cl^- in the Palaeozoic limestone aquifer of Georgina Basin, Northern Territory, Australia. (Redrawn after Verma and Jolly 1992)

Table 11.4 Chemical analysis data of two representative groundwater samples. (After Hem 1989)

Chemical	Sample 1			Sample 2		
Constituent	mg l^{-1}	meq l^{-1}	meq l^{-1} (%)	mg l^{-1}	meq l^{-1}	meq l^{-1} (%)
Ca^{2+}	37	1.85	6.0	49	2.45	21.8
Mg^{2+}	24	1.97	6.5	18	1.48	13.2
$Na^+ + K^+$	611	26.58	87.5	168	7.29	65.0
HCO_3^-	429	7.03	23.2	202	3.31	29.6
SO_4^{2-}	1010	21.03	69.2	44	0.92	8.2
Cl^-	82	2.31	7.6	246	6.94	62.2
TDS	1980	–	–	649	–	–

meq l^{-1}. These include Collins bar diagram, Stiff diagram, radial diagram and circular diagram. Different shapes and patterns of these diagrams/plots show variation in water quality.

The chemical analysis data of two water samples as given in Table 11.4 have been plotted in various hydrochemical diagrams, e.g. Figs. 11.3, 11.4, 11.5, 11.6 and 11.8).

Collins' Bar Diagrams *These* are vertical bar diagrams. Each sample is represented by two bars, one for cations and the other for anions (Fig. 11.3). The height of each bar is proportional to the total concentration of cations or anions in meq l^{-1}. As the sum of the cations should be equal to the sum of the anions, both expressed in meq l^{-1}, the height of the two bars for each sample should be the same. The concentration of cations and anions can be plotted either in absolute values (Fig. 11.3) or as the percentage of the total epm.

Stiff's Diagrams In Stiff's diagram, the analytical data is plotted on three or four horizontal parallel axes equidistant from each other (Fig. 11.4). These diagrams like other pattern diagrams have a disadvantage as separate diagrams are to be prepared for each analysis. However, they are useful in visualising the differences in the distribution of cations and anions based on their varying patterns.

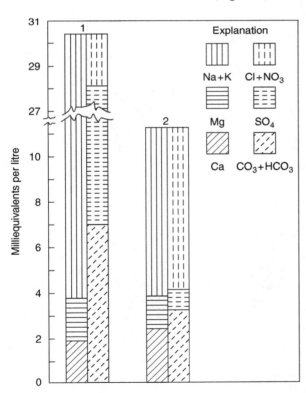

Fig. 11.3 Representation of chemical analysis data (meq l^{-1}) by Collins bar diagrams. (After Hem 1989)

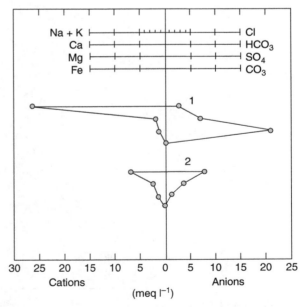

Fig. 11.4 Representation of chemical analysis data (meq l^{-1}) by Stiff diagram. (After Hem 1989)

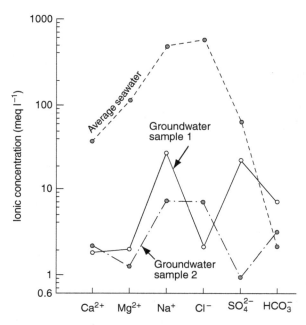

Fig.11.5 Chemical analyses (meq l^{-1}) represented by Schoeller's semilogarithmic plotting

Schoeller's Diagram A semi-logarithmic plot, in which on the abscissa (on airthmetic scale), the various cations and anions are arranged in the order—Ca, Mg, Na, Cl, SO$_4$ and HCO$_3$, from left to right at equidistance. The concentration of each of these ions in mg l^{-1} or meq l^{-1} is plotted along the ordinate in logarithmic scale. The plots are joined by straight lines (Fig. 11.5). These diagrams have an advantage over the other diagrams described earlier, as more than one analysis can be plotted in the same diagram and hydrochemical characteristics of different samples can be compared. It is also possible to compare the ratios of the various elements from the slope of the lines joining the plots of the adjacent cations or anions, viz. γMg/γCa, γSO$_4$/γCl and others (Schoeller 1959). Waters of similar composition will plot as near-parallel lines.

Hill–Piper Diagram The Hill–Piper tri-linear diagram, which is used extensively, was first conceived by Hill and later improved by Piper (1953). It has two triangular fields, one for the cations and the other for anions, and a central diamond shaped field (Fig. 11.6).

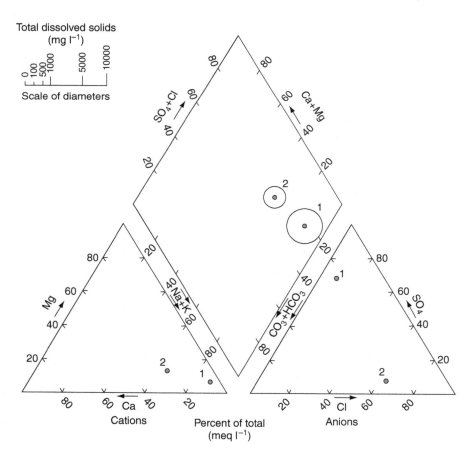

Fig.11.6 Hill–Piper diagram showing plotting of analytical data. (After Hem 1989)

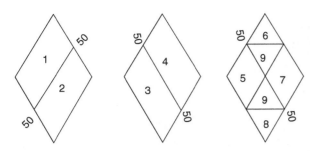

Fig. 11.7 Subdivisions of the diamond shaped field of Hill–Piper diagram showing various classes of water

In the two triangular fields, the concentration of cations and anions is plotted as percentages in meq l^{-1} so that the total of cations (Ca^{2+}, Mg^{2+}, and $Na^+ + K^+$) and anions (Cl^-, SO_4^{2-} and $CO_3^{2-} + HCO_3^-$) are each considered as 100%. The respective cation and anion plots are projected into the central diamond shaped field showing the overall chemical characteristics of the water sample in terms of certain combination of cations and anions. The total salinity can be depicted by drawing circles with radii in proportion to the logarithmic values of total dissolved solids (Fig. 11.6).

The Hill–Piper system has an advantage of depicting analytical data of a large number of samples in one diagram. It is also used for the classification of water samples into various hydrochemical types depending on the relative concentration of major cations and anions (Fig. 11.7, Table 11.5). It also has an advantage of demonstrating changes in water quality due to mixing of water of two extreme types, viz. mixing of fresh groundwater and seawater in coastal aquifers and also effect of base-exchange. One of the main disadvantages of Hill–Piper diagram is that it shows the relative concentration of different ions and not their absolute concentration.

Durov's Diagram This is also a type of tri-linear diagram suggested by Durov (Zaporozec 1972; Lloyd and Heathcote 1985). In this diagram, the concentration of major cations and anions in percentage meq l^{-1} is plotted in two separate triangles. The sample points on the two triangles are projected to a central square field which represents the overall chemical character of the sample. The concentration of any other chemical characteristics, viz. TDS, EC, pH etc. can also be shown by extending the point from the central square field to the adjacent one or two scaled rectangular fields as shown in Fig. 11.8. Durov's diagram can also be used for plotting the concentration of minor ions. Burdon and Mazloum (1961) and Lloyd and Heatheote (1985) have given expanded versions of the Durov's diagram which have an advantage over the Hill–Piper diagram in that they provide a better display of different types of water and also some hydrochemical processes such as ion exchange, simple dissolution and mixing of waters of different qualities. Al-Bassam et al. (1997) have given a computer program for plotting hydrochemical data in Durov's diagram.

Hydrochemical Sections These are plotted to show the variation in concentration of several hyrochemical characteristics in the direction of groundwater movement. Such type of sections can also be of help in depicting the changes in water quality due to mixing of sea-water and fresh water in coastal aquifers. In Fig. 11.9 is given one such section from the carbonate rock aquifer in coastal parts of Saurashtra, India.

Software packages which are now available have greatly facilitated the plotting and interpretation of hydrochemical data.

Table 11.5 Chemical characteristics of groundwater plots in different areas of the Hill–Piper diagram. (After Piper 1953)

Area 1	Alkaline earths exceed alkalis
Area 2	Alkalis exceed alkaline earths
Area 3	Weak acids exceed strong acids
Area 4	Strong acids exceed weak acids
Area 5	Carbonate hardness (secondary alkalinity) exceeds 50%—chemical properties of the groundwater are dominated by alkaline earths and weak acids
Area 6	Non-carbonate hardness (secondary salinity) exceeds 50%
Area 7	Non-carbonate alkalinity (primary salinity) exceeds 50%—chemical properties are dominated by alkalis and strong acids, ocean waters and many brines plot in this area, near its right-hand vertex
Area 8	Carbonate alkali (primary alkalinity) exceeds 50%—here are plotted groundwaters which are inordinately soft in proportion to their content of dissolved solids
Area 9	No one cation–anion pair exceeds 50%

Fig. 11.8 Durov's diagram showing plotting of analytical data

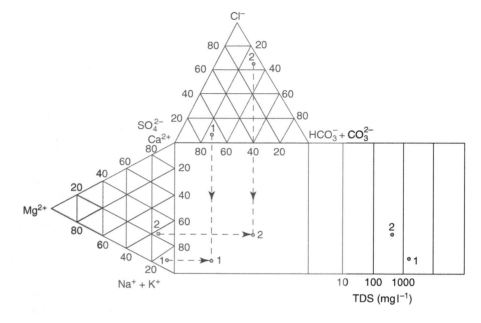

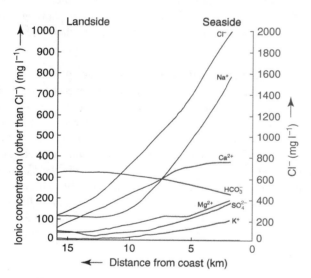

Fig. 11.9 Hydrochemical section of the Miliolite limestone across the coast of Saurashtra, India. (After Nair 1981)

11.6 Modification of Chemical Characters of Groundwater

In addition to dissolution of rock material, the other processes which change the chemical composition of groundwater are ion exchange, membrane filtration and sulphate reduction. Climate and length of flow trajectory also affect the water quality.

11.6.1 Ion Exchange

Groundwater on coming in contact with rock matrix and organic material has a tendency to exchange ionic constituents. These reactions are known by a variety of names, i.e. sorption, adsorption, absorption and desorption. Sorption involves mass transfer from the solution to the solid by adsorption or absorption. In case of adsorption, the molecule or the ion is held at the solid surface, while in absorption the sorbate enters the mineral structure. Desorption is the release of molecules and ions from the solid phase to the solute. Another reaction of interest to hydrogeologists is ion or base exchange, when an ion in solution gets sorbed to a surface and a similar charged ion is released leading to adsorption of cations on mineral surfaces. Most of the exchange reactions involve cations and therefore it is referred as cation or base-exchange. The common natural materials which act as the most important adsorbants are clay minerals. In addition glauconite, rock-forming silicates like zeolites, micas, feldspathoids and organic material like humus also form adsorbants.

The cation-exchange capacity (CEC) of a material is defined as the number of exchangeable ions in milliequivalents per $100\,g$ of solids at $pH=7$. The CEC of some commonly occurring minerals and other natural materials is given in Table 11.6.

The cation exchange capacity increases with decrease in the particle size. Colloidal sized par-

Table 11.6 Cation exchange capacity of some natural materials. (After Matthess 1982)

Mineral/Rock	Cation exchange capacity (meq/100 g)
Tuff	32–49
Kaolinite	3–15
Glauconite	10–200
Montmorillonite	70–100
Zeolites	100–600
Feldspathoids	400–1000
Organic substances in soils	150–500

ticles (10^{-3}–10^{-6} mm) have a greater tendency to ion exchange because they have a large electrical charge relative to their surface areas. As the nature of the surface charge is a function of pH, the tendency of adsorption of cations or anions will also depend on the pH of the solution. The exchange capacity also depends on the ion characteristics. The affinity for adsorption is greater in ions with higher valencies, i.e. it is more difficult to remove ions of higher valency than that of lower valency. For ions of the same valency, the affinity for adsorption increases with increase in the atomic number and the ionic radius. The normal order of preference for adsorption is:

$$Ba^{2+} > Sr^{2+} > Ca^{2+} > Mg^{2+} > C_s^{+}$$
$$> K^{+} > Na^{+} > Li^{+}$$
$$\text{Stronger} \rightarrow \text{Weaker}$$

The order indicates that at similar concentrations, ions will replace those ions which lie to their right. The concentration of the adsorbed mole fractions at the initial condition and the concentration ratio of the two ions in solution are also of importance to determine the direction in which an ion exchange reaction will take place. The equilibrium between the quantity of the substance (S) adsorbed on the solids and the concentration of this substance in solution (C) can be described by Freundlich's isothermal Eq. 11.3.

$$S = K_d C^b \tag{11.3}$$

where K_d and b are coefficients that depend on the solute species and nature of the medium. Equation 11.3 shows that an increase in the concentration of a solution will raise the adsorbed quantity, and a decrease in the concentration will lead to desorption.

The effect of cation exchange on groundwater quality due to rock lithology and movement of contaminants has been cited widely in groundwater literature

(Back and Hanshaw 1965; Freeze and Cherry 1979; Hem 1989; de Marsily 1986; Kehew 2001).

Base exchange also changes soil characteristics. The replacement of Ca^{2+} by Na^{+} due to exchange process in clays decreases the permeability of the soil drastically due to swelling. This could happen in areas where saline or brackish water is used for irrigation.

11.6.2 Membrane Effect

The sedimentary rock aquifers are often interbedded with relatively impermeable clay and shale beds which serve as semi-permeable membrane for the selective movement of ions under higher confining pressures. A semi-permeable membrane is defined as one that restricts or prevents the passage of charged species (cations and anions) while allowing relatively unrestricted flow of water molecule (Back and Hanshaw 1965). The concentration of solutes on the input side of the membrane therefore increases relative to the concentration on the output side. Thereby the input side gets saltier and the receiving aquifer becomes fresher. This ion expulsion effect is commonly referred to as salt filtering, ultrafiltration or hyperfiltration (Freeze and Cherry 1979; Matthess 1982). Divalent cations are usually filtered more easily than monovalent cations due to differences in ionic size and charge. Temperature also influences the rate of salt filtering. Clays with higher cation exchange capacity like montmorillonite have higher ion-filtering efficiencies as compared with kaolinite. Laboratory experiments also indicate that salt filtering is more effective in sedimentary formations at depths greater than 500–1000 m. The formation of deep seated brines and occurrence of saline waters in non-marine sediments and those devoid of evaporite deposits has been explained by the process of salt filtering (Back and Hanshaw 1965; Neuzil 1986).

11.6.3 Sulphate Reduction

The reduction of SO_4^{2-} to HS^{-} is a common feature in groundwater which takes place in the presence of bacteria (Eq. 11.4).

$$SO_4^{2-} + CH_{4(bacterial)} \rightarrow HS^{-} + H_2O + HCO_3^{-}$$

$$\tag{11.4}$$

Therefore, waters which have undergone sulphate reduction are characterized by the presence of HS^- and high content of HCO_3^- as is reported from several oil fields. Micro-organisms (bacteria) play an important role in bringing out changes in the chemical quality of groundwater as is revealed from recent researches, although earlier it was believed that such organisms may not exist in the aquifers, a few metres below the ground surface (Hasan 1996). The study of these micro-organisms is important from the point of view of remediation of contaminated groundwater (also see Sect. 12.4.3).

Sulphate reducing bacteria are reported from the shallow groundwater in coastal parts of Orissa, India. The H_2S released from the reduction of SO_4^{2-} gives bad odour and also attacks iron pipes and other parts of well assembly thereby increasing the concentration of ferrous ion in water. Sulphate reducing bacteria have been also reported from even greater depths (>500 m) in petroleum deposits. Some bacteria can withstand pressures above 1700 bars (Matthess 1982). Sulphate reduction can also take place by hydrogen released through the decomposition of organic matter by anaerobic bacteria (Eq. 11.5).

$$CaSO_4 + 4H_2 \rightarrow 4H_2O + CaS \qquad (11.5)$$

11.7 Hydrochemical Zoning and Hydrochemical Facies

11.7.1 Hydrochemical Zoning

On a regional scale, groundwater exhibits both lateral and vertical variation in its chemical characteristics which is a result of variation in the lithology of the aquifer, climatic variations and length of trajectory. Schoeller (1959) has identified three types of zonations namely geological zonation, vertical zonation and climatic zonation. The geological zonation is a result of variation in the mineralogical composition of the aquifer material.

Vertical and lateral variation (zonation) in a groundwater flow system is a result of greater dissolution leading to an increase in total dissolved solids and different ions along its flow paths. Therefore, the groundwater in recharge areas and at shallow depths will have lesser concentration of dissolved solids than water in the deeper zones. Such a variation is described from

Russia by Siline-Bektchourine (Schoeller 1959) and by Chebotarev (1950) and Domenico (1972) from Australia. As per this scheme, following changes in the anion species in groundwater are reported with increase in the length of trajectory and increase in residence time:

$$HCO_3 \rightarrow HCO_3^- + SO_4^{2-} \rightarrow SO_4^{2-} +$$
$$HCO_3^- \rightarrow SO_4^{2-} + Cl^- + SO_4^{2-} \rightarrow Cl^-$$

The decrease in redox potential of groundwater along its flow path is due to decrease in dissolved oxygen at greater depths (Freeze and Cherry 1979).

As the climate influences the rainfall and evaporation etc., the chemical evolution of groundwater will also depend on these climatic factors. This is the reason that groundwater in arid and semi-arid regions is more saline than in humid and temperate regions. Temperature alongwith humidity also controls the activity of soil micro-organisms and the organic matter which produce CO_2. This is an essential factor in the dissolution of silicate and carbonate minerals in the rocks. Schoeller (1959) traced the evolution of the chemistry of groundwater in the former USSR, from the temperate down to the equatorial regions, and has shown variations in total dissolved solids and various cation and anion species.

The lateral and vertical sequence of various types of water is also modified by the palaeo-climatic conditions. There are several examples where due to palaeoclimatic conditions, fresh groundwater occurs at varying depths especially in arid and semi-arid regions. Certain chemical processes like base-exchange and sulphate reduction also cause reversal in the above sequence. Vertical hydrochemical zoning in crystalline rocks and its importance in storing radioactive waste is discussed in Sect. 13.7.2.

Hydrochemical zoning can also result due to mixing of water of different qualities. For example, in the coastal dunes of the Netherlands a zonation in hydrochemical facies is noted due to ion exchange reactions between fresh recharging water and the intruded seawater (Kehew 2001) (Fig. 11.10).

11.7.2 Hydrochemical Facies

The concept of hydrochemical facies as given by Back (1961) is based on a similar concept of lithofacies used commonly in geology which states that there will be a definite trend in the formation of minerals

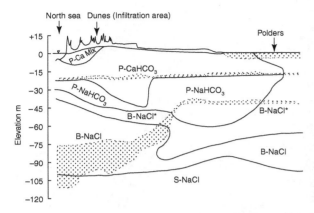

Fig. 11.10 Cross section through coastal dunes area in Netherlands showing chemical facies of groundwater. *P* fresh, *B* brackish, and *S* saline. Low permeability layers shown by shading. Ion exchange during artificial recharge has refreshened the groundwater. (After Kehew 2001)

under a given geological environment. Similarly, the development of a particular hydrochemical facies is controlled mainly by the lithology of the aquifer and its distribution is controlled by the groundwater flow pattern.

Piper's diagram forms the basis of classification of waters into various hydrochemical facies (Fig. 11.11). The characteristics of various hydrochemical facies is given in Table 11.7. The difference between the Piper's diagram and the hydrochemical facies diagram is that instead of giving equal increments to various variables, the hydrochemical facies are distinguished into 0–10, 10–50, 50–90 and 90–100 percent domains. The areal distribution of various facies can be shown with the help of maps, fence diagrams and cross sections (Back 1961). The concept of hydrochemical facies has been used for explaining the changes in water quality from recharge to discharge areas and along the flow paths in a given lithology (Back and Hanshaw 1965; Back 1966; Toth 1966). Some of the main conclusions are:

(a) Bicarbonate content is low in recharge areas and high in discharge areas;
(b) Sulphate content decreases in the direction of flow and bicarbonate increases, as a result of sulphate reduction; and
(c) The ratio of sulphate to chloride decreases in the direction of flow.

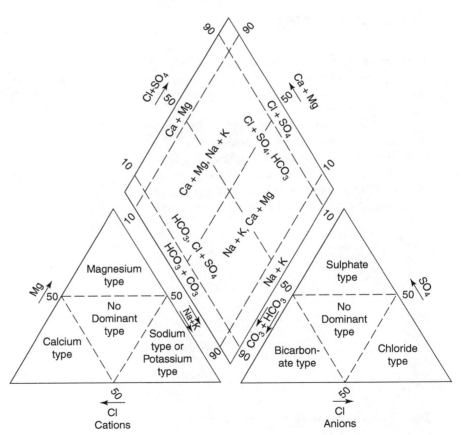

Fig. 11.11 Hydrochemical facies diagram. The ionic concentration is in percent of total meq l^{-1}

Table 11.7 Classification of hydrochemical facies. (After Back 1966)

	Percentage of constituents (meq l^{-1})			
	Ca+Mg	Na+K	HCO₃+CO₃	Cl+SO₄
Cation facies				
Calcium–magnesium	90–100	0 < 10	–	–
Calcium–sodium	50–90	10 < 50	–	–
Sodium–calcium	10–50	50 < 90	–	–
Sodium–potassium	0–10	90 < 100	–	–
Anion facies				
Bicarbonate	–	–	90–100	0 < 10
Bicarbonate-chloride–sulphate	–	–	50–90	10 < 50
Chloride–sulphate-bicarbonate	–	–	10–50	50 < 90
Chloride–sulphate	–	–	0–10	90–100

11.8 Quality Criteria for Various Uses

Groundwater is mainly used for drinking, irrigation and industrial purposes. Therefore, quality criteria depend on the use of water for a particular purpose. Quality standards have to be maintained in water supply for different uses to avoid deleterious effects.

11.8.1 Domestic Use

Groundwater forms an important source of water for drinking and other domestic purposes. Therefore, groundwater, in general, is safer for use than surface water especially from the point of view of bacterial pollution but the chemical composition of water is also important, as certain chemical constituents become toxic beyond a particular concentration although they may be beneficial in lower amount. This can be expressed by the dose response curve (Fig. 11.12). The threshold value will vary for different constituents.

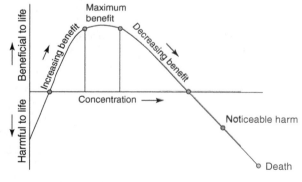

Fig. 11.12 Dose-response curve for human consumption

Prescribed standards for drinking water vary from country to country, depending upon economic conditions, climate, food habits and geographic location. There are also conflicting evidences with respect to safe limits for certain constituents.

Quality criteria for drinking water normally are based on a water intake of two litres per person per day. The optimum desirable concentration of various constituents depend on the age of the person, food habits and the climate. Therefore, exact drinking water standards could vary from one individual to another. Drinking water standards as prescribed by European Union, US-Environmental Protection Agency (USEPA), WHO, and Bureau of Indian Standards (BIS) are given in Table 11.8 (also see websites: http://www.who.int/water_sanitation_health/dwq/guidelines2/en/, and http://www.epa.gov/safewater/contaminants/index.html

The importance of some constituents in water for drinking purpose is discussed below.

TDS: The potability of water in terms of TDS is suggested as follows (WHO 1984):

Excellent	<300 mg l^{-1}
Good	300–600 mg l^{-1}
Fair	600–900 mg l^{-1}
Poor	900–1200 mg l^{-1}
Unacceptable	>1200 mg l^{-1}

Hardness: The standards of hardness for drinking water supply vary, viz. the WHO suggested an upper limit of 500 mg l^{-1} of CaCO₃. The American Water Works Association has prescribed the upper limit of 80 mg l^{-1} of CaCO₃ for an 'ideal' water (Hem 1989). As per the standards prescribed by the European Economic Community, the minimum required concentra-

Table 11.8 Upper limits of various constituents ($mg\,l^{-1}$) in drinking-water

Parameter	European Union (1998)	USEPA (2002)	WHO (1984)	BIS (1991)
TDS	–	500	1000	500
pH	–	6.5–8.5	6.5–8.5	6.5–8.5
Total hardness ($CaCO_3$)	–	–	500	300
Ca	–	–	75–200	75
Mg	–	–	–	30
Na	200	–	200	No limit
Cl	250	250	250	250
SO_4	250	250	400	150
NO_3	50	44	50	45
Fe	0.2	0.2	0.3	0.3
F	1.50	4.00	1.5	0.6–1.2
Pb	0.01	0.01	0.01	0.1
Hg	0.001	0.002	0.001	0.001
Zn	–	5.0	5.0	5.0
Cu	–	–	1.0	0.05
Cd	0.005	0.005	0.003	0.01
As	0.01	0.01	0.01	0.05
Cr (hexavalent)	0.05	–	0.05	0.05
CN (Cyanides)	–	0.2	0.05	0.05
Overall α radioactivity	–	$15\,pCi\,l^{-1}$	$3\,pCi\,l^{-1}(0.1\,Bq\,l^{-1})$	$1\,pCi\,l^{-1}$
Overall β radioactivity	–	–	$30\,pCi\,l^{-1}(0.1\,Bq\,l^{-1})$	$10\,pCi\,l^{-1}$
Coliforms	–	–	No fecal coliforms	100 coliforms per litre

USEPA Environmental Protection Agency, USA, *WHO* World Health Organization, *BIS* Bureau of Indian Standards

tion of hardness is $2–10\,meq\,l^{-1}$ of $CaCO_3$. Studies in USA, Canada and Japan indicated that incidence of cardiovascular diseases is more in areas where drinking water is extremely soft (Hem 1989), However, WHO (1984) showed no such definite relationship.

Chloride (Cl): The limits of chloride have been put more from the point of view of taste rather than its adverse effect on human health. Water with chloride concentration more than $250\,mg\,l^{-1}$ will be saline in taste. For persons suffering from hypertension, the chloride concentration should be less.

Iron and Manganese: These may not have direct effect on human health but they are of importance from aesthetic considerations. Certain type of bacteria, such as crenothrix change the dissolved iron to an insoluble slime which causes plugging of pipes, valves and water meters. Iron also imparts dirty reddish colour, stains clothes and fixtures and bad taste and odour. It also clogs the well screen and adjoining formation. These problems arise when the iron concentration approaches $0.3\,mg\,l^{-1}$. Therefore, the guideline value for iron is chosen as $0.3\,mg\,l^{-1}$. The use of chlorine is effective in stopping crenothrix growth. The common

method of removal of iron from water is by aeration followed by sedimentation.

Fluoride (F): is beneficial when present in small concentration ($0.8–1.0\,mg\,l^{-1}$) in drinking water for calcification of dental enamel but it causes dental and skeletal fluorosis if present in higher amount (Table 11.9). Higher concentration of fluoride in drinking water is also linked with cancer (Smedley 1992). The permissible limit of F depends on temperature; higher intake of fluoride can be permissible in colder climate (Hamill and Bell 1986).

Table 11.9 Impact of fluoride in drinking water on health. (Dissanayake 1991)

Concentration of fluoride ($mg\,l^{-1}$)	Impact on health
Nil	Limited growth and fertility
0.0–0.5	Dental caries
0.5–1.5	Promotes dental health
1.5–4.0	Dental fluorosis (mottling of teeth)
4.0–10.0	Dental and skeletal fluorosis (pain in back and neck bones)
>10.0	Crippling fluorosis

In areas of higher fluoride concentration, defluoridation can be adopted to supply safe drinking water. The most common method, adopted in India, is the Nalgonda process developed by the National Environmental Engineering Research Institute (NEERI). The Nalgonda technique comprises addition of lime and aluminum sulphate or aluminum chloride which helps in the removal of fluoride by flocculation. This method is quite simple and can be used in villages at domestic level and also for small communities.

Mercury (Hg): is a toxic element and serves no beneficial physiological function in human beings. The main effect of mercury poisoning is in the form of neurological and renal disturbances. It also disturbs the cholesterol metabolism. There is no evidence that high dose of mercury is carcinogenic. The recommended guideline value of mercury in drinking water is $0.001\,mg\,l^{-1}$.

Lead (Pb): concentration in drinking water supply is kept low ($10–20\,\mu g\,l^{-1}$) after treatment, but due to lead in water supply pipes and storage tanks, the lead in drinking water may be higher ($>300\,\mu g\,l^{-1}$). Lead is not essential for the human beings, therefore the intake of lead should be kept to minimal. Lead in high doses acts as metabolic poison causing anaemia, behavioural and mental diseases. A guideline value of $0.05\,mg\,l^{-1}$ of lead in drinking water is recommended (WHO 1984).

Arsenic (As): is toxic and carcinogenic (Smedley 1992). Arsenic intake by humans is greater from food (e.g. seafood) and inhalation than from drinking water. Earlier, the WHO recommended limit for As in drinking water was $50\,\mu g\,l^{-1}$, but it is recently reduced to $10\,\mu g\,l^{-1}$ (Smedley and West 1995). See Sect. 12.3.1 for more details.

Iodine (I): The daily iodine requirement for human diet is about $100–200\,\mu g\,l^{-1}$ (Smedley 1992). Deficiency of iodine causes goitre. As sea-water is the main source of iodine, it is natural that it is deficient in areas away from the coast. This is the reason that goitre has been mainly reported from Alpine, Himalayan and Andean regions. Iodine concentration of less than $1\,\mu g\,l^{-1}$ is reported in goitrous areas in Nepal (Smedley 1992). As high Ca in water suppresses the solubility of iodine, goitre is also observed in limestone areas.

Nitrate (NO_3): Nitrate (NO_3) is one of the most common groundwater contaminants in the world and its presence in higher concentrations poses human health and ecological risks. Studies in different parts of USA show that agricultural fertilizers and urban septic leachates are the primary sources of large nitrate concentrations in surface and groundwaters (see Sect. 12.3.2). In many countries, water supplies having high levels of nitrate have been responsible for bluebaby disease in infants (infantile methaemoglobinaemia) and death. This problem does not arise in adults. Maximum prescribed limit of NO_3 in municipal water supplies is $10\,mg\,l^{-1}$.

Pesticides: From among the pesticides, DDT and aldrin are important from water quality point of view. The main effect of DDT and aldrin is on the nervous system and liver. There is no evidence of their being carcinogenic. The guideline values for DDT is $1\,\mu q\,l^{-1}$ and for aldrin is $0.03\,\mu q\,l^{-1}$ (WHO 1984). However, as DDT accumulates in fatty tissues and causes reproductive defects, its use was banned several decades ago.

Radionuclides: Radionuclies emit radiations in the form of alpha particles, beta particles, and gamma rays. Among the naturally occurring radionuclides, ^{238}U is both alpha and beta emitter and ^{226}Ra and ^{234}U are the typical alpha-emitting radionuclides while ^{90}Sr is among the man-made beta—emitters of interest.

Among the alpha-emitting radionuclides, ^{226}Ra, ^{224}Ra, ^{210}Po, ^{232}Th, ^{238}U and ^{238}U have high toxicity and ^{90}Sr, ^{89}Sr, ^{134}Cs, ^{137}Cs, ^{131}I and ^{60}CO are the toxic beta-emitting radio nuclides. From water quality point of view, gross alpha and beta activity is of importance rather than of individual radionuclides. The radioactivity of water is expressed as Curie (Ci) per litre or Becquerel (Bq) per litre. As Curie is a larger unit, therefore picoCurie (pCi) is usually used ($1\,pCi = 2.7 \times 10^{-12}\,Ci$). Bq is the unit of radioactivity in the Internationl system (SI) of units ($1\,Bq = 30\,pCi$).

WHO (1984) has recommended guideline value of $1\,Bq\,l^{-1}$ ($30\,pCi\,l^{-1}$) for gross beta activity in drinking water (Table 11.8). Guideline values represent values below which water is considered potable without any further radiological examination.

Micro-organisms: Micro-organisms (viruses, bacteria) are important as particulate contaminants in groundwater. Viruses due to their small size, have a greater mobility, even in fine grained rocks with small pore size. However, the mobility of bacteria in groundwater systems is slower due to their comparatively larger size except in rocks with wide fractures and solution conduits in carbonate rocks.

11.8.2 Quality Criteria for Other Uses

Water Quality Standards for Use by Livestock - Fundamentally, the same standards can be applied as for human being. However, animals can have a greater tolerance of total dissolved solids ($1000 \, mg \, l^{-1}$). Standards also depend upon livestock type, age and food habits, (Bouwer 1978; Lloyd and Heathcote 1985).

Industrial Use The quality criteria of water for industrial purposes depend on the type of industry, processes and products. The quality criteria of water for use in boilers is important as water used in high pressure boilers should be free from suspended matter; should have low total dissolved solids and no acid reaction. Low pressure boilers can use water with total dissolved solids up to $5000 \, mg \, l^{-1}$ and $CaCO_3$ hardness up to $80 \, mg \, l^{-1}$, while in high pressure boilers total dissolved should be less than $50 \, mg \, l^{-1}$ and $CaCO_3$ hardness less than $1 \, mg \, l^{-1}$. Therefore, water treatment is necessary before use in high pressure boilers.

For the manufacture of pharmaceuticals and high grade paper, water approaching or equalling the quality of distilled water is required. Very pure water is also desirable in nuclear reactors to keep the radioactivity caused by neutron activation of dissolved solids as low as possible (Hem 1989). In construction industry, sulphate content of water is important to avoid deterioration of concrete. To take care of such problems, sulphate resisting cements are developed to suit a range of sulphate concentrations in the soil and water (Lloyd and Heathcote 1985).

Irrigation Water The quality standards for irrigation water are based on (1) total dissolved solids which may effect the intake of water and other nutrients by plants through osmosis, (2) the relative concentration of alkalies and alkaline earths which effect the soil texture due to cation exchange and thereby its permeability and drainage characteristics, and (3) the concentration of specific ions, viz. boron, selenium, cadmium etc., which are toxic to the growth of plants beyond certain levels. In addition to chemical characteristics

of applied water, other factors which decide the suitability of a particular quality of water for use in irrigation are: (a) soil characteristics, (b) type of crop and its stage of growth, (c) climate, (d) depth to water-table, (e) method of irrigation and (f) drainage characteristics of soil. The quality standards for irrigation water cannot be applied very rigorously as in the absence of water of required quality, poor quality water can also be used in farming by making necessary adjustments and water management practices.

One of the earliest system of classification of water for use in irrigation was given by Wilcox (1955) which is based on electrical conductivity (EC), percent sodium (% Na) and boron concentration. Percent sodium (% Na), is expressed as

$$\% Na = \frac{Na^+ + K^+}{Ca^{2+} + Mg^{2+} + Na^+ + K^+} \times 100 \tag{11.6}$$

where, all the ionic concentrations are expressed in $meq \, l^{-1}$. Boron is useful in very small concentration for the normal growth of plants, but it becomes toxic in greater concentration depending upon the type of plants (Todd 1980).

The US Salinity Laboratory of the US Department of Agriculture used salinity hazard and sodium hazard as the two important criteria for the classification of irrigation waters. Salinity hazard is a measure of EC and sodium hazard is expressed in terms of Sodium Adsorption Ratio (SAR) which is defined as

$$SAR = \frac{Na^+}{\left[\left(Ca^{2+} + Mg^{2+}/2\right)\right]^{\frac{1}{2}}} \tag{11.7}$$

where all the ionic concentrations are expressed in $meq \, l^{-1}$ Adjusted SAR has been suggested as another criteria to assess the suitability of water for irrigation (Ayers 1975). Reader may refer to Todd (1980) and Lloyd and Heathcote (1985) for necessary details regarding various systems of classification of irrigation waters.

Summary
Groundwater quality is very important for its possible use for various purposes. The water quality depends on the climatic conditions,

lithology of aquifer, sources of recharge and its residence time. The concentration of major ions in water is expressed in $mg\,l^{-1}$ or parts per million (ppm) and of minor and trace elements in $\mu g\,l^{-1}$ or parts per billion (ppb). The chemical analysis data can be plotted in the form of maps or pattern diagrams. The various hydro-chemical process which modify water quality are ion exchange, rock-water interaction and reduction-oxidation. On a regional scale, both lateral and vertical variation in water quality occurs.

Quality criteria of water for domestic use have been proposed by various national and international agencies depending upon the food habits, water intake and its availability. Quality criteria of water for irrigation, industrial and other purposes have also been prescribed.

Further Reading

Freeze RA, Cherry JA (1979) Groundwater. Prentice Hall Inc., New Jersey.

Kehew AE (2001) Applied Chemical Hydrogeology. Prentice Hall Inc., New Jersey.

Kresic N (2007) Hydrogeology and Groundwater Modeling. 2nd ed., CRC Press, Boca Raton.

Schwartz FW, Zhang H (2003) Fundamentals of Ground Water. John Wiley & Sons Inc., New York.

Groundwater Contamination

12

12.1 Introduction

The term contamination is used for addition of any solute into the hydrological system as a result of man's activity while the term pollution is restricted to a situation when the contamination attains levels that are considered to be objectionable (Freeze and Cherry 1979). There could also be deterioration in water quality due to natural reasons namely dissolution of rock material. Contamination of groundwater can take place from either a wider source like percolation from agricultural fields on account of application of fertilizers and pesticides or from a point source like waste disposal sites. Atmospheric composition will also affect the composition of precipitation and thereby causes water pollution. In a polluted atmosphere, many oxidizing compounds (e.g. ozone), acid-forming gases (oxides of nitrogen and sulfur) and particulate material will be higher than in an unpolluted atmosphere. This will be area specific depending on the anthropogenic activities. For example, combustion of fossil fuels results in acid rains containing H_2SO_4 and HNO_3, which reduces the pH of source water. This induces greater solubility of aquifer material (see effect of Climate Change in Sect. 20.10). Overexploitation of groundwater, especially in coastal areas is also responsible for contaminating fresh water aquifers due to sea-water intrusion. This aspect is discussed in Sect. 20.7.

During the last few decades, due to the increased industrialization, urbanization and agricultural activities, the quality of groundwater as well as surface water has deteriorated considerably in several areas. This has caused great concern both in developed and developing countries. It is comparatively easier to detect the contamination of surface water sources like rivers and lakes, but as underground pollution cannot be observed directly, it is detected only after a long time has elapsed and then it becomes a difficult task to rehabilitate the aquifer for safe water supplies. However, as the groundwater moves with a much lower velocity and so also the plume of pollution, one can plan ahead and avoid the outcome of pollution by taking advance measures.

12.2 Movement of Contaminants

The movement and interaction of contaminants with aquifer material and groundwater is a complex phenomenon. As the contaminant moves vertically downward towards the water-table, it passes through different zones in the aquifer. In the unsaturated zone, attenuation of contaminant will take place as some chemicals are adsorbed on organic material and clay minerals, some are decomposed through oxidation and bacterial activity and some are taken up by plants or released into the atmosphere.

After reaching the water-table, the dissolved contaminant will move along with the groundwater in the direction of its hydraulic gradient mainly by advection (Sect. 7.4.1). As the pore spaces in the zone of saturation are devoid of oxygen, there is no likelihood of oxidation. Further, due to very low groundwater velocity ($<0.03\,md^{-1}$), the contaminant generally takes several years to move from the source to the well point. The concentration of the contaminant in groundwater also decreases with increasing distance of flow due to hydraulic dispersion and other attenuation effects.

As discussed in Sect. 7.4.1, two types of hydrodynamic dispersion are prominent; the longitudinal dispersion which occurs in the direction of flow, and

B. B. S. Singhal, R. P. Gupta, *Applied Hydrogeology of Fractured Rocks*,
DOI 10.1007/978-90-481-8799-7_12, © Springer Science+Business Media B.V. 2010

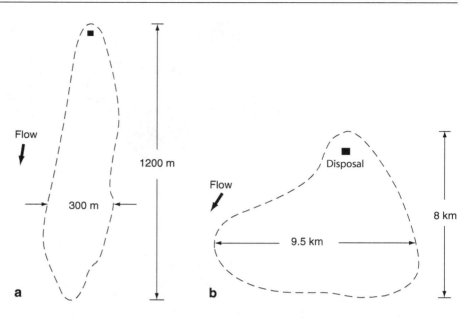

Fig. 12.1 Effect of differences in longitudinal and transverse dispersivities on shapes of contaminant plumes. **a** chromium plume, Long Island, USA, after 13 years, **b** chloride plume in fractured basalt, Idaho, USA, after 16 years. (After Miller 1985)

the transverse dispersion which occurs normal to the direction of flow. Due to hydrodynamic dispersion, the concentration of contaminant will decrease with distance from the source. The spread of the solute in the direction of flow will be greater than in the direction perpendicular to flow as longitudinal dispersivity is generally more than the transverse dispersion (Fig. 12.1a). However, in fractured basalts at Idaho, USA, as the transverse dispersivity is more than the longitudinal dispersivity, the lateral spread of the chloride plume was found to be longer than its spread in

the direction of groundwater flow (Fig. 12.1b). The effect of rock discontinuities on the spread of the contaminant from a uranium mill tailing pond is illustrated in Fig. 12.2. This figure shows that the baseline natural uranium activity follows a linear pattern following the strike of the fractures. After 11 years of operation of the tailings pond, the uranium activity increased several folds and the plume extended to the northwest following the strike of fractures. A case history illustrating the importance of structural and karst features in carbonate rock on contaminant migration from a

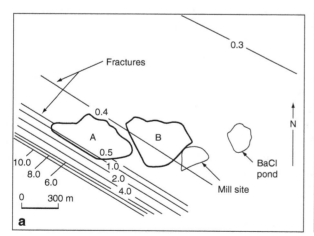

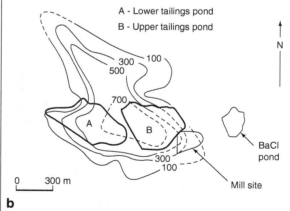

Fig. 12.2 Effect of fractures on the spread of uranium activity from two unlined uranium tailing ponds in sandstone formation, Utah, USA. **a** Baseline natural uranium activity (μ Ci ml$^{-1} \times 10^7$) before the tailing ponds were put into service. This was due to

natural leaching of uranium bearing rocks. **b** Elevated activity of natural uranium in the summer of 1981, i.e. after about 11 years of operation. (After White and Gainer 1985, reprinted with permission of Ground Water Publishing Co., C. 1985)

chemical recycling facility in US Midwest is described by Hasan (1996).

Adsorption and cation exchange reactions, cause attenuation of solutes as discussed in Sect. 11.6. Laboratory and field studies show that fractured shales in the unsaturated zone act as barriers for the migration of Cd^{2+} and Pb^{2+} obtained by leaching from sanitary land fills, fly-ash piles and other concentrated sources of these metals. These metals are retained in shales through a combination of precipitation reactions, cation exchange and adsorption reactions (Angino and Wilbur 1990).

12.3   Sources of Groundwater Contamination

Groundwater contamination could be from geogenic or anthropogenic sources.

Major sources of groundwater contamination are listed in Table 12.1.

12.3.1   Geogenic Sources

The concentration of certain toxic or other substances beyond acceptable limits for human consumption can take place due to the dissolution of aquifer materials. Increased concentration of fluoride, arsenic, iron and other heavy metals in groundwater is usually due to this type of contamination.

Fluoride problem has received much attention in India and in several other countries (Handa 1988). Fluoride-rich water in East African Rift Valley, e.g.

Ethiopia is associated with calcium deficient volcanic rocks and calcium poor granitic rocks (Smedley 1992). The hot spring in this area also have high fluoride ($90\,mg\,l^{-1}$). In India, high fluoride ($2–10\,mg\,l^{-1}$) water is reported mainly from granites and pegmatites which have F-bearing minerals, e.g. fluorite (CaF_2) and fluor-apatite (Sinha and Kakar 1974). Excessive concentration of fluoride in water is also reported from the eastern part of Sri Lanka, Thailand, China and several African countries.

Iron in groundwater is mainly due to the dissolution of iron oxides in laterites and other iron-bearing minerals especially in high rainfall areas, viz. coastal Orissa, Assam and Kerala in India. In these areas, Fe^{2+} is in the range of $0.4–46\,mg\,l^{-1}$ and total iron is in the range of $6.83–55\,mg\,l^{-1}$. High concentration of iron is also reported from the crystalline basement rocks in Malawi (Chilton and Foster 1993) and other places.

Arsenic (As)   Arsenic in drinking water has received much attention in recent years as evidence of its detrimental health effects has grown (Welch and Stollenwerk 2003). As per the World Health Organization, the upper limit for As in drinking water is $0.01\,mgl^{-1}$. In India, the permissible limit of As in drinking water is $0.05\,mg\,l^{-1}$ High concentration of As in groundwater is reported from several countries world-over, viz. Bangladesh, India, China, Argentina, Chile, Mongolia, Taiwan and United States. People consuming high levels of arsenic are prone to develop skin, bladder and lung cancers.

Dissolved arsenic may exist in groundwater either as arsenite in trivalent form (As^{3+}), or as arsenate in pentavalent form, (As^{5+}). Both anions are capable of adsorbing to various subsurface materials, such as ferric oxides and clay particles. Arsenate strongly adsorbs to

Table 12.1 Major sources of groundwater contamination

Source of pollution	Type	Principal contaminants causing concern
Natural: Dissolution of minerals and rocks	Diffuse	Cl, F, Fe, As, Pb, Hg, Cd
Industrial	Point source	Hydrocarbons, chlorinated solvents, heavy metals, etc. depending on type of industry
Agricultural practices	Diffuse	Cl, NO_3, pesticides and herbicides
Mining and other associated activities	Point source	Cl, SO_4, trace metals, e.g. Fe, As, Cu, Pb, Zn, Hg
Municipal wastes, e.g. landfills and other waste piles	Point source	TDS, CH_4, and various other organic and inorganic contaminants
Radioactive waste disposal	Point source	Radioisotopes
Miscellaneous, e.g. sea-water intrusion, deforestation, de-icing salts application, acid rain etc.	Diffuse	Cl, SO_4, trace metals

these surfaces in acidic and neutral waters. An increase in the pH to alkaline condition may cause both arsenite and arsenate to desorb and they are expected to be mobile in an alkaline environment (Kresic 2007).

Several geochemical mechanisms, including oxidation of arsenic-bearing sulfide minerals (e.g. pyrite, arsenopyrite), reduction of arsenic bearing iron and manganese hydroxides, release of sorbed arsenic from mineral surfaces, evaporative concentration, and contamination from fly ash, and acid mine drainage can be responsible for higher concentration of arsenic in groundwater.

Origin of arsenic in groundwater could be both geogenic and anthropogenic Arsenic is present in many rock forming minerals especially sulfide minerals, viz. pyrite and arsenopyrite. In aerobic (oxygenated) conditions, dissolution of sulfide minerals, viz. pyrite and arsenopyrite contributes arsenic to groundwater and surface water. As a result, arsenic rich water is common in mineralized areas where mining activity has catalysed the oxidation of sulfide minerals (Leblanc et al. 2002). High arsenic concentration is reported from acid mine drainage in Obusi gold mines in Ghana and peat deposits in The Netherlands. In the Netherlands, oxidation of pyrite in peat deposits is regarded to be the source of arsenic in water.

Arsenic in groundwater may also be a result of human activities viz., fossil fuel combustion (fly ash) and from industrial products such as pesticides, pharmaceuticals, glassware and dye manufacturing. Irrigation in some parts of USA has also liberated As. It can also be derived from use of pesticides and herbicides.

In the western United States, high ($>50\,\mu g\,l^{-1}$) concentration of arsenic in groundwater is associated with areas of sedimentary rocks derived from volcanic rocks, geothermal systems and gold deposits (Fetter 1993).

In the Indian subcontinent, high concentration of arsenic ($0.3-1.1\,mg\,l^{-1}$, highest being $5\,mg\,l^{-1}$) in groundwater of shallow alluvial aquifers of Bangladesh and West Bengal, India, has posed serious problems to millions of people. In Bangladesh, more than 50 million people are currently exposed to the risk of As poisoning. About 44% of the total population of West Bengal is suffering from arsenic related diseases including skin cancer (Chandrasekahram 2005). In these areas arsenic rich water is restricted to shallow alluvial aquifers, within a depth of 20–80 m, while the deeper aquifers (below 100 m), separated from shallow aquifers by clay layers, are arsenic-free.

The origin of arsenic in groundwater in the alluvial tracts of West Bengal and Bangladesh is still speculative. Earlier, the high concentration of arsenic was attributed to the oxidation of pyrite present in clay minerals, due to the lowering of water-table by pumping. However, recent opinion attributes it to the release of adsorbed arsenic from the Fe-hydroxide and Mn-hydroxide surfaces under reducing conditions (Nath et al. 2006). While discussing origin of arsenic in the groundwater of Bangladesh, Kresic (2007) concludes that "The pumping may affect arsenic concentrations, but not by the oxidation of sulfides, or by slow reduction of iron hydroxides or sorbed arsenate by detrital organic carbon, as has been previously reported."

Arsenic can be removed from water by chemical precipitation, membrane separation and ion exchange. The membrane separation process is expansive. The use of iron oxide-coated sand and activated alumina filtration are reported to be alternative emerging technologies for arsenic removal from drinking water (Ramakrishna et al. 2006). Leblanc et al. (2002) have suggested the use of oxidizing bacteria for the removal of As from heavily contaminated acid mine drainage system.

Chromium (Cr) Chromium contamination is significant problem world wide. In groundwater it may be both from anthropogenic and natural source. Cr contamination in soil and water is mainly due to its use in numerous industrial processes viz. chromium plating and alloying, hide tanning and chemical manufacturing. It is also present in minerals and rocks. The common ore of Cr is chromite ($FeCr_2O_4$) which is associated with ultramafic rocks (dunites and serpentinites). Therefore, it may enter the surface water and groundwater from acid mine drainage also.

Cr may occur in groundwater as Cr^{3+} or Cr^{6+}. Cr^{3+}, as trace elements, is regarded to be useful for living organisms but Cr^{6+} is toxic for biologic system. Therefore, it is quite important to distinguish between Cr^{3+} and Cr^{6+} rather than reporting it as "total chromium". Cr^{6+} may be reduced to Cr^{3+} thereby reducing the toxic impact of chromium contamination. Cr^{6+} has a deleterious effect on the liver, kidney and respiratory organs. WHO has prescribed $50\,\mu g\,l^{-1}$ as the guideline value for drinking water.

Cr^{6+} compounds are more soluble and mobile compared with Cr^{3+}, the latter makes only a small percentage of the total concentration of Cr in groundwater. Further, due to greater solubility and mobility, Cr^{6+} can persist in groundwater for longer duration and also transported to longer distances compared with Cr^{3+}. Presence of ferrous iron or natural dissolved organic carbon can serve as reducing agent to immobilize Cr^{3+} by sorption to hydroxide surfaces. In situ biodegradation is effective in converting Cr^{6+} into less toxic form Cr^{3+}.

A chromium plume from chrome plating wastes is reported to extend nearly 1300 m down gradient in the glacial deposits of Long Island, New York (Kehew 2001). Cr contamination in groundwater is reported in India due to mining activities of chromite in Sukinda area in Orissa (Tiwary et al. 2005), and also from tanneries and chemical plants in Tamilnadu.

Other heavy metals viz. Pb, Zn, Cd, and Hg could also be a source of groundwater contamination derived from aquifer sources.

Radon Radon gets dissolved in water from either the source rocks and soil or leakage through cracks in building foundations. High concentration of radon is commonly present in areas underlain by rocks which have significant uranium concentration viz. granites and similar rocks. The Rn content in groundwater depends on the uranium content, grain size and permeability of host rocks and the nature and intensity of fracturing. High concentration of Rn of the order of several hundred or even thousand $Bq\,l^{-1}$ has been reported from the granites and gneisses in Norway and from the Archean rocks of Russia. Rn is one of the main causes of colon and lung cancer and leukaemia. At the same time, radon water from the thermal springs in Russia has been used since long for therapeutic purposes (Voronov and Vivantsova 2005).

US EPA (US Environmental Protection Agency) has prescribed the Maximum Contaminant Level (MCL) for Rn to be $300\,pCi\,l^{-1}$ (picocuries per litre). In Russia, Norms of Radioactivity Safety (NRS) prescribe a maximum allowable concentration of radon of $60\,Bq\,l^{-1}$, while in Sweden it is $100\,Bq\,l^{-1}$. More information about the permissible limits of Rn in drinking water in USA can be seen at the web sites: http://www.epa.gov/iaq/radon, and http://sedwww.cr.usgs.gov.8080/radon.

12.3.2 Anthropogenic Sources

Industrial Activities The waste water discharged from industries carries with it a variety of dissolved and suspended impurities, the composition of which depends on the type of industries and the processes used. At many places, the effluents are discharged without proper treatment thereby causing contamination of surface water and groundwater. For example, high concentrations of Cr, Ni, Cd, Pb and Zn in surface water and groundwater are reported from several places due to industries in north India (Kakar 1988, 1990; Handa 1994). Pollution due to organic compounds is discussed in Sect. 12.4.

Mining industry is responsible for groundwater pollution mainly due to infiltration of leachate from mine waste and tailings. Dewatering of mines during operation causes entry of oxygen which results in the oxidation of sulphide ores resulting in strongly acidic (H_2SO_4) waters, known as acid mine drainage. Bacteria speed up the process. The acid mine drainage water from coal mines is characterized by high amount of suspended and dissolved solids, particularly iron and sulphate and low pH. Acidity (pH) depends on the amount of reactive pyrite present in the coal. The problem of acid mine drainage is quite acute from high organic sulphur Tertiary Coals of Assam in northeastern India (Singh 1988). In this area, the problem is aggravated due to high rainfall of the order of 400 cm per year. The mine drainage water is highly acidic (pH = 2.3–4.0) and contains more than $3000\,mg\,l^{-1}$ of sulphate and about $300\,mg\,l^{-1}$ iron. It also has undesirable concentration of heavy metals like As, Cd, Cr, Pb and Zn and therefore is a major source of pollution of surface water and groundwater. In People's Republic of China, coal has a high content of sulphur and mine water usually has a pH of 2–3. The mine water contaminates surface water leading to a decline in fish and shrimp catches and also the yield of farm crops irrigated with contaminated water (UNESCO/UNEP 1989). Compared to colliery waters, drainage waters from metal mines have lower concentration of sulphate but higher amount of heavy metals (Hg, Pb, Cd, Cu etc.) due to lower pH. Acid mine drainage emanating from Rand gold mine in South Africa contains uranium which is toxic to aquatic life (Humphreys 2009).

Water from polymetallic (Zn–Pb–Cu) mines viz. Rangpo mine in Sikkim, India, has low pH (around

5.0) and higher amount of TDS (3616 mg l^{-1}), Pb (277 mg l^{-1}), Zn (35 mg l^{-1}) and Cu (1.80 mg l^{-1}) (Rai 1994). Acid mine drainage is also a significant problem in eastern coal mining and some western base metal mines in USA. The subsurface migration of acid mine water is generally controlled by fractures (NRC 1996).

Electrical resistivity method is useful in determining both the extent and degree of acid mine drainage and also for monitoring of contamination owing to high conductivity of acidic water (Rogers and Kean 1981).

Arsenic and mercury contamination of stream water, reported from the gold-mining belt of Ghana, appears to be a result of both mine effluent discharges and atmospheric emissions from the ore-roasting plant. Some of the local streams in the area have up-to 170 μg l^{-1} of arsenic (Smedley and West 1995). Arsenic contamination (>10 μg l^{-1}) of water is also reported from tin mines due to the presence of arsenopyrite from southern Thailand, Taiwan, and the southern Andes (Smedley 1992). Discharge of mercury from gold mining activities has contaminated streams and basement rock aquifers in northern and central Precambrian shield in Brazil and Guiana creating serious health risks for settlements (Reboucas 1993).

Leachate from mine tailing dumps, which may contaminate surface as well as groundwater can be minimized by providing a cover of clay or some other impervious material to reduce the entry of water and oxygen in the mine tailing dumps (Herbert 1992).

Municipal Waste Most of the municipal solid waste is disposed in sanitary landfills. At some places, hazardous waste viz. organic chemicals and pesticides are also disposed in these landfills. The sanitary landfills are made in natural depressions or in man made excavations like abandoned quarries etc. The leachate from landfills may contain large number of contaminants; the total dissolved solids may be as much as 30 000 mg l^{-1}. Therefore, it can cause severe damage to groundwater quality. The problem is more acute in humid climate, especially in permeable and fractured formations where water-table is at shallow depths. The leachate plume may extend to distances of several hundred metres in permeable formations which may last for many decades. In addition to the formation of leachate, a variety of gases such as CH_4 and CO_2, are also produced due to the biochemical decomposotion

of organic matter. CH_4 can be very dangerous as it is combustible and can cause severe damage. Therefore, landfill sites should be selected by necessary hydrogeological investigations to minimize the groundwater contamination by the leachate. In permeable formations, it is advisable to provide impermeable lining of asphalt, clay or membrane over the bottom of the landfill to reduce infiltration of leachate. It is also necessary to construct cut-off drains around the landfill to prevent the flow of surface water into the landfill area. A good account of site selection and design of various types of landfills is given in Hamill and Bell (1986), Fetter (1988), Mather (1995) and Keister and Repetto (2007). In fractured rocks, detailed characterization involving surface mapping of fracture density and orientation in addition to hydraulic properties of the substrata through exploratory boreholes, geophysical logging and hydraulic tests is necessary.

In India, with increasing population, the leachates from landfills pose a major environmental problem. The NO_3 content in groundwater from a landfill located in fractured quartzite in Delhi is reported to be more than 110 ppm which necessitates abatement solutions to protect groundwater quality (Dey et al. 2003).

Agricultural Activities These are one of the most important factors which cause contamination of groundwater due to greater use of inorganic fertilizers and pesticides for obtaining higher crop yields. This is particularly the case in the developing countries (Fig. 12.3). The inorganic fertilizers are the source of

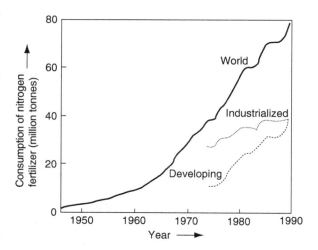

Fig. 12.3 Consumption of nitrogen fertilizers (1946–1989). (After Chilton et al. 1995)

nitorgen, phosphorous and potassium. Nitrogen, in the form of NO_3^-, is more soluble in groundwater as compared with phosphorous while K^+ has low mobility due to cation exchange. High amount of NO_3^- in groundwater can also be due to contamination from human and animal waste excreta. This is more common in fractured crystalline rocks with thin soil cover. If the NO_3^- concentration exceeds $50\,mg\,l^{-1}$, it becomes undesirable for drinking purposes (Table 11.8).

NO_3^- concentration is usually higher in shallow aquifers but with time and greater input, deeper aquifers are also likely to be contaminated. Studies in UK show that the nitrate front moves in the saturated zone at 1 m per year in Chalk to 2 m per year in sandstones with varying amounts of dispersion. A rise of NO_3 in groundwater from $14\,mg\,l^{-1}$ in 1959, to between 26 and $52\,mg\,l^{-1}$ in 1970 is reported from East Yorkshire, UK due to the application of fertilizers (Hamill and Bell 1986).

Nitrate concentration of groundwater due to the use of synthetic and organic fertilizers poses serious problems in USA especially in areas where corn and wheat is grown. In Jaffna Peninsula of Sri Lanka, groundwater in Miocene limestones contains in excess of $90\,mg\,l^{-1}$ of NO_3, which is mainly due to the increased use of inorganic fertilizers (Skinner 1985). Higher concentration of NO_3^- ($>100\,mg\,l^{-1}$) is also reported from groundwater in the shallow unconfined aquifers of the semi-arid regions of western parts of India.

The contamination of groundwater by phosphorous is less than nitrogen, as PO_4^{3-} is easily adsorbed on soil particles and has low mobility in groundwater. However, surface water may have comparatively higher amount of phosphorous which causes excessive vegetative growth in lakes and reservoirs (Schwartz and Zhang 2003).

In addition to fertilizers, pesticides and herbicides are also being extensively used. The annual growth rate world-over reached 12% in the 1960s but later decreased to around 3–4% (Chilton et al. 1995). The use of pesticides is more rapid in many developing countries than in the developed world. In India, the average rate of application of pesticides to agricultural land is about 180 grams per ha (Handa 1994). Canter et al. (1987) have given a good account of the effect of pesticides on groundwater quality, based on field, laboratory and model studies. The contamination of groundwater by pesticides is dependent on the properties of the soil, and application factors. The chance

for pesticide contamination is higher in surface water than groundwater due to the adsorption of pesticides by clay mineral grains and organic matter degradation by bacteria. Some pesticides with higher solubility in water have significant mobility in some type of aquifer materials, such as clean sands and gravels (Freeze and Cherry 1979). Preferential flow paths, e.g. macropores and fissures, in the unsaturated zone will also lead to fast movement of pesticides to the water-table.

The adverse effects of agriculture on groundwater quality can be reduced by sound management practices and optimum use of fertilizers, pesticides and irrigation water. LeGrand (1970) suggested that chances of groundwater contamination are reduced at those sites that have (1) deep water-table, (2) high clay content, (3) water supply wells located on the upgradient side from the source of contamination, and (4) large distances between wells and contamination sources.

12.3.3 Radioactive Waste Disposal

Radioactive waste is generated in all the stages of the nuclear fuel cycle. This includes mining, milling and refining of uranium ore, fuel fabrication and fuel consumption in reactors, waste solidification and burial of solidified waste. The large amount of solid material obtained from mining and milling is a low-level radioactive waste. The uranium refining process also produces small quantities of solid or semisolid low-level radioactive waste.

Large volume of high-level radioactive waste is produced from nuclear reactors for power production, weapons manufacturing and research. This is known as reactor waste. Radionuclides with short half-lives can be disposed of on or in the ground depending on their sorption by the mineral grains and organic matter till they decay to safe levels. The reactor waste is currently temporarily stored in iron, concrete or steel containers which are put underground, preferably in the unsaturated zone above the water-table. Ultimately these will be stored in the underground repositories.

A list of environmentally important radioactive isotopes produced from various industries and at different stages from the nuclear fuel cycle, nuclear weapons test, research and medical waste is given

Table 12.2 Sources of environmentally important radioactive isotopes. (After Bedient et al. 1993)

Source	Radionuclides
Mining waste, uranium, phosphate, coal	^{222}Rn, ^{226}Ra, $^{230,\ 232}$Th
Industrial wastes, e.g. nuclear power plant	$^{59,\ 63}$Ni, ^{60}Co, ^{90}Sr, $^{93,\ 99}$Zr, ^{99}Tc, ^{107}Pd, ^{129}I, ^{137}Cs
Nuclear weapons tests	^{3}H, ^{60}Co, ^{90}Sr, ^{137}Cs, $^{239,\ 240}$Pu
Weapons manufacturing, research and medical waste	$^{134,\ 137}$Cs, ^{60}Co, ^{151}Sm, $^{152,\ 154}$Eu, ^{237}Np, $^{239,\ 240,\ 242}$Pu, $^{241,\ 243}$Am

in Table 12.2. Nuclear and health physics data for some selected radionuclides is listed in Table 12.3. The wastes from mining, milling and refining processes usually contain ^{226}Ra, ^{230}Th and ^{238}U. ^{226}Ra is a common contaminant in surface and groundwater associated with uraniferous rocks. It has a half-life of 1600 years and poses the biggest environmental problem as even small amounts of ^{226}Ra leached from waste rock or tailings can make groundwater unfit for drinking purposes.

Most of the reactor waste from nuclear power plants and nuclear weapons consists of ^{90}Sr, ^{137}Cs and ^{60}CO which have half-lives of 29, 30 and 5 years respectively and will decay to safe levels in about few hundred years. Longer containment is required for Pu and Am which alongwith their radioactive daughters may require 500 000 years of isolation from the environment. In a time frame of one million years, isotopes of ^{99}Tc (half-life 2.1×10^5 year), and ^{237}Np (half-life 2.2×10^6 year) pose a greater potential to escape from repositories due to their long half-lives.

The high-level radioactive waste containing various radionuclides may enter groundwater flow systems either by intentional release or accidental escape from atomic/nuclear reactors and radioactive waste management systems. One such example is from the Chernobyl disaster in Ukarine (former USSR), on 26 April 1986, which caused widespread contamination of air, soil and groundwater from radioactive nuclides i.e. ^{137}Cs, and ^{90}Sr, coming to the soil surface from the atmospheric fallout. A plume of radionuclide also

reached Sweden, among other countries. The analytical data show that ^{137}Cs and ^{90}Sr concentrations in groundwater samples taken during 1992–1997 from wells in Kyiv and Chernobyl regions reached $100\,\mathrm{Bq\,dm^{-3}}$ indicating preferential flow of contaminated water into the subsurface up to depths of 100 m and more. This indicates the necessity of assessing the vulnerability of not only shallow aquifers but also of deep confined aquifers (Shestopalov et al. 2006).

The safe disposal of nuclear waste is a very difficult problem from the perspective of contamination. The migration of radionuclides from a repository will vary significantly depending on the rock and site characteristics and the properties of the radioincludes. Most of the repositories are located in rocks of very low permeability, viz. crystalline rocks, shales and clays, so that the infiltration of water is slow, and thereby there will be sufficient time for the substantial decay of radionuclides having short half-life.

Storage of radioactive waste in underground fractured rocks repositories, will cause thermo-mechanical effects induced by the heating of fluids and the rock material. This may lead to further fracturing or reduction in porosity depending upon the degree of saturation and interconnection between existing fractures. Decrease in the mechanical properties of rocks and triggering of earthquakes are some of the other associated problems (Domenico and Schwartz 1998).

The movement of dissolved radionuclides in groundwater is retarded by several processes such as diffusion, adsorption, membrane filtration, chemical

Table 12.3 Nuclear and health physics data for selected radionuclides. (After UNESCO 1980)

Radionuclide	Half-life (yr)	Major radiation[a]	Critical organ	Biological half-life	MPC[b] (μci ml^{-1})
^{3}H	12.3	β	Total body	12 days	3×10^{-3}
^{90}Sr	28.1	β	Bone	50 days	3×10^{-6}
^{129}I	1.7×10^7	β, γ	Thyroid	138 days	3×10^{-8}
^{137}Cs	30.2	β, γ	Total body	70 days	3×10^{-3}
^{226}Ra	1600	α, γ	Bone	45 days	3×10^{-8}
^{239}Pu	24 400	α	Bone	200 days	3×10^{-6}

[a] α alpha particle, β beta particle, i.e., electron, γ gamma ray
[b] *MPC* maximum permissible concentration for water consumed by the general public without readily apparent ill effects

precipitation and radioactive decay (Sect. 7.4.1). For example, zeolites and montmorillonite have considerable influence on the migration of radioactive nuclides (Domenico and Schwartz 1998). All nuclides have some degree of sorption except ^{14}C and ^{129}I. Sorption coefficient for different radionuclides varies significantly for different host materials such as granite, basalt, shale, clay and salt which form the common repositories for radioactive waste disposal (see Table 7.2). The degree of adsorption is also related to the pH of the solution. Readers may refer to Matthess (1982) for case studies of relative distribution of radioisotopes below some radioactive waste disposal sites.

Fracture characteristics of host rock (aperture, length and interconnections) and fracture infilling will have a considerable influence on rate of transport (Sect. 7.4). It is also important to visualize possible changes in the permeability of the fractures with time due to their opening or closure by processes of removal or deposition of clay material, thermal changes and seismic activity as we are thinking of safety measures for the disposal of radioactive waste in terms of thousands of years. The temporal and spatial distribution of concentration and flux through fractures for some typical high level waste (HLW) radionuclides is given by Krishnamoorthy et al. (1992) (see Sect. 7.4.1).

As the high level nuclear wastes have to be isolated from the accessible environment over long time period ($>10^3$ years), it has to be demonstrated that the disposal of high level radioactive waste will not have any adverse effect on the environment and human health so that such programmes are acceptable to the public. Therefore, both site characterisation (SC) and performance assessment (PA) of nuclear waste repositories are important. Several studies on modelling of groundwater and solute transport are being carried out to have an improved understanding of radionuclide transport in fractured rocks. These studies are quite challenging as they involve a precise understanding of the transport processes for predicting the possible migration of radionuclides with groundwater over long distances and time inspite of the fact that experiments can be made over much shorter times and distances. For details readers may refer to Hodgkinson et al. (2009) and references therein. The importance of hydrological and ecological modelling on the assessment of nuclear waste assessment is emphasised by Berglund et al. (2009).

Underground laboratories for such studies are of two types, those associated with existing mines and other underground cavities and those constructed specifically for testing purposes. The repository sites are provided with drifts for inspection purposes to monitor the migration of contaminants after waste emplacement and for assessing the safety. Sub-seabed disposal as an alternative method is also suggested. Initial studies considered salt deposits and granite to be more favourable due to the high thermal conductivity of these rocks in order to combat the problem of long term production of heat due to decaying radionuclides (Domenico and Schwartz 1998). The heat generated will change the pre-emplacement conditions such as alteration of minerals, and hydraulic and thermo-mechanical properties including triggering of seismicity.

Some of the well cited examples of field studies on land are from crystalline (granite) rock terrains, e.g. Stripa iron ore mine and the Aspo Hard Rock Laboratory (HRL) in Sweden, Underground Research Laboratory (URL) in Manitoba, Canada, the Grimsel Rock Laboratory in Switzerland, ONKALO in Finland and Beishan area of Gansu province, China. Similar studies are also done in Columbia basalts at Hanford, in Washington state and volcanic tuffs of Yucca Mountain, Nevada, USA. Bae et al. (2003) have given an account of hydrogeological and hydrochemical studies for the HLW waste in a granitic terrain in South Korea. Other low permeability formations viz. bedded and domal salt deposits in Germany and USA and clay formations in Belgium, Switzerland and Italy also form potential repositories for radioactive waste (Neerdael et al. 1996; Put and Ortiz 1996; NRC 1996). The study involves surface and borehole techniques for characterization of the subsurface geological, hydrogeological, and hydrochemical environment and to assess the changes in the physical and chemical conditions in the rock mass and groundwater caused by excavation and disposal of nuclear waste. Subsurface geophysical methods are of much help in characterizing the fractures. Special hydraulic tests viz. crosshole tests and tracer injection tests have contributed to a better understanding of the flow regime and interaction of solutes with the rock material (Simmons 1985; Abelin and Birgersson 1985, Put and Ortiz 1996 and Walker et al. 2001). Ilman (2001), Ilman and Neuman (2003), and Ilman and Taratovsky (2005, 2006) have described details of crosshole pneumatic tests from the Apache Leap Research Site (ALRS) near Superior, Arizona, USA (see Sect. 9.2.4).

At the Yucca Mountain, Nevada, USA, the repository for the disposal of High Level Waste (HLW) is proposed in the unsaturated (vadose) zone within bedded volcanic tuffs of Tertiary age in a desert environment. The tuffs contain fractures formed during cooling of the magma and are confined to individual beds. A number of large faults cut across these beds. The potential repository is located at a depth of approximately 350 m beneath the ground surface and 225 m above the water-table. Climate was an important reason, for choosing Yucca Mountain as the leading candidate for the permanent nuclear waste storage in USA. The area gets about 15 cm of rain a year. Infiltration is less than 1 mm per year and water-table is deep (>600 m). Faults and fractures will play an important role in the movement of contaminants. Larger fractures filled with gas or air, above the water-table, will provide an effective barrier to fluid flow. On the other hand, fracture zones and faults may serve as conduits for the transport of water from the surface to the water-table below the repository (NRC 1996). In the unsaturated fractured rock, the contaminants would migrate both laterally and upward by capillary action in vapour form to the atmosphere and also downward to the zone of saturation (Evans and Nicholson 1987). This would contaminate the biosphere as well as the groundwater. The rate of migration of the contaminants to the biosphere and to the zone of saturation would depend on fracture characteristics, degree of saturation, temperature conditions and air circulation in the vadose zone. The development of conceptual models for flow in variably saturated fractured rocks as at the Yucca Mountains is still incomplete. Recent studies have created doubts on the suitability of this repository as there are indications of fast seepage of rainwater through the vadose zone. It was earlier proposed that this repository will be ready to accept waste in 2010 but this is delayed due to technical difficulties, public protests and resistance from the host state.

In India, the deep abandoned gold mine in the Precambrian crystalline rocks in the Kolar Gold Field is regarded to be a potential repository for the disposal of HLW waste (Bajpai 2004).

12.4 Organic Contaminants

Contamination of groundwater by organic compounds due to inadvertent release of fuel and industrial compounds is of great concern and poses safety hazards to drinking water especially in the industrialized countries of Western Europe and North America. The developing countries are also facing similar problem due to rapid industrialization. The subsurface migration of these contaminants depends on their physical properties viz., vapor pressure, solubility and density, and the hydrogeological characteristics of surface and subsurface formations.

Petroleum hydrocarbons are the most common organic contaminants but they do not pose significant risk because of their lower aqueous solubility and toxicites. However, halogenated hydrocarbons e.g. carbon tetrachloride and ethylene dibromide, where certain hydrogen atoms are replaced by chlorine, bromine and fluorine atoms are of greater concern due to their persistence and toxicity (Kehew 2001). Halogenated hydrocarbons are used for making plastics, as solvents and many types of manufacturing processes. Halogenated hydrocarbons may cause brain disorder and liver damage etc. It is however difficult to establish which compounds are more toxic due to insufficient data from clinical tests. Their permissible limit in drinking water is less than one part per billion.

Most of the organic compounds are immiscible in water with limited dissolution between aqueous and organic phases. Such liquids are known as non-aqueous phase liquids (NAPLs). If the density of NAPL is less than that of water ($<1\,g\,l^{-1}$), the liquid is classified as a light non-aqueous phase liquid (LNAPL), viz., acetone, gasoline, kerosene and benzene etc., i.e. mainly petroleum products. If the density of NAPL is greater than that of water, it is classified as dense non-aqueous phase liquid (DNAPL), common examples being chlorinated hydrocarbons, phenol, coaltar and chloroform etc.

12.4.1 Transport of LNAPLs

The transport of LNAPL depends upon the quantity of LNAPL released. If the quantity is small, it will flow through the unsaturated zone. The infiltrating water will dissolve the soluble components of LNAPL and transport them to the water-table forming a plume while the volatile components will be transported to the upper parts of the aquifer by molecular diffusion causing secondary contamination in the unsaturated zone by volatilization or by dissolution in groundwater (Fig. 12.4). The composition of the dissolved plume is

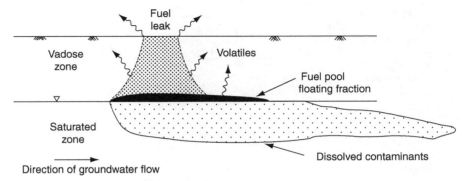

Fig. 12.4 LNAPL dissolved into groundwater forms a contaminant plume flowing in the direction of groundwater flow. (After Schwartz and Zhang 2003)

a complex mixture of many compounds that change in composition along the flow path (Kehew 2001). Depending on the permeability characteristics of the saturated zone, the volatiles can migrate up to several tens of meters in a few weeks and may either escape into the atmosphere or retained in significant high concentrations in the soil mass (Schwartz and Zhang 2003). If the volume of released LNAPL is large, it will move through the unsaturated zone to the top of the capillary fringe, floating above the water and moving in the direction of groundwater flow (Fig. 12.5). Continuous input of LNAPL may cause depression of the NAPL—water-table interface. Further, water-table fluctuations will cause complexities in the distribution of NAPL as a rising water-table is likely to entrap the NAPL below it. A fluctuating water-table will result in a more complex horizontal and vertical distribution (redistribution) of different component phases in groundwater and vapor phase in the unsaturated zone and smear the NAPL compounds over a thicker vertical range of the aquifer (Kehew 2001).

A blanket of LNAPL with sufficient thickness floating on the water-table can be detected by electri-

cal resistivity and GPR (Ground Penetrating Radar) methods due to higher electrical resistivities. However, biodegradation may lead to free ions which causes reduced resistivities (Kirsch 2006; Atekwana et al. 2006). Therefore, the resistivity anomalies of a LNAPL spill will change from high resistivities of a fresh spill to low resistivities after biodegradation (Kirsch 2006; Atekwana et al. 2006).

12.4.2 Transport of DNAPLs

The most common DNAPLs are halogenated hydrocarbons (viz. trichloroethylene-TCE), coaltar, wood treating oil, and polychlorinated biphenyls (PCB). Pure chlorinated solvents are generally more mobile than coaltar and PCBs due to their relatively high density and viscosity ratios.

The migration paths of DNAPL plumes in the subsurface are complicated. They are influenced by the gravity and groundwater flow, and also by the permeability structure of the underground matrix, dip direction

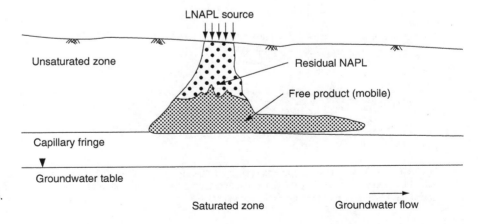

Fig. 12.5 Movement of LNAPL in the unsaturated zone and the capillary fringe. (After Hasan 1996)

Fig. 12.6 Conceptual distribution of pure DNAPL, a vapor in the soil zone and a dissolved phase in the groundwater. (After Domenico and Schwartz 1998)

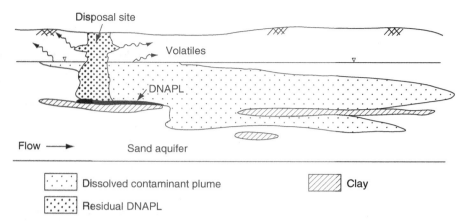

of underlying confining layers, and by the interaction with clay and other minerals. Lithologies influence the movement of DNAPL as clay rich horizons are likely to be more water saturated thereby forming barriers for the downward movement of DNAPL (Fig. 12.6). DNAPLs due to their high specific gravity and low viscosity, as compared with water, have a tendency to move downwards to the base of the aquifer displacing groundwater in a pattern known as viscous fingering.

After reaching a low permeability layer, DNAPL will move laterally in the aquifer spreading in the dip direction which may be different than the direction of groundwater flow (Fig. 12.7). DNAPLs while moving downwards in the aquifer will be partially dissolved in water forming large dissolved plumes moving in the direction of groundwater flow. Enhanced DNAPL concentration can be found in structural traps like paleochannels, fault zones or depressions of the aquifer bottom.

As the main body of DNAPL moves through the rock, a part of it will be retained infilling pores and fracture spaces known as residual. The residual remains heterogeneously distributed (Fig. 12.8). It may be partly reduced by process of dissolution, diffusion, advection and volatilization but the remaining part may remain in the unsaturated zone for years or decades. In granular aquifers, due to higher porosity, residual DNAPL can occupy up to about 8–10% of the rock volume. However, in fractured hard rocks, the residual DNAPL will occupy a much smaller fraction of the rock volume (Lawrence et al. 2006).

In fractured rocks, DNAPL will migrate through fracture networks under the influence of gravity and viscous forces. Capillary forces will resist the forward

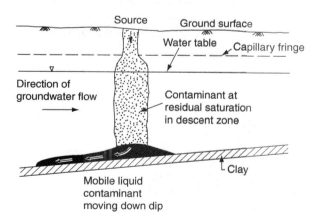

Fig. 12.7 The flow of DNAPL pool is opposite to the groundwater flow and is controlled by the dip of the underlying impervious (clay) layer. (After Domenico and Schwartz 1998)

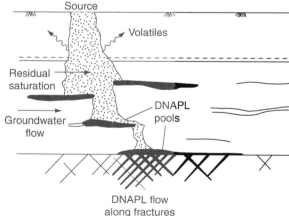

Fig. 12.8 Complex distributions of DNAPL pools controlled by the rock heterogeneties and fractures. (After Domenico and Schwartz 1998)

movement of DNAPL mass which causes separation and immobilisation in the form of ganglia from the DNAPL mass.

In fractures, DNAPL movement will be rapid and localized controlled by fracture aperture, interconnectivity and orientation resulting in complex patterns of DNAPL distribution (Fig. 12.8). Travers et al. (2001) have shown that in fractured rocks, the DNAPL plume can migrate along fractures which may be at an angle to the piezometric gradient. In vertical fractures, if present, DNAPL will move relatively easily.

In a double-porosity fractured sandstone aquifer, the movement of DNAPL in fractures will be different than in the matrix block. As the fracture porosity is usually lower than that of matrix blocks, therefore the contaminant would be in contact with a greater area of matrix blocks than fractures. The DNAPL mass from the fracture will also have a tendency to diffuse into the porous matrix and therefore one could anticipate in fracture sites where little or no DNAPL exist although it was there earlier (Freeze and McWhorter 1997).

The migration of DNAPL through fractures primarily depends upon the capillary pressure (P_c) of the invading fluid and the entry pressure (P_e). The invasion front will progress successively through interconnected fractures, where the capillary pressure of the invading fluid exceeds the threshold (entry) pressure (P_e) of the given fracture (Wealthall et al. 2001). The capillary pressure (P_c) and entry pressure (P_e) are defined by Eqs. 12.1 and 12.2.

$$P_c = (\rho_{nw} - \rho_w)\, gh \qquad (12.1)$$

$$P_e = 2\sigma\cos\theta/e \qquad (12.2)$$

where ρ_{nw} is the density of non-wetting fluid, ρ_w is the wetting phase density, g is the gravitational constant, h is the height of the DNAPL pool, σ is the interfacial tension between the DNAPL and water, θ is the contact angle on the fracture walls, and e is the fracture aperture. Equation 12.2 shows that entry pressure is directly proportional to the interfacial tension between the liquids of concern and inversely proportional to the fracture aperture.

Numerical simulation shows that the DNAPL will enter the fracture at points of largest aperture (Kueper and Mcwhorter 1992). The ability for DNAPL to enter smaller apertures depends upon whether or not the cap-

illary pressure at the advancing front exceeds the local entry pressure of the fracture. This ability increases as a function of the depth of penetration. The implication of this finding is that one can expect migration of DNAPL to greater depths in fractured rocks although the frequency and fracture aperture tends to decreases with depth. The above simulation studies further show that the time taken for DNAPL to traverse a fractured rock is inversely proportional to the height of the DNAPL pooled above the fracture, and the aperture and dip of the fracture.

Accurate prediction of DNAPL movement is fractured rocks is difficult due to extreme heterogeneities and problems in aquifer characterization (Kueper and McWhorter 1992; Freeze and McWhorter 1997; Wealthall et al. 2001; Lawrence et al. 2006). Estimation of spatial variation in fracture aperture and interconnectivity is important to study the movement of these contaminants. While discussing the contaminant migration studies in the fractured Triassic Sherwood sandstone in northern England, Lawrence et al. (2006) have outlined necessary investigation methodologies including sampling density for understanding the contaminant pathways and distinguishing between aqueous and non-aqueous phases of DNAPL.

12.4.3 Remediation

In homogenous porous aquifers, it is easier to delineate and control halogenated hydrocarbons as the plumes are well defined. However in heterogeneous aquifers, the problems of delineation and remediation are multiplied. The problem of groundwater contamination is further aggravated due to the continuing dissolution of trapped DNAPL which could be a continuous source of contamination for years or even decades (Kresic 2007).

A number of approaches are available for the remediation of contaminated groundwater, viz. (1) Pump-and- treat, (2) Permeable reactive barriers (PRBs), and (3) In situ biodegradation.

In *pump-and-treat system,* contaminated groundwater is extracted from the ground and finally discharged or re-injected after treatment. It is however difficult to remove the contaminant if it has entered in the rock matrix. Horizontal wells are preferable over conventional vertical wells for extracting LNAPLs floating on

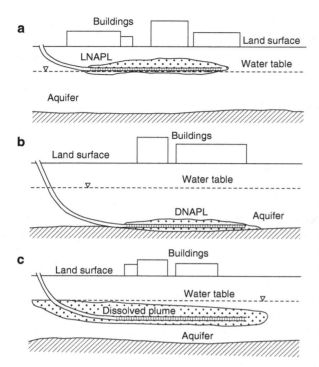

Fig. 12.9 Use of horizontal wells in groundwater remediation. **a** Recovery of LNAPL floating on water table, **b** recovery of DNAPL resting on impermeable aquifer base, **c** removal of dissolved contaminant plume. (After Kresic 2007)

water-table, for the recovery of DNAPLs resting over impervious formation, and removal of dissolved plume in pump-and teat method (Fig. 12.9).

Permeable reactive barrier (PRB) is an in situ method. A trench is made downstream of the contaminant source, and filled with some reactive material which allows the treatment of contaminated groundwater passing through the PRB. The PRBs are normally installed vertically to treat contamination plumes moving horizontally, but can also be put horizontally to check the vertical movement of the contaminant towards water-table. It is a cost-effective remediation technique and has low demand of energy. However, it is often difficult to remove the diffused dissolved contaminants from the rock matrix by conventional remediation techniques like pump-and-treat method. Chemical oxidation technologies have also proved as an alternative and cost effective method to remediate DNAPL contaminated groundwater (Tsai et al. 2008).

Biodegradation/Bioremediation Bioremediation of both inorganic and organic compounds is regarded

as a likely long-term and cost-affective solution for many types of subsurface contamination which needs greater integration of hydrogeology and microbiology (Chapelle 1993; Kehew 2001). Bioremediation degrades hazardous substances into less toxic or nontoxic substances, both under aerobic and anaerobic conditions. Many organic substances, viz. petroleum and chlorinated solvents like tri-chloroethylene (TCE), can be biodegraded by micro-organisms. Bioremediation can also transform inorganic contaminants to less toxic or immobile forms e.g. nitrate to nitrogen gas or hexavalent chromium to trivalent chromium. However, it is difficult to biodegrade certain synthetic organic compounds, such as chlorinated ethenes or chlrofluorocarbons (Houlihan and Berman 2007).

Although geologic record indicates the presence of microbes on Earth as early as the Precambrian (Atekwana, et al. 2006), very little was known earlier about the micro-organisms in groundwater. However, recent research has contributed significantly to the knowledge of microbial life in the subsurface. Micro-organisms which are of importance in controlling water quality are protozoa, viruses, fungi and bacteria. Micro-organisms most responsible for bioremediation are bacteria. Bacteria are ubiquitous in groundwater and can exist up-to about 2.5 km depth (Chapelle 1993; Kresic 2007).

Bacteria cause biodegradation of organic compounds by processes such as hydrolysis, oxidation, reduction and dehalogenation. Reductive anaerobic reduction of chlorinated solvents involves the substitution of H^+ for Cl^- in the chlorinated solvent structure, $(R-Cl)$ as given in eq. below:

$$R - Cl + H^+ + 2e \rightarrow R - H + Cl^-$$

$$(12.3)$$

Such a series of dechlorination reactions are known as sequential degradation. During each of these transformations, the parent compound $(R-Cl)$ release one Cl ion and gains one hydrogen ion. During this process, two electrons are transferred which provide energy to the micro-organisms.

Recently, a new technology known as phytoremediation using vegetation has been also suggested for bioremediation of both inorganic and organic contaminants in groundwater and surface water (Yang 2008).

12.5 Miscellaneous Sources of Groundwater Contamination

In addition to the sources of contamination mentioned above, there are several other causes of deterioration of water quality viz. (1) deforestation, over-irrigation especially in arid and semi-arid regions, (2) oil leaks and spills, (3) road salts, (4) sewage sludge, (5) urban runoff, (6) thermal power plants, (7) deep well disposal of liquid waste, and (8) sea-water intrusion etc.

Deforestation may cause rise in water-table and subsequent increase in the salinity of groundwater (Ruprecht and Schofield 1991). In arid and semi-arid areas intensive irrigation leads to rise in water-table, thereby increasing the soil and groundwater salinity due to evapotranspiration losses. This problem is quite severe in flat alluvial plains viz. Pakistan, Egypt, India and China (Greenman et al. 1967; Shahin 1985; Tyagi 1994; Chen and Cai 1995). A case study of soil salinization due to rise in water-table from San Joaquin Valley in California, USA is described by Kehew (2001) (Sect. 20.5.8). Deterioration in groundwater quality due to sea-water intrusion is dealt in Sect. 20.7.2. Readers may also refer to Bouwer (1978), Freeze and

Cherry (1979), and Watson and Burnett (1993), among others, on this subject.

12.6 Evaluation of Contamination Potential and Hazard

There is no precise method for evaluating the contamination potential and hazard from waste disposal sites. A precise knowledge of hydrogeological environment and chemical and biological aspects of the various types of wastes is necessary for such studies.

An empirical point count system for evaluating the contamination potential was suggested by LeGrand (1964). The system was designed for contaminants that attenuate or decrease in potency with time or by oxidation, chemical or physical sorption, and dilution through dispersion. Thus, it is suitable for a quick initial appraisal of sites, where hydrogeological data are scarce. Three rating charts were given for different geological environments; one such rating chart applicable for radioactive waste disposal in fractured rocks and unconsolidated alluvial formations is given in Fig. 12.10. It

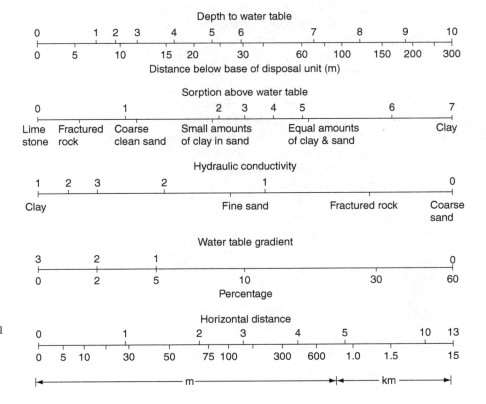

Fig. 12.10 Rating chart for pollution assessment due to radioactive waste-disposal in both fractured rocks and unconsolidated sediments. (After LeGrand 1964; reprinted from Journal AWWA, Vol 56, No.8, by permission, copyright 1964, American Water Works Association)

Table 12.4 Estimation of pollution potential by LeGrand method. (After Le Grand 1964)

Total points	Probability of pollution
0–4	Imminent
4–8	Probable or possible
8–12	Possible but not likely
12–25	Very improbable
25–35	Impossible

uses parameters of depth to water-table, sorption above the water-table, hydraulic conductivity, hydraulic gradient and horizontal distance from the disposal site. A numerical value is read above the line from the chart for each of the above five factors, corresponding to the data below the line. The sum of the points for all the given factors is a measure of the pollution potential of the site (Table 12.4). Based on the above method, the US Environmental Protection Agency (1978) developed the Surface Impoundment Assessment (SIA) method for evaluating groundwater contamination potential. This method has been applied to a large number of waste water ponds, septic tank systems and landfills (e.g. Canter 1985).

A distinction is also between pollution potential and pollution hazard. Pollution potential is considered as a natural risk index, not involving the effect of man-made features while groundwater pollution hazard is considered in terms of the probability of groundwater pollution in the area. Therefore, pollution hazard is based on pollution potential together with the society's pollution contribution in the area, given in terms of land use/cover, and hazardous material sites etc. The DRASTIC model developed by Aller et al. (1987) has been more frequently used to assess groundwater pollution potential and hazard, which has been described in Sect. 6.4.5. Denny et al. (2007) have modified the DRASTIC methodology for mapping the vulnerability of fractured bedrock aquifers by incorporating the structural (fracture) characteristics. This modified methodology has been termed DRASTIC-Fm (See Sect. 6.4.5).

Summary

Groundwater contamination is usually due to either mineral dissolution of the aquifer or as a result of human activity. This problem has become more acute in the last few decades as a result of increased industrialization, urbanization and agricultural activities which have seriously deteriorated the quality of both surface water and groundwater.

Mineral dissolution may result in undesirable concentration of arsenic, fluoride, and several heavy metals, viz. iron, chromium, lead, zinc and mercury. Radon (^{222}Rn) which is a decay product of uranium also deteriorates water quality in crystalline rocks making it undesirable for human consumption. Disposal of radioactive waste material from mines, and atomic reactors also poses a serious problem of groundwater contamination.

Contamination of groundwater by organic compounds, viz. petroleum hydrocarbons and halogenated hydrocarbons is also of serious concern. The movement of dense nonaquous phase liquids (DNAPLs) in the aquifer is a complex process depending upon permeability structure of rocks. The movement of DNAPLs in fractured rocks is controlled by fracture aperture, interconnectivity and orientations resulting in complex distribution of organic compounds. Some of the methods for the remediation of contaminated aquifers from organic compounds are pump-and-treat, permeable reactive barriers, and in situ biodegradation

Further Reading

Delleur JW (ed) (2007) The Handbook of Groundwater Engineering. 2nd ed., CRC Press, Boca Raton, FL.

Domenico PA, Schwartz FW (1998) Physical and Chemical Hydrogeology. 2nd ed., John Wiley & Sons, New York.

Fetter CW (1998) Contaminant Hydrogeology. 2nd ed., Printice Hall, New Jersey.

Kehew AE (2001) Applied Chemical Hydrogeology. Prentice Hall, New Jersey.

Schwartz FW, Zhang H (2003) Fundamentals of Ground Water. John Wiley & Sons Inc., New York.

Hydrogeology of Crystalline Rocks 13

13.1 Introduction

Crystalline rocks include plutonic igneous rocks (granites, diorites, etc.) and metamorphic rocks (gneisses, granulites, quartzites, marbles, schists and phyllites, etc.). The plutonic igneous rocks, viz. granites, usually occur as large size intrusive bodies (plutons) while some other rocks, viz. dolerites and pegmatites, are of comparatively small size in the form of dykes and veins. The hydrogeological characters of dykes are described in Chap. 14 as they are more commonly found in volcanic rock terrains. Like other hard rocks, the crystalline rocks are characterized by negligible primary porosity and permeability. However, weathering and fracturing can impart significant secondary porosity and permeability which is highly variable.

The crystalline rocks form important Precambrian shield areas in different parts of the world, mainly in Canada, Scandinavian countries, northeastern United States, India, Sri Lanka, China, Australia, Russia and many African countries. Less extensive but important outcrops of crystalline rocks are also found outside the above shield areas, as in Spain, France and the Bohemian massif in Central Europe. The total extent of shield rock outcrops is estimated to be about 20% of the present land surface, i.e. about 30 million km^2 (Fig. 13.1).

Earlier, the crystalline rocks, like other hard rocks, were not given due attention for groundwater development on account of low permeability and also difficulties in water-well drilling. However, in the last few decades, due to the need for safe drinking water for vast rural populations, especially in the developing countries, crystalline rocks are being investigated in detail for groundwater development (Wright and Burgess 1992; Lloyd 1999; Ahmed 2007). Faster and more efficient method of water-well drilling, like down-the-hole hammer (DTH), have also greatly promoted the construction of wells in crystalline rocks. Crystalline rocks also form a potential host rock for the safe disposal of high-level radioactive waste (Sect. 12.3.3).

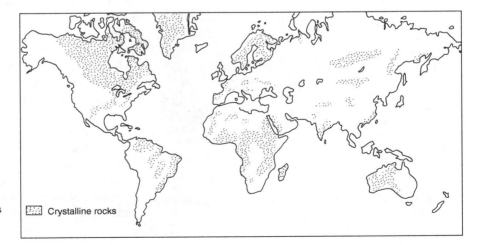

Fig. 13.1 Major occurrences of the basement crystalline massifs on the continents

☒ Crystalline rocks

B. B. S. Singhal, R. P. Gupta, *Applied Hydrogeology of Fractured Rocks*, DOI 10.1007/978-90-481-8799-7_13, © Springer Science+Business Media B.V. 2010

13.2 Landforms and Drainage

The commonly developed landforms in crystalline rock terrains are structural hills, inselbergs, pediments, buried pediments, erosional valleys and valley fills (Fig. 13.2). An aerial stereopair of metamorphic rock terrain in Central India is given in Fig. 13.3a, b and its interpretation map in Fig. 13.3c showing different landforms. The landform interpretation map of a granitic terrain derived from an IRS-image (Fig. 4.9) is given in Fig. 13.4. The characteristic features including groundwater potential of these landforms are listed in Table 13.1.

Structural Hills The morphology of structural or residual hills is controlled by large-scale rock structures and lithology. They evolve by the combined processes of tectonism and denudation. Since the rocks in this unit are hard and compact, they act as run-off zones; limited infiltration can take place along the weak planes like faults, fractures and joints. The groundwater may discharge as springs or seepages along the narrow valley portions, and the groundwater availability in general is very poor.

Inselbergs These are small residual hills which stand in isolation above the general level of the surrounding erosional plains. Similar to the structural hills, they have negligible groundwater potential due to the small recharge area, steep slopes and low permeability of rocks.

Erosional Valleys These occur within the structural hills as narrow valleys, and the bedrock in the valleys is usually weak like phyllites. They have limited thickness of unconsolidated material and hence their groundwater potential is meagre.

Pediment A broad, flat or gently sloping, rock-floored erosional surface or a plain of low relief. Such features are typically developed due to the processes of denudation by the subaerial agents including running water in an arid or semi-arid region at the base of an abrupt mountain front or plateau escarpment. At many places, the pediment is covered by a thin veneer of colluvial material; the pediment has low moisture content. Groundwater potential of this unit is limited due to the thin depth of weathered material and greater water-table fluctuation. Dugwells may not be very successful but borewells may yield small quantities of water. However, pediments developed along lineaments may form potential sources of groundwater.

Buried Pediments Formed when the sloping surface of the pediment gets gradually covered with a thick mantle of soil and colluvial material. The thickness of overburden may reach 20–100 m. The buried pediments have greater moisture and denser vegetation than pediments. The water-table fluctuation is relatively less and the recharge area is large. Hence, they form potential zones for groundwater development by dugwells and borewells.

Valley Fills A type of channel deposit developed by the process of deeper pedimentation, in an erosional environment in hard rock terrains. These are characterized by gentler slopes, greater moisture and denser vegetation. Therefore, the valley fills are the most important landforms for groundwater development in a crystalline rock terrain. It has been shown in Chap. 4 that enhancement of multispectral data could lead to better discrimination of such landform features.

Besides the above, there could be special landform features like buried channels and palaeochannels which also form potential areas for groundwater development. As these are predominantly comprised of clastic fluvial sediments, they are discussed in Chap. 16.

The drainage network shapes depend on surface geology and permeability characteristics of rocks. A dendritic drainage pattern is characteristic of massive crystalline rocks while rectangular/angular patterns are indicative of fractured rocks. Parallel patterns indicate steep slopes and often higher surface runoff.

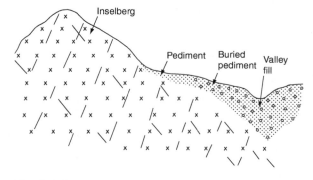

Fig. 13.2 Schematic profiles of landforms in a crystalline rock terrain

Fig. 13.3 a and **b** Represent
an aerial stereopair. The area
comprises a typical metamorphic rock terrain exhibiting
lithological bandings, folds,
faults and fractures. **c** Is the
interpretation map showing
landforms, such as denudo-
structural hills (*ridges*),
erosional valleys, pedi-
ment, buried pediments and
structural features, viz. faults.
(a and b, courtesy B. White)

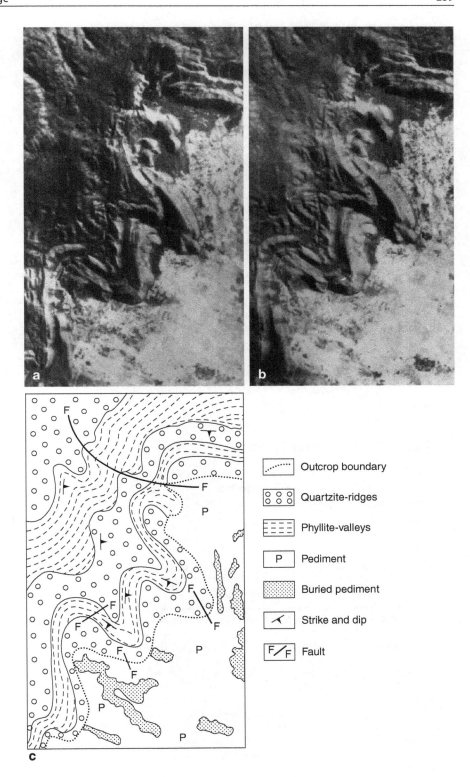

Fig. 13.4 Landform interpretation map showing inselbergs/hills, pediments, buried pediments and valley fills in a granitic terrain. The corresonding remote sensing (IRS-LISSII) FCC is shown in Fig. 4.9

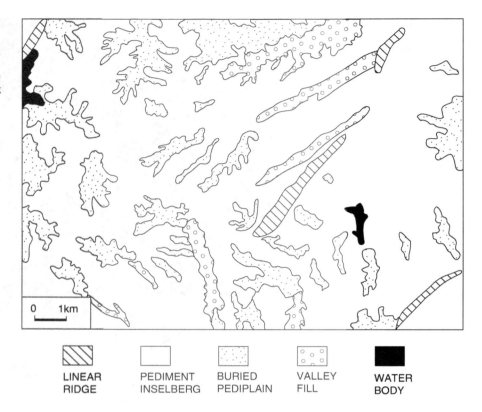

| LINEAR RIDGE | PEDIMENT INSELBERG | BURIED PEDIPLAIN | VALLEY FILL | WATER BODY |

13.3   Groundwater Occurrence

The weathered layer (regolith) and fractures are the main sources of groundwater supply in crystalline rocks.

13.3.1   Weathered Profile

The weathered layer, also called regolith, developed on crystalline basement rocks is an important source of groundwater, especially for the rural water supply

Table 13.1 Landforms in crystalline rock terrains. (Modified after Philip and Singhal 1992)

Landform	Description	Shape	Lithology	Groundwater potential
Structural hills	Ridges and erosional valleys	Ridges and valleys	Competent rocks (quartzites and granite gneiss)	Poor
Inselbergs	Isolated residual hills in a vast plain	Island-like	Quartzites, granites etc.	Poor
Pediments	Low-lying rock outcrops, some-times dissected by gullies	Irregular	Crystallines with thin veneer of soil cover	Moderate, suitable for borewells
Buried pediments	Low-lying bed rock covered with in situ material generally with high moisture	Irregular	In situ weathered material	Good, suitable for dug wells
Erosional valleys	Elongated depression occurring between two adjacent hills	Irregular generally elongated	Bedrock is of incometent rocks viz. phyllites etc.	Poor
Valley fills	Marked by presence of palaeo-drainage pattern in pediments	–	Coarse sediments and rock debris	Good, suitable for dug and borewells

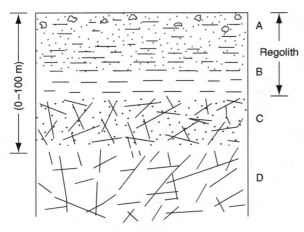

A = Soil cover with sand and clay

B = Saprolite

C = Saprock

D = Parentrock

Fig. 13.5 Idealized weathered profile in a crystalline rock terrain

in developing countries. Thick and a really extensive weathered layer can form a potential aquifer. However, if there is a good and perennial source of recharge, even a thin layer of 5–7 m thickness can be a good source of water supply.

A typical weathered profile consists of the following zones from top to bottom (Fig. 13.5).

Zone A   Sandy clay or clay-sands, often concretionary, usually a few metres thick.

Zone B   Saprolite layer having mainly massive accumulation of secondary clay minerals, and characterized by high porosity and low permeability. Thickness may reach up to 30 m

Zone C   Saprock, the parent rock progressively altered upwards with relict rock structures. Thickness may range from a few metres to about 30 m

Zone D   The parent rock with fractures, low porosity but moderate permeability.

The thickness, areal extent and physical characteristics of the weathered layer vary from one region to another depending on climate, topography, lithology, and vegetative cover.

1. Climate
This is an important factor in the development of regolith. In arid and semi-arid regions, the weath-

ered layer is usually thin (<1 m). In sub-humid areas and humid tropic regions where annual rainfall exceeds 1000 mm, the weathered layer is very thick (65–130 m), e.g. in the upper regions of Ghana, the Ivory Coast and Togo (UNESCO 1984a). Further, in some present-day arid areas, thick weathered zones may be found, for example in arid parts of Rajasthan (India), where although the present average annual rainfall is barely 380–460 mm, the weathered layer in places is found to be 25–30 m thick. Similarly, in Sudan and Nigeria, the weathered layer attains a thickness of 50 m (UNESCO 1984a). Such thick regolith zones in arid regions must be related to past pluvial climate.

2. Topography
This also influences the development of a weathered layer. Extensive thick weathered layers are developed in erosional peneplain areas of low relief, as in Brazil, Sub-Saharan Africa, Sri Lanka, Peninsular India and parts of Australia (UNESCO 1984a).

3. Lithology and texture
Lithology and texture of the parent rock influence the thickness and permeability of the weathered layer. Coarse-grained salic rocks such as granite and orthogneiss give rise to thick and permeable weathered layers. In basic rocks, like gabbro and dolerite, although the weathered horizon may be equally thick, it is more clayey and therefore less permeable.

A generalized section of the weathered layer illustrating the groundwater flow regime in an African weathered basement aquifer is illustrated in Fig. 13.6. Water-level in general represents and follows the ground topography in a subdued fashion. At shallow depths, in the interfluve areas, the water movement is vertically downwards and at deeper levels there is a slower lateral movement towards the topographic depressions forming groundwater discharge areas (*dambo*). The flow pattern also influences water quality as water at shallow depths is of low mineralization but the deeper waters are strongly mineralized.

Figure 13.7 shows possible depth-wise variation in the hydraulic properties of regolith based on data from Zimbabwe and Malawi. The overall transmissivity is low (1–5 $m^2 d^{-1}$) with occasional values of an order of several magnitudes lower or higher for the weathered material. Specific capacity also shows inconsistent values (Chilton and Foster 1993).

Fig. 13.6 Generalized section of a weathered crystalline rock aquifer from Central Africa showing groundwater flow regime. (Modified after Chilton and Foster 1993)

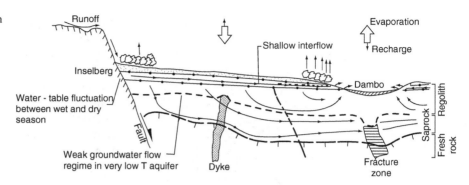

13.3.2 Fractured Media

The occurrence and movement of groundwater in massive crystalline rocks is mainly controlled by fractures and other discontinuities, which have been described in Chap. 2. Figure 13.8 shows an example where distribution of surface drainage, soil moisture and vegetation (which are groundwater indicators) is controlled by fractures and discontinuities. Characterization of the fracture system is important in order to have a proper understanding of fluid flow and distribution of permeability. This involves identification of different sets of fractures based on their orientation (strike and dip), frequency, interconnectivity, aperture, nature of filling, continuity and shape etc.

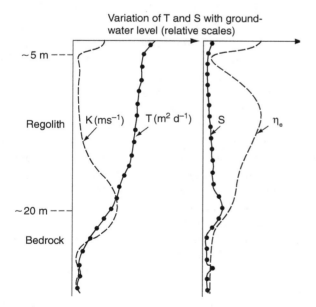

Fig. 13.7 Conceptual hydrogeological model showing variation of hydraulic properties (K, T, S, η_e) in a typical African weathered basement rock aquifer. (After Chilton and Foster 1993)

Open joints and fractures which are not filled with weathered or broken rock material form potential passages of groundwater movement but their permeability is greatly reduced when filled with clayey material such as smectite or montmorillonite. The extent of alteration and type of clay minerals which are formed depend on the original lithology and composition of the parent rock, tectonic history and composition of the circulating groundwater. The effect of fracture characteristics on rock permeability is discussed in Sect. 8.2.

Although crystalline rocks extend over large areas and to greater depths, due to structural controls it is unlikely that the flow system extends over long distances. However, there are a few exceptions as observed at the Underground Research Laboratory (URL) in Manitoba, Canada, and in Stockholm, Sweden. At URL, sub-horizontal fractures with good permeability extend over a distance of about 1 km (Fig. 13.9).

13.4 Hydraulic Characteristics

As mentioned above, in crystalline rocks, the extent of weathering and fracture characteristics decide their hydraulic conductivity and other properties.

The massive crystalline rocks have a very low porosity and permeability. The porosity of unweathered crystalline rock mass usually varies between 0.1% and 1% while weathered rocks may have as much as 45% porosity. The range of porosity and specific yield in metamorphic rocks is shown in Fig. 13.10. The hydraulic conductivity varies in the range of 10^{-6}–10^{-3} m s^{-1} depending on the extent of weathering and fracturing. The range of hydraulic conductivity and permeability values for some sites in Europe and the USA are given in Fig. 13.11.

Fig. 13.8 Example showing control of fractures and discontinuities on distribution of surface drainage, soil moisture and vegetation in crystalline rocks of Andhra Pradesh, India

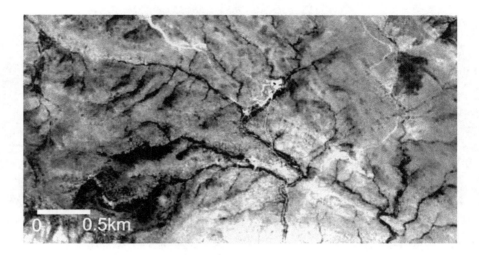

As in other rocks types, hydraulic properties of crystalline rocks can be estimated from laboratory as well as *in situ* field tests (Chap. 9). The *in situ* values of permeability determined from field tests in fractured rocks are significantly greater and more representative than laboratory values, because in the laboratory permeability measurements are usually made from compact and unfractured core samples while in the boreholes a greater thickness of rock with fractures is sampled. Permeability estimates from boreholes tests may show greater variability depending on the aperture and frequency of fractures intercepted by the borehole. On the other hand,

regional values show comparatively less variation due to proximity to REV (representative elementary volume) (Fig. 13.12).

Pumping tests including slug tests are commonly used for estimating hydraulic characteristics of crystalline rocks (Chap. 9) in areas where fractures are interconnected so that the medium is considered as a continuum. Slug tests which involve injection of water may modify the hydraulic properties due to a decrease in the effective apertures of fractures. This is attributed to blocking of fractures by particulate material, and the precipitates formed due to chemical differences in the quality of naturally occurring water

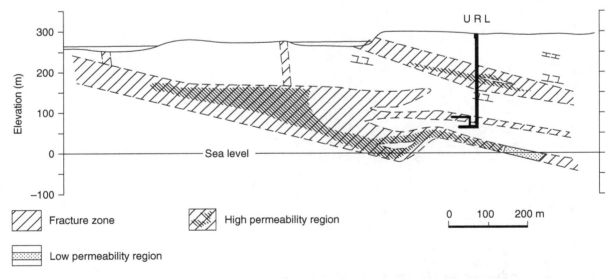

Fig. 13.9 Geological cross-section across the Underground Research Laboratory (*URL*), Monitoba, Canada, showing the regional extension of sub-horizontal fracture zones. (After Davison 1985)

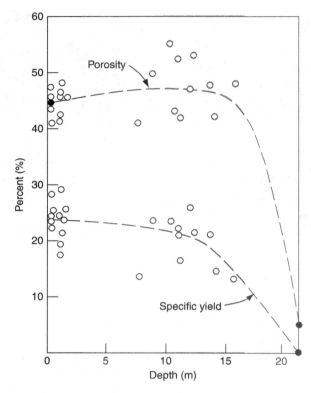

Fig. 13.10 Range of porosity and specific yield values as estimated from laboratory tests on core samples in weathered metamorphic rocks from the Georgia Nuclear Laboratory area Georgia, USA. (After Davis and DeWiest 1966)

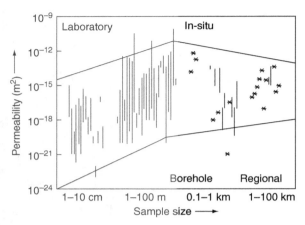

Fig. 13.12 Variation in permeability values of crystalline rocks as a function of scale of measurement; bars mark the permeability range based on several reported values, and stars represent single values. (After Clauser 1992)

and the injected water. Therefore, necessary care is required to remove the suspended material from the injected water and its chemistry should also be known (Heath 1985).

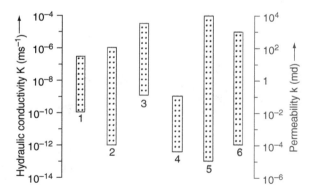

Fig. 13.11 Range of hydraulic conductivity (K) and permeability (k) in some crystalline rocks estimated from *in-situ* borehole tests. (Based on data from Brace 1984 and Black 1987) *1* Granite batholith, Monticello, SC, USA; *2* Altnabreac, Scotland; *3* Carynnen, Cornwall, UK; *4* HDR, Cornwall, UK; *5* Deep drilling in northern Switzerland; *6* Four sites in Sweden

Pumping tests in hard rocks are performed either on large diameter dugwells which tap the regolith, or on borewells in underlying fractured rocks. In most fractured rock masses, local flow is controlled by the fracture system and is usually anisotropic. Therefore, configuration of multiple observation wells for determining anisotropic permeabilities of fractured rocks from the pumping test is necessary. In areas where weathered material (regolith) is underlain by fractured rocks, and groundwater development is from both these horizons, it is also necessary to have both shallow and deep observation wells to estimate the extent of interconnection between the two aquifer systems. The aquifer response could be similar to either a confined, leaky or unconfined type depending on the relative permeabilities of the constituent horizons. Various types of drawdown and build-up curves, depending on the nature of the fractures, boundary conditions and well characteristics are reported in literature (Davis and De Wiest 1966; UNESCO 1979; Kruseman and de Ridder 1990; Boehmer 1993). These aspects are discussed in Chap. 9.

Pumping test data can be analysed by the conventional Theis and Jacob methods (Eagon and Johe 1972; Wesslen et al. 1977; Kresic 2007, among others), based on the assumption of continuum. Leaky aquifer and unconfined aquifer models are also used for aquifer parameter determination in crystalline rocks (CGWB 1980; Sridharan et al. 1990; Sekhar et al. 1993; Boehmer 1993; Levens et al. 1994; Kaehler et al. 1994). Double porosity approach is used by

Table 13.2 Range of hydraulic parameters in fractured crystalline rocks

Country	Rock type	Borehole numbers	Mean transmissivity (m^2d^{-1})	Transmissivity (m^2d^{-1})	Hydraulic conductivity (m^2d^{-1})	Testing method
Zimbabwe	Mobile Belt gneiss	228	4.2	0.5–79	0.01–2.3	Pumping test[1]
	Younger granite	209	3.6	0.5–71	0.01–1.9	Pumping test[1]
	Older gneiss	392	–	0.5–101	0.01–2.8	Pumping test[1]
Malawi	Biotite Gneiss	2	–	–	0.1–0.2	Packer test[1]
United States	Granite	58	–	–	0.0–17	Injection test[1]
Europe, USA	Varied	8 site	–	–	1×10^{-8}–0.9	Packer test[1]
India	Granite and gneiss	–	–	5–50	–	Pumping test[2]

[1] Wright (1992)
[2] CGWB (1980)

Singhal and Singhal (1990), and Sidle and Lee (1995), among others, to analyse pumping test data from fractured crystalline rocks. Methods for the analysis of test data from wells tapping a single vertical or horizontal fracture, and from fractured dyke systems (dykes and the adjoining fractured rocks) are described in Chap. 9. Specific capacity values are also used to estimate transmissivity (Sect. 9.2.5). A range of values of hydraulic properties of crystalline rocks are given in Table 13.2; values estimated at specific locations are shown in Table 13.3. These data may act as representative values and may be useful for groundwater resource estimation. A comparison of transmissivities of some crystalline rocks with those of alluvium is shown in Fig. 13.13.

In geotechnical and contaminant transport investigations, where permeability of discrete fractures and flow geometry is more important than the overall rock-mass permeability, borehole packer (Lugeon) tests and cross-hole tests are more advantageous rather than conventional pumping tests (e.g. Gale 1982a; Banks et al. 1992; Levens et al. 1994). Cross-hole and pneumatic injection tests which are specially designed to study the distribution and connectivity of fractures are useful in determining the hydraulic properties of the unsaturated zone (see Sect. 9.2.5). As pumping tests are expensive, spring discharge data can also be used to estimate hydraulic conductivity (see Sect. 15.5).

Tracer injection test in boreholes is used to study the flow regime including estimation of groundwater velocities and permeability of rock mass. Special tracer injection (forced-gradient) tests are also designed to estimate directional permeabilities and to study the mechanism and migration of solutes in fractures (see Chap. 10).

Table 13.3 Hydraulic characteristics of crystalline rocks as estimated from pumping and packer tests

Rock type	Country	T (m^2d^{-1})	K ($m\,s^{-1}$)	Source
Granite	Portugal	0.2–160	–	Carvalho (1993)
Granite-gneiss (fractured)	Zimbabwe	2–5	–	Chilton and Foster (1993)
Granite-gneiss (regolith)	Malawi	2–10	–	Chilton and Foster (1993)
Granite-gneiss	Sweden	$1–2 \times 10^{-1}$–10^{-2}	9×10^{-7}–6×10^{-8}	Carlsson and Carlstedt (1977)
Granite	Sweden	50	2×10^{-4}	Wesslen et al. (1977)
Granite-gneiss	Norway	–	10^{-8}–10^{-6}	Banks et al. (1992)
Granite	USA	–	10^{-13}–10^{-9}	Sidle and Lee (1995)
Granite-gneiss and charnockites	India	3.65–7.2	–	Perumal (1990)
Granites and gneiss	India	5–50	10^{-4}–10^{-3}	CGWB (1980)
Granodiorite	Canada	–	2×10^{-9}–5×10^{-8}	Gale (1982b)
Chert	USA	–	1.84×10^{-12}	DeWiest (1969)
Chert	Australia	0.8–20	–	Pearce (1982)
Quartzite	India	54	–	Kittu and Mehta (1990)
Slates	USA	–	1×10^{-10}	UNESCO (1975)
Schist	USA	–	1×10^{-10}–2×10^{-6}	UNESCO (1972)
Mica schist	Czech Republic	2×10^{-2}–7×10^{-3}	–	Carlsson and Carlstedt (1977)

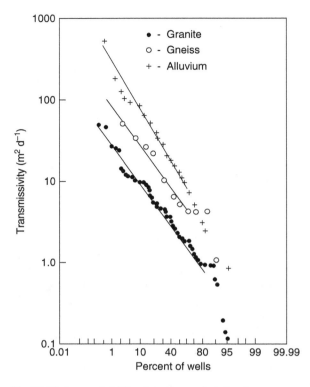

Fig. 13.13 Log probability plot of transmissivity of some crystalline rocks and alluvium from Keonjhar district, Orissa, India. (After Das et al. 1996)

It may be emphasized that for a realistic appraisal of hydraulic characteristics in crystalline rocks, an integrated study involving structural mapping and controlled hydraulic *in situ* tests is necessary. The selection

of a particular method would depend on the accuracy needed and purpose of study.

13.5 Water Wells

The common groundwater structures for water supply in crystalline rocks are dugwells, borewells, dug-cum-borewells and collector wells. Infiltration galleries and horizontal boreholes are also constructed in suitable situations (Chap. 17).

13.5.1 Well Yields

Well yields in crystalline rocks are usually low being less than $20\,m^3 h^{-1}$. Only a small percentage of wells have high yields (Table 13.4). High yields of the order of $2000\,m^3 d^{-1}$ are reported from some areas in southern Norway and central Sweden (Gustafson and Krasny 1994). Specific capacities of wells range from about $10\,m^2 d^{-1}$ in weathered crystalline rocks to about $18\,m^2 d^{-1}$ in fractured rocks (Houston 1992). Data from India on well productivity from some crystalline rocks and alluvium are given in Fig. 13.14.

Wells for a rural water supply are usually located in proximity to the demand site. A number of factors such as climate, lithology, topography, landforms, depth of waethering, distance to the nearest water

Table 13.4 Yield of borewells in crystalline rocks

Rock type	Region	Well yield ($m^3 h^{-1}$)	Well depth (m)	Source
Granite-gneiss	Southern and eastern India	15–80	3–150	UNESCO (1984a), CGWB (1995b)
Granite and Quartzite	Mt Lofty, Australia	3–300	20–200	Lloyd (1999)
Granite-gneiss, Charnockite	Tamilnadu, South India	6–18	30–70	Perumal (1990)
Granite and gneiss	Kwara, Nigeria	9.5	50–75	Houston (1992)
Granite and gneiss	Victoria, Zimbabwe	4.1	20–30	Houston (1992)
Gneiss (fractured)	Maharashtra, India	48	164	CGWB (1995c)
Granodiorite and gabbro (fractured)	Lee Valley, USA	3.5–23	60–150	Kaehler and Hsieh (1994)
Pegmatite (fractured)	Karnataka, India	110	42	CGWB (1995c)
Quartz vein (fractured)	Bihar, India	55	137	CGWB (1995c)
Marble	Sri Lanka	2.4–24	–	Jayasena et al. (1986)
Quartzite	Sri Lanka	1.6–28.8	–	Jayasena et al. (1986)
Schist	Connecticut, USA	4.16	33.4	UNESCO (1972)
Slate	Maine, USA	3.42	–	UNESCO (1979)

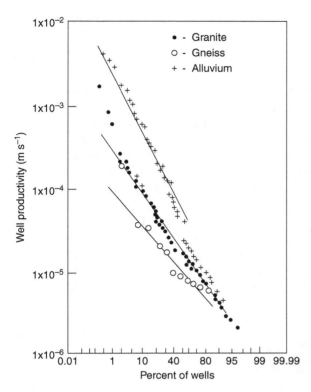

Fig. 13.14 Frequency graphs of well productivity in fractured crystalline rocks and alluvium from Keonjhar district, Orissa, India. (After Das et al. 1996)

body, well proximity to lineaments, size of the drainage area upgradient of a well, etc. influence the well yield (viz. Moore et al. 2002; Henriksen 2003; Neves and Morales 2007).

Climate has an important bearing on well yields as it influences the depth of weathering, nature of weathered material, and the recharge potential.

Lithology and rock texture also influence well yields. Generally, wells in quartz-rich and coarse-grained rocks like granites, pegmatites, quartz veins and quartzites have higher yields than other crystalline rocks as these rocks are more competent, and they develop and preserve good open joint systems. Higher yield from these rocks compared to slates and schists is due to their being more brittle, hence greater fracturing, and the coarse-grained characters of their weathered product. High yields from wells tapping pegmatites and quartz veins are reported from several places, worldwide, for example from India (Karanth 1992; CGWB 1995c), USA (Florquist 1973) and Zimbabwe (Owaode 1993). Well yields in phyllites, schists and slates is usually low (Table 13.4) as these rocks

have micaceous minerals which along with clay minerals, obtained as weathering products, tend to close the fractures. This is also the reason for their lower transmissivity as compared with granites and gneisses (Table 13.3). However, statistical studies of data from Burkina Faso, in western Africa are inconclusive about the influence of rock lithologies on well yields (Lloyd 1999). Therefore, as compared to lithology, the climate, topography, landform and rock structures are more important to account for differences in well yields.

The palaeo-weathered horizons can also form potential sources of water, if opportunities of recharge are available. For example, Henriksen (1995) reported high yielding wells with median specific capacity of $5 \times 10^{-6}\,\mathrm{m}^2\,\mathrm{s}^{-1}$, terminating at the contact of tectonized phyllite and underlying weathered granite gneiss in Norway. Similarly, wells tapping weathered granite underlying the Deccan basalt in central India have higher yields.

Topography and landform have a strong influence on well yield, especially of shallows wells, as they influence the thickness of the weathered zone (LeGrand 1967; McFarlane et al. 1992; Henriksen 1995). Wells located in valleys have greater yields than those located on steep slopes, sharp ridges and interfluvial areas. Therefore, many drillers use only the topographic considerations in predicting well yields.

LeGrand (1967) distinguished 18 types of topographic characteristics from the point of view of well yield. However, a recent reinterpretation of LeGrand's data shows that geomorphology alone is not the guiding factor in controlling well yields, but that fracture traces have a greater influence (Yin and Brook 1992). As topographic depressions usually follow fracture traces/faults, and have a greater thickness of weathered material, the probability of well yields being greater in valleys is quite logical. Further, these depressions are also occupied by streams thereby providing a greater opportunity of recharge. The effect of distance from valleys on well yields is shown in Fig. 13.15.

In Central Malawi, higher well yields are supported from areas of low relief as compared with areas with high relief adjacent to inselbergs (McFarlane et al. 1992). Statistical analysis of well yield data from Norway also indicates the importance of geomorphological controls on well yield. Wells in valley bottoms and flat lands have the highest median yields, due to the greater thickness of superficial cover and source of recharge from surface water bodies (Henriksen 1995,

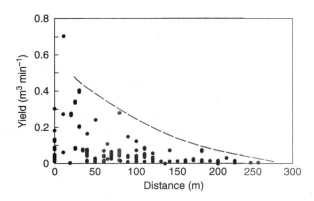

Fig. 13.15 Well yields in relation to distance from valleys. (After Hansmann et al. 1992)

2003). In Sri Lanka, the average well yield in the crystalline rocks from valleys is reported to be $3.6\,m^3h^{-1}$, but it is only $0.78\,m^3h^{-1}$ from interfluvial areas (Hansmann et al. 1992). In Uganda, out of 14 boreholes drilled in valley bottoms, 13 were productive with an average yield of $14.4\,m^3h^{-1}$, while those located on watersheds in hilly country had only 25% success, and the average yield was as low as $1.8\,m^3h^{-1}$ (UNESCO 1984a).

The influence of landforms on well yields is also demonstrated by Perumal (1990) from a study of granite-gneiss and charnockite formations in the Arthur Valley of Tamil Nadu. Fault controlled buried pediments, where the weathered horizon is thicker, give higher well yields as compared to other landforms (Table 13.5).

Rock Structures and Folding Antiformal folds develop intense fracturing at crests which is responsible for higher secondary porosity and permeability, and hence greater well yields. However, in the case of open synformal folding, groundwater is more controlled by the dip of strata. In such open large-scale structures doubly plunging synclinal folds (i.e. tectonic basins) may coincide with groundwater basins. For example, in the Precambrian basement in Southeastern Botswana, successful boreholes appear to be

Table 13.5 Yield of borewells in favourable landforms. (After Perumal 1990)

Landform	Yield (m^3s^{-1})
Valley	0.003–0.006
Shallow buried pediment	0.0006
Buried pediment (fault controlled)	0.020–0.026

preferentially situated in areas where large scale synforms intersect (Dietvorst et al. 1991).

Fractures and Lineaments Fractures and lineaments form important loci for groundwater. Lineaments generally represent vertical or near vertical fractures along which the rocks, in certain situations, may be also deeply weathered (Fig. 4.16). This may be revealed by resistivity survey (Sect. 5.2). Lineaments are also characterized by a higher degree of saturation and thick vegetative cover. The importance of lineaments for locating successful wells has been reported from several crystalline rock terrains, e.g. Kaehler et al. (1994) and Sander et al. (1997). The yield of wells located along lineaments in phyllites and quartzites in Andhra Pradesh, India, was four-orders of magnitude higher than elsewhere in similar rock types (Waters et al. 1990). Tensile fractures are more productive than shear fractures as the latter have smaller apertures and are often filled with mylonite. High well yields (50–$80\,m^3h^{-1}$) are reported from several sites located at the intersection of major tensional fractures (Waters et al. 1990).

Near-horizontal sheet joints, formed by erosional unloading, when interconnected by sub-vertical fractures, form an important source of water for shallow borewells (Carruthers et al. 1993). Such sheet joints are more open and closely spaced near the ground surface but decrease in frequency and aperture lower down.

Lineaments are best mapped as aerial photographs, and satellite images (Sect. 4.8.10). Digital image processing of remote sensing data can be carried out to enhance lineaments and also landforms. A collective study of lineaments and landforms can be extremely helpful in locating target zones (Fig. 4.19). Further, GIS-based modelling relating yields of existing wells with various spatial characteristics like distance from valleys, lineaments, and surface reservoirs may be used for potential groundwater zoning (Hansmann et al. 1992). Lineament density maps can be combined with drainage density maps to define target zones (Fig. 13.16). An example of the importance of integrated study involving geophysical, structural and geomorphological investigations for locating high-yielding wells is given in Sect. 17.2. However, in places there may be limitations to using lineaments for locating successful wells because of their variation in dip, width and possible tightening at depth due to shearing (Greenbaum 1992; Carruthers et al. 1993).

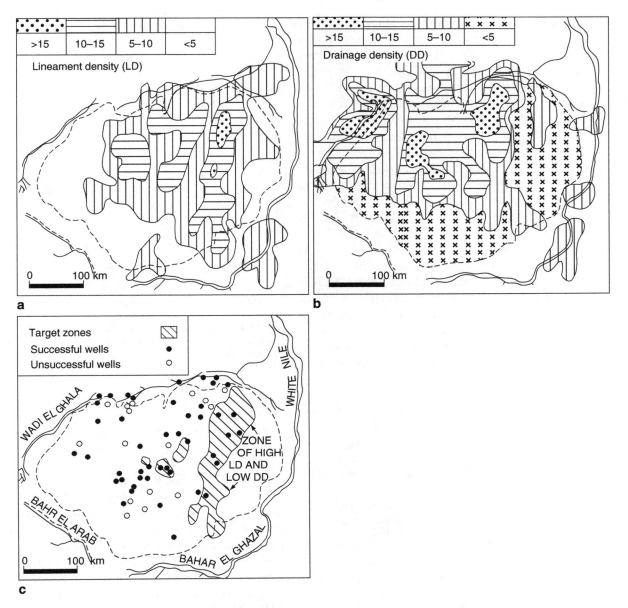

Fig. 13.16 Delineation of target zones for groundwater development in Neuba Mountain, Sudan. **a** Lineament density map showing number of lineaments per 20×20 km area, **b** drainage density map showing number of channels per 20×20 km area, **c** target zones for groundwater development selected on the basis of overlap of areas with high lineament density and low drainage density. Distribution of successful and unsuccessful wells drilled in the area are also shown. (After Ahmed et al. 1984)

13.5.2 Optimum Depth of Wells

The optimum depth of well is the depth at which an unsuccessful well should be abandoned, so as to give the lowest average cost per successful well. The concept of optimum depth of wells holds good in crystalline frac- tured rocks as the permeability usually decreases with depth (Sect. 8.2.3). While deciding the optimum depth of wells, it is necessary to consider what makes a well successful or unsuccessful in a region. The cost of well construction, water use, well yield and water quality together decide the success or failure of the well. Even a relatively small well yield (about $1\times10^{-4}\,\mathrm{m^3\,s^{-1}}$) of good quality water may make a well, fitted with a hand pump, successful for a rural domestic water supply (Houston 1992). On the other hand, for a piped water supply to small communities and for minor irrigation, the yield requirement from a well with an electric submersible pump should be higher ($>0.002\,\mathrm{m^3\,s^{-1}}$).

Read (1982) suggested that a curve between the function $-\ln[1-S(d)]$, where $S(d)$ is the estimated cumulative success rate to depth (d) when plotted against depth (d) gives a better idea of the optimum depth of a well for a given expected discharge. A smooth curve is drawn through the resulting step-graph. The optimum depth is indicated by the point where a straight line through the origin is tangential to the curve (Fig. 13.17).

An overall decrease in well yield with depth is reported from various crystalline rock terrains in different parts of the world (Davis and Turk 1964; UNESCO 1979; Woolley 1982; Henriksen 1995). Based on well productivity data, Davis and Turk (1964) suggested the optimum depth of wells in crystalline rocks to be between 50 m and 60 m (Fig. 13.18). Similar conclusions are drawn by several other workers (Table 13.6).

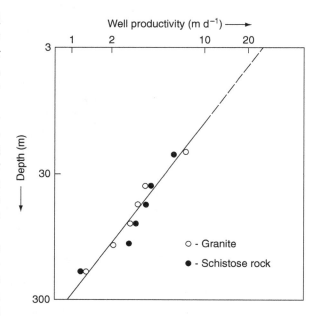

Fig. 13.18 Plot showing decrease in well productivity with depth in crystalline rocks of Eastern United States. (After Davis and Turk 1964)

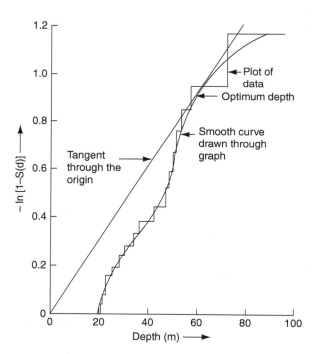

Fig. 13.17 Plot of $-\ln[1-S(d)]$ against depth of 27 wells drilled in gneisses and schists of Haast Bluff area, NT, Australia. $S(d)$ is cumulative success rate to depth 'd' where $5 \times 10^{-4}\,\mathrm{m^3\,s^{-1}}$ is classified as success. The straight line drawn through the origin, tangential to the curve, indicates an optimum depth of about 60 m. (After Read 1982)

There are also cases where well yields do not necessarily decline with depth. For example, in a recent study, Loiselle and Evans (1995) did not find any evidence of either decrease in fracture density or facture yield up to a depth of about 180 m in the crystalline rocks of Maine, USA. High yields from deep fracture horizons ($>100\,\mathrm{m}$) are also reported from Sri Lanka and India. In Sri Lanka, deep fresh-water flow is reported along the Trincomalee submarine canyon indicating a considerable loss of fresh groundwater into the sea. Data on well drilling in parts of the Indian shield also show a considerable improvement in well productivity with an increase in depth up to 300 m (Jagannathan 1993; CGWB 1995c). High discharge from deep gold mines in Karnataka, India, and Rand mines in South Africa also indicate high potentiality of deep fractured aquifers (Das 2007). Two deep wells ($>2\,\mathrm{km}$) drilled in crystalline basement of Germany under a continental deep drilling project have revealed unexpectedly high K values of the order of $4 \times 10^{-8}\,\mathrm{m\,s^{-1}}$ at a depth of 4 km and $K = 1.32 \times 10^{-9}\,\mathrm{m\,s^{-1}}$ at 4.4 km depth (Stober and Bucher 2005). These wells were pumped for 1 year and the total amount of pumped water was $23\,000\,\mathrm{m^3}$ indicating high hydraulic connectivity of deep fractures even at deeper levels under the prevailing tectonic conditions (Table 13.7).

Table 13.6 Data on optimum depth of wells in different regions and rock types

Region	Rock type	Optimum depth (m)	Source
Cyprus	Gabbro	170–200	Read (1982)
Satpura hills, India	Granite	45–60	Uhl (1979)
	Gneiss and schist	45–75	Read (1982)
North Carolina, USA	Granite	75–90	Read (1982)
Karamoja, Uganda	Gneiss and schist	30–92	Faillace (1973)
Zimbabwe	Greenstone and gneiss	40–80	Wright (1992)
Ghana	Granite and gneiss	40–50	Darko and Krasny (2000)
Norway	Granite and gneiss	40–60	Henriksen (1995)

From the above it can be summarized that although there is a general decrease of well yield with depth in crystalline rocks, the variation is case specific and may depend upon the type of formations, type of fractures and other variables. In the case of sheet joints, formed due to near surface processes, one would expect a decrease of well yield with depth. However, lineaments and other deep-seated fractures may be in an open state even at greater depths.

13.6 Assessment of Groundwater Recharge

Estimation of recharge is of importance for the assessment of groundwater resources of an area. In temperate and tropical regions, rainfall is the main source of groundwater recharge. In addition to this, other important sources of recharge are influent seepage from rivers and canals, inter-basin groundwater flow and return flow from irrigation. Various methods of estimation of groundwater recharge are described in Sect. 20.1.3. The estimation of groundwater recharge in crystalline and other hard rocks is faced with greater problems as compared with sedimentary rock aquifers, on account of their heterogeneity. Estimates of recharge from rainfall by using different methods in some crystalline rock terrains are given in Table 13.8; there is a wide variation in recharge depending upon local geology and climatic conditions. In most cases the recharge is around 5–15% of rainfall.

Isotopic and hydrochemical data are also helpful in understanding the flow mechanism and sources of recharge in fractured crystalline rocks (Fontes et al. 1989; Mazor 1991; Offerdinger et al. 2004) (Sect. 10.5.1).

The groundwater residence time in crystalline rocks may vary from only a few years in shallow weathered horizons to as much as several thousand years in deep fractures, as revealed from ^{226}Ra, ^3H and ^{14}C data (Gale et al. 1982; Sukhija and Rao 1983; Silar 1996). This indicates fast movement of groundwater in the weathered horizon but very retarded movement and lack of interconnection in deep fractures.

In areas where sources of natural recharge are limited, artificial recharge may be practiced, depending on the availability of water. An account of various methods of artificial recharge under different hydrogeological conditions is given in Sect. 20.8.2. In crystalline rocks, on account of their typical geomorphic setting and hydrogeological characteristics, the common methods of artificial recharge are: (a) percolation tanks (infiltration ponds), and (b) subsurface dykes/dams. Soil conservation methods also promote natural groundwater recharge.

Table 13.7 Deep-fluid data from the crystalline basement of central Europe. (After Stober and Bucher 2005)

Drill site	KTB-VB	Urach
Depth (m b. s.)	3850–4000	3320–4444
Permeability ($m\,s^{-1}$)	4.07×10^{-8}	1.32×10^{-9}
Porosity (%)	0.7	0.5
Temp. (°C)	120	175
TDS ($g\,l^{-1}$)	62	72
Ca	15800	1230
Mg	3	164
Na	6415	25000
K	212	905
Sr	265	111
Cl	38700	42700
HCO_3	34	294
SO_4	336	790
Br	516	475
Ca/(Na+Ca) (mol)	0.59	0.03
Cl/Br (ppm)	75	90

All concentrations in $mg\,l^{-1}$ unless noted otherwise.
KTB-VB: pilot-hole of German continental deep drilling project
Urach: HDR test site in Black Forest basement, Germany.

Table 13.8 Estimated values of groundwater recharge (mm) from rainfall

Area/Country	Annual rainfall (mm)	Method of estimation					
		Base flow measurement	Water-table fluctuation	Chloride balance	Soil moisture	Isotope method	Reference
Malawi	750–900	30–200		95–234	–	–	Wright (1992)
Zimbabwe	750–1400	50–200 (5–20%)		80–115	–	22–33	Chilton and Foster (1993)
Kwara, Nigeria	1700–2000	–	–	–	(15%)	–	Houston (1992)
Vedavati Basin, India	590–620		17–90 (3–15%)			(6.2%)	Athavale (1985)
Ponnani and Rao	800–1300				(13–21%)		Sukhija and Rao (1983)
Noyil basin, India	750–900		(5.7–12.20%)		(4.5–9.8%)		UNESCO (1984a) and Athavale (1985)

Values in brackets give recharge as percentage of rainfall

13.7 Groundwater Quality

In crystalline rocks, the groundwater flow mainly takes place through fractures. Therefore, the contact area between water and the rock matrix is small, as compared to rocks possessing primary porosity. Further, the crystalline rocks contain mainly silicate minerals which weather very slowly. Therefore, in general, groundwater in metamorphic and plutonic igneous rocks has relatively small amounts of dissolved solids (TDS) and the water is of Ca–HCO_3 type. However, in arid and semiarid regions, due to low rainfall and excessive evaporation, groundwater may have high salinity.

13.7.1 Effect of Lithology

Igneous rocks are classified as acidic, basic and ultramafic types. Acidic igneous rocks (e.g. granites) and their metamorphic equivalent (gneisses) have appreciable amounts of quartz and other aluminosilicate minerals such as feldspars and micas. Quartz has very low solubility under existing pH and temperature conditions. Other silicate minerals, e.g. feldspars and micas, dissolve incongruently in water having dissolved CO_2. This causes release of common cations like Ca^{2+}, Mg^{2+}, Na^+ and K^+ into water, leaving behind a clayey residue which is rich in aluminosilicates (clay). It is observed that nearly all groundwaters from igneous rocks plot in the kaolinite field of stability diagrams, suggesting that alteration of feldspar and micas

to clay (kaolinite) is quite common (Freeze and Cherry 1979). This is accompanied by a rise in pH and HCO_3 concentration of water.

The amount of total dissolved solids in water from crystalline rocks is usually less than $300\,mg\,l^{-1}$. The silica content is low ($10–30\,mg\,l^{-1}$), its solubility increases with temperature. There is dominance of Na^+ over K^+, as sodium is more soluble than potassium, and the latter is more easily fixed on clay minerals in the rock matrix. Ca^{2+} is also dominant over Mg^{2+} and HCO_3^- more than Cl^- and SO_4^{2-} (Table 13.9). The ratio of Na^+/Ca^{2+} in aqueous extracts is generally much lower as compared to the mole ratio in the parent rocks, indicating greater solubility of Ca^{2+} to that of Na^+. Granitic rocks usually have moderately acidic waters (pH=6.0–6.8).

At places, the fluoride content of groundwater in acidic igneous rocks (granites and pegmatites), and mica schists may be greater than the drinking-water permissible limits ($<1.5\,mg\,l^{-1}$), due to the presence of fluoride-bearing minerals like apatite, fluorite and biotite. Among other trace elements, groundwater in granitic and metamorphic rocks may have a high concentration of Pb, Cr, Fe, Mn and low I. For example; Smedley (1992) reported high Pb (around $0.15\,mg\,l^{-1}$) and Cr concentrations in acidic waters from granitic terrains in Ghana.

Crystalline basement rocks (granites and gneisses) which have high concentration of uranium and radium are reported to have radon (^{222}Rn) concentration of several hundred or even more than few thousands $Bq\,l^{-1}$ which is much above the permissible limits for drink-

Table 13.9 Chemical characteristics of groundwater from different types of crystalline rocks (mg l^{-1})

Rock Type	TDS	Ca	Mg	Na	K	Cl	SO$_4$	HCO$_3$	Fe	SiO$_2$	Location
Granite	223	27	6.2	9.5	1.4	5.2	32	93	1.6	39	Maryland, USA[a]
Granite gneiss	137	28	1.9	6.8	4.2	1.0	1.4	121	2.7	31	Baltimore, USA[b]
Granite	340	11	1.22	98	0.59	1.51	25.2	39.7	0.02	8.3	Czech Republic[c]
Diorite	347	72	4.1	10	2.8	6.5	115	114	0.04	22	N. Carolina, USA[a]
Gabbro	359	32	16	25	1.1	13	10	203	0.06	56	N. Carolina, USA[a]
Olivine-tuff (ultramafic)	281	20	42	19	–	7	22	279	–	31	Arizona, USA[b]
Quartzite	52	1.6	5.8	2.8	–	9.9	2.0	18	–	8	Transvaal, S. Africa[a]
Schist	221	27	5.7	16	0.7	2.5	9.6	138	0.11	21	Georgia, USA[a]

[a] Hamill and Bell (1986)
[b] Hem (1989)
[c] Jezersky (2007)

ing water. However groundwater in overlying Quaternary deposits derived from the weathering of basement rocks, do not have such problem due to the leaching of uranium bearing minerals viz. micas (Morland et al. 1998; Veeger and Ruderman 1998).

The basic igneous rocks (e.g. gabbros) are rich in Ca and Mg minerals, e.g. plagioclase feldspars and pyroxenes. The groundwater in these rocks, therefore, has high ratios of Ca/Na and Mg/Ca. The total dissolved solids are usually less than 400 mg l^{-1} (Table 13.9). The total iron content is moderate but silica is high (>20 mg l^{-1}). Groundwater from fissured basic igneous rocks has considerably lower radon concentration while limestone, shale and sandstone have normal radon concentration (50–150 Bq l^{-1}). The radon concentration in surface water may be as low as 2 Bq l^{-1} due to its short half-life (3.8 days).

The ultramafic igneous rocks (e.g. dunites and peridotites) are made up mainly of magnesium and iron silicate minerals (olivines and pyroxenes). Although mafic and ultramafic rocks contain little or no quartz and have lower amounts of total silica than acidic rocks (granites), the amount of dissolved silica in groundwater may be high, indicating greater solubility of silica in the presence of magnesium. The Mg content in these rocks is quite high; therefore, the spring water from these rocks is reported to be of Mg–HCO$_3$ type (Table 13.9).

Low-grade metamorphic rocks (phyllites and slates) have comparatively high porosity but low permeability. Therefore, on account of greater contact of groundwater with rock matrix, the total dissolved solids and concentration of cations and anions is greater than in high-grade metamorphic rocks like granulites and gneisses. Groundwater in pyrite bearing slates

has higher SO$_4$ and lower pH. In mica schists, which are equivalent to granite in chemical composition, the groundwater is alkali-rich but in amphibolites, which are equivalent to basic igneous rocks, groundwater is rich in alkaline earths (Matthess 1982).

13.7.2 Hydrochemical Zoning

Like sedimentary basins, the chemical composition of groundwater in crystalline rocks also varies with the climatic conditions and depth of occurrence resulting in both lateral and vertical variation in water quality. In temperate areas, the groundwaters in crystalline rocks are usually of CaHCO$_3$ type with low TDS (100–500 mg l^{-1}). Also in the humid areas of African continent (viz. Congo basin) and in areas of high altitude, groundwater has low TDS. TDS generally increases upto 3–5 g l^{-1} northwards and southwards from the equatorial zone and the groundwater quality varies from HCO$_3$ to SO$_4$ or Cl type. In desert areas (Sahara and Kalahari) TDS is high being of the order of 10 g l^{-1}. Groundwater with short residence time, viz. from ephemeral springs has low concentration of various ions while groundwater from perennial springs with longer residence time has higher concentration (Schwartz and Zhang 2003).

In Precambrian shields and massifs of crystalline rocks, vertical zoning in the chemical composition of groundwater is reported from several areas (Fig. 13.19). Near-surface groundwaters are dilute and are of Ca–Na–HCO$_3$ type. With increased depth and length of flow path, at greater depths (150–500 m), the groundwater becomes brackish and tends towards

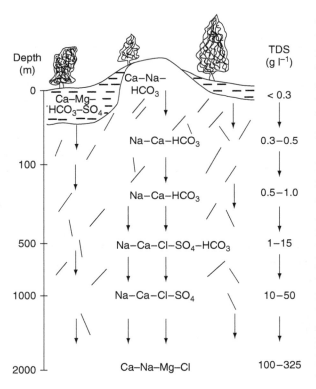

Fig. 13.19 Schematic representation of the chemical evolution of groundwater with depth in plutonic igneous rocks. (After Gascoyne and Kamineni 1993)

Na–Ca–HCO$_3$–Cl type. At increasingly greater depths (>500 m), the water is usually saline having 5000–50000 mg l^{-1} TDS and the composition is dominated by Na–Ca–Cl and Ca–Cl types. Such a general pattern has been reported viz. from the Canadian shield (Gascoyne and Kamineni 1993; Gascoyne et al. 1996), granitic suite of rocks in Sweden (Smellie et al. 1995), Jurassic granites in Korea (Bae et al. 2003), and Bohemian Massif in Czech Republic (Krasny 1999). Salinities of the order of 32000 to as high as 72000 mg l^{-1} are reported from deep boreholes at depths of 2–4 km from several shield areas of the world (Gascoyne and Kamineni 1993; Edmunds and Savage 1991; Kessels and Kuck 1995; Stober and Bucher 2005) (Table 13.7). Br/Cl ratios and isotopes (viz. ^{6}Li, ^{129}I, ^{3}H, ^{2}H and ^{18}O) have been used to study the origin of hypersaline Ca–Cl brines in the shield areas of Canada and northern Europe. ^{129}I data suggest a residence time of 80 million year. in some parts of the Canadian shield (Bottomley et al. 2003).

The saline waters in crystalline rocks are typically depleted in Na$^+$ relative to Cl$^-$ which makes them different from the saline waters formed either due to entrapment of sea-water or dissolution of evaporite deposits. The change from Ca–Na–HCO$_3$ type to Na–Ca–HCO$_3$ type with increased depth has been explained by the hydrolysis of plagioclase feldspars which liberates Na and Ca coupled with loss of Ca by precipitation and/or ion exchange (for Na) on clay minerals in the fractures (Gascoyne and Kamineni 1993). Several other explanations have been given for the origin of deep thermal saline waters including migration of sedimentary formation brines (Couture et al. 1983; Kessels and Kuck 1995), marine transgressions, residual hydrothermal fluids, breakdown of fluid inclusions, radiolytic decomposition of water and water-rock interaction (Edmunds and Savage 1991). Fontes et al. (1989) based on hydrochemical and isotopic data of deep groundwater in granites in Stripa mine, Sweden, suggested mixing of local infiltrating water, which has been circulating in the fracture system since 10–1000 years, with a residual brine of sedimentary origin entrapped in the micropores of granite. It appears that no single origin can hold good in all terrains (Horta 2005).

The composition of groundwater in deep-seated crystalline rocks is not only of geochemical interest but is also important for the disposal of radioactive waste and tapping of geothermal power. In this context two theories i.e. autochthonous and allochthonous have been suggested for the origin of deep saline brines in the basement rocks of Canada and northern Europe (Stober and Bucher 2005). Autochthonous origin supports in situ origin which is a result of extensive leaching of rocks while allochthonous theory indicates an external origin from the brines in the ancient marine deposits and therefore indicates hydraulic continuity of the deep aquifers with rocks at shallow depths. Areas where the deep waters show allochthonous character, may not be suitable for the disposal of HLW waste.

13.7.3 Groundwater Contamination

Crystalline rocks are more vulnerable to contamination from anthropogenic activities, as the vadose zone is usually thin and rapid flow takes place through fissures and joints. In arid regions, although the water-table is usually deep as the weathered layer is coarse-grained and devoid of clayey material, the contamination may move faster. High nitrate in groundwater is reported from several crystalline rock aquifers (Reboucas 1993). Contamination from pesticides is likely to be present in fractured rock aquifers and coarse-grained regolith with a shallow water-table. Movement of leachates from landfills along fractures also contaminates groundwater in crystalline rocks. Faecal contamination of shallow wells and boreholes in the weathered basement aquifers of Malawi, Botswana and elsewhere is widespread. In such cases borewells with casing pipes should be preferred to conventional dugwells in order to reduce chance of pollution.

Acid rain affects groundwater quality especially in those hard rock terrains which consist of non-carbonate rocks, as reported from the granitic rocks of Scotland and metamorphic and granitic rocks of the Bohemian Massif in the western part of the Czech Republic (Mather 1993). Crystalline rocks, due to low permeability, form potential repositories for high-level radioactive waste. These aspects are discussed in Sect. 12.3.3.

The main source of groundwater in these terrains is the weathered horizon (regolith). Groundwater potential of the regolith depends upon its thickness, lithology, permeability and climatic conditions. The potential of deeper massive rocks depends upon intensity of fracture characteristics including degree and interconnectivity.

Yields of wells in crystalline rocks is usually low which depends on rock weathering, topography and rock structures (fractures and lineaments). Water quality depends on rock compositions; TDS is usually less than $300\,mg\,l^{-1}$. Vertical zonation in water quality is reported from several shield areas of the world. Near surface water has low TDS but the salinity generally increases with depth.

The crystalline rocks form potential repositories for the deposal of radioactive waste which requires detailed hydrogeological investigations.

Summary

Crystalline rocks of Precambrian age which include plutonic igneous rocks and crystalline rocks viz. granites etc. form important shield areas in different parts of the world covering about 20% of the land surface. These rocks develop different landforms, and drainage characteristics depending on their structural and lithological characteristics and climatic conditions.

Further Reading

Krasny J, Sharp JM (eds) (2008) Groundwater in Fractured Rocks. IAH Selected Papers Series, vol. 9, CRC Press, Boca Raton, FL.

Lloyd JW (1999) Water resources of hard rock aquifers in arid and semi-arid zones. Studies and Reports in Hydrology 58, UNESCO, Paris.

Wright EP, Burgess WG (eds) (1992) The Hydrogeology of Crystalline Basement Aquifers in Africa. Geol. Soc. Spl. Publ. No. 66, The Geological Society, London.

Hydrogeology of Volcanic Rocks

Volcanic rocks are formed by the solidification of magma at or near the ground surface. The most common volcanic rocks are basalts, which are of basic composition. Acidic and intermediate rock types such as rhyolites and andesites have comparatively very limited occurrences. Basalts are formed due to the eruption of lava either on the ground surface (subaerial eruption) or on the sea floor (submarine eruption). Eruption on the land surface could be of fissure type (*plateau basalts*), covering large areas on the continents or of *central type*, which is of limited distribution mostly forming volcanic cones (Fig. 14.1).

14.1 Weathering, Landform and Drainage

Volcanic rocks possess characteristic landforms, weathering and drainage patterns. Generally, they are highly susceptible to weathering. Older extrusive rocks exhibit deeper weathering implying a thicker soil and vegetation cover; younger volcanic rocks are generally barren of vegetation. In the tropical climate, basalt weathers into 'chernozem', also known as 'black cotton soil' which being rich in clay (montmorillonite) is quite impervious but has high fertility. Under good leaching conditions, the ultimate weathering products of basalts are laterite and bauxite.

A variety of landforms develop in volcanic rock terrains which are linked to the composition of lava and type of eruption. They also govern the hydrogeological characteristics of the terrain. Acidic lava is viscous, and therefore restricted in extent, often forming steep-sided bulbous domes. Basic lava has a relatively lower

viscosity and is commonly of a ropy type. It exhibits flow structures and possesses an oblate outline with flat topography.

The central type of eruption is marked by sloping cones, calderas, vents and conical landforms. Eruptions of the fissure type are flatter near the center and become serrated along the periphery, the flows generally have a rough surface topography and discordant contacts with the bedrock. The lava flows are often interbedded with weathered and pyroclastic material and other neo-volcanic sediments. From a distance, on scarp faces, a number of flows may collectively impart the impression of rough sub-horizontal bedding. The landscape looks like a series of giant steps, each flow producing a steep escarpment and a flat top (mesa) (Fig. 14.2). This step-like topography is known as *trappen* in German from which the word trap is derived. Due to such a landscape, the basaltic flows which cover large parts in Central India are known as Deccan Traps (Fig. 14.3). These lava flows are associated with dykes which acted as feeding channels and now occur as extensive ridges (Fig. 14.4), which may form barriers for surface water and groundwater flow.

Lava tubes and tunnels are locally important subsurface features. A lava tube or tunnel is formed when the surface of a lava flow has cooled and hardened, but the interior remains more fluid and happens to drain out from beneath the solidified crust, leaving behind a tunnel-like void.

Basalts typically exhibit columnar jointing (Fig. 14.5). Some of the columns have 4, 5, 8 or 10 sides, but the majority of them are near-perfect hexagons, about 30–50 cm in diameter. They are formed due to contraction of the cooling lava, forming prismatic

B. B. S. Singhal, R. P. Gupta, *Applied Hydrogeology of Fractured Rocks*,
DOI 10.1007/978-90-481-8799-7_14, © Springer Science+Business Media B.V. 2010

Fig. 14.1 World distribution of major Mesozoic and younger plateau basalts and active volcanic arcs. *1* Siberian Traps; *2* Karoo Province; *3* Parana Volcanics; *4* Deccan Traps; *5* Thulean Province; *6* Ethiopian Basalts; *7* Columbia River Basalts

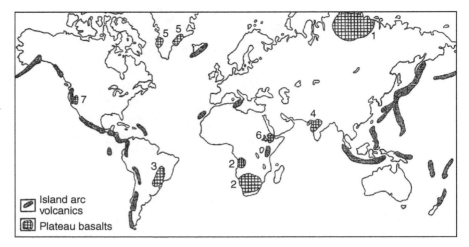

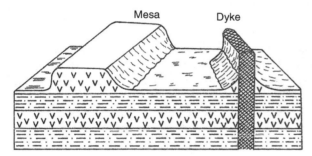

Fig. 14.2 Formation of 'mesa' and dykes in volcanic terrain

patterns in the solidifying rock. As shrinkage continues, the joints extend through the rock mass resulting in the network of vertical joints causing anisotropy and strong vertical hydraulic conductivity in basalts.

The lava flows may obliterate the pre-existing water divides, resulting in changes in surface and groundwater flow regimes, as is exhibited in the Snake River

Plain in Idaho, USA (Stearns 1942). Lakes may be formed either due to closed depressions, viz. crater lakes, and also due to damming of river courses by lava flows. In areas of tectonic activity, volcano-tectonic lakes are formed as in New Zealand and Indonesia (Ollier 1969). The drainage pattern is influenced by the type of eruption and rock characteristics; radial and annular drainage patterns are associated with a central type of eruption.

Surface drainage, in volcanic rocks, increases with weathering and age; older lavas display high drainage density and a dendritic pattern. Surface drainage in young volcanic rocks, however, may be almost absent owing to high porosity and permeability. For exam-

Fig. 14.3 Extensive layering produced by successive eruptions in Deccan Traps, Western India. (Courtesy H. Kulkarni)

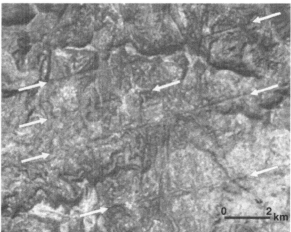

Fig. 14.4 Dolerite dykes striking nearly ENE–WSW and running for several tens of kilometre in the Deccan Trap region; the fissures corresponding to the dykes formed feeding channels for eruption of the plateau basalts. (Image source: GoogleEarth)

favour a high infiltration rate. Here, drainage basins are of a small size, and watersheds are characterized by steep slopes and steep valley walls with little channel storage. Therefore, in such areas surface storage of water is difficult due to the high rate of infiltration (Peterson 1972). Stearns (1942) has highlighted the problems of construction of reservoirs and dams in basaltic terrains.

14.2 Hydrogeology

14.2.1 Plateau Basalts

Plateau basalts, also known as continental flood basalts, are widespread in various parts of the world. They are of different geological ages, but the most common ones on the continents are of late Mesozoic and younger ages. Some of the important occurrences of plateau basalts from the hydrogeological point of view are listed in Table 14.1.

The plateau basalts usually consist of a number of flows of varying thickness superimposed over each other. The thickness of the individual layer ranges from less than 1 m to more than 30 m, most being between 10 and 30 m. The individual layer is much thicker in continental food basalts as compared to basalts of oceanic islands (Fig. 14.6). This could be due to greater original surface gradients of oceanic islands, as compared to that of continental plateau areas (Davis 1974).

The lateral extent of individual flow units in plateau basalts depends upon the composition and viscosity of lava, rate of supply and loss of volatiles during cooling. Basaltic lava with low silica and low viscosity spreads over larger areas, forming thin and extensive sheets as compared to acidic lavas such as rhyolite and andesite which are more viscous (Fig. 14.7). The lat-

Fig. 14.5 Columnar joints in basalts, N. Ireland. (After Reader's Digest Association 1980)

ple, an area of about 25 000 km² in southern Idaho, USA, underlain by basalt, has no surface run-off (Stearns 1942). Also in tropical volcanic islands such as Hawaii, despite heavy rainfall, which may exceed 750 cm per year, most of the streams are flashy and are seasonal. This is primarily due to a high rainfall and peculiar watershed characteristics which favour a high rate of flashy surface run-off; and also the high permeability of the basalts and soil cover which

Table 14.1 Major plateau basalt provinces on the continents

Country	Basalt province	Age	Area (km²)	Maximum thickness (m)
India	Deccan Traps	Upper Cretaceous to Eocene (65–60 Ma)	500 000	1500
Brazil	Parana Volcanics	Lower Cretaceous (140–120 Ma)	900 000	2000
USA	Columbia and snake River basalt	Miocene to Quaternary (17.5–6 Ma)	200 000	1500
Russia	Siberian Traps	Middle Paleozoic to early Mesozoic	1.5×10^6	700
South Africa	Karoo Province	Early Jurassic (190 Ma)	3×10^6	

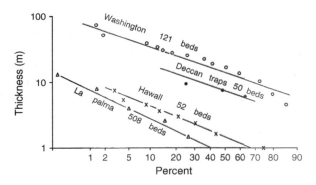

Fig. 14.6 Log-probability plot of bed-thickness of basalt flows. Data for Washington, Hawaii and La Palma are from Davis (1974) and data for Deccen Traps are from Krishnan (1949)

eral extent of individual flow may vary from a few tens of meters to as much as hundreds of kilometers. Many basalt flows occupy ancient river valleys and therefore locally enclose alluvial sediments of varying lithologies and thicknesses, which are important sources of water supply.

Many continental lava flows develop an autobrecciated top surface which is also progressively pushed beneath the flow as it advances (Fig. 14.8). This tends to produce a relatively permeable blocky top and base to many lava flow units (Mathers and Zalasiewicz 1994). Description of some important plateau basalt provinces is given below.

Columbia-Snake River Basalts These are of Miocene to Quaternary age and occupy an area of more than 200000 km^2 covering the states of Washington, Oregon and Idaho in the USA. The total thickness is more than 1500 m. The individual flows are very thick (50–150 m), which can be traced for as much as 250 km

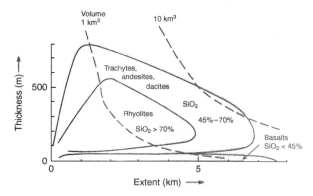

Fig. 14.7 Variation in thickness and lateral extent of lava flows depending on chemical composition (SiO$_2$%). (After Walker 1973)

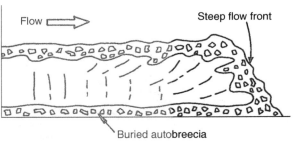

Fig. 14.8 Development of autobrecciated material on the top surface of lava flow and its burial due to downward sliding along the steep moving front

indicating high fluidity of lava. The basalt flows are either flat or slightly dipping with an angle of 1–2° to the southeast. The interflow surfaces and sedimentary layers are highly permeable (K = 10–10000 md^{-1}); deeper aquifers are more productive than shallower ones (Fetter 1988).

Pillow structures formed due to rapid cooling under marine conditions have very high permeabilities supporting large springs on the banks of the Snake River Canyon at Thousand Springs, Idaho (Table 14.4). The Columbia Snake River basalts also exhibit typical karst features like other plateau basalts with openings being as wide as 3.5 m and extending to few kilometre (Krishnaswamy 2008).

Deccan Traps The Deccan Traps cover an area of about 500000 km^2 in the western and central parts of India (see Fig. 20.3). Originally they would have covered an area of more than 1 × 10^6 km^2. The age of the Deccan Traps, based on argon data, is estimated to be 65–60 Ma (Duncan and Pyle 1998). Based on deep seismic-sounding surveys, the greatest thickness of basalt is reported to be about 1500 m near the western coast (Kaila 1988). The thickness decreases towards the east, and it becomes only a few tens of metres in the eastern and southeastern parts. The eruption took place through wide fissures as is evidenced by the presence of dolerite dykes. These dykes extend in length from a few hundred metres to 70 km or more (Fig. 14.4) with an average width of 1–10 m. The dykes have a NW–SE trend on the western coast and an E–W trend in the central region. Lava tubes having diameter of about 100 m and lava channels more than 300 m wide are also reported (Misra 2002).

In the Deccan Traps as many as 29 lava flows are reported from a borehole at Bhusawal in Maharashtra. The thickness of individual flow units varies from a

Fig. 14.9 a Field photograph showing layering within a flow in Deccan Traps (Courtesy H. Kulkarni), **b** interpreted profile

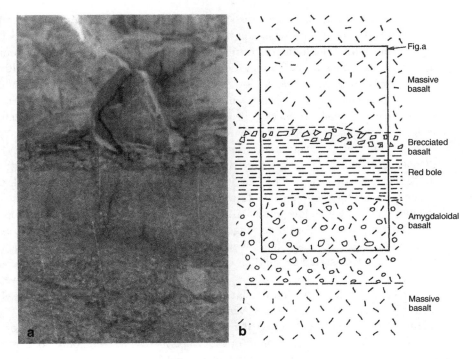

few metres to 50 m. The individual flows are separated from each other by residual soil locally called *bole* and other interflow sedimentary deposits (intertrappeans). The lower part of an individual flow is usually massive, and it becomes vesicular and amygdaloidal towards the top (Fig. 14.9). The *bole* is usually of a red colour but occasionally it may be green coloured also, the thickness being usually less than 1 m. It is rich in clay minerals and is believed to be a product of atmospheric weathering and/or hydrothermal alteration of amygdaloidal basalt or of pyroclastic material (Inamdar and Kumar 1994; Wilkins et al. 1994). Being rich in clay, the bole layer usually serves as a confining or semi-confining layer. However, when fractured it forms an aquifer as is seen in many dug well sections. Natural gamma ray logging is useful in the detection of red boles (Versey and Singh 1982).

14.2.2 Volcanic Islands—Submarine Eruptions

Submarine basalts show distinct hydrogeological characteristics as compared to subaerial continental types, due to differences in the physical conditions of solidification of lava. The submarine basalts are characterized by a pillow structure in which the intervening spaces between the pillows form easy channels for the movement of water. Pillow structures can also develop under continental conditions when lava is erupted into lakes or in water-saturated sediments. In comparison to plateau basalts, the submarine basaltic flows show greater heterogeneity and are commonly associated with pyroclastic materials with low porosity and permeability.

Volcanic islands are of two types: (1) oceanic islands, viz. Hawaii and French Polynesia in the Pacific Ocean and Mauritius in the Indian Ocean, which are of basaltic composition and therefore belong to the Basaltic Province; and (2) island-arc islands, viz. Indonesia, New Guinea and the Philippines, which are formed of mainly andesitic lavas and hence belong to the Andesitic Province. The basaltic oceanic islands can further be classified into two types—high islands and low islands. Some of the high islands, e.g. Hawaii islands rise to a height of more than 5000 m above the floor of the Pacific Ocean. They are occupied mainly by young basalts and may be partly flanked by marine and terrestrial sediments. Low islands e.g. the Cook islands in the Pacific and Laccadive (Lakshadweep) Islands in the Arabian Sea, are only a few metres above sea level which may be partly or fully covered with coralline limestone reefs forming atolls underlain by volcanic rocks (CSC 1984).

The basaltic flows on volcanic islands are usually thin (6 m or less) but form main aquifers due to their high permeability. The high permeability is primarily due to clinker zones in the *aa* type flows, lava tubes and gas vesicles in the *pahoehoe* flows, columnar vertical joints and the irregular openings between the lava flows (Peterson 1984). In places, the flows are interbedded with ash beds which form confining layers.

Pahoehoe lava generally grades into *aa* lava with increasing distance from the source. In aa lava flow, the clinker zones which have high permeabilities occur at the top and bottom of the flow while the central part is massive with low permeability. The hydraulic conductivity of clinker zone ranges from several hundred to several thousand md^{-1} which is similar to that of coarse well sorted gravel (Schwartz and Zhang 2003).

The central part of the Hawaii islands is characterized by a swarm of closely spaced vertical or steeply inclined dykes which cut across the gently dipping lava flows. The dykes, being more or less impermeable, serve as barriers against lateral movement of groundwater and thereby form water compartments in which groundwater may occur at different heights (Fig. 14.10).

In the Indian Ocean, Mauritius is an important volcanic island. The island is totally composed of basalts varying in age from 7 Ma to Recent. Younger flows are vesicular and scoriaceous with high permeability (Rogbeer 1984).

14.2.3   Dykes and Sills

Dykes are vertical or steeply inclined intrusive igneous bodies which cut across the pre-existing rocks. They are usually of basic composition. Dykes vary in thickness from a few decimeters to hundreds of metres, but widths of 1–10 m are most common. In length they may be from a few metres to several kilometers long. They represent feeders for the lava flows. As dykes are usually more resistance to erosion than the country rock they stand out prominently as wall-like ridges. Massive and unweathered dykes form barriers for lateral movement of water, thereby confining large volumes of surface and groundwater (Fig. 14.11). On the other hand, fractured dykes may form good aquifers. One of the dykes in the Palaghat Gap (western coast of India)

extends for a strike length of about 14 km and is highly fractured, forming a potential source of groundwater. The discharge from some of the wells in this dyke is of the order of 240–840 m^{3}d^{-1} (Kukillaya et al. 1992). Sometimes intrusion of dykes may cause fracturing of the adjacent country rock due to thermal effects and differences in the mechanical properties resulting in formation of effective conduits for groundwater flow parallel to the contact (Gudmundsson et al. 2003). In such cases, drilling into the country rock close to the dyke is recommended, e.g., in the Precambrian crystalline basement rocks of South Africa, Western Australia and north-east Brazil (Boehmer and Boonstra 1987). Pumping tests in dyke areas from Botswana indicate that dykes that are thicker than 10 m serve as groundwater barriers, but those of smaller width are permeable as they develop hydraulic continuity with the country rock through cooling joints and fractures (Bromley et al. 1994).

Sills are nearly horizontal tabular bodies which are commonly concordant and follow the bedding of enclosing sedimentary rocks or the older lava flows. Some of the sills are very thick and extend over large areas, as in the Karoo System of South Africa. Due to their low permeability, except when fractured, sills form perched water bodies, as in the Hawaiian Islands. However, studies in the Palisades sill in Newark Rift Basin, New York show that the main transmissive zones are located within the dolerite—sedimentary rock contact zones characterized by chilled dolerite. This is mainly a result of thermal cracking and fracturing of both formations resulting in higher permeability along the contact zone (Matter et al. 2006).

14.3   Groundwater Occurrence

Groundwater in volcanic rocks occurs under perched, unconfined and confined conditions. Perched water occurs above the regional water-table due to the presence of impervious formations, viz. sills, ash beds and dense basalt flows. Water under confined conditions is reported where the previous lava beds are confined between impervious sedimentary beds (Stearns 1942; Fetter 1988). Confined conditions are also created when vesicular or fractured basalt is sandwiched between massive units (Singhal 1973). Free-flowing artesian conditions are reported from

Fig. 14.10 Different types of groundwater structures used in volcanic islands and coastal areas. (Modified after UNESCO 1987)

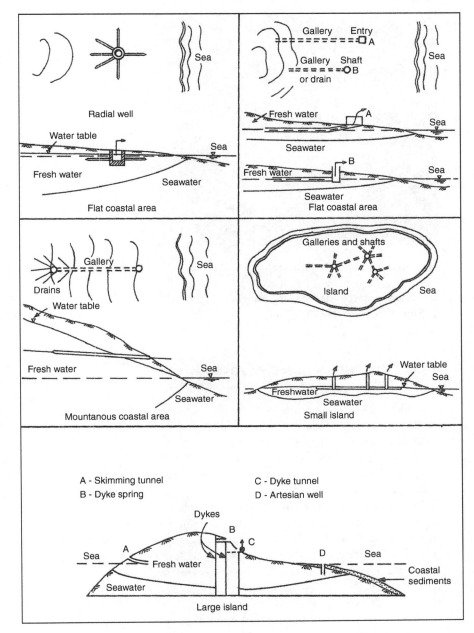

some places, viz. Honolulu (Stearns 1942) and Java island in Indonesia (Krampe 1983). Water under artesian conditions is also reported from the Deccan Traps (Kittu 1990). Dykes also cause lateral confinement of water forming compartments as described earlier.

The water-table gradients are low (<1:1000) in areas where volcanic rocks have high permeabilities, as in Columbia-Snake River basalts. The depth to the water-table varies depending upon the recharge and permeability of volcanic rocks, and the depth to the water-table is considerable (>150 m) in areas where the basalts are highly permeable (Ollier 1969). In the Deccan Traps, which are less permeable than Columbia River basalts, the water-table is generally shallow (<10 m) and it follows the ground topography.

The general velocity of groundwater in basalts is high, but not as high as in carbonate aquifers. The

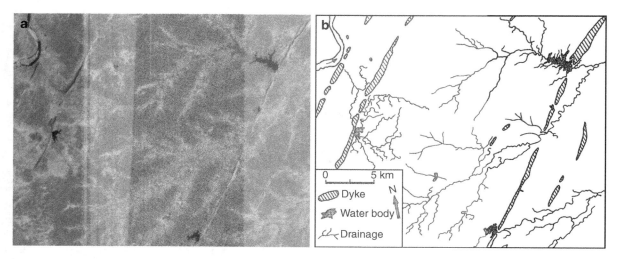

Fig. 14.11 a LANDSAT TM image showing damming of surface streams by dykes, **b** interpretation map

velocities are reported in the range of about 3.6–255 md^{-1} (Shelton 1982).

14.4 Hydraulic Characteristics

Primary porosity and permeability in volcanic rocks depend on the rate of cooling, viscosity of magma and degassing during cooling. Acidic volcanic rocks, like rhyolites and trachytes, are usually more massive than basalts and therefore have lower porosity and permeability, with some exceptions.

The various openings which impart porosity and permeability to basaltic rocks are (a) scoariae, (b) breccia zones between flows, (c) cavities between pahoehoe lava flows, (d) shrinkage cracks, parallel to the flow surfaces or columnar joints, (e) gas vesicles, (f) lava tubes, and (g) fractures and lineaments (Stearns 1942; UNESCO 1975). The order of importance may, however, vary in different areas. *Pahoehoe* flows and pillow lavas are more permeable than the thick dense *aa* flows. The lower and upper brecciated parts of flows develop good permeability (Fig. 14.8). Other interflow spaces may also impart permeability. Sheet and columnar joints formed due to cooling, and other fractures and lineaments produced as a result of later tectonic activity, sometimes impart high permeability. Table 14.2 gives the range of values of porosity and hydraulic conductivity of volcanic rocks, indicating that the variation in permeability is almost nine

orders of magnitude. Such a variation indicates their importance both from the point of view of water supply as well as for the selection of suitable sites for the disposal of radioactive waste (Dietz 1985; Evans and Nicholson 1987; Flint et al. 2002).

The hydraulic conductivity of pyroclastic deposits depends on the degree of consolidation and welding. The unconsolidated pyroclastic deposits, especially those which are reworked, show higher conductivities, as in Indonesia, Japan and in central California, USA (Stearns 1942; Davis and De Weist 1966). Welded tuffs which are formed at high temperatures by the fusion of rock fragments have low porosity in the range of 5–20% and very low conductivity (10^{-9}–10^{-7} ms^{-1}) depending on the degree of welding (Sharp et al. 1993).

Volcanic rocks also show anisotropy. Horizontal hydraulic conductivity is several times greater than

Table 14.2 Porosity and hydraulic conductivity values in volcanic rocks. (Based on data after Davis 1969; Freeze and Cherry 1979; UNESCO 1975; Sharp et al. 1993)

Rock type	Porosity (%)	Hydraulic conductivity (ms^{-1})
Basalt		
Dense	0.1–1	10^{-11}–10^{-8}
Vesicular	5–11	10^{-9}–10^{-8}
Fractured, weathered	10–17	10^{-9}–10^{-2}
Pumice		
Pyroclastics, (tuffs)	87	
Zeolitized	39	4×10^{-9}
Welded	5–20	10^{-9}–10^{-7}

vertical hydraulic conductivity due to the presence of interflow spaces and horizontal fractures. In Gran Canaria, Spain, the ratio of K_h/K_v is reported to be between 20 and 100 (Custodio 1985). Therefore, the rate of horizontal flow may be several orders of magnitude greater than the vertical flow. However, at places, columnar jointing may impart high vertical conductivity which may be of the order of $10^{-1}\,\mathrm{m\,s^{-1}}$.

A decreasing trend in porosity and hydraulic conductivity of volcanic rocks with increasing geological age is reported from several places. This is attributed to the filling of vescicles and fractures by secondary minerals such as zeolites, calcite and secondary silica. For this reason pre-Tertiary basaltic rocks of Brazil, Deccan Traps of India and basalts of eastern United States form generally poor aquifers, as compared to the Tertiary basalts of Columbia-Snake River area and Hawaiian islands (USA), and Canary Islands (Spain). Pillow lavas of Triassic age in Connecticut (USA) and the island of Guam also show lower conductivity due to the infilling of interstices with secondary minerals, as compared with the younger Tertiary pillow lavas from Snake River Canyon, USA (Stearns 1942) and Pleistocene and Holocene basalts of Mexico and Mauritius (Niedzielski 1993; Rogbeer 1984) (Table 14.3).

The above generalized relationship of hydraulic conductivity with age holds good mainly for primary poros-ity and permeability. However, at times the older basalts can aquire even higher secondary porosity and permeability due to fracturing, as is reported from some places in the Deccan traps, and even older Rajmahal Traps of Jurassic age, from India (Khan and Raja 1989).

The groundwater flow characteristics in basalts are similar to those in other hard rocks. Accordingly, basalts may have discrete fracture flow, or may form a continuum medium when fractures, lava tubes and vesicles are interconnected. Therefore, estimation of aquifer parameters in volcanic rocks is besieged with problems similar to those in other fractured rocks (Chap. 9). In literature there are many examples where the conventional methods of Theis, Jacob and Thiem were used to determine the aquifer properties from pumping test data (viz. Walton and Stewart 1961; Davis 1969; Adyalkar and Mani 1974; Jalludin and Razack 1993). Further, packer tests have been used for estimating permeability of discrete zones. The Papadopulos model has been considered for analyzing pumping test data from large diameter dug wells (Rao 1975). The double-porosity method has also been used for estimating transmissivity of matrix blocks and fractures (Singhal and Singhal 1990). Attempts are also made to estimate hydraulic conductivity from specific capacity data. viz. Rotzoll and El-Kadi (2008) estimated hydraulic conductivity ranging from 3 to $8200\,\mathrm{md^{-1}}$

Table 14.3 Hydraulic properties of volcanic rocks

Rock type	Age	Location	$T\ (\mathrm{m^2\,d^{-1}})$	S	Source
Dacite (weathered)	–	Red Hill, Australian Capital Territory	0.5–1.2	5×10^{-5}	Jacobson (1982)
Andesite (vesicular and fractured)	Lower Permian	Banana area, Queensland	1.5–200, average 65	–	Pearce (1982)
Basalt (fractured)	Jurassic, Cretaceous	Rajmahal, India	90–330	1.4×10^{-4}	Khan and Raja (1989)
Basalt	Cretaceous-Eocene	Deccan Traps	10–700	10^{-3}–10^{-1}	Adyalkar and Mani (1974); Versey and Singh (1982); Kittu and Mehta (1990)
Basalt	Miocene	Columbia Snake River area, USA	202–22780, average 5310	2×10^{-2}–6×10^{-2}	Walton and Stewart (1961)
Basalt	Miocene Quaternary	Gran Canaria Spain	50–300	–	Fernandopulle et al. (1974); Custodio (1985)
Basalt (fractured)	Pleistocene–Holocene	Mexico	605–865	–	Niedzielski (1993)
Basalt	Pliocene	Republic of Djibouti	34–5100	10^{-3}–10^{-4}	Jalludin and Razack (1993)

for the basalts of Hawaii islands. This helped in the preparation of hydraulic conductivity maps for use in groundwater management.

Step-drawdown test data have been analysed by Rao (1975), and Mishra et al. (1989) using Jacob's and Rorabough methods for estimating well and formation loss coefficients as well as transmissivities. Methods of analysis of pumping test data from dykes and adjacent country rocks are described in Sect. 9.2.4. Acidic rocks, e.g. dacites and andesites, usually have low transmissivities as compared with basalts. Hydraulic characteristics of some volcanic rocks from different parts of the work, as estimated from pumping tests, are given in Table 14.3.

In recent years, a greater interest has developed in the hydrogeological characteristics of basalts and tuffs as they could be potential host rocks for the disposal of radioactive waste, due to their low permeability. Therefore, detailed in situ hydraulic tests have been conducted in Columbia River basalts at Hanford, USA (Dietz 1985) and in fractured tuffs at the Nevada test site in the USA (Russell et al. 1987), to assess their hydraulic parameters (see Sect. 12.3.3).

14.5 Groundwater Development

Springs are an important source of water in volcanic terrains. They usually show high discharge, quite comparable to those in carbonate rocks (Table 14.4). In the USA, out of the 66 first magnitude springs (discharge $2.83\,m^3\,s^{-1}$ or more), 36 issue from basalts (Stearns 1942). Springs from the Quaternary basalts in Idaho, USA contribute more than 90% of the total flow of Snake River and are an important source of irrigation. In eastern Australia, some springs show a discharge as much as $5\,m^3\,s^{-1}$ (UNESCO 1975). In volcanic islands, dykes and ground topography control the location of springs (Fig. 14.10). The spring discharge depends on the size and interconnection of fractures and other passageways like volcanic tubes etc. Relatively constant discharge is

expected where the fractures are narrow and the groundwater body is large, while greater variability is observed when flow is through wide fractures and other openings which are well-developed and abundant.

Dug wells and borewells are constructed to tap groundwater. Horizontal boreholes, tunnels and galleries are used to tap perched water on hills or on impermeable ash or tuff layers between flows and interappeans in plateau basalts and also in volcanic islands (Fig. 14.10).

Yield of large-diameter dugwells in the Deccan Traps of India is reported to be $0.01-0.05\,m^3\,s^{-1}$ for moderate drawdown (Adyalkar and Mani 1974). Specific capacity of these wills is in the range of $1\times10^{-4}-5\times10^{-3}\,m^2\,s^{-1}$ depending on the permeability of basalts (Deolankar 1980). Borewells in the Deccan Traps are usually of 10–20 cm diameter and 40–50 m depth having a yield of $1-2\times10^{-3}\,m^3\,s^{-1}$. The well productivity generally decreases beyond a depth of 70 m. Hydro-fracturing helps to increase well yield. Borewells located in intersections of lineaments have higher yields ($0.02-0.04\,m^3\,s^{-1}$) for a drawdown of 3–5 m (Kittu 1990; Subramanian 1993). Borewells in older Rajmahal traps of Jurassic age in eastern India have an yield of 2×10^{-3} for a drawdown of less than 1 m from fractured horizons (Khan and Raja 1989). In Columbia River basalts of the USA, typical well yields from confined aquifers located at depths between 150 and 300 m are $0.06-0.12\,m^3\,s^{-1}$ (Fetter 1988). Specific capacity of wells in Columbia basalts is reported to vary from 2×10^{-6} to $8\times10^{-2}\,m^2\,s^{-1}$ with an average value of $2\times10^{-3}\,m^2\,s^{-1}$ (Maxey 1964).

In coastal areas and volcanic islands, basal groundwater which floats over sea-water, is mostly developed by radial wells (Maui type), tunnels, galleries, shafts and a combination of these to minimize the chances of sea-water intrusion (Fig. 14.10). The type of structure depends upon topography, rock structure and their permeabilities. Confined aquifers in coastal sediments are also an important source of water supply tapped by screened well. Such wells in Honolulu, Hawaii, have an yield of $0.025\,m^3\,s^{-1}$ (UNESCO 1975).

Table 14.4 Discharge values of some major springs in basalts

Location	Discharge ($m^3\,s^{-1}$)	Source
Datta Spring, USA	1.4–3	UNESCO (1975)
Thousand Springs, Idaho, USA	15–20	UNESCO (1975)
Kalauaa Spring, Pearl Harbour, Oahau Island, USA	0.4–1	UNESCO (1975)
Oaxtepec and Aqua Hedrionda Springs, Mexico	0.7	Niedzielski (1993)

The *Maui* type of tunnels may produce up to $2\,m^3\,s^{-1}$ depending on their length and nature of rocks (Fetter 1988). In places, the tunnels dip slightly towards the entrance so that water can flow out by gravity. Sometimes they are also provided with vertical bores (well shafts) to tap simultaneously deeper aquifers, as in the Canary Islands. These tunnels or galleries are of a few kilometers in length and are located at a depth of $100\,m$ below ground level (Custodio 1985). Water confined in the dyke compartments is either discharged as springs or is tapped by tunnels.

14.6 Groundwater Recharge

The recharge in volcanic rocks can be estimated by the same techniques as in other rocks (Chap. 20). The main source of recharge is the direct infiltration from precipitation, seepage from streams and return flow from irrigation. Fog-drip could also be an important source or recharge as reported from the Maui Island in Hawaii (Stearns 1942) and La Palma island of Spain (Veeger et al. 1989).

Recharge in old volcanic rocks is generally low which is attributed to weathering effects and low permeability of rocks resulting in high run-off. In the Deccan Traps, based on water balance and water-table fluctuation methods, groundwater recharge is estimated to vary from 10 to 20% of the rainfall (Adyalkar and Rao 1979). Tritium data also gave similar values of recharge. Even in high rainfall areas, recharge is low due to the high relief and impervious nature of the Deccan Traps. In arid regions of different parts of the world, the groundwater recharge in basaltic terrains is reported to be about 10% of the annual rainfall (UNESCO 1975). Recharge in the fractured tuffs at the Nevada nuclear test site, USA, located in an arid climate, was estimated to be approximately 8% of precipitation (Flint et al. 2002). These estimates are of importance in the selection of potential repository sites for high level waste disposal. At this site, a comparison of the travel time of water with the pre-nuclear testing period indicated increased permeability due to induced fracturing as a result of nuclear testing (Russell et al. 1987). In the Hawaiian islands, where annual rainfall varies from 500 to 6300 mm, the recharge in older basalts is 6–10% and in younger volcanics 30–36% of the annual rainfall (Wright 1984).

In Mauritius, where basalts are of Quaternary to Recent age, and average annual rainfall is 1200 mm, the recharge is computed to be 24% of the rainfall (Rogbeer 1984).

Deeper confined and semi-confined aquifers may be recharged by leakage from shallow horizons through vertical fractures, and partly by lateral subsurface flow from outcrop areas. In Columbia River basalts, good recovery of water-levels after pumping from deeper aquifers, and other hydraulic data indicate high vertical hydraulic conductivity and interconnection between aquifers. On the other hand, the age of groundwater as determined by the ^{14}C method ranges from modern to as old as 32 000 year BP; age increasing with increasing depth. This indicates lack of interconnection between shallow and deeper aquifers (Fetter 1988). Low tritium content and greater salinity of water in the deeper aquifers of the Deccan Traps also indicates very limited or negligible vertical movement of water (Versey and Singh 1982).

As in carbonate rock aquifers, spring discharge data are also used to determine the recharge pattern and the groundwater potentiality of volcanic rocks. Analysis of spring flow data from fissured basalts of Mexico shows a double recharge mechanism as in karstic terrains (Sect. 15.5). In this case, the main recharge takes place immediately within a period of 1–3 days after rainfall through open fractures, but the recharge due to diffused flow is slow through fine fissures and is delayed by 1–3 months with respect to a rainy period. This diffused flow is mainly responsible for dry weather discharge (Niedzielski 1993).

14.7 Groundwater Quality

Basalts mainly consist of plagioclase and ferromagnesian minerals rich is Ca and Mg which weather rapidly. Release of Mg from olivine is several times faster than the release of Ca and Na from plagioclases. Therefore, groundwater in basalts has a higher amount of alkaline earths (Ca^{2+} and Mg^{2+}) and a lower amount of alkalis (Na^+ and k^+). In olivine bearing rocks, Mg may be a dominant cation. The TDS is usually less than $500\,mg\,l^{-1}$ (Table 14.5). Among the anions, HCO_3 is dominant over Cl^- and SO_4^{2-}. The pH is usually in the range of 6.7–8.5 and the SiO_2 content is high ($>30\,mg\,l^{-1}$). In general, the groundwater in basalts is

Table 14.5 Chemical composition of groundwater (in mg l^{-1}) from volcanic rocks

Rock Type	TDS	Ca	Mg	Na	K	Cl	SO$_4$	HCO$_3$	Fe	SiO$_2$	Location
Rhyolite	234	14	5.8	20	5.2	4.0	7.7	112	–	62	Oregon, USA[a]
Columbia River Basalt	280	24	15	12	5.3	15	1.6	156	0.43	50	Oregon, USA[a]
Deccan Trap	505	62	28	24	–	37	30	294	–	30	Hyderabad, India[a]
Basalt	–	24	15	12	5.3	15	1.6	156	0.43	50	Oahau and Hawaii, USA[b]

[a] White et al. (1963)
[b] Morrison et al. (1984)

of the Ca–Mg–HCO$_3$ type. In the case of acidic volcanic rocks (rhyolties), groundwater has a higher concentration of SiO$_2$ and Na$^+$ and the waters are of the Na–HCO$_3$ type (Table 14.5).

In areas where volcanic racks have high permeability, there are greater chances of groundwater contamination. Pollution of groundwater in volcanic islands, due to sewage and increased use of pesticides, has been reported from some places (CSC 1984). However, in areas where permeable basalts are covered with some impervious material, the chances of downward migration of contaminants are minimized, as in the Columbia-Snake River area (Davis and DeWiest 1966). Overexploitation of groundwater in oceanic islands can also lead to sea-water intrusion.

Summary

The most common volcanic rocks are basalts formed as a result of eruption of magma at or near the ground surface. The basaltic lava flows are often associated with dykes which acted as feeding channels and now occur as extensive ridges, often forming barriers for surface water and groundwater. Submarine basalts show distinct hydrogeological characteristics as compared to subaerial continental types. Groundwater in volcanic rocks occurs under perched, unconfined and confined conditions. Water under confined conditions may occur where the pervious lava beds are confined between impervious sedimentary beds or when the vesicular or fractured basalt is sandwiched between massive units. Acidic volcanic rocks, like rhyolites and trachytes, are usually more massive than basalts and therefore have lower porosity and permeability though some exceptions may occur.

The various openings which impart porosity and permeability to basaltic rocks are scoariae, breccia zones between flows, cavities between pahoehoe lava flows, shrinkage cracks, parallel to the flow surfaces or columnar joints, gas vesicles, lava tubes, and fractures and lineaments. Volcanic rocks also show anisotropy. Horizontal hydraulic conductivity is several times greater than vertical hydraulic conductivity. In recent years, a greater interest has developed in the hydrogeological characteristics of basalts and tuffs as they could be potential host rocks for the disposal of radioactive waste, due to their low permeability. Recharge in old volcanic rocks is generally low. Groundwater in basalts has a higher amount of alkaline earths (Ca^{2+} and Mg^{2+}) and a lower amount of alkalis (Na$^+$ and K$^+$), the pH being usually in the range of 6.7–8.5.

Further Reading

Custodio E (1985) Low permeability volcanics in the Canary Islands (Spain). Mem. IAH. vol. 17(2), 562–73.

Fetter CW (2001) Applied hydrogeology. 4th ed., Prentice Hall Inc., New Jersey.

Stearns HT (1942) Hydrology of volcanic terraines, in Hydrology (O.E. Meinzer ed.), Dover Publ., Inc., New York, pp. 678–703.

UNESCO (1987) Groundwater Problems in Coastal Areas. Studies and reports in hydrology 45, UNESCO, Paris.

15.1 Introduction

Carbonate rocks are sedimentary type of rocks containing more than 50% carbonate minerals which are mainly calcite, $CaCO_3$, and dolomite, $CaMg(CO_3)_2$. The carbonate minerals could be a result of chemical precipitation, organic processes or may occur as detrital material. Some of the dolomites may be a result of diagenesis known as dolomitization which involves replacement of calcite by dolomite to a varying extent. The term limestone is used for those rocks which contain more than 90% carbonates. If the rock contains more than 50% but less than 90% carbonates, it is termed as arenaceous limestone or an argillaceous limestone, depending upon the relative amounts of quartz and clay minerals. Chalk is a type of limestone which is soft and white in colour and is rich in shell fragments. The carbonate rocks occupy about 10% of the Earth' surface and supply nearly a quarter of the world's population with water. Locally, their cumulative thickness may be 10 000 m (UNESCO 1984b). In the geological past, most limestones and dolomites were deposited during the intermediate phases of Caledonian, Hercynian and Alpine tectonic cycles and their ages are mainly Palaeozoic and Mesozoic. One of the largest carbonate aquifers in the world is the Floridan aquifer system of Palaeocene to Miocene age in the southeast United States consisting of gently dipping thick sequences of carbonate sediments separated by less permeable clastic sediments.

Carbonate rocks have attracted greater attention of hydrogeologists in the past few decades, both from the point of view of water supply as well as potential sites for waste disposal. Carbonate rocks are also of great economic value having large deposits of oil, gas and ore minerals.

15.2 Weathering and Landforms—The Karst

Carbonate rocks (limestones and dolomites) are soluble to a large degree in water, rich in carbonic acid giving rise to a characteristic landscape known as *karst*. The term karst is of Serbian origin as karst features were first described from the Dalmatian coast of the Mediterranean area in the former Yugoslavia.

The development of karst landforms is governed by a number of factors such as, lithology, porosities (matrix and fracture), hydraulic conductivity, chemically active groundwater, proper drainage, palaeoclimates, tectonic movements and eustatic changes, mainly during Pleistocene and Holocene times. Rock discontinuities facilitate the development of solution cavities (Fig. 15.1). Karst development in dolomites is generally more subdued than in limestones due to slower dissolution kinetics as a result of which conduit systems are generally less developed in dolomite aquifers (White 2007). Higher rainfall and humidity along with a higher concentration of CO_2 promote karstification (Eq. 15.1).

$$2H_2O + CaCO_3 + CO_2$$
$$\Leftrightarrow H_2O + Ca^{2+} + 2HCO_3^- \quad (15.1)$$

The presence of soil cover will promote dissolution, as partial pressure of CO_2 in the soil zone is higher than in the atmosphere due to the weight of the overburden, and additional CO_2 is also derived from the decay of organic material. Conversely, a decrease in CO_2 pressure will cause precipitation of calcite where the groundwater either discharges on the surface through spring outlets or when the CO_2 escapes through the cave openings. The carbon cycle, therefore, plays an

B. B. S. Singhal, R. P. Gupta, *Applied Hydrogeology of Fractured Rocks*,
DOI 10.1007/978-90-481-8799-7_15, © Springer Science+Business Media B.V. 2010

Fig. 15.1 Solution cavities (karren) along discontinuities in Miliolite limestone of Saurashtra, India. (After Nair 1981)

important role in karst development, as the carbonate rocks being a huge storehouse of carbon, form an important sink or source of atmospheric CO_2 due to the processes of dissolution and deposition (Yuan 1996).

The sea level changes influence the potential hydraulic gradient between the recharge area and the outlet zone. Therefore, a lowering of sea level will promote solution activity, viz. higher transmissivities in Floridan aquifer, USA are attributed to lower sea levels during glacial times (Schwartz and Zhang 2003). The direction of groundwater movement is also an important factor in the development of karst. There is also a distinct relationship between the degree of karstification and the permeability, lithology and the thickness of the overlying stratigraphic unit as the solution activity depends upon the circulation of groundwater. Therefore, the solution activity will be most prominent in unconfined carbonate aquifers, which readily allow recharge and discharge from the aquifer (Domenico and Schwartz 1998). In the Floridan aquifer, the karst is best developed where either the carbonate rocks outcrop on the surface or the overlying confining layers are thin. The transmissivities in the central and northern Florida are often greater than $93\,000\,m^2\,d^{-1}$ while in southern Florida the transmissivities are much less because the aquifer is confined by thick beds (Schwartz and Zhang 2003). A rule of thumb is: the thicker the limestone sequence, the more important and dominant will be the effect of karstification on the hydrological regime of the area. If the carbonate rocks are thin, then only a superficial karst may develop (UNESCO 1984b).

Dissolution of limestones and development of cavities become more pronounced due to the mixing of waters of different compositions, viz. fresh and sea-

water in coastal aquifers. This situation can also occur anthropogenically, due to the disposal of waste water of a composition different to that of the resident water (Deike 1990). In areas where the potential of dissolution is low, mechanical erosion dominates giving rise to cliffs instead of karst as in the arid southwestern United States.

Karst is well developed in all geographic latitudes, viz. karst of the Urals is in part at latitude 70°N and that of South Australia at latitude 35°S (Komatina 1975). Karst covers large areas in different parts of the world, namely CIS (former USSR), USA, Central America, South China, North Africa (Atlas Mountains), and southern Iran. The karst of Central America, namely Jamaica, Cuba, Mexico and Guatemala together with that of Florida and Texas in the USA constitutes one of the largest karst belts in the world, covering an area of about $500\,000\,km^2$. In China, karst landforms are well developed in the carbonate rocks ranging in age from Proterozoic to Quaternary. The bare karst terrain of south China is one of the largest and most densely populated tropical karsts in the world covering an area of about $9 \times 10^5\,km^2$. It is developed in Palaeozoic carbonate rocks characterized by low porosity and thin soil cover. The total karst water resources in China are estimated to be $2029 \times 10^3\,m^3$ out of which $1847 \times 10^3\,m^3$ is in south China and only 182×10^3 in north China (Chen and Cai 1995).

Another classic karst area is the Dinaric karst in the Balkans bordering the Adriatic Sea in Europe which has some of the largest springs viz. Buna in the erstwhile Yugoslavia (Table 15.1). Other examples of potential karst aquifers are Jaffna limestone in Sri Lanka, in Indonesia and some low lying coral islands of the Indian Ocean, viz. the Maldives.

Table 15.1 Discharge values of some of the major karst springs. (After Ford and Williams 1989; Korkmaz 1990; Bonacci 1995)

Spring location	Discharge ($m^3 s^{-1}$)			
	Mean	Maximum	Minimum	Basin areas (km^2)
Buna, Yugoslavia	40	440	2	–
Ljubljanica, Yugoslavia	39	132	4	1100
Chingshuli, China	33	390	4	1040
Vaucluse, France	39	260	4.5	2100
Ombla, Croatia	25	106	4.0	800–900
Silver, USA	23	36	15.3	1900
Kirkgoz, Turkey	16	19	12	1800

In England and France, the chalk of Cretaceous age forms highly productive aquifers. In England, chalk accounts for 70% of groundwater supply. The chalk is a soft microporous fractured limestone which has three components of porosity and permeability (1) high intergranular porosity (25–45%) but low intergranular permeability due to small pore size (<1 µm), (2) primary fractures that have low porosity (0.1–1%) but may have high permeability depending on their frequency, and (3) secondary fractures developed due to solution and enlargement of primary fractures which account for the bulk of flow in the saturated zone (Goody et al. 2007).

The karst processes have been widespread not only in the Recent but also in the past geological periods. The karst process is quite rapid in terms of geological times, and only a few thousand years, generally less than 50 k-years may be required for developing a karst network (Bakalowicz 2005).

Karst Characteristics and Types The karst is characterized by a number of features, viz. sinkholes and other depressions, long dry valleys, sparse streams, internal drainage, springs and bare rocks. Sometimes due to weathering, a red-coloured iron-rich material known as *terra-rossa* covers the land surface. The various karst features can be identified on aerial photographs, multispectral images and SAR images (Figs. 15.2 and 15.3). On panchromatic stereo aerial photographs, fresh limestone surfaces appear light toned, whereas terra-rossa is dark-toned; variation in moisture usually leads to a mottled texture. Sudden or gradual disappearance of streams (sinking creeks) or dry river beds may be observed (Fig. 15.4a, b). Similarly, springs and a sudden increase in stream discharge is also characteristic of karst terrains (Fig. 15.5). On SAR images, the radar return is related to the variation in surface moisture and surface roughness due to subsidences, vegetation etc. (Fig. 15.3).

Fig. 15.2 Stereo-photographic pair showing limestone terrain characterized by depressions, dry valleys and mottled surface due to variation in surface moisture. (Courtesy Aerofilms Ltd.)

Fig. 15.3 SAR image of karst terrain in Indonesia characterized by numerous pits and depressions. (After Sabins 1983)

The characteristic subsurface features are the caverns and solution channels; the solution activity in carbonate rocks results in the lowering of the ground surface. In zones of intense fracturing, this effect is more pronounced giving rise to topographic depressions known as *dolines, uvalas* and *poljes*. The surface drainage is directed underground. In the old stage of landform development, due to the collapse of the land surface, the underground caverns which were earlier filled with water and air are destroyed, and the water-table comes closer to the ground surface.

Although karst phenomenon is not restricted to any particular climatic zone, there is no doubt that the

degree of solution is related to climate, as it depends on the temperature regime and atmospheric precipitation. Therefore, attempts have been made to identify different types of karst landforms that develop in temperate and tropical regions. Temperate karst regions are usually characterized by broad depressions, e.g. dolines, uvalas and poljes with subordinate development of positive forms such as pinnacles and cones. Tropical karst, on the other hand, is dominated by positive forms (Fig. 15.6).

The common landforms of karst areas developed in humid climates are briefly described below. For more detailed description, the reader may refer to Burger and Dubertret (1984) and Ford and Williams (1989).

Doline is a Yugoslav term for a depression or valley in karstic terrain. Dolines are circular in plan and vary from a few metres to about one kilometer in diameter (Fig. 15.7). The mechanism for the development of dolines is shown schematically in Fig. 15.8. Dolines are called 'sinkholes' in North America and 'swallow holes' in England. Formation of sink holes can cause serious safety problems to surface structures viz. dams, reservoirs highways, pipe lines and vehicles. The process of subsidence gets accelerated by dewatering of mines. In the limestone area of Florida, USA, due to solution and subsidence, lakes are formed which are also called 'sinkholes'. Several dolines coalesce to form *uvalas*. A *karst cave* is a solution opening in carbonate rocks that is large enough for human entry. The

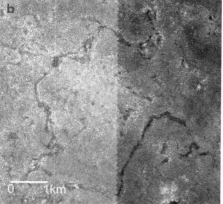

Fig. 15.4 Influent seepage and sinking streams. **a** Field photograph of a sinking stream Rak at the entrance to Tklaca cave in classic karst of Slovenia (after Kresic 2007). **b** Karst in the Nullarbor Plain, Australia. It is a flat almost treeless, arid or semi-arid terrain and the world's largest single piece of limestone. Due to

the absence of vegetation cover, karst structures can be readily observed. The image shows very scanty vegetation, and whatever is present is aligned along drainage channels, controlled by fractures. Influent drainage can be clearly interpreted. Possible dolines are present at the intersection of fractures. (Source: GoogleEarth)

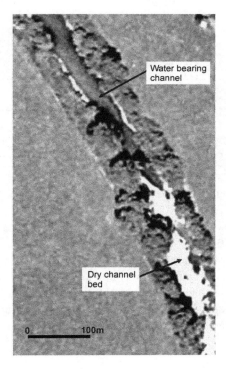

Fig. 15.5 Effluent groundwater seepage in the karst terrain near Immendingn, Germany. The Danube River flows from SE to NW. Note the dry channel in the upstream region (SE) and the water-bearing channel in the downstream (NW). Image generated in black and white from GoogleEarth

formation of caves and cavern passages are facilitated by the dissolution of carbonate rocks along the various systems of joints (Fig. 15.9). The caves may be formed in the vadose zone or in the deep phreatic region. On the basis of analogue and numerical modeling, Cas-

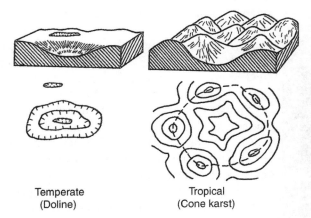

Fig. 15.6 Comparison of karst terrains in temperate and tropical climates. Doline are characteristic of temperate climate and cone karst of tropical areas

Fig. 15.7 Doline-a typical circular shaped depression formed due to dissolution in carbonate rocks. Field photograph from the Nam La River valley, Son La Province, Vietnam. (Courtesy: Okke Batelaan, Belgium)

ini et al. (2006) have shown that tensile stresses promote jointing causing greater dissolution of carbonate rocks.

Tower karst, also known as *stone forest,* is one of the most spectacular features of the karst terrains in South China which is characterized by the development of residual hills or peaks separated by large depressions (Fig. 15.10). It covers an area of 500 000 km^2 in South China. The formation of tower karst within Upper Devonian limestones and dolomitic limestones is related to intense fracturing of these rocks during Mesozoic and Tertiary episodes. Stone forest aquifers have very large lateral permeabilities but very small storage capacities as the aquifers are thin. This is responsible for large fluctuations in karst water discharge and alternating drought and flood hazards which are characteristics of these naked karst areas. Therefore, tower karst has rather poor development potential, though it is an important source of local water supply.

Palaeokrast refers to karst developed largely or entirely in ancient geological times during the periods of emergence above sea-level. As palaeokrasts are decoupled from the present hydrogeochemical system, they are also termed as *fossil karst.*

Palaeokarsts are of different geological ages and have been described from various parts of the world (Bosak et al. 1989). The earliest plaeokarsts are of Preterozoic age as reported from China and Canada. Palaeokarsts are of importance both from the point of view of supply of good quality water, as well as for temporarily subsurface storage of imported water

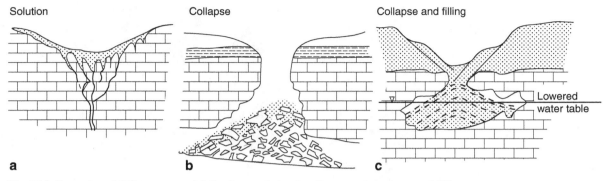

Fig. 15.8 Formation of different types of doline by: **a** solution, **b** collapse, and **c** collapse and filling

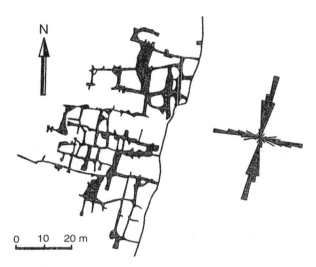

Fig. 15.9 Map of Herron Cave, Indiana, USA, showing the pattern of cavern passageways in relation to joint orientations. The rose diagram includes 136 joint measurements. (After Powell 1977)

Fig. 15.10 Tower karst of Guilin area, China. (Courtesy Yuan Daoxian 1996)

(Burdon and Al-Sharhan 1968). Palaeokarsts are also of great economic importance as they contain at places economic deposits of oil and gas, silver, lead and zinc, bauxite, phosphate and other minerals, such as lead-zinc deposits in Morocco, Algeria and Tunisia, and oil and gas in the USA (Bosak et al. 1989).

15.3 Springs

Springs are characteristic features of karst terrains, some of which have high discharge (Table 15.1). They are an important source of water supply especially in hilly terrains. Karst springs show a tremendous variation in their physical form and rates of discharge. Springs with high discharge can also be a potential source for the generation of hydropower. Location of springs is mainly controlled by structural and stratigraphic features. Karst springs can be classified based on: (a) morphology, (b) variability of discharge, (c) hydrodynamic conditions, (d) water quality (temperature and geochemistry), and (e) spring biota, etc. A review of various systems of spring classification is given in Springer and Stevens (2009).

Springs may emerge on the land surface and also on the bottom of a lake or sea. On the land surface, springs are formed by stream erosion or other denudational processes. Spring locations are usually controlled by geological discontinuities like faults and impervious beds (Figs. 15.11, 15.12). Springs occurring at different altitudes can be tapped by tunnels; one such tunnel was constructed for water supply to the city of Innsbruck, Austria, having a discharge of $0.6–1.7 \, \text{m}^3 \, \text{s}^{-1}$ (Graziadei and Zotl 1984).

Fig. 15.11 Types of karst springs. **a** fresh water spring controlled by lithologic layering and fissures, **b** submarine spring, **c** brackish water spring formed due to mixing of fresh groundwater and seawater

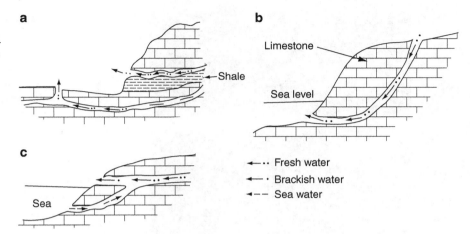

Brackish springs above sea level are formed due to flow of sea-water inland through the sub-sea conduits due to the decrease in the head of fresh water or changes in water density. (Fleury et al. 2007; Fleury et al. 2008) (Fig. 15.11c). Such type of brackish karst springs are common along the eastern coast of the Adriatic sea in Croatia, parts of Serbia and also in the coast of Greece, Israel and Florida. Spring water may be fresh or brackish. The flow of water in the conduits open to sea below sea level depends on the hydraulic head gradient between the aquifer and the sea and therefore is a function of water density and head losses in the aquifer.

Submarine springs are formed when the discharging water in the bottom of the lake or sea is under sufficient hydrostatic pressure to overcome the weight of the overlying surface water (Fig. 15.11b). Sometimes, due to differences in hydrostatic pressure, reversal of flow may take place when sea-water moves landwards

through the solution cavities, making fresh groundwater brackish (Fig. 15.11c); the proportion in which fresh water and sea-water get mixed will vary seasonally. Submarine karst springs are most common along the eastern coast of the Adriatic Sea in the former Yugoslavia. The depth of submarine springs varies from 0 m up to 150 m. Spring water may be fresh water or brackish.

Submarine springs are important for groundwater supply to meet local needs. The springs result in local eddies and currents in sea-water and hence may be detected by photography (Fig. 15.13). Further, they frequently cause thermal anomalies in surface water due to mixing. Hence, thermal IR sensing can enable detection of such freshwater discharge sites (Fig. 15.14) and this can be corroborated by field

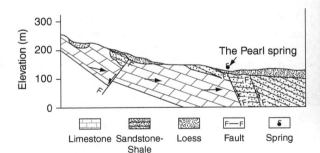

Fig. 15.12 Hydrogeological profile of the Pearl spring, Henan province, PR China. High discharge from the spring is controlled by a fault in karstified limestone. (After Zhengzhou et al. 1996)

Fig. 15.13 Freshwater submarine spring in karst formations in the Adriatic Sea. The spring discharges under high hydrostatic pressure caused by the high elevation of the recharge area in the hinterland. (After Alfirevic 1966)

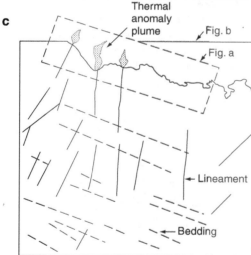

Fig. 15.14 Submarine coastal springs in limestone. **a** thermal IR aerial scanner image (predawn) showing discharge of cooler (darker) groundwater into warmer (lighter) seawater, **b** NIR-band aerial photograph of the same area acquired at noon-hours, **c** interpretation map. A number of fractures can be delineated on the aerial photograph controlling freshwater discharge into the sea. (Courtesy J. Bodechtel)

measurements of density, salinity and temperature distribution in the sea-water. The remote sensing surveys for delineation of submarine freshwater springs are timed to enable detection of perennial freshwater discharges. The thermal IR data (pre-dawn) provides zones of thermal anomalies. The land in the night is cooler than the sea and therefore, freshwater which has a temperature corresponding to that of the land is cooler than the sea-water. After eliminating thermal patterns due to surface drainage, the remaining sites of thermal plumes can be attributed to subsurface discharge. Lee (1969) has given nomograms for estimating spring discharge based on data about the area of thermal plume and temperature of air, sea-water and groundwater. Remote sensing images and photographs may be interpreted for lineament-geological structures, and areas of higher surface moisture and healthy vegetation. Superimposition of all the above data on to a common base map may yield valuable information about the fracture zones which control the groundwater movement (Fig. 15.14).

Considering the variability of spring discharge, the following types can be recognized:

1. Fluctuating springs: These show changes in discharge related to precipitation. The degree of fluctuation also depends upon the size of the groundwater reservoir; if the storage is large, fluctuations may be less, otherwise large fluctuations may take place. For example, the Sreid spring, emerging from Jurassic and Cenomanian dolomitic limestones in Lebanon, shows a time lag of 3 months between maximum precipitation in the catchment area during January, and maximum discharge ($3\,m^3\,s^{-1}$) which is in April (Khair and Haddad 1993). Springs in the stone forest region of the Guilin area in South China also show great fluctuation in discharge from $0.001\,m^3\,s^{-1}$ to $7\,m^3\,s^{-1}$ which indicates both uneven precipitation and also the heterogeneity of the system (Yuan et al. 1990).

2. Rhythmic springs: These show periodic (regular) variation in discharge, unconnected with seasonal or tidal influence. These springs are also referred to as periodic or ebb- and flowing springs. They are usually dammed springs with a siphoning reservoir system, controlling their discharge (Ford and Williams 1989).

3. Springs of changing outlet where the discharge point shifts from time to time.

4. Springs of constant discharge that have more or less constant discharge.

Hydrodynamically, karst springs can be classified as descending, ascending, or combined type. Descending springs drain water from an unconfined aquifer and show seasonal variation in discharge. An analysis of their recession curve can yield data about the size of their catchment and volume of water in storage. Ascending springs are fed by a confined aquifer. They show less variation in their discharge, chemical composition and temperature. Truly ascending springs are rare. Most of the springs are of combined type being fed by aquifers with varying hydrodynamic conditions.

15.4 Hydrogeological Characteristics

Karst terrains have their own characteristic distribution of porosity and permeability, both in the unsaturated and saturated zones. The hydraulic characteristics (K, T and S) of carbonate rocks depend on lithology, intensity of fracturing and solution activity. The lithology controls the mechanical and dissolution properties of rocks and hence the development of fractures and solution cavities. Limestone and dolomite usually have well-developed fractures as compared with clayey marl and chalk formations. A greater amount of matrix material and clay will also lower rock permeability of unfractured limestones and dolomites. Fracturing and solution impart higher permeabilities to CO_3 rocks. The hydraulic conductivity of the fractured rock mass is considerably greater than the massive part as the main movement of groundwater is along fractures and joints (Table 15.2). For this reason, the transmissivities are largely anisotropic, as is evidenced from the drawdown ellipse obtained from the pumping test and cave passageway orientation (Greene and Rahn 1995). In chalk, the vertical hydraulic conductivity of $5 \times 10^{-8}\,\mathrm{m\,s^{-1}}$ is about 500 times greater than that attributed to intergranular conductivity (Hamill and Bell 1986).

A matured karst shows heterogeneous spatial distribution of hydraulic conductivities ranging from 10^{-10} up to $10^{-1}\,\mathrm{m\,s^{-1}}$; the rock matrix has low conductivities ($10^{-10}\,\mathrm{m\,s^{-1}}$) while high conductivities are characteristic of conduits. Three main types of porosity are primary (matrix), fracture, and conduit porosity. Primary or intergranular porosity is usually very small. Fracture porosity is due to interconnected fractures and conduit porosity includes solution channels and pipes. Variation in total bulk porosity is reported as being from near zero to as high as 45% (Table 15.3). Chalk and calcareous tuff have high porosity; in chalk, inspite of high porosity, since the pores are small (1.0–0.012 µm), primary permeability is low and specific retention is high. However, the rock may acquire high secondary porosity and permeability, to varying extents, depending on fracturing and dissolution of mineral matter. Dolomites are usually more porous than limestones due to reduction in volume as a result of dolomitization. On the other hand, diagenetic processes, namely cementation etc. tend to reduce the porosity and permeability.

The karst areas generally show a variation with depth in hydrogoeological characteristics. As the porosity, hydraulic conductivity and storativity depend on fracturing and dissolution, they generally reach a maximum value close to the lower position of

Table 15.2 Comparison of hydraulic conductivities (K) of unfractured and fractured carbonate rocks. (After UNESCO 1984b)

Rock type	K values of unfractured rocks (m s^{-1})	K values of fractured rocks with one fracture per metre	
		Opening (mm)	K along the fractures (m s^{-1})
Limestone	10^{-13}–10^{-12}	0.1	0.7×10^{-6}
	10^{-11}–10^{-9}	4.0	0.5×10^{-1}
Dolomite	10^{-10}	6	1.6×10^{-1}
Chalk	10^{-7}–10^{-6}	–	10^{-5}–10^{-4}

Table 15.3 Total porosity values of some carbonate rock. (After UNESCO 1984b)

Lithology	Location	Total porosity (%)
Compact limestone	Italy	0.2–14
Fissured limestone	Buxton, UK	14
Poorly fissured limestone	Dundee, USA	2.2–9.4
Poorly fissured limestone	UK	1.4–1.6
Oolitic limestone	Monk's Park, USA	20
Chalk	France	30–45
Dolomite	Micheldeau, UK	9–22
Marble	–	0.4–2.1
Calcareous tuff	–	20–32

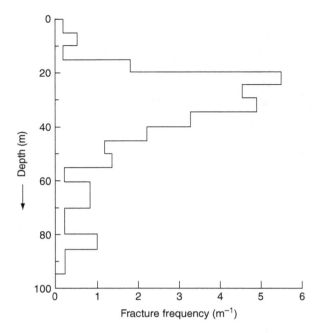

Fig. 15.15 Fracture distribution with depth in the Cretaceous Chalk aquifer of London Basin, UK. (After Owen 1981)

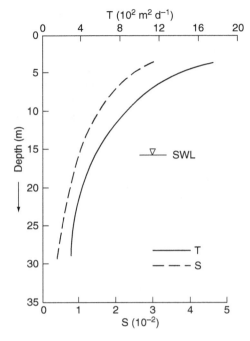

Fig. 15.16 Variation in transmissivity (T) and storativity (S) with depth in the Cretaceous Chalk aquifer of London Basin, UK. (After Owen 1981)

water-table and then decrease with depth (Fig. 15.15). A decreasing trend in transmissivity (T) and storativity (S) with depth can also be due to decreasing intensity of fracturing, as is demonstrated from a study in the Cretaceous chalk aquifer from England (Fig. 15.16 and 15.17). Giusti (1977) also reported a decrease in K with depth from Puerto Rican karst ranging from about $1\,m\,s^{-1}$ in the upper parts to $10^{-4}\,m\,s^{-1}$ in the basal parts. Such a trend is also reported from many other carbonate rock terrains (Otkun 1977; Rushton and Raghava Rao 1988; Arihood 1994). The infilling of solution cavities by clastic sediments will also result in the lowering of hydraulic conductivity at deeper levels (Huntoon 1992a).

The flow characteristics in carbonate rocks can be explained by discrete fracture flow, equivalent porous medium and double porosity models. A 'triple porosity' model has also been conceptualized in well developed karst aquifers consisting of matrix, fracture and conduit porosities (Kresic 2007). For the purpose of regional groundwater resource assessment and management, the continuum character of the aquifer (equivalent porous medium) can be an acceptable generalization, especially when fractures and solution cavities are closely spaced and well interconnected (Maslia and Prowell 1990; White 2007). The flow through the matrix is slow as per Darcy's

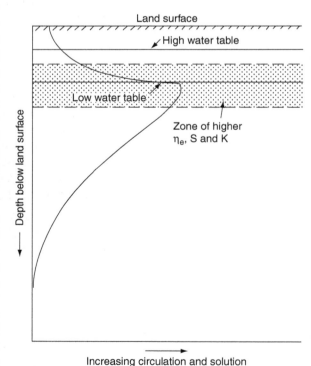

Fig. 15.17 Relationship between hydraulic properties and depth in carbonate rocks. Note that the maximum values of η_e, S and K are close to the water table. (After UNESCO 1984b)

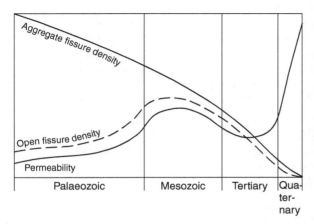

Fig. 15.18 Variation in hydraulic conductivity and fissure density with geological age in carbonate rocks. (After UNESCO 1984b)

law while flow through fractures and conduits can be non-Darcian when the aperture exceeds about 1 cm (Gabrovsek and Dreybrodt 2001; White 2002). The reader may refer to LaMoreaux et al. (1975), LaMoreaux (1986), Bogli (1980), UNESCO (1984b), Paloc and Back (1992), White (2007) for a more detailed account of the hydrogeology of carbonate and karstic formations.

Generally, the hydraulic conductivity of carbonate rocks shows a decreasing trend with increasing geological age (Fig. 15.18). Often higher values of hydraulic conductivity are observed in the carbonate rocks of Quaternary age than in older carbonate rocks. This figure also shows that although older limestones have high aggregate fissure density, hydraulic conductivity is lower, possibly due to the healing and infilling of fractures by calcite in older rocks. Therefore, K is directly related to the density of open fractures rather than the total number of fractures. In some cases, however, higher hydraulic conductivity is also observed in Precambrian limestones and dolomites.

Matrix permeability decreases roughly exponentially with age. For example, the matrix permeability of Paleozoic limestone at Manmouth cave in Kentucky, USA is $10^{-17} m^2$ in comparison to that of Holocene rocks in Florida and Bahamas ($10^{-10} m^2$).

In old carbonate rocks of Palaeozoic and Mesozoic ages, which are well cemented and recrystallized, flow is often primarily through conduits and fractures. In contrast, the younger aquifers of Tertiary and Quaternary ages, have greater primary porosity and permeabilities than recrystallized carbonates, the matrix permeability averages about four orders of magnitude greater in Floridan aquifer (Palaeocene to Miocene) in USA than in old carbonate rocks.

The usefulness of water-table maps in karst aquifers is limited due to the differences in the hydraulic conductivity of the conduit system on the one hand and of matrix and fractures on the other. Water-table can be mapped satisfactorily only when there are sufficient number of wells tapping the fractures and the matrix. In such cases, location of conduits is generally indicated by the presence of troughs (White 2007).

In view of the complexity of groundwater flow, the hydrogeological characterization of karstic aquifers can be made from (1) geomorphological features, (2) fracture analysis, (3) tracer studies, (4) hydraulic tests, (5) groundwater temperature, (7) biomonitoring, and (8) spring flow analysis.

Geomorphological Features The geomorphological features and rock permeabilities control the position of the water-table to a large extent. A deep water-table is observed under high relief and in thick permeable formations, while a shallow water-table develops in low topographic conditions and in areas where impervious beds underlie permeable karstic horizons at shallow depth. Groundwater movement is fast through bigger cavities and is slow and of diffuse type thorough small fractures and intergranular spaces.

Groundwater flow in karst limestones may be divided mainly into two components, (1) diffuse (Darcian) as in granular aquifers, and (2) turbulent (conduit) flow through fractures and conduits. In areas with prominent karstification, conduit flow will be most prominent. This is reflected in the trend of water-level fluctuations and spring discharges. Seasonal water-level fluctuations in karst aquifers are usually high, being of the order of 20–80 m. This is due to the large infiltration capacity of karstic regions leading to quick inflow of water into solution cavities.

Mature karst areas with very high porosity due to dissolution generally have very high infiltration. For example in the karst areas of Montenegro, in erstwhile Yugoslavia, recharge rates of the order of over 80% of total precipitation have been recorded resulting in very high temporary discharge of over $300 m^3 s^{-1}$, which happens to be one of the largest in the world (Kresic 2007). The high infiltration capacity of young CO_3 rocks is also reflected by few streams or rivers. In such areas, groundwater may be the only source of water supply.

Fracture Analysis Fracture and lineament studies are particularly useful in groundwater investigations in carbonate rocks. Details of techniques and parameters for describing fractures are given in Chap. 2. Their utility in exploration of groundwater and locating successful wells in carbonates rocks has been well elucidated (Lattman and Parizek 1964) (see Sect. 4.8.10).

Fractures and discontinuities control dissolution of carbonate rocks and development of various karst landforms. Figure 15.19 gives an interesting insight into the movement of groundwater in carbonate rocks in Vietnam region. The Nam La river in NW Vietnam flows through a highly karstified limestone terrain disappearing underground several times and reappearing on the surface again. Tam et al. (2004) attempted a correlation of the lineament density map and possible flow pathways, and confirmed observations with limited tracer tests. The dark cells in the figure correspond to high lineament-fracture density areas and appear to preferentially control the surface as well as inferred groundwater flow channels.

Tracer Studies On account of fracturing, rock heterogeneities and solution cavities in carbonate rocks, tracers (fluorescent dyes, spores, salt, CFCs, and noble gases viz. helium and neon) give a better idea of groundwater flow than water-table or potentiometric surface data (Thrailkill 1985; Long et al. 2008). Fluorescent dyes have been more commonly used. Noble gases (helium or neon) are preferred for tracing the groundwater movement through comparatively tight fractures (White 2007). Using tracer tests, linear

velocities through conduits are estimated to vary from less than $200\,m\,d^{-1}$ to more than $20\,km\,d^{-1}$ (Domenico and Schwartz 1998). Tracer studies also provide useful information on pollution and the conduits' structure in structurally complex karst formations. ^{3}H, ^{14}C and ^{36}Cl data have suggested four levels of water circulation in karst aquifers with residence time <20 years in shallow aquifers to ~100 000 years at depths of ~2500 m in the Pyreenes Mountains of southern France (Aquilina et al. 2003).

Hydraulic Tests The hydraulic properties of carbonate rocks can be estimated in the laboratory and in the filed from slug tests, pressure injection (Lugeon) tests, pumping tests and tracer injection tests, as described in Chap. 9. However care may be taken that the pumping well does'nt tap a conduit as it would show unrealistic high yield and too little drawdown. On the contrary, if the well penetrates the matrix part, the discharge will be less and drawdown will be large (White 2007). Spring discharge and groundwater-level data are also used to estimate hydraulic conductivity (Sect. 15.5). The conventional Theis-Jacob drawdown method has often been used for estimating aquifer characteristics from pumping test data in carbonate rocks (Zeizel et al. 1962; Csallany 1965; Eagon and Johe 1972). Leaky aquifer and double-porosity models have also been used to estimate aquifer properties in carbonate rocks (e.g., Motz 1982; McConnell 1993; Marechal et al. 2008). Values of T and S as estimated from borehole tests in some carbonate rocks are given in Table 15.4.

Fig. 15.19 Correlation of lineament density map and possible flow pathways of ground water in Nam La river catchment, Vietnam. The dark cells in the figure correspond to high lineament-fracture density areas and appear to preferentially control the surface as well as inferred groundwater flow channels. (After Tam et al. 2004)

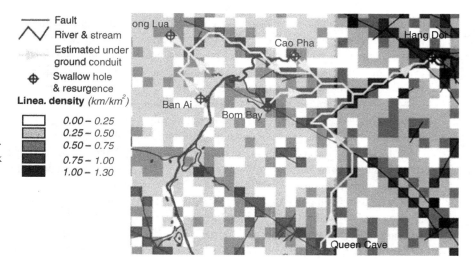

Legend:
— Fault
/\/ River & stream
⋯ Estimated under ground conduit
⊕ Swallow hole & resurgence

Linea. density *(km/km²)*
- 0.00 – 0.25
- 0.25 – 0.50
- 0.50 – 0.75
- 0.75 – 1.00
- 1.00 – 1.30

Table 15.4 Transmissivity and storativity values of some carbonate rock aquifers

Location	Rock type (age)	T (m²d⁻¹)	S	Source
Ohio, USA	Limestone and Dolomite (Silurian-Devonian)	4–128	2×10^{-4}	Eagon and Johe (1972)
Georgia, USA	Dolomites (Upper Floridan)	5907	3.5×10^{-4}	Maslia and Prowell (1990)
Israel	Fractured limestone (Jurassic)	86–11000	0.01	UNESCO, (1984b)
Lebanon	Fractured limestone	86–5000	0.001–0.024	UNESCO, (1984b)
Witwarterstand, S. Africa	Karstified dolomite	3300–4600	–	Boehmer (1993)
Mt.Larcom, Queensland, Australia	Limestone (Middle Devonian)	10–3000, average 500	–	Maslia and Prowell (1990)
Damman, Saudi Arabia	Limestone (Palaeocene)	55720	–	Otkun (1977)
London Basin, UK	Chalk (Lias)	15–370	–	Water Res. Board (1972)
Sanibel Island, Florida, USA	Limestone and dolomite (Miocene)	119	2.7×10^{-5}	Motz (1982)
Coastal Saurashtra, India	Miliolite limestone (Pleistocene)	200–900 175–950	– 0.12	Nair (1981) Rushton and Raghava Rao (1988)
Ghataprabha, India	Kaladgi limestone (Precamabrian)	150–500	–	Angadi (1986)
Chattisgarh, India	Charmuria Limestone (Precambrian)	400	–	Khare (1981)

As in other fractured rocks, due to anisotropy and heterogeneity, the measured value of hydraulic conductivity is a function of the reference volume of the rock mass, which may vary by more than three orders of magnitude, depending on the scale of measurement (Fig. 15.20). The greatest difference is in karstified rocks (Rovey and Cherkauer 1995). In chalk from the London Basin, laboratory values of K using permeameters were $5 \times 10^{-9} \, m \, s^{-1}$ as compared with the values of K ($2 \times 10^{-6} – 7 \times 10^{-5} \, m \, s^{-1}$) estimated from pumping tests (Water Res. Board 1972). Therefore permeability values

estimated from pumping tests may be useful for small scale problems, such as well head protection or local water supplies but may not be applicable at the aquifer scale to predict contaminant transport etc. (Mace 1997). In such cases tracer studies are more useful.

Groundwater Temperature Groundwater temperature data are also used to study the residence time and groundwater velocities (Screaton et al. 2004). This also helps in estimating the groundwater flow from matrix into the conduit system.

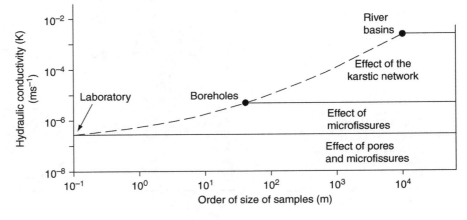

Fig. 15.20 Schematic representation of the effect of scale on the estimated hydraulic conductivity of karst. Laboratory values obtained from core samples are several orders of magnitude lower than those obtained from borehole tests and regional assessment. (After Kiraly 1975)

Biomonitoring Biomonitoring which is commonly used to evaluate the quality of groundwater and surface water has also been used to monitor the flow mechanism of groundwater in karst aquifers (Humphreys 2009). A distinction can be made between diffuse percolation via shallow holes (allochthonous origin), flow though fractures and fissures, and the saturated zone (autochthonous origin) by studying bacterial components (Pronk et al. 2009).

15.5 Assessment of Groundwater Recharge

A model for groundwater flow, recharge and storage in karst formations is given by Smart and Hobbs (1986) (Fig. 15.21). Usually, karst rocks have a high rate of infiltration which at places may be even 80% of precipitation (Boni et al. 1984). Khair and Haddad (1993) have reported infiltration rates to be 40–44% of the annual rainfall (1600 mm) in fractured carbonate rocks of Jurassic and Cenomanian age in the Lebanon. Healy and Cook (2002) have reported water-level rise of approximately 15 m in response to rainfall of approximately 50 mm over 24 h in fractured Cretaceous limestones and marly limestones in southern France. Even in arid areas, recharge rates can be high due to high permeability of the surface. For example, in the karst areas of Saudi Arabia, recharge is about 45% of the yearly rainfall (145 mm) (Al-Saafin et al. 1990).

The following four methods have been used for assessing groundwater recharge in karst regions.

1. Conventional water balance method.
2. Chloride method. This is suitable for the estimation of groundwater recharge in carbonate rocks because runoff is very low, and the rocks contribute very little to the chloride content of groundwater (UNESCO 1975).
3. Isotope techniques using tritium (^{3}H) and stable isotopes (^{2}H and ^{18}O) are well-suited for hydrogeological studies in karstic aquifers. Measurement of ^{3}H, ^{2}H and ^{18}O helped in determining the relative recharge from rainfall and snowmelt in the Alpine karstic aquifers (Long 1995). Nair et al. (1993) used ^{3}H and ^{14}C for studying the recharge pattern in the Vindhyan limestone of the Jodhpur district in India. Jones and Banner (2000) found dissolved chloride and ^{18}O data useful in providing insight into the spatial and seasonal distribution of recharge in a tropical karst aquifer in Barbodas.
4. Spring hydrograph analysis.

The first two methods are described in Chap. 20. The isotope techniques are given in Chap. 10. Spring hydrograph analysis is described below. The hydrograph of a spring in karst aquifers will depend on the interplay of three independent processes of recharge, storage and flow (Fig. 15.21). The importance of hydrograph analysis in karst hydrology is emphasized by White (2002).

The hydrograph of a spring discharge shows a shape similar to that of a surface water stream, i.e.

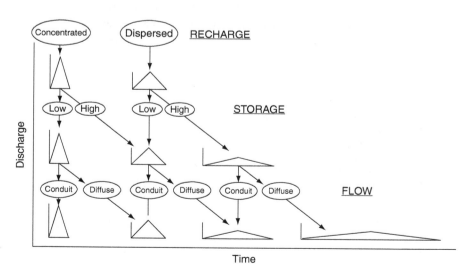

Fig. 15.21 Hypothetical effects of variations in recharge, storage and discharge upon spring hydrographs. (From Smart and Hobbs 1986; National Ground Water Association Copyright)

a rising limb, peak and a recession limb (Sect. 1.2). It also shows a time-lag between precipitation and peak of the hydrograph. In permeable, formations, like karstic limestones, the peak of the spring hydrograph will be broader and flatter with a greater time-lag. In contrast to this, in less permeable formations, the hydrograph has a sharper peak due to small aquifer storage. In karstic aquifers, the peak and the recession limb of the spring hydrograph may also have several bumps indicating complex recharge and flow mechanism (Fig. 15.22). A concentrated recharge through a conduit with low storage will result in spiked hydrograph while diffused recharge in an aquifer of long-term storage and diffuse flow will show an extended spring discharge with low peak (Kresic 2007). The importance of matrix (diffuse) flow to spring discharge is shown by several workers, viz. Atkinson (in White 2002) concluded from the analysis of spring hydrographs that 50% of the spring discharge was by quickflow and 50% by slowflow i.e. diffuse flow. Recent studies also demonstrate that in addition to conduit flow, matrix flow also contributes significantly to spring flow (Florea and Vacher 2006).

A quantitative analysis of spring hydrograph recession is generally based on Maillet's formula (Eq. 15.2).

$$Q_t = Q_0 \, e^{-\alpha \, t} \qquad (15.2)$$

where Q_t is the discharge rate at time t, Q_0 is discharge rate at any previous time during recession, t is the elapsed time between Q_t and Q_0, e is the base to the Napierian logarithm, and α is the recession (discharge) coefficient of the dimension (T^{-1}). The coefficient, α will be a function of the aquifer transmissivity, storage coefficient or specific yield and the catchment geometry. Larger values of α indicate higher hydraulic conductivities and therefore is also a measure of degree of fracturing. Amit et al. (2002) used the recession constant values to evaluate the hydrological and flow characteristics of a group of springs in fractured chalk and dolomite aquifer in Galilee Mountains, northern Israel.

In logarithmic form to the base 10, Eq. 15.2 is written as

$$log \, Q_t = log \, Q_0 - 0.4343 t \, \alpha \qquad (15.3)$$

The recession coefficient, α may be obtained directly from Eq. 15.3 as

$$\alpha = \frac{log Q_0 - log Q_t}{0.4343 \, (t_2 - t_1)} \qquad (15.4)$$

The recession curve usually plots as a straight line on a semi-log paper with time on the arithmetic scale. Sometimes these semi-logarithmic plots consist of two or more segments (Fig. 15.23) with different values of α indicating complex hydrogeological characteristics (permeability and storativity) of the aquifer. Higher values of α, at the earlier time of recession indicate

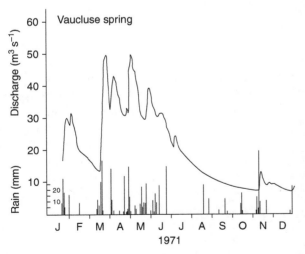

Fig. 15.22 Hydrograph of Vaucluse spring in karst formation, France. Rainfall data is also shown. (After Ford and Williams 1989)

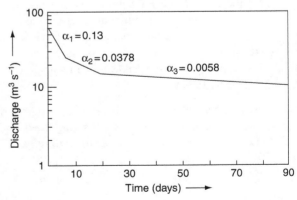

Fig. 15.23 Composite hydrograph recession of Ombla spring, Croatia. Note that the recession coefficient (α) decreases with increase in time of recession. (After Milanovic 1981)

fast drainage (quick flow) from conduits and large fractures, while the smaller values towards the later part of recession point (slow flow) is due to diffuse flow through smaller fractures and pore spaces into the conduits. In such cases, spring discharge can be expressed as

$$Q_t = Q_{01}e^{-\alpha_1 t} + Q_{02}e^{-\alpha_2 t} + \ldots Q_{0n}e^{-\alpha_n t} \quad (15.5)$$

The concept of a half-flow period, $t_{0.5}$, has also been used to evaluate the recession constant value. The concept of a half-flow period is similar to the half-life concept in nuclear physics, which is defined as the time required for the base flow of the stream or spring to halve i.e. $2Q_{t0.5}=Q_0$. By substituting this into Eq. 15.2

$$Q_{t_{0.5}} = 2Q_{t_{0.5}}\,\beta^{t_{0.5}} \quad (15.6)$$

Hence

$$\frac{1}{2} = \beta^{t_{0.5}} \quad (15.7)$$

and

$$t_{0.5} = constant/\log\beta \quad (15.8)$$

Like recession coefficient, α, half-flow period, $t_{0.5}$ is a direct measure of the rate of β recession and can be used as a means of characterizing exponential base-flow recessions of a spring. It is sensitive to changes in the spring flow domain and can have values in the range of zero to infinity.

With large values of β and small values of $t_{0.5}$, the recession curve will be steep, indicating fast drainage of the aquifer and little groundwater storage, i.e. the conduit function is prominent. Small values of β and large values of $t_{0.5}$ are characteristic of slow drainage from an extensively fractured or porous aquifer, with large storage capacity and limited recharge (Ford and Williams 1989).

Estimation of Groundwater Recharge Groundwater recharge can be estimated from spring flow data using Eq. 15.9 based on the principle of continuity.

$$AR = \frac{Q_2}{\alpha} - \frac{Q_1}{\alpha} + \int_{t_1}^{t_2} Qdt \quad (15.9)$$

where A is the recharge area (L^2), R is the recharge rate (LT^{-1}), t_1 and t_2 are the instance of time at the end

Table 15.5 Recharge rate as estimated from spring discharge data in some Mediterranean countries

Country	P (mm)	Q_w (mm)	$R_p=Q_w/p\times100$
Tunisia*	463–633	108–264	23–42
Greece*	1150–1400	583–723	50.7–51.6
Israel*	600–750	229–335	30.5–53.0
Turkey**	–	270	–

Source: *UNESCO (1975); **Korkmaz (1990)
P rainfall (mm), Q_w recharge; spring discharge divided by the intake area, R_p rate of recharge in terms of percentage of rainfall

of the one dry season and the beginning of the next one, Q_1 and Q_2 are the spring discharge rate (L^3T^{-1}) at time t_1 and t_2 respectively, and α is the recession coefficient, as defined earlier. Based on this approach, Korkmaz (1990) analysed the spring discharge data of the Kirkgoz spring in the Mediterranean karst of Turkey. The average annual recharge was estimated to be $493\times10^6 m^3$, i.e. $0.27\,m\,year^{-1}$ (Table 15.5.). Bhar (1996) used monthly discharge data from the Kirkgoz spring to estimate recharge on a monthly basis. Annual values of recharge obtained by Bhar (1996) are in close proximity to those obtained by Korkmaz (1990). Some values of recharge based on spring flow data are given in Table 15.5.

Evaluation of Hydraulic Parameters Considering the relationship between the recession coefficient, α and aquifer characteristics, Domenico (1972) provided Eq. 15.10 for estimating the transmissivity, T

$$T = \frac{4\,\alpha\,L^2 S}{\pi^2} \quad (15.10)$$

where L is the groundwater path length, i.e. the average distance from a catchment divide to the spring, and S is the coefficient of storage or specific yield. Equation 15.10 is valid when the thickness of the saturated zone is high enough so that the change in the saturated thickness during the drainage period can be neglected. Further, in fractured and karstic formations, variation in the value of α due to rock heterogeneity and inaccuracy in the estimation of L in the field, may not provide realistic estimates of T. However, in some cases the values of T estimated from Eq. 15.10, and from pumping tests, are quite comparable (Stasko and Tarka 1996).

Hydraulic conductivity, K can also be estimated from spring discharge and groundwater level data

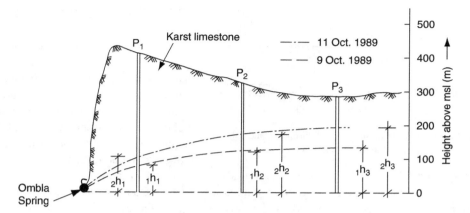

Fig. 15.24 Section across the karst massif at Ombla spring, Croatia, showing the position of three piezometers (P_1, P_2 and P_3) at distances of 1.62, 4.66 and 7.40 km respectively from the spring outlet. Groundwater levels in the three piezometers on 9 October 1989 and 11 October 1989 are also shown. (After Bonacci 1995, reprinted from *Journal of Hydrology*, with kind permission of Elsevier Science-NL, Amsterdam, the Netherlands)

using Eq. 15.11 based on Dupuit's assumption for steady-state flow in an unconfined aquifer.

$$K = \frac{2.3 \; Q \; log \; r_2/r_1}{\pi \; \left(h_2^2 - h_1^2\right)} \qquad (15.11)$$

where, h_1 and h_2 are the groundwater levels in two observation wells at distances r_1 and r_2, from the spring's outlet, respectively. Equation 15.11 is the well-known Thiem's equilibrium equation which is used for estimating K from pumping tests in unconfined aquifers under steady-state conditions.

Hydraulic conductivity, K of the karstic limestone at the Ombla spring, Croatia, was estimated using Eq. 15.11 by Bonacci (1995). The layout of the observation wells is shown in Fig. 15.24. K was computed to be in the range of 2×10^{-3}–5×10^{-3} m s^{-1} for the two sets of water-level observations. Bhar (1996) used a search technique, for estimating transmissivity and storativity of the confined karstic aquifer feeding the Kirkgoz spring, Turkey, based on monthly spring discharge data. T was estimated to be 686 m^2 d^{-1} and S was 0.0013.

15.6 Groundwater Quality

The water quality in CO_3 aquifers is influenced by the rock composition, structures and geomorphological set-up which control the groundwater recharge, movement, and storage processes and thereby the residence time.

Extensive work on the thermo-dynamic equilibrium of dissolution-precipitation processes in $CaCO_3$ and $MgCO_3$ species have helped in predicting whether water is unsaturated, supersaturated or is at equilibrium (Garrels and Christ 1965; Back and Hanshaw 1965; Hem 1989). The measurement of pH, temperature, calcium and carbonate concentrations, and ionic strength will provide this information.

Groundwater in carbonate rocks contains more alkaline earths and carbonate ions, relative to alkalis, chloride and sulphate. Therefore, groundwater is mainly of Ca–Mg–HCO$_3$ type. The pH of the groundwater is generally above 7. As the contact area between groundwater and rock matrix is limited due to fissure flow, groundwater in carbonate rocks usually has low total dissolved solids. However, where these rocks are interbedded with shale and other low permeability formations, there may be development of Na–HCO$_3$ and Na–Cl waters due to ion-exchange and greater residence time (Kehew 2001).

The Mg^{2+}/Ca^{2+} ratio provides information on lithology of rock units through which the groundwater has moved. In dolomites, the ratio of γMg/γCa in groundwater is less than in the rock due to preferential dissolution of calcium carbonate. Intermediate values imply mixed lithology. Mg^{2+}/Ca^{2+} ratio is also used commonly as an indicator of the residence time of water; an increase in the Mg/Ca ratio indicates longer residence time.

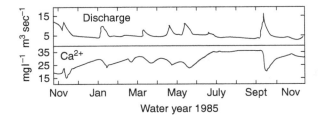

Fig. 15.25 Relationship between spring discharge and calcium concentration in groundwater. (After Delleur 2007)

Chemographs Chemical variation in the composition of spring water can be shown by plotting variation in the chemical constituents with time which depends on the hydrogeology of the specific spring. Chemographs can be superimposed on spring hydrograph to study the relation between the spring discharge and the water composition (White 2002). Generally, specific conductance, temperature, hardness, and Ca^{2+} (Figs. 15.25, 15.26) are used for this purpose. A sudden decrease in specific conductance and hardness is an indication of dilution of spring water by storm water from the surface. The conduit spring will show a large seasonal variation in hardness and temperature as compared with diffuse-type spring. The temperature of a diffuse spring remains more or less constant throughout the year but on the other hand wide variation in the temperature of a conduit spring is noted which is attributed to the large seasonal variations in the temperature of recharging water and shorter residence time, which is not normalized due to rapid flow (Fig. 15.26).

15.7 Groundwater Development

Water supplies in karst are obtained from springs and wells. Downhole video cameras can be used to locate open conduits and other productive horizons in karst aquifers. Springs are the main source of water supply in hilly regions. Springs draining conduit systems have highly variable discharge and high turbidity and risk of pollution. On the other hand, springs fed by fracture system, due to their uniform discharge and small chemical variation, are a more reliable source of water supply. Conduits pose greater uncertainty of water supply through wells due to their rapid drainage after the storm and low storavity during inter-storm period.

Dug wells and boreholes are common structure constructed for water supply. Location of wells in karst is challenging except in young limestones, viz. Floridan aquifers in USA, where matrix permeability is high. In younger carbonate rocks, primary porosity, permeability and specific yield is greater than the recrystallized older aquifers, therefore main supply of groundwater in young CO_3 rocks appears to be from rock matrix than conduit flow (Screaton et al. 2004).

Dug wells of 5–10 m diameter are common in the stone forest aquifers of China (Huntoon 1992b). The yield of such wells varies from $2 \times 10^{-4} m^3 s^{-1}$ to $2 \times 10^{-3} m^3 s^{-1}$ with small drawdown. Very high discharges of the order of $0.13 m^3 s^{-1}$ with practically no drawdown is reported from the Precambrian (Vindhyan) limestones of the Borunda-Bilara area in western

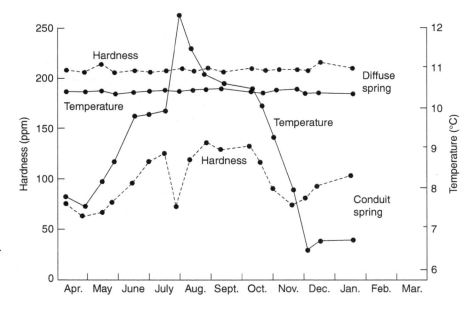

Fig. 15.26 Plots of temperature and hardness vs. time for a diffuse-flow spring and a conduit-flow spring. (After Kehew 2001)

India. In the chalk of South-east England, a common practice is to sink a well and provide adits to intercept the maximum number of vertical fractures. The yield of such wells is in excess of $0.15\,m^3\,s^{-1}$. However, suitably located smaller diameter wells in chalk aquifers are more successful (Hamill and Bell 1986).

The yield of borewells from rocks with good secondary porosity and permeability can be very high. For example, the usual discharge from wells in the Ocala limestone of Eocene age in Florida and Georgia, USA, is $0.063\,m^3\,s^{-1}$ and one well gave a yield of $0.475\,m^3\,s^{-1}$ with a drawdown of 3 m only (Maslia and Prowell 1990). The sustained yield of borewells from fissured chalk aquifers of unconfined and confined zones in England is reported to be 0.05–0.10 and $0.002–0.05\,m^3\,s^{-1}$ respectively (Owen 1981). The yield of borewells from partly karstified Pakhal and Cuddaph limestones of Precambrian age in central India ranges from 0.01 to $0.02\,m^3\,s^{-1}$ for 3 m drawdown.

Specific capacity and well productivity data are useful in comparing the relative performance of wells in different hydrogeological environments. Well productivity value is the specific capacity per unit saturated thickness of the aquifer adjusted to a common radius and pumping period. The well productivity in carbonate rocks depends on: (a) fractures, lineaments and other structural features, (b) solution features, (c) topography, (d) depth, and (e) palaeoweathered horizon.

The importance of fracture traces, lineaments and solution features has been demonstrated by several workers. Figure 15.27 indicates that wells located on fracture traces have higher yields than nonfracture trace wells. A study of well productivities in limestones from Maryland, USA, suggested that specific capacity of well s located within a distance of 30 m from the lineaments was about 15 times greater than those located at longer distances, indicating the importance of lineaments for locating high-yielding wells (LaRiccia and Rauch 1977).

Solution features, which are mainly controlled by rock solution openings developed along horizontal bedding planes are important for the water supply from wells, as there is a greater probability of verticals wells striking horizontal openings rather than vertical joints. In folded carbonate rocks, wells drilled in the crest of the anticlines are likely to be more productive than those in synclinal areas as the intensity of fractur-

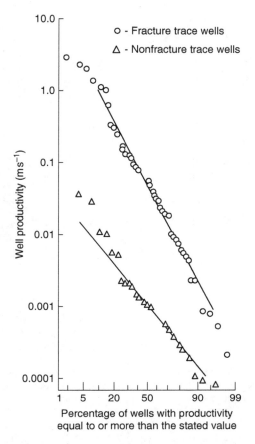

Fig. 15.27 Production-frequency data for water wells in relation to their location with respect to fracture traces (lineaments) in carbonate rocks. (After Siddiqui and Parizek 1971)

ing and solution will be more pronounced in the former case. The well productivity usually decreases with depth due to decrease in fracturing and solution (Walton 1970). A similar trend is observed from Charmuria limestone in the Raipur basin, India (Fig. 15.28).

Topography influences transmissivity and thereby well productivity. For example, in the Upper Crtaceous chalk aquifer of the Seine valley in France, tansmissivity was estimated to be $16\,000\,m^2\,d^{-1}$ along the valleys and less than $20\,m^2\,d^{-1}$ in the interfluves (UNESCO 1984b). Palaeoweathered horizons and unconformities form a good source of water to wells as is illustrated from a study in the Darwin Rural Area, NT, Australia (Fig. 15.29). In addition to the above factors, methods of well construction also influence well yields (Knopman and Hollyday 1993). Acid treatment is very effective in improving well discharge. This aspect is discussed in Chap. 17.

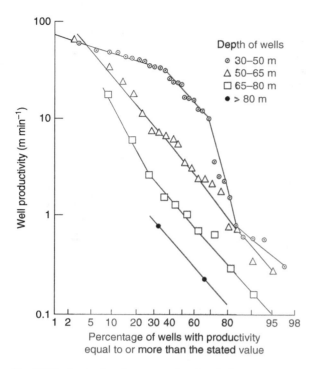

Fig. 15.28 Production-frequency plots in relation to depth of water wells in Charmuria limestone, Raipur Basin, India. Note the decreasing trend of well productivity with depth. (After Khare 1981)

The reliable yield to wells in CO_3 aquifers depends upon the base flow component of recharge and discharge, as most of the 'quick flow' through conduits has a too short residence in the aquifer to be exploited (Atkinson 1977; Screaton et al. 2004).

Although karst areas are an important source of water supply, they may also pose several hydrological and environmental problems listed below:

Scarcity and poor predictability of groundwater supplies due to great variation in permeability, both laterally as well as vertically. Further, zones of high permeability need not form potential sources of groundwater supply as the water may rapidly drain out from these places and the water-table may be very deep.

Rapid drainage of groundwater causing alternate flooding and drought conditions. For example, in the karst area of South China, due to high infiltration, the water-table comes close to the ground surface in the rainy season but in the dry period the water-table recedes considerably resulting in drought conditions (Huntoon 1992b). Leakage from surface reservoirs can also be high. *Perennial streams are scarce* and there is variability of flow due to large infiltration.

Poor soil cover which is infertile and easily eroded, resulting in poor agricultural production and reduction in spring discharge.

Unsatisfactory waste disposal environment due to high permeability and thin soil cover, causing erratic and fast movement of the pollutant to the zone of saturation without getting purified. It is also difficult to predict the movement of contaminants due to rock heterogeneity. The common contaminants are fecal material from septic tanks, pesticides and insecticides etc, from agriculture fields. Organic contaminants viz., LNAPLs (gasoloine and diesel fuel) due to leakage from underground tanks and DNAPLs (viz. chlorinated hydrocarbons) also occur as contaminants. Suspended clayey sediments with adsorbed heavy metals viz. Cr, Cd and As are also reported from carbonate aquifers. Swallow holes are the major source of nitrate, organic carbon and faecal bacteria, particularly after

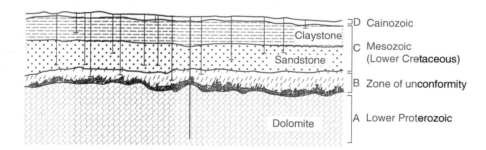

Fig. 15.29 Hydrogeological section in Darwin Rural Area, NT, Australia. B is the hydro-stratigraphic unit formed due to unconformity. This horizon consisting of chert, dolomite and other weathered materials forms the most productivie horizon with well yields upto $0.7\,m^3\,s^{-1}$. (After Verma and Qureshi 1982)

heavy rainfall. Therefore, sanitary landfills and other waste disposal sites should be located with great care and regular monitoring of surface and groundwater quality is essential. However, as the residence time of water in karst aquifers is small due to fast drainage, groundwater is easily renewed and thereby effects of accidental pollution do not persist for long durations and therefore chances of diffuse pollution are also less as compared with porous rocks (Bakalowicz 2005). Biomonitoring of karst water is useful in assessing the groundwater contamination and flow mechanism (Humphreys 2009).

Instability of the ground surface and subsidence due to the development of subsurface solution cavities. This also creates land use problems such as construction of roads, mining, agricultural and other civilian activities.

Summary

Carbonate rocks are comprised of dominantly calcite and dolomite and occupy more than 10% of the earth's surface. These rocks are largely soluble in water rich in carbonic acid giving rise to characteristic karst landforms exhibiting sinkholes, cavities, dry valleys and internal drainage. Springs of different types and varying discharge are important features which form an important source of water supply. The carbonate areas exhibit variation in hydrogeological characteristics owing to differences in porosity, fracturing, thickness, composition, geological age, paleoclimate, proximity to water-table, and sea-level changes. Various models such as equivalent porous medium, double-porosity and triple-porosity models have been conceptualized to represent the hydrogeological behaviour. Fractures and lineaments are particularly useful in groundwater exploration in carbonate rocks. Downhole video cameras are used for detecting producing fractures and open conduits. Tracer studies can provide an idea of subsurface groundwater flow. Hydraulic properties can be estimated in the field from slug test, pressure injection test, pumping test, and tracer injection tests. Usually, karst areas have high infiltration (>80% of precipitation). Groundwater in carbonate rocks contains more alkaline earths and bicarbonate ions ($Ca–Mg–HCO_3$) type; pH being >7.

Further Reading

Ford DC, Williams PW (2007) Karst Hydrology and Geomorphology. 2nd ed., John Wiley & Sons Inc., London, 567 p.

Kresic N (2007) Hydrogeology and Groundwater Modeling. 2nd ed., CRC Press, Boca Raton, FL, 805 p.

White WB (2007) Groundwater Flow and Transport in Karst, in The Handbook of Groundwater Engineering (Delleur JW ed.), CRC Press, Boca Raton, FL, pp. 18–1 to 18–36.

Hydrogeology of Clastic Formations

16

Under clastic formations, we have included both unconsolidated and consolidated sediments. Unconsolidated sediments include various admixtures of boulder, sand, silt and clay deposits. These on consolidation form clastic sedimentary rocks, e.g. sandstone, siltstone and shale.

16.1 Unconsolidated Sediments

Most of the unconsolidated sediments were deposited during the last few million years, and are of Quaternary–Recent age. They have been formed under different sedimentary environments, viz. fluvial, glacial, eolian and marine. The coarse grained sediments, (gravel and sand) form potential aquifers due to their high hydraulic conductivity and storativity. Being incoherent in nature, unconsolidated or semi-consolidated sediments are largely unfractured, except some glacial deposits like clay tills.

16.1.1 Fluvial Deposits

The fluvial deposits are characterized by typical landforms, viz. alluvial fans, flood plain, and terraces etc. which can be identified on aerial photographs and space imageries. Fluvial deposits form ideal aquifers as they occur along river valleys and in areas of even topography with adequate recharge. Some of the river valleys are deep and narrow, e.g. the Jordan river valley along the border of Jordan and Israel, and the Owens river valley in California, USA. Others form broad plains, e.g. Indo-Gangetic basin in Indian subcontinent, and North Plain in China.

Buried valleys and palaeochannels are of special interest for groundwater development in hard rock terrains. They contain thick deposits of gravel and sand and are a result of fluvial and fluvio-glacial processes involving shifting of river courses due to either tectonic or climatic reasons. Palaeo-channels are identified by their typical sinuous serpentine shape and form. In an arid or semi-arid terrain, the palaeochannels may posses anomalous moisture and vegetation, in comparison to the adjoining areas. Table 16.1 gives a summary of physical conditions of palaceochannels and the corresponding spectral characters on remote sensing data.

An interesting example of palaeochannel is furnished by the '*lost*' Saraswati river in India, which is said to have been a mighty river in the olden *Vedic or pre-vedic* times (about 5000 years BC). It used to flow in the area now occupied by the Thar desert in western India. Here, the palaeochannels can be identified in most cases on the basis of higher moisture content and vegetation patterns (Fig. 16.1). Using Landsat images, the ancient course of the river Saraswati has been delineated for a distance of about 400 km (Yashpal et al. 1980; Baklliwal and Grover 1988). Some of the buried segments of river Saraswati are a potential source of groundwater (Venkateswarlu et al. 1990). In arid and hyper-arid regions, the buried channels are of added importance for water supply, owing to acute water scarcity conditions. The SIR-A data in Eastern Sahara has demonstrated the applicability of SLAR data for delineation of palaeochannels in hyper-arid regions (McCauley et al. 1982; Elachi et al. 1984). In the sand-covered limestone terrain of Rajasthan desert, India, Mehta et al. (1994) interpreted presence of relict rivers on the basis of ERS-SAR images (Fig. 16.2).

B. B. S. Singhal, R. P. Gupta, *Applied Hydrogeology of Fractured Rocks*,
DOI 10.1007/978-90-481-8799-7_16, © Springer Science+Business Media B.V. 2010

Table 16.1 Physical conditions of palaeochannels and their spectral characters on remote sensing data

Physical conditions associated with the palaeochannel	Spectral character
Higher surface moisture	Medium to dark in visible bands; dark on NIR and SWIR images; dark on SLAR images
Preferential denser vegetation cover	Dark on visible bands; light toned in NIR; light on SLAR images
Dry sand cover; no surface anomaly in terms of vegetation; bedrock occurs at a greater depth	Darker on longer wavelength SLAR images
Alluvial cover, thick gravelly river bed resulting in relatively less surface moisture and poor vegetation	Very light on visible and NIR bands

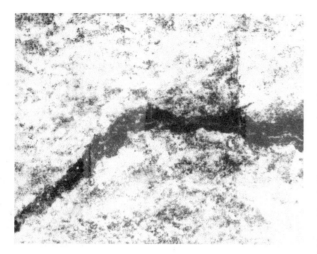

Fig. 16.1 Landsat MSS band-2 image showing the presence of palaeochannel of the 'lost' Saraswati river in the north-western India. The bed is about 6–8 km wide and is marked by dense vegetation

16.1.2   Glacial Deposits

Unconsolidated glacial deposits occur mainly in North America, northern Europe and Asia. Isolated occurrences are also observed on major mountain chains such as the Alps, the Andes, and the Himalayas, while close to the equator, glaciation was restricted to Mt. Kenya and Kilimanjaro in East Africa. Glacial deposits in North America and elsewhere have been studied in detail both as a source of water supply and also as host rock for the disposal of waste (Cherry 1989; Nilsson and Sidle 1996).

Glacial tills are unstratified and typically poorly sorted ice transported sediments in which the grain size varies from clay fraction to boulder size material. In some areas tills are thick and unfractured but wherever fractured, they provide active hydraulic connections and potential contaminant pathways (Hendry 1982; Ruland et al. 1991). Fracture spacing usually increases with depth thereby affecting vertical distribution of permeability (Fig. 16.3).

Fractures are often filled with calcite and gypsum. In some areas, thin rootlets are observed along the fractures to depths of 5–10 m below the ground surface (Freeze and Cherry 1979). Shallow fractures are regarded to be a result of alternate cycles of wetting and drying and freezing and thawing. The origin of deep open fractures in the unweathered clays is more problematic. These are attributed to stress changes caused by post-Pleistocene crustal rebound or to stress changes caused by advance and retreat of the last Pleis-

Fig. 16.2 a Palaeochannel as delineated on the ERS-SAR image of Rajasthan, Thar desert, India. The darker tones in the palaeochannel area are due to local thicker sand cover, as compared to the adjacent area (Courtesy: ESA). **b** Schematic interpretation. The lower radar backscatter is attributed to the thicker sand cover in the palaeochannel area

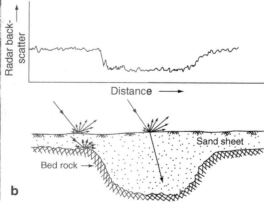

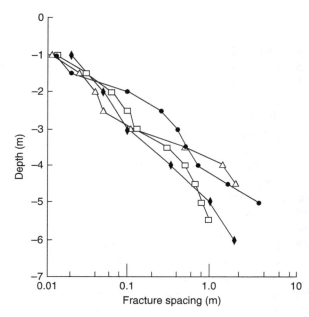

Fig. 16.3 Variation of fracture spacings in till with depth. (After Ruland et al. 1991)

this will promote fast movement of pesticides and contaminants from landfills and other waste disposal pits. The anisotropy ratio (K_v/K_h) can have a wider range depending on the relative role of horizontal layering and vertical fracturing. Laboratory values of hydraulic conductivity measured from core samples is two to four order of magnitude smaller than that determined from field tests (Jones et al. 1992).

Detailed pumping tests with an array of observation wells, distributed radially, can provide better idea of aquifer parameters including anisotropy (Edwards and Jones 1993). Tracer injection tests are also carried out for the estimation of hydraulic parameters. Conservative tracers like chloride were used to estimate the potential of groundwater contamination in fractured tills in Denmark (Nilsson and Sidle 1996). Tritium profiles indicate active groundwater circulation in fractured tills upto a depth of 5–10 m (Ruland et al. 1991). The probability of monitoring the movement of contaminants in fractured clayey till can be enhanced by constructing large diameter horizontal or angled wells as vertical wells may not be able to intercept the fractures (Jorgensen et al. 2003).

Water in glacial aquifers is generally very hard and has high concentration of total dissolved solids and sulphate. In areas of active groundwater circulation, water quality is usually good. High concentration of SO_4 (1000–10 000 mg l^{-1}), reported from weathered clays, is attributed to oxidation of organic sulphur or pyrite (Cherry 1989). In such cases, the amount of SO_4^{2-} can be used as a natural tracer to study the downward solute migration through fractures in the unweathered clays.

The chemical evolution of groundwater in the fractures is influenced by the composition of the matrix and of pore water as well as the diffusion gradients. Studies in fractured glacial till indicate that different ions, e.g. Ca^{2+} and Cl^- behave differently due to

tocene glacier (Cherry 1989). Volume changes due to geochemical processes, such as cation exchange, have been also suggested for the development of fractures (Freeze and Cherry 1979).

Glacial tills have porosities in the range of 25–45%; clay tills have higher porosities. Due to their unsorted character, the matrix (intergranular) permeabilities of glacial tills are low (10^{-10}–10^{-9} m s^{-1}). However, due to weathering and fracturing, they acquire higher hydraulic conductivities (K) of the order of 10^{-9}–10^{-6} m s^{-1} (Table 16.2). The upper parts of the till deposits usually have higher hydraulic conductivity due to greater intensity of fracturing and weathering (Ruland et al. 1991; Jones et al. 1992). The higher values of hydraulic conductivities in weathered and fractured tills will facilitate efficient drainage of agricultural lands but

Table 16.2 Hydraulic properties of glacial deposits

Type of deposit	Location	T(m²d⁻¹)	K(ms⁻¹)	Sy	Source
Till (fractured clayey)	Manitoba and Montreal, Canada	–	10^{-10}–10^{-9}	–	Cherry (1989)
Till (fractured)	Alberta, Canada	–	5×10^{-9}–2×10^{-7}	–	Jones et al. (1992)
Till (weathered)	Iowa, USA	–	4×10^{-4} (av.)	0.04 (av.)	Edwards and Jones (1993)
Glacial outwash	Central Illinois, USA	43–1 66488 (av.)	2×10^{-6}–7×10^{-6}	0.5	Walker and Walton (1961)
Glacio-fluvial	Michigan, USA	1440–1920	10^{-5}–10^{-4}	0.04–0.35	Kehew et al. (1996)

variation in their diffusion coefficients and adsorption by the matrix. Laboratory studies show that Ca^{2+} passes through the fractures more rapidly than Cl^- due to smaller diffusion coefficient of Ca^{2+} and adsorpition of Ca^{2+} within porous matrix (Grisak et al. 1980).

16.2   Consolidated Sediments

The common clastic sedimentary rocks are sandstone, siltstone and shale. These are formed in almost all environments including marine, fluvial, deltaic, lacustrine and eolian. Usually, sandstones and fine-grained clastic rocks, e.g. shales and siltstones, occur as alternating beds with varying thicknesses in most sedimentary sequences. Although sandstones being more permeable, are of main interest as a source of water supply, but their potentiality as well as water quality is greatly influenced by the composition of intergranular cementing material and properties of the interbedded shales and siltstones.

Sandstones being more resistant to erosion usually form hills, ridges and scarps while shales erode easily forming hill slopes and valleys. For the same reason, sandstones have low to medium drainage density but in shales drainage density is high. In sandstone, drainage pattern is rectangular or angular due to rock discontinuities. Shales show typical dendritic pattern. A comparison of landform and drainage characteristics in these two rock types is illustrated in Fig. 16.4.

16.2.1   Sandstones

The hydrogeological properties (porosity and permeability) of sandstones depend on their textural characteristics, which in turn is influenced by the depositional environments and subsequent changes due to cementation, consolidation and fracturing. The porosity, hydraulic conductivity and specific yield of sandstones is less than sands due to compaction and cementation. Common cementing materials are clays, carbonates, silica and iron oxides. As the degree of compaction depends on depth of burial, the porosity of sandstones decreases systematically with depth and age. Poorly cemented sandstones may have porosity of about 35%. Hydraulic conductivity of sandstones may vary from 1×10^{-6} to $1 \times 10^{-4} \, m \, s^{-1}$ (Table 16.3). Geologically, older sandstones have lower porosity and hydraulic conductivity due to greater compaction and cementation, but there are several exceptions also. Stratification imparts anisotropy; hydraulic conductivity along the bedding plane is usually higher than across it. In Berea Sandstone, USA, permeability parallel to the bedding was found approximately four times higher than that normal to the bedding (Lee and Farmer 1993). At other places, fracturing may impart higher vertical conductivities.

Younger sandstones of Mesozoic and Tertiary ages form most productive aquifers covering large areas, e.g. Nubian sandstone in North Africa, Dakota sandstone in USA, Ranmark and Murray Group of sandstone in Australia, the Jurassic sandstone in Great Artesian Basin (GAB) in eastern Australia and Cud-

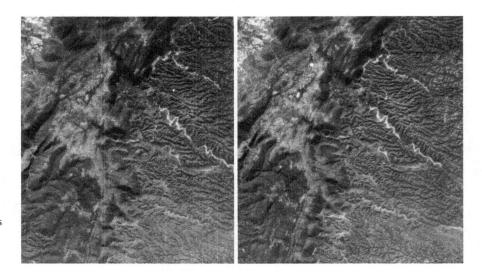

Fig. 16.4 Stereo aerial photographs showing sandstones and shales marked by differences in landform and drainage. (Courtesy A. White)

Table 16.3 Hydraulic properties of sandstones and argillaceous rocks

Rock type	Age	Location	$T(m^2 d^{-1})$	S	$K(ms^{-1})$	Source
Charmuria sandstone (fractured)	Algonkian	Raipur, India	3–13	–	3×10^{-7}–3×10^{-6}	Khare (1981)
Berea sandstone	Mississipian	USA	–	–	3×10^{-6}	Read et al. (1989)
Shewood sandstone (fractured)	Triassic	UK	600 (av.)	0.05–0.25	–	Kimblin (1995)
Karoo sandstone (fractured)	Permo-Triassic	Botswana, South Africa	200	10^{-4}–10^{-2}	–	Bromley et al. (1994)
Nubian sandstone	Cretaceous	Egypt	100–15 000	3.5×10^{-4}	–	Shata (1982)
Renmark Group sandstone	Lower Tertiary	Victoria, Australia	800–900	1.8×10^{-4}	–	Lawrence (1975)
Cuddalore sandstone	Miocene	Tamilnadu, India	1000–5000	2×10^{-4}–5×10^{-4}	–	Gupta et al. (1989)
Gunderdehi shale (fractured)	Algonkian	Raipur, India	2–18	–	10^{-8}–10^{-7}	Khare (1981)
Shales and siltstones (fractured)	Upper Permian	Queensland, Australia	1–21 12 (av.)	–	–	Pearce (1982)
Siltstones (fractured)	Devonian	Appalachian Plateau, New York, USA	–	–	10^{-8}–10^{-6}	Merin (1992)
Brunswick shale (fractured)	Triassic	New Jersey, USA	708–933	10^{-4}	10^{-5}–10^{-4}	UNESCO (1972) Michalski (1990)
Pierre shale	Cretaceous	USA	–	–	10^{-14}–10^{-10}	Neuzil (1986)
Opaline clay	Mesozoic	Mt. Terri, Switzerland	–	–	10^{-11}	Neerdael et al. (1996)
Boom clay	Neogene	Belgium	–	$S_s = 1.3 \times 10^{-5} m^{-1}$	10^{-12}	Put and Ortiz (1996)

dalore sandstone (Miocene–Pliocene age) in India. They form some of the best known confined aquifer systems extending over several thousand square kilometre and possess high transmissivities (Table 16.3). A case study of numerical modelling to estimate the effect of pumping of groundwater in Cuddalore sandstone (India) to facilitate open cast mining of overlying lignite is described in Sect. 19.5.5.

Groundwater in the Nubian sandstone is dated to be old (about 35 000 years) which was probably recharged during pluvial climate in Pleistocene time. Therefore, groundwater extraction in such areas should be planned carefully to avoid groundwater mining. Similar situation may exist in several other arid areas also.

Fractured Sandstones In firmly cemented sandstone, intergranular permeability is negligible but secondary permeability due to fracturing is of significance. Fracturing in sandstones could develop due to unloading, tectonic movements and at their contact with dolerite dykes. Fractures developed due to expansion, as a result of unloading, will be more prominent only upto a shallow depth (50 m or so), but tectonic fractures related with folding and faulting can be deep seated. Fracturing due to both tectonic movements and intrusion of dolerite dykes is reported from the Permo-Triassic sandstones of Karoo Supergroup in Botswana and other places in South Africa. As a result of fracturing, sandstones have acquired good transmissivity and specific yield (Table 16.3) (Bromley et al. 1994; Sami 1996). Triassic sandstones in UK also show much lower integranular hydraulic conductivity than the bulk aquifer conductivity as determined by pumping tests which is explained due to dominant fracture flow component (Hamill and Bell 1986). Similarly, other fractured sandstones also show moderate to high transmissivities (Table 16.3).

16.2.2 Shales and Siltstones

Fine-grained argillaceous clastic rocks, e.g. shales and siltstones are formed by compaction and lithification of clay and mud deposits. Porosity of freshly deposited clays and muds is high (50–80%), but due to compaction on burial, porosities are reduced to less than 30%. Shales usually have porosities in the range of 1.5–2.5%. Intergranular permeabilities in shales are low (10^{-13}–10^{-9} m s^{-1}) so that groundwater cannot move faster than few centimetre per century through intact shales (Table 16.3). Permeabilities of shales depend upon porosity, clay mineralogy, clay contents and texture. Permeabilities also depend upon fluid composition which affects clay swelling.

Even if the primary permeabilities of shales and siltstones are low, they are capable of transmitting large quantities of water and solutes over large contact areas by leakage across lithologic boundaries. For example, a 30 m thick siltstone bed with a hydraulic conductivity of 10^{-9} m s^{-1} having a hydraulic head difference of 3 m perpendicular to the bedding, will transmit about 3153 m^3 of water each year through each square kilometre of its surface areas.

Low permeability argillaceous rocks (clays and shales) have attracted greater attention of hydrogeologists in recent years from the point of view of disposal of high-level radioactive waste. With this in view, detailed investigations are in progress in the Underground Research Laboratories in the Boom Clay, Belgium and opaline clay/shale at Mt. Terri in Switzerland (Put and Ortiz 1996; Neerdael et al. 1996). Special *in situ* interference tests were designed in these underground laboratories to assess the hydraulic conductivity and storativity (Table 16.3).

Fractured Shales and Siltstones Fracturing can impart good hydraulic conductivity (10^{-7}–10^{-4} m s^{-1}) to shales, siltstones and other fine grained clastic rocks which are otherwise impervious (Table 16.3). In USA, notable examples include the Brunswick shale

of Triassic age in New Jersey and adjoining states of New York and Pennsylvania, shales of Pennsylvanian age in the Central States and fractured siltstones of Devonian age in the Appalachain basin of New York state. It is opined that fractured shales represent a double porosity system, in which the fractures control the major amount of flow while the intervening porous blocks contribute slowly by transient flow (Neuzil 1986). Michalski and Britton (1997) have emphasised the importance of bedding-plane partings, particularly those enlarged by stress relief, to provide principal groundwater flow pathways and groundwater contamination in the Raritan unit of Newark Basin, USA. A similar conclusion is drawn from the Maquoketa aquitard (dolomitic shale of Ordovician age) in the southeastern Wisconsin where the hydraulic conductivity due to bedding plane fractures, extending to several kilometre, is about five orders of magnitude higher than the vertical permeability. Water quality also indicates the isolated nature of vertical fractures (Eaton et al. 2007). In a heterogeneous sedimentary sequence, thin incompetent beds, such as shales will be more intensely fractured as compared with thicker units of strong and resistant formations such as mudstones. This is evidenced from the Brunswick Formation, in USA, where mudstone horizons at shallow depths have lower values of K (10^{-6} m s^{-1}) while shales at deeper levels (20–45 m) are fractured forming aquifer horizons with hydraulic conductivity (K) varying from 2×10^{-5} to 5×10^{-4} m s^{-1} (Michalski 1990). Devonian siltstones of Appalachian plateau, USA exhibit two sets of fractures-subhorizontal bedding plane fractures and vertical fractures. Unlike vertical fractures, the spacing of bedding fractures increases with depth due to greater confining stress at deeper levels. Slug tests in these siltstones indicate that closer spacing of bedding plane fractures at shallow depths (<7 m) impart hydraulic conductivities (2×10^{-8}–2×10^{-6} m s^{-1}) which are 10–100 times greater than those obtained from deeper wells (>50 m) (Merin 1992).

16.2.3 Groundwater Development

Groundwater from clastic consolidated rocks is tapped both by dugwells and drilled wells, the latter being more common. The common method of water-well drilling in open textured sandstones is the hydraulic rotary method. In hard and well cemented sandstones, percussion and DTH methods are used (Chap. 17). Geologically younger sandstones are less cemented and therefore are more productive. The yield of 100–3000 m deep wells, in the Jurassic sandstones of the Great Artesian Basin in eastern Australia, is 0.05–0.1 $m^3 s^{-1}$. Here water from deeper wells has high temperature (1000 C). Well yields vary from less than $3 \times 10^{-4} m^3 s^{-1}$ in compact sandstones to $3 \times 10^{-2} m^3 s^{-1}$ in open textured rocks. Specific capacity of wells in Cambrian—Ordovician sandstones of northern Illinois, USA is in the range of 53–70 $m^2 d^{-1}$ indicating the hydraulic conductivity to be in the range of 1×10^{-5}–$1.8 \times 10^{-5} m s^{-1}$ (Walton and Csallany 1962). Well yields from fractured Charmuria sandstone of Precambrian age and Athgarh sandstone of Jurassic ages in India are of the order of $1.5 \times 10^{-3} m^3 s^{-1}$ and 10^{-3}–$10^{-2} m^3 s^{-1}$ respectively depending upon the degree of compaction and fracturing (Khare 1981; CGWB 1995b). In compact sandstones, well yields can be increased by shooting using explosives and to a less extent by acid treatment, viz. shooting increased the specific capacity of sandstone wells in northern Illinois by an average of 22–38% (Walton and Csallany 1962).

Integrated geophysical and geological studies have proved very valuable in the location of high yielding fracture zones in sandstone formation of Karoo supergroup especially in areas with complex geology obscured by overburden (Bromley et al. 1994). Figure 16.5 gives a comparison of the specific capacity of wells sited on fractures mapped by the above approach as compared with those in unfractured aquifer. This figure also indicates that major subvertical fractures identified by VLF survey have highest well yield. Studies by Sami (1996) in the Karoo sandstone of South Africa also show that wells located in the fracture zone in the vicinity of dolerite dykes and also those located along stream channels of higher orders have higher yields. Well yield in fractured shales will be low to moderate. However, in the absence of other good aquifers, wells in fractured shales may yield some water. Well yield from Lower Proterozoic fractured shales in Darwin Rural Area, NT, Australia is reported to be 1×10^{-3}–$3 \times 10^{-3} m^3 s^{-1}$.

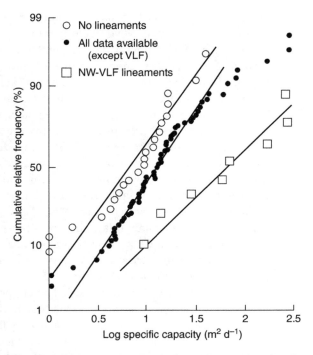

Fig. 16.5 Cumulative frequency of specific capacity of wells in fractured and non-fractured locations in Karoo sandstones, Botswana. Note that NW-VLF lineaments are most productive. (After Bromley et al. 1994, reprinted by permission of journal of Ground Water)

16.3 Water Quality

The chemical composition of groundwater in clastic rocks, as in other rock types, depends on their lithology and texture. In purely siliceous sandstones, the rocks consist almost wholly of quartz and therefore there are very few elements that can be dissolved in water. The groundwater in such rock types has low pH (about 6 or 5) and the HCO_3 content is also low; Ca, Mg, Cl and SO_4 are also in small concentrations; Na may exceed Ca. In rocks having lime or gypsum, the water has greater concentration of Ca and HCO_3. In arid and water logged regions, shallow groundwaters may have higher contents of salts due to evaporation (Greenman et al. 1967). In coastal areas, groundwater may show complex chemical characteristics due to mixing of fresh groundwater and sea-water as well as cation exchange. We return to this subject in Sect. 20.7.

Sandstone aquifers in large sedimentary basins show both horizontal (lateral) and vertical hydrochemical zoning (see Sect. 11.7.1). Groundwater in the recharge area is mainly of HCO_3 type which changes to SO_4-Cl type in the direction of flow. A decrease in SO_4 in the direction of flow is also reported from some places, e.g. groundwater in the Fox Hills sandstones in Dakotas, USA shows an increase in salinity but decrease in SO_4 from recharge to discharge areas. The source of SO_4^{2-} in recharge area is presumably due to pyrite and/or gypsum dissolution. The decrease in SO_4^{2-} appears to be a result of sulphate reduction in the presence of reduced carbon (lignite) while increase in chloride is attributed to cross flow of saline water from underlying Pierre shale (Thorstenson et al. 1979). Leakage from interbedded aquitards (clay, sillstone deposits) causing an increase in Cl^- content of groundwater is also reported from the Triassic Sherwood sandstone aquifer in northwest England (Kimblin 1995).

Shales and silstones being very fine and porous provide an enormous contact area with groundwater. The permeabilities are also very low, thereby increasing the residence time and greater opportunity for dissolution of rock material. Therefore, groundwater in these formations has high total dissolved solids ($>1000\,mg\,l^{-1}$) and the amount of SO_4^{2-} and Cl^- is more than HCO_3^- Both Ca–Mg and Na-rich waters are reported. In shales, SO_4^{2-} is the dominant anion while Cl^- is dominant in both shales and sandstones. Cation exchange is a characteristic phenomena in modifying the chemical characteristics of groundwater due to the presence of clay minerals. Clay and shales also play an important role in acting as semi-permeable membrane (osmotic filters) inspite of their low permeability as described in Sect. 11.6. Such a process is believed to be responsible for the occurrence of saline waters in non-marine sediments (Back and Hanshaw 1965; Neuzil 1986).

Movement of tracers and contaminants in fractured sandstone and shale will behave like in other dual porosity aquifers with rapid movement in fractures and diffusion into low permeability but porous matrix. Fracture skins formed due to the chemical alteration of sandstones into oxides of iron and manganese consid-erably lower the matrix rate of diffusion. These aspects have been discussed in Sect. 7.4.1.

Summary

Clastic formations include porous sedimentary rocks (sandstones, shales etc.) which have undergone consolidation and cementation to varying degree. The consolidated and cemented clastic formations have low porosities and permeabilities. Therefore their hydrogeological characteristics are similar to those of crystalline rocks. Glacial deposits viz. tillites have been studied in detail both as a source of water and also as host rocks for the disposal of waste. Sandstones, silstones and shales are characterised by bedding planes and are also fractured exhibiting varying degree of primary/secondary porosities and permeabilities. Younger sandstones of Mesozoic and Tertiary ages generally form more productive aquifers. Older sandstones of Precambrian age are often less productive. Shales and siltstones have low permeability, but fracturing can impart greater hydraulic conductivity. The water quality depends on lithology and permeability. Groundwater in sandstones is usually of good quality but in shales and siltstones, TDS is high on account of large residence time of water. Hydrochemical zoning is reported from large sedimentary basins.

Further Reading

Davis SN, DeWiest RJM (1966) Hydrogeology. John Wiley & Sons, Inc., New York.

Kresic N (2007) Hydrogeology and Groundwater Modeling. 2nd ed., CRC Press, Boca Raton, FL.

Walton WC (1970) Groundwater Resource Evaluation. McGraw-Hill, New York.

Water-wells are vertical shafts or holes used by mankind since time immemorial for obtaining groundwater for drinking purposes and other uses. The type of well to be constructed depends on the nature of the geological formation, depth to water-table, yield requirement and economic consideration. Sometimes horizontal or sub-horizontal wells or galleries are also constructed under favourable geohydrological conditions to tap subsurface water. Wells serve other purposes also, such as subsurface exploration, artificial recharge and waste disposal.

17.1 Types of Wells

For large-scale groundwater supplies, especially in unconsolidated and semi-consolidated formations of high permeability, tubewells are recommended. These wells are usually 100–300 m deep, or even more, depending on the depth of aquifer horizons, and water quality. Tubewells are provided with screens and are often gravel packed, creating a zone of high permeability around the well which also increases its effective diameter.

In hard rocks, due to their low permeability and depth limitations, the common structures for groundwater withdrawal are dugwells, borewells and a combination of these types. The technology of well drilling, design and construction is, however, more developed for unconsolidated formations (Campbell and Lehr 1973; Driscoll 1986).

17.1.1 Dugwells

Traditionally in most of the developing countries of Asia and Africa, large diameter dugwells are popular in low permeability hard rock aquifers. They can be constructed by local labour and are ideally suited when the weathered layer is thick and the water-table is shallow. There are hundreds of thousands of such wells in tropical, sub-tropical and temperate regions of the world.

Dugwells are about 3–10 m in diameter and have a depth of 10–15 m (Fig. 17.1a). Sometimes they extend in depth to 40 m or even more but the diameter is reduced telescopically with depth (Fig. 17.1b). In weathered rock mass, these wells are circular in cross-section, but in fractured rock, they are made rectangular so that the strike of joints, foliation and other discontinuity planes in the country rock are at right angles to the longer axis of the rectangle (Fig. 17.2). Such a well gives higher yield as compared with a circular well, as it is likely to intersect greater number of fractures.

Dugwells are usually unlined, except with a masonry curbing in the upper 5 m or so. Sometimes, they are also provided with perforated concrete rings in the water bearing zone.

Dugwells of large diameter in low permeability rocks as in crystalline formations, also serve as storage reservoirs, in which water seeps-in slowly during recuperation period after pumping. These wells are pumped for 3–4 h and are left idle, preferably overnight, for full recuperation which may take about 6–8 h. Shallow dugwells have the risk of getting dry in summer season or during drought. To ensure an adequate supply, dugwells must extend at least 2–3 m below the water-table.

In earlier years, large size dugwells were at places provided with steps for direct accessibility into the well. Another ancient mode of abstraction of water has been through 'mhot' (leather bag) or persian wheel driven by animals. In recent years, centrifugal pumps,

B. B. S. Singhal, R. P. Gupta, *Applied Hydrogeology of Fractured Rocks,*
DOI 10.1007/978-90-481-8799-7_17, © Springer Science+Business Media B.V. 2010

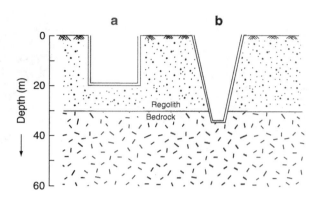

Fig. 17.1 Types of dugwells: **a** shallow dug well, and **b** deeper dug well with gradual reduction in diameter

Table 17.1 Yield of dug wells in crystalline rocks of India. (After Subramanian 1992)

Rock type	Well yield ($m^3 d^{-1}$)		
	Recharge area	Discharge area	Canal command area
Granite	29–78	48–128	300–772
Charnockite	9–53	42–96	60–360
Schist	9–32	18–54	–

which are coupled to a diesel engine or electric motor, have become more popular.

The yield characteristics of dugwells depend upon several factors, namely,

1. Landform—whether located in pediment, buried pediment or valley fill areas.
2. Regolith—its thickness and permeability.
3. Fracture characteristics of bedrock.
4. Local groundwater regime: whether the well is located in groundwater recharge or discharge area (Table 17.1).
5. Depth of water-table and its fluctuation.

A dugwell in unconsolidated formation can yield 10–30 $m^3 h^{-1}$ but in hard rocks, the yield is usually less than 5 $m^3 h^{-1}$. The yield may change with season, e.g. in West Africa yield is 5 $m^3 h^{-1}$ in rainy season and only 1 $m^3 h^{-1}$ in dry periods (UNESCO 1979).

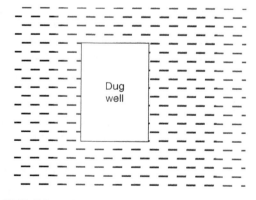

Fig. 17.2 Schematic plan of a rectangular dug well. The longer axis of the rectangle is kept at right angle to the strike of foliation or fracture planes to intercept maximum number of discontinuities

Dugwells are used for irrigation, livestock and domestic water supplies. When used for drinking purposes, the well should be protected by providing a parapet wall and cement sealing around the well to protect against surface water contamination, and a tin shade to inhibit contamination from airborne material.

17.1.2 Borewells

Small diameter drilled wells (borewells) are being increasingly used in several countries in Asia, Africa and South America for domestic water supply and minor irrigation. These wells in hard rock terrains were earlier drilled by conventional percussion (cable tool) or rotary methods. However, due to the introduction of down-the-hole air hammer (DTH) drill, the construction of such wells in hard rocks has become faster and economical. By this method in hard rocks, a well say of 155 mm diameter and of 60 m depth can be drilled in 1 or 2 days.

Borewells have the advantage of tapping a greater thickness of aquifer in the weathered layer and also the underlying fractured rocks (Fig. 17.3). Sometimes, saturated clay horizons within the weathered layer create problems in well construction due to caving. This problem is usually taken care of by providing a blank casing.

Depth of borewells is usually 30–70 m while the diameter depends on the type of pump. Wells which are provided with handpumps are of 100 mm diameter but for electric submersible pumps the minimum diameter should be 150 mm. Mark II and Mark III handpumps developed in India are being successfully used for rural water supply in several countries.

Borewells are either of open-end type in massive hard rock or they are provided with screen and packed with gravel when they tap the weathered layer

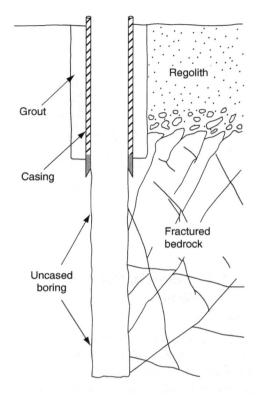

Fig. 17.3 Borewell in fractured bedrock overlain by regolith

(Fig. 17.3). A steel casing or PVC pipe is provided to protect the well against caving, and cement grout is used to protect from pollution.

The yield from a borewell depends mainly on fracture characteristics and varies considerably from one region to another and within short distances. The various factors which affect yields of borewells in crystalline rocks are: (a) climate, (b) lithology,

(c) geomorphologic setting, and (d) rock structures (see Sect. 13.5.1). A minimum yield of $1.5\,m^3h^{-1}$ is usually required for wells fitted with hand pumps and $7.5\,m^3h^{-1}$ for wells fitted with submersible pumps. However, the yield criteria would vary from one country to another and also from one region to another depending on economics. In several countries of Africa, e.g. Malawi, Zimbabwe and Nigeria, a successful well fitted with hand pump may have a yield of $0.72\,m^3h^{-1}$ (Wright and Burgess 1992). Typical well yields of borewells in different rock types are given in respective chapters on hydrogeology (Chaps. 13–16).

Borewells are preferred over dugwells under certain geohydrological conditions viz. insufficient thickness of weathered horizon, deep water-table, larger seasonal fluctuations of water-table, and for supplying safe drinking water.

Relative merits of dugwells and borewells are listed in Table 17.2.

17.1.3 Horizontal Borewells

Borewells may also be oriented horizontally or sub-horizontally in areas of high relief where fractures and joints are vertical or where groundwater occurs as perched water above an impermeable outcropping bed or is trapped behind dykes or other vertical barriers. The *horizontal borewells* are preferably aligned normal to the fractures for intercepting maximum number of such fractures for a given length of borehole. These wells are free flowing, and are provided with valves or

Table 17.2 Relative merits of dug wells and borewells

Dug wells	Borewells
1. Difficult to protect against pollution	1. Easy to protect
2. May require about 6 months for construction	2. A well of 150 mm diameter and 60 m depth can be constructed in a 1 or 2 days by DTH machine
3. Wastage of agricultural land as each well occupies quite a large area	3. Well occupies very small space
4. May go dry during prolonged drought	4. Less affected by seasonal fluctuation; therefore more dependable for water supply in drought period
5. Pumped by centrifugal pumps which are cheap and easy to maintain	5. Submersible pumps are required which are expensive and more difficult to maintain
6. Can be pumped only intermittently, at higher rates of about 5–$10\,m^3h^{-1}$	6. Can be pumped constantly with an average rate of about $5\,m^3h^{-1}$
7. The unit cost of water, considering all the items of input and benefit, is higher	7. The unit cost of water is lower considering all the items of input and benefit

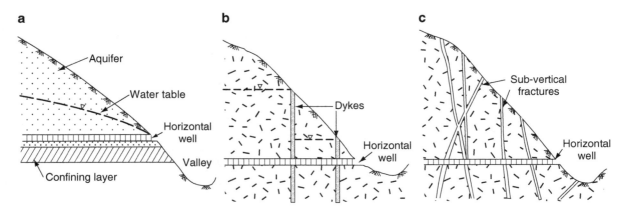

Fig. 17.4 Horizontal wells tapping **a** perched water from above an impervious layer, **b** confined water from dyke compartments and **c** confined water from steeply inclined fractures

other flow control devices at the outlet (Fig. 17.4). The horizontal or inclined borewells may have problems of stability when the rocks are highly fractured (Banks et al. 1992), and may also have problems of clogging of well screens and limitations of gravel packing. Horizontal borewells generally have an yield of 10–50 $m^3 d^{-1}$; high yields (300–1000 $m^3 d^{-1}$) are reported from weathered Tertiary andesites in the Lake Tahoe Basin, California, USA (Fig. 17.5).

Horizontal wells are also preferred over vertical wells for fluid recovery as part of groundwater remediation program, viz. recovery of light nonaqueous phase liquids (NAPLs) floating on groundwater and of dense nonaqueous phase liquids (DNAPLs) over a shallow impermeable layer (See Sect. 12.4.3).

Fig. 17.5 Example of horizontal wells tapping weathered andesite at a height of 2400 m above m.s.l. in the Lake Tahoe Basin, California, USA. (After Matthews 1985)

17.1.4   Dug-cum-Borewell and Collector Well

Dug-cum-borewells and collector wells are more or less similar in principle as both of them have a large size open well with extension bores for tapping a greater thickness or surface area of the aquifer (Fig. 17.6).

The term dug-cum-borewell are very popular for rural water supply in the hard rock terrains of India. In such a well one or two small diameter (50–90 mm) vertical bores are drilled at the bottom of the existing dugwell, to a depth of about 20 m for tapping deeper fractured aquifer (Fig. 17.6a). Horizontal or inclined boreholes may also be drilled to tap vertical or steeply inclined joints or permeable layers, e.g. inter-trappean formations in basalts (Singhal 1973). The yield of dug-cum-borewells in crystalline rocks of south India is usually in the range of 5–15 $m^3 h^{-1}$ (Singhal et al.

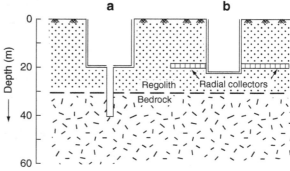

Fig. 17.6 a Dug-cum-bore well in crystalline rocks. **b** Collector well in weathered crystalline rock

1987), i.e. an increase of about 50–100% compared to that of a dugwell. Dug-cum-borewells also recuperate faster and may provide a perennial source of water round the year.

A typical collector well, also known as Ranney well is constructed in the river bed/or terrace deposits (predominantly sand and gravel). It consists of a central caisson of about 4 m diameter and 20–40 m depth, and is provided at the bottom with radiating screen pipes to facilitate recharge from the river (Fig. 17.7). The yield of such wells may range between 4000 and $100000 \, \text{m}^3 \, \text{d}^{-1}$ with an average of about $25000 \, \text{m}^3 \, \text{d}^{-1}$ (Hamill and Bell 1986).

The first radial collector well in unconsolidated formation was built by an American engineer, Leo Ranney in 1934 for water supply of London in U.K. Radial collector wells are also quite popular in Europe. For example, Belgrade, capital of Serbia, gets its water supply from about 90 Renney-type collector wells, located in the flood plains of Danube and Sava rivers. The yield of individual wells is more than $400 \, \text{l} \, \text{s}^{-1}$ (Kresic 2007).

In India, collector wells have been recently constructed at a number of places in the flood plain of river Yamuna for water supply to the cities of Delhi, Agra and Mathura.

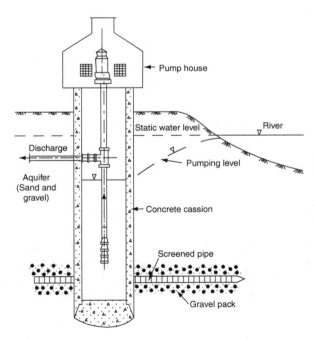

Fig. 17.7 A Ranney type collector well in alluvium adjacent to the river bed

Collector wells although traditionally constructed in alluvial formation, have also been found to be quite successful in the weathered crystalline rock terrain of Zimbabwe, Malawi and Sri Lanka (Wright 1985, 1992). These wells are usually of 2–3 m diameter and 20 m depth. They are provided with radial collectors of 90–150 mm diameter up to 40 m length at the base of the well. They show less drawdown and fast recovery in comparison to borewells and dugwells. The yield from collector wells in the basement rocks in South Africa is reported to be of the order of 1×10^{-3}–$8 \times 10^{-3} \, \text{m}^3 \, \text{s}^{-1}$ with drawdowns of 2–3 m. In comparison, the typical well yield from slim boreholes is 1×10^{-4}–$1 \times 10^{-3} \, \text{m}^3 \, \text{s}^{-1}$ with a drawdown of 30 m or even more (Wright 1992). Safe yield of collector wells in the crystalline rocks of South Africa and Sri Lanka is in the range of 1.1–$8.1 \, \text{l} \, \text{s}^{-1}$ with drawdown of 2–3 m as compared to the yield of borewells in the range of 0.1–$0.7 \, \text{l} \, \text{s}^{-1}$ (Wright 1992).

17.1.5 Infiltration Galleries (QANAT)

Infiltration galleries are constructed in thin aquifers which are in hydraulic continuity with some surface water body as in such cases vertical wells may not provide sufficient quantities of water. *Qanat* is an Arabic term for the canal. The *qanat* principle of intercepting a gently sloping water-table was conceived in ancient Persia in the seventh century and is still practiced in many countries in the Middle East. It was adopted and perpetuated by Spanish colonialists in South America. The *qanat* system consists of a number of vertical shafts linked by a tunnel which intercepts the water-table; the slope of the tunnel is less than that of the water-table. Figure 17.8 gives an outline of an infiltration gallery built in the river alluvium in central India during the Moghul period in the 15th century. A percolation tank was also constructed to augment recharge to the alluvial aquifer. *Qanat* is regarded as an elegant solution of water management as it makes over-exploitation of water resources virtually impossible and thereby avoids mining of water. Although infiltration galleries are usually constructed in permeable unconsolidated formations, their use for groundwater extraction in weathered and fractured crystalline rocks has also been advocated (Ruden 1993). They could form a potential method of tapping the thin weathered rock mantle.

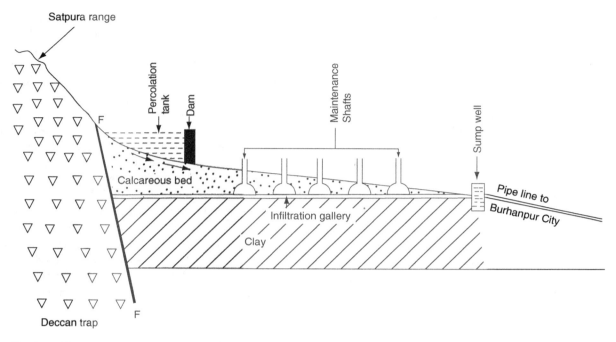

Fig. 17.8 The hydrogeological setting of an infiltration gallery (qanat) for supplying water to Burhanpur City, Madhya Pradesh, India. (Courtsey S. Romani)

Galleries and horizontal tunnels are well known from basaltic terrains in USA and Spain. These are common mode of groundwater abstraction in oceanic islands for skimming fresh water floating over seawater (Sect. 14.5). Many such structures are built on the basaltic formations in the Canary Islands in the Atlantic and the Hawaii islands in the Pacific, and also in islands made of coral limestones (Falkland 1984).

17.2   Criteria for Siting Wells

The wells for domestic and minor irrigation are to be located within reasonable distance from the place of use. This puts a severe constraint on the site selection of such wells.

Based on the experience of early high-yielding wells, target areas may be defined followed by mapping of significant lineaments using satellite imageries and aerial photographs and ground reconnaissance. This should be followed by geophysical surveys at identified sites, and finally drilling. Criteria for the site selection of common types of wells in hard rocks are briefly described below.

Dugwells These are constructed in areas with thick (>10 m) and permeable weathered layer. They are more successful in groundwater discharge areas like valleys and buried pediments where the depth and seasonal fluctuation of water-table is limited. Canal command areas where perennial irrigation is practiced and sites in proximity to surface water tanks, are also ideal for dugwells (Table 17.1). Remote sensing data can help to delineate areas of relatively deeper weathering, buried pediments and valleys. Vertical electrical sounding is useful in ascertaining the saturated thickness of weathered layer. Seismic refraction technique is also usually very appropriate to assess the nature of the weathered material and identify permeable zones within a practical range of 10–20 m (Sect. 5.6).

Borewells Fractures and lineament are important for locating borewells. The relation between fracture systems and streams is significant as the streams may recharge the aquifer due to flow along fractures. Therefore, drilling may be done on such fractures which are hydraulically connected to the streams (UNESCO 1972; Moore et al. 2002).

LeGrand (1967) has emphasized the importance of sheet joints for groundwater supplies from shallow wells in granitic and gneissic rocks. In Australia, such types of joints are often very open near the ground surface and upto a depth of about 100 m.

Borewells tapping these fractures have an yield of about 0.01 m³ s⁻¹ (UNESCO 1972) Similar inference is also drawn from studies in the Archean granite of SE Zimbabwe (Carruthers et al. 1993).

Studies in the shield rocks of a number of regions (India, Africa, Scandinavia, etc.) show that tensile fractures are more productive than shear fractures. Broad criteria for making distinction between the two are given in Sect. 2.2.3.

It may be emphasized that the mere presence of fractures is not sufficient for locating borewells in crystalline rocks, as fractures when filled with clay minerals will not serve any useful purpose. It is also difficult to distinguish between clay filled and water saturated fractures by any geophysical or remote sensing method.

In addition to the study of fractures and lineaments by remote sensing, the resistivity, seismic refraction and EM techniques are regarded as potential exploration tools for siting of wells. Resistivity azimuth surveys can be helpful in ascertaining the fracture orientation and their relative saturation and hydraulic conductivities in different directions (Sect. 5.2.2).

The data from African countries indicate that success rate in water-well construction incrases by using both EM and resistivity techniques (Table 17.3). Radon gas survey has also proved useful in locating water bearing fractures (Pointet 1989; Wright 1992; Reddy et al. 2006). Buried fracture zones which are targets for drilling of borewells can be identified by a careful interpretation of EM, resistivity and radon surveys (Wright 1992). Figure 17.9 shows the case of a possible erroneous siting of a borewell which could be avoided by- a close and judicious interpretation of both EM (horizontal loop) anomaly and radon anomaly data. In the example shown, the positive

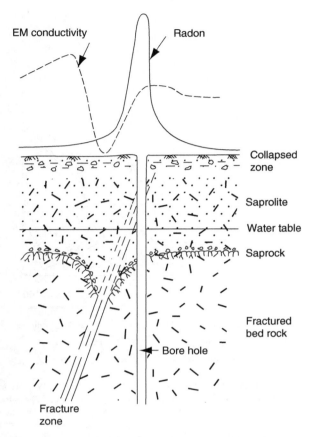

Fig. 17.9 Example of the radon anomaly and EM (horizontal) loop anomaly across a dipping fracture zone and a possible erroneous siting of borewell which could be avoided by careful interpretation of data. For details see text (modified after Wright 1992)

peak of radon anomaly, indicating higher moisture circulation, is in broad agreement with the negative peak of EM anomaly. This would normally tempt one to drill a vertical borewell close to the radon peak, which would miss the dipping fracture zone at depth. A closer look reveals the asymmetry in the EM anomaly on either side of the minimum value and also a slight shift of the EM minimum with respect to the radon peak. These features indicate the dipping nature of the fracture, which ought to be considered for the correct siting of borewells. Further, the importance of integrated study using satellite images, aerial photographs and geophysical data for the siting of high yielding wells in the crystalline rocks is also demonstrated from another study in West Africa (Fig. 17.10).

Table 17.3 Success rate for various siting techniques. (After Barker et al. 1992)

Technique	Percentage of successful boreholes
EM + resistivity	90
Resistivity	85
EM	82
Hydrogeology	66
Air photo	61
Logistical	50

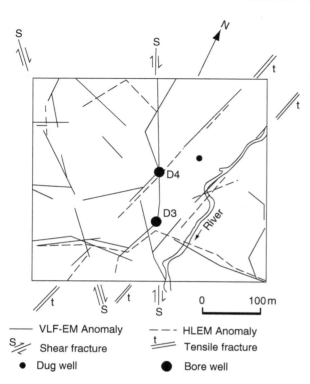

Fig. 17.10 Map of a part of crystalline basement of West Africa, showing EM anomaly axes and traces of tensile and shear fractures. The borewells sited at the intersection of important tensile and shear fractures coincident with EM anomalies were found to be highly successful. The borewell D4 gave an exceptionally high yield of $7\,m^3h^{-1}$ due to its proximity to the flood plain. (After Boeckh 1992)

In some areas, lineaments of a particular orientation may be more potential than other. For example, Bromley et al. (1994) observed that W–NW-trending lineaments deciphered from EM–VLF surveys gave the best results in identifying such potential lineaments in the buried Karoo sandstones in Zimbabwe (Sect. 5.3.2).

Horizontal Borewells and Infiltration Galleries Horizontal borewells are effective in areas of high relief where rocks are steeply dipping or have other vertical discontinuities in the form of dykes or fractures. Infiltration galleries are preferred to tap thin pervious horizons like intertrappean sediments and vesicular horizons. They are of additional advantage in volcanic islands to check deterioration of water quality due to sea-water encroachment. For successful siting of different type of wells, the various factors, viz. topography, structure, landform, lithology, lineaments,

source of recharge etc. need to be collectively considered, and this can be accomplished in a GIS environment (Sect. 6.4).

Dug-cum-Borewells These are preferred in areas with comparatively lesser thickness of overburden and greater water-table fluctuation. The orientation of boreholes in such wells is controlled by fracture pattern so that boreholes are aligned perpendicular to the fracture orientation. Instead of constructing a new dug-cum-borewell, an existing dugwell is usually revitalised by extension drilling.

Collector wells with radial strainers are preferred in local depressions in bedrock surfaces with adequate thickness of regolith. In volcanic islands, such type of wells (also known as '*Maui*' wells) also check deterioration in water quality due to sea-water encroachment.

17.3 Methods of Well Construction

The method of well construction depends on the type of well and nature of geological formation. Dugwells in weathered rock are excavated by hand using pick and shovel. In hard rocks, power driven equipment and explosives are used to facilitate excavation. Tubewells and borewells in unconsolidated sand and gravel formations, are drilled by rotary and reverse rotary circulation methods. In bouldery formations, cable tool method is more effective. In crystalline rocks, air rotary and down-the-hole (DTH) hammer methods are recommended for drilling small size boreholes of 30–100 m depth.

In volcanic rocks, e.g. basalts, the suitability of a particular method depends on the texture and permeability of the various flow units. In Deccan Traps, where basalts are generally massive, down-the-hole hammer method is widely practiced for drilling 30–50 m deep and 100–150 mm diameter borewells. On the other hand, on some volcanic islands, viz. Hawaii and French Polynesia, where there are thick permeable horizons, deep (upto 400 m) and 200 mm diameter high yielding $(0.1\,m^3\,s^{-1})$ wells are drilled by a combined air hammer and rotary machine (Wright 1984).

In fractured and karstified limestones, the cable tool method is more advantageous as it would avoid problems of lost circulation. If lost circulation is not a

Table 17.4 Comparison of common water well drilling techniques

Drilling method	Formations for which best suited	Usual range of possible depth and diameter	Typical penetration rates ($m\,h^{-1}$)	Advantages	Disadvantages
Cable tool (percussion)	Unconsolidated clays and, gravel, and boulders; consolidated and hard rocks, viz. sandstones, limestones, crystalline rocks	Depth from 100 to 600 m; diameter from 80 to 600 mm	Clay and shale (0.2–0.6); sandstones (0.61–1.25); hard rocks (0.04–0.12)	Low cost, simple operation; good sampling of rock cuttings & water; low water requirement	Slow penetration, casing is required in loose material
Direct circulation hydraulic rotary drilling	Unconsolidated sand, silt, soft consolidated rock	Depth from several hundreds to few thousand metres; diameter 80 to 300 mm	Unconsolidated material (6.5); consolidated rocks (0.4–0.6)	Fast method, capable of drilling deep wells; casing usually not required	High cost of rig, requires large quantities of water, samples contaminated with mud
Reverse circulation hydraulic rotary drilling	Unconsolidated sand, silt, clay and other soft rocks	Depth from 80 to 200 m, diameter 400 to 1500 mm	About 10	Capable of drilling large diameter wells; suitable for gravel packed wells; no contamination with mud	High cost of rig, requires large quantities of water; not effective in large-size gravel and boulder formation
Air rotary and down-the-hole hammer drilling	Very hard dry rock, viz. crystalline rocks, basalts and massive limestones	Depth upto 200 m–diameter 50 to 400 mm	3 or more	Fast rate of drilling and most effective in hard rock; water or mud not required; good sampling of rock cuttings	Not effective below water-table, casing required in unconsolidated overburden

major problem, the choice of the rig would depend on the number of holes to be drilled and the urgency of work. Cable tool method has an additional advantage as it facilities interpretation of lithological and water quality changes during drilling.

Table 17.4 gives the comparative merits of various drilling methods; the suitability of these methods in the common geological formations is indicated in Table 17.5. Reader may refer to Driscoll (1986) for the details of various drilling methods. A brief description of some common methods of drilling is given below:

17.3.1 Cable Tool Method

In the cable tool method, also known as percussion method, drilling is done by alternate lifting and

Table 17.5 Relative performance of drilling methods in some common types of geological formations

Type of formation	Drilling method		
	Cable tool	Hydraulic rotary	Air rotary DTH hammer
Loose sand and gravel	Difficult	Rapid	NR
Loose boulders	Suitable casing required	NR	NR
Firm shale	Rapid	Rapid	NR
Sandstone	Slow	Slow	NR
Limestone and dolomite	Rapid	Rapid	Very rapid
Limestone, fractured	Rapid	Difficult	Rapid
Limestone, karstified	Rapid	Difficult	NR
Massive igneous and metamorphic rocks	Slow	NR	Rapid

NR: Not Recommended

dropping of a heavy string of tools. The drill bit is massive and heavy with relatively sharp chisel edge which crushes the rock by impact. In the zone above the water-table, water is added to a the hole to make a slurry of the crushed rock so that it can be removed easily by a bailer. There is usually no need of providing a casing in consolidated formation but in unconsolidated and loose rock materials casing is necessary to check the caving of the hole.

Cable tool method is a more versatile method for drilling in a variety of rock formations like boulders and fissured and cavernous rocks. It is capable of drilling holes of 60–600 mm diameter to depths of 100–600 m. If the hole is of small diameter, if can be drilled relatively to greater depth.

The main advantages of cable tool method are that representative data about the rock formations, water quality as well as static water-levels in different formations can be obtained. Further, it also requires less quantity of water for drilling which is advantageous in arid and semi-arid regions. The main drawbacks are the slow rate of drilling and limitations about the diameter and depth of drilling. Driving of casing along with drilling in unconsolidated rocks and also maintaining the plumbness of the hole is often problematical.

17.3.2 Direct Circulation Hydraulic Rotary Drilling

Direct circulation hydraulic rotary method is suitable for drilling deep wells of upto 30 cm diameter or even more, in unconsolidated formations. The hole is drilled by rotating a hollow bit attached to the lower end of a drill collar and a string of drill pipes. Drilling mud is circulated into the hole through the drill pipe, which brings out the rock cuttings.

Direct rotary method is used widely for drilling oil wells and water-wells. This technique is advantageous due to rapid rate of drilling in a variety of geological formations. Two types of bits are generally used—the drag (fish tail) and the roller cone (rock) bit. It is difficult to use this method in highly jointed and cavernous formations due to heavy loss of drilling fluid and also in loose bouldery formations as the bit may be unable to crush such material. In some areas the availability of adequate quantity of water for drilling may also be problematical.

17.3.3 Reverse Circulation Rotary Method

As the name indicates the reverse circulation method has a reversed circulation of drilling fluid so that the drilling fluid alongwith rock cuttings is pumped out through the drilling pipe. This method is advantageous in drilling comparatively larger diameter (400–1500 mm) high capacity wells to depths of 200 m, especially in situations where gravel packed wells are to be constructed. This method is more suitable for soft unconsolidated clay and sand formations. If the rock consists of cobbles and boulders of size larger than the drill pipe, or openings in the bit, drilling is not possible as such a material cannot be brought out through drill pipe.

17.3.4 Air Rotary Method

In air drilling methods, instead of water air is used as the drilling fluid. There are two such methods—direct rotary air and down-the-hole hammer methods. In both these methods, a large compressor is used to force the air through the drill pipe and bit into the hole which brings the cuttings to the ground surface, The air also cools the bit.

Air drilling methods are most effective for drilling small diameter (100–400 mm) holes to depths of 200 m in consolidated sedimentary formations and other hard rocks where chances of caving of the hole are rare. These methods are also advantageous for drilling in fissured rocks in areas where availability of water for drilling is limited. Large diameter holes can also be drilled by using foams and special additives to the air to reduce loss of air to the formation (Driscoll 1986). Use of foams also helps in speedy removal of cuttings.

Down-the-hole Hammer Method The equipment consists of tungsten-carbide bit attached to the hammer and drill pipe (Fig. 17.11). The hammer is operated by compressed-air. It is capable of delivering 500–1000 blows per minute. A rotation speed of about 40 rev min^{-1} is usually satisfactory, though in abrasive rocks, such as quartzites, the rotation may be at about 10 rev min^{-1}. These machines, are capable of drilling 4″ (102 mm) and 6″ (152 mm) diameter wells upto a depth of about 50 m. Large diameter (380–760 mm) holes can also be drilled by a set of six or eight down-the-hole

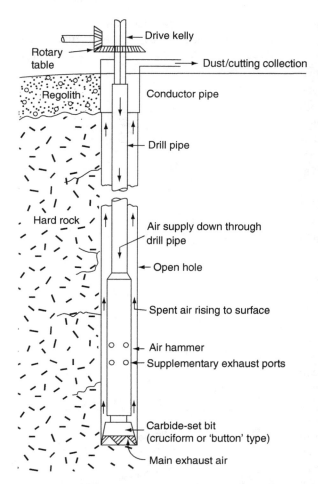

Rotary table — Drive kelly

Dust/cutting collection

Regolith

Conductor pipe

Drill pipe

Hard rock

Air supply down through drill pipe

Open hole

Spent air rising to surface

Air hammer
Supplementary exhaust ports

Carbide-set bit (cruciform or 'button' type)

Main exhaust air

Fig. 17.11 Down-the-hole hammer drill. (After Cruse 1979, reprinted from Journal of Hydrology, with kind permission of Elsevier Science-NL, Amsterdam, The Netherlands)

hammers, arranged concentrically around a leading (pilot) hammer which are rotated as one unit to give a large size hole. Down-the-hole hammer method is more efficient than the cable tool method as in the former the rock cuttings are continuously removed by air.

Compressor forms an important component of air drilling as it has to provide adequate air volume at a certain operating pressure in order to reduce drilling time. Compressed air is supplied to the hammer at a pressure of 100–110 psi (690–758 kPa). The upward velocity of air through the annular space between the drill pipe and the sides of the hole should be about 90 m s^{-1} or more for the effective removal of rock cuttings (Driscoll 1986).

The advantages of down-the-hole method are: (a) Rapid removal of rock cuttings and therefore rate of

drilling is fast; (b) The method is especially suited for fractured hard rocks such as basalts, granites, quartzites, limestones and dolomites; (c) Water requirement for drilling is negligible. Further, yield characteristics from a particular horizon can be estimated during drilling. The main disadvantages of the down-the-hole hammer machine is the damage caused to the bits by abrasion with continued use. Further, down-the-hole hammer method is also unsuitable for unconsolidated formations like loose sand, clay and cavernous limestones.

17.4 Well Development

After a well is completed, it is developed to increase its yield and efficiency. In unconsolidated formations, this is achieved by removing finer material from around the well screen thereby leaving an envelope of coarse grained material. In gravel packed wells also, well development is recommended to increase its efficiency.

17.4.1 Development Techniques in Unconsolidated Formations

In uncosolidated rocks, well development methods include pumping, surging, and hydraulic jetting (Todd 1980; Driscoll 1986).

Pumping is the simplest method where well is pumped at rates higher than its design capacity. This causes higher velocities which help in removing finer material. *Overpumping* method of well development is more effective in non-stratified formation where gravel packed wells are constructed.

Well development by *surging* is achieved by either using a plunger or compressed air. The downward motion of a plunger (surge block) forces the water from the well into the aquifer through the screen while its upward motion causes back washing. This results in removal of fine material into the well which should be periodically removed with the help of bailers. Surging is more effective in aquifers having clay or mica as these minerals are likely to clog the screen. In hard rocks, well development by surging can be done in the casing above the open hole. In air surging, compressed air is injected into the well through an air pipe which causes surging and also lifts water with washout material to the surface.

High Velocity Jetting by water is an effective method of well development both for open end wells in hard rocks and also in screened wells. Water with high velocity jet action enters the formation through the screen openings agitating the aquifer material adjacent to the screen. The fine grained aquifer material is dislodged and enters into the well by the turbulent flow of water.

Deflocculants like polyphosphates are also used for well development as they disperse the clay particles. This method is effective both in unconsolidated and consolidated formations.

17.4.2   Development of Wells in Hard Rocks

In hard rocks, wells are either provided with screen or they are open end type without screen. The well face and the fractures are often clogged by the fine rock material produced during drilling. Removal of this material improves the well yield. In addition to methods described above, the other techniques which are used for increasing well yields, especially in hard rocks are actually directed towards increasing the rock permeability rather well development. These are therefore more appropriately called as aquifer development methods or aquifer stimulation methods. The common aquifer development techniques are acidizing, blasting or shooting, and hydraulic fracturing.

Acidizing In this method hydrochloric acid is used in carbonate aquifers and other calcareous formations to dissolve carbonate minerals deposited on the wall of the well screen as encrustation or those within the rock fractures or voids to improve the well yield. In a limestone aquifer near Adelaide, South Australia, acidization by hydrochloric acid increased the well diameter from 200 mm to 300–600 mm in the depth range of 130–160 m as revealed by calipper logging (Fig. 17.12). Aquifer test also indicated an improvement in well efficiency as the drawdown was reduced by about 8 m over an extended period of pumping (Fig. 17.12). It is however preferred that the deposition of $CaCO_3$ is minimized by keeping low drawdown.

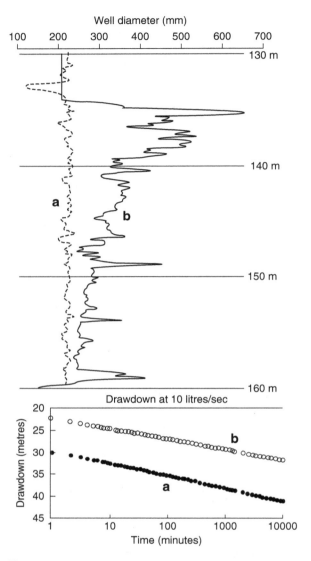

Fig. 17.12 Effect of acidisation of limestone aquifer on well diameter and well production efficiency at The Paddocks aquifer storage and recharge site near Adelaide, South Australia **a** before acidisation, and **b** after acidisation. (After Barnett et al. 2000)

In siliceous rocks, hydrofluoric acid can be used for similar purpose. The amount of hydrochloric acid required to treat an encrustated screen depends on the screen diameter and its length. Other acids which are used to remove carbonate encrustations are sulfuric acid and hydroxyacetic acid (Driscoll 1986).

In an advanced method of pressure-acidizing, larger volume of muriatic acid (commercial grade of HCl), is pumped into the aquifer at high pressure. This will result in developing larger volume of aquifer due to

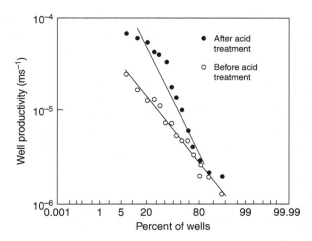

Fig. 17.13 Effect of acid treatment on well productivity in carbonate rocks. (After Csallany and Walton 1963)

deep radial penetration of acid (Hamill and Bell 1986). Experience has shown that acidizing is more effective in fractured carbonate rocks (Fig. 17.13). Treatment with sulfamic acid (NH_2SO_3H) is suggested when encrustations are high in organic and clayey substances. Bacterial slimes (especially those formed due to iron bacteria) are the most common cause of clogging of bore screen. Bacterial slimes can be removed by using chlorine which has greater advantages then using acids.

Blasting or Shooting Explosives (viz. nitroglycerine—liquid or solid) are used to shoot rock wells to enhance well yields which is accomplished by an increase in well diameter, development and enlargement of fractures and removal of fine-grained material including well face encrustations. It is necessary to use proper blasting procedure and explosives in a given situation depending on the rock type, size of well and depth of water. Care should also be taken when blasting is done in populated areas to avoid any damage to life and property. Earlier small explosive charges, usually 14–45 kg, were used but recently larger charges of 400–900 kg are used in igneous rocks for better results (Driscoll 1986).

Excellent results in improving well yields by blasting in hard rocks is reported from several places. Explosives have also been used to rehabilitate old wells, as specific capacities of wells may sometime decrease with time due to clogging. Shooting of wells in sandstone aquifers of Cambrian-Ordovician age in northern Illinois, USA, improved the specific capacities from a few percent to 71% (Walton and Csallany 1962). Shooting of wells by using mild explosives in limestone rocks of Raipur basin, India improved the well discharge by more than ten times (Adyalkar 1983).

Hydraulic Fracturing or Hydro-fracturing This is an advanced technique to enhance the well yields in hard rocks by opening existing fractures and providing better interconnection between them in the vicinity of the well. This method owes its origin to oil reservoir engineering where it is employed to create reservoir fractures for increasing oil productivity during secondary recovery. More recently, this method is used to create new fractures in HDR systems.

For hydrofracturing, an inflabable packer is lowered into the well which is inflated, a little above the production zone, thereby isolating this zone. Sometimes the production zone is isolated by two packers (Fig. 17.14). Water is pumped into the isolated zone under pressure exceeding the pressure exerted by the overburden.

The overburden pressure for every 1 ft (0.3 m) depth will be about 1 psi (6.9 kPa). Therefore, at a depth of about 100 ft (30 m) the overburden pressure can be overcome if water is injected under a pressure of more

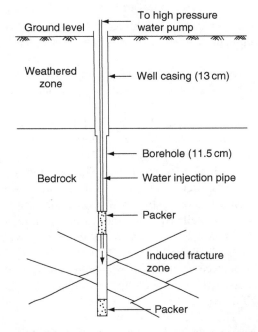

Fig. 17.14 Schematic set-up of hydrofracturing in a hard rock aquifer. (After Klitten 1991)

than 100 psi (690 kPa). Usually water is pumped into the isolated zone at pressures 3500–70 000 kPa depending on the depth of this zone and the nature of the rock (Driscoll 1986). Higher pressure is required in massive rocks with few fractures. Hydraulic fracturing is caused due to rock failure when fluid pressure increases to the point that effective stress becomes negative and exceeds the tensile strength of the rock. It is preferred that the wells for hydrofracturing are drilled normal to the existing fractures so that the induced fractures are developed at right angles to the existing ones to provide interconnectivity.

Before hydrofracturing, it is necessary to clean the borehole walls of any clay coating or drill cuttings either mechanically or using chemical additives such as diluted hydrocholoric acid to remove carbonate deposit. Experience has shown that the productive zone to be subjected to hydrofracturing should be identified beforehand by say electrical logging, and also the diameter of the hole by caliper logging. This will help in placing the packer of the right size at the proper depth. It is also necessary to determine safe injection pressure from step-rate tests. If the temperature of the injected water is different than formation water, the pressure required to extend the hydrofracture will change. The geometry of hydrofractures is not well understood as it depends on existing fractures and matrix permeability. Pressure and temperature changes during hydrofracturing may also cause precipitation of minerals in fractures and thereby may reduce the permeability.

As the newly created fractures are likely to heal with time, sometimes propping material such as sand, special beads or small particles of high-strength plastics are introduced alongwith injected water to keep the cracks open, after the water pressure is removed. This is called '*sand fracing*' (Hamill and Bell 1986).

Hydraulic fracturing may increase the permeability of aquifer for about 100 m radius or so around the well, by about 200%. A distinct improvement in the yield of wells by hydraulic fracutring in a variety of rocks is reported from many countries (Williamson 1982; Sheila and Banks 1993). Experience shows that it is more economical to improve the yield of failed wells by hydro-fracturing rather than to drill new wells (Talbot et al. 1993).

The effect of hydrofracturing on improving the transmissivity is greater in hard crystalline rocks (gneisses, granites and migmatites) as compared

with soft rocks like schists and low grade amphibolites. Klitten (1991), based on studies in Tamilnadu, India reported an increase of 100% in the transmissivity of hard crystalline rocks while in softer rocks the increase was only 20–30%. Results from Australia show that the specific capacity of shallow low yielding wells improved by about six times in granites and schists but in phyllites the ratio was 0.95 indicating deterioration instead of improvement (Williamson 1982).

17.5   Hydraulic Characteristics of Wells

These include well loss (C), well efficiency (E_w), specific capacity (S_c) and well productivity.

Well Loss (C) The value of well loss coefficient C, is a measure of the degree of well deterioration due to clogging or other reasons. Values of C between 0.5 and $1.0 \, min^2 m^{-5}$ indicate mild deterioration while clogging is severe when C is between 1.0 and $4.0 \, min^2 m^{-5}$. It is then difficult to restore the original capacity (Walton 1962). If a well does not show any well loss, it is a perfect or ideal well. Step-drawdown tests are useful both in porous and fractured rocks for determining flow regime and in discriminating linear and non-linear head losses around a pumping well (Mackie 1982; Mishra et al. 1989).

Several graphical (Walton 1962) and numerical methods (Avci 1992) are developed for the analysis of step-drawdown tests data for the estimation of aquifer parameters and well loss constant. Walton (1962) suggested a graphical method in which the time-drawdown data for successive steps are plotted on a semilog paper; t on the logarithmic scale and s_w on the arithmetic scale (Fig. 17.15). Based on step-drawdown test (Sect. 9.2.6), well loss, C can be expressed as

$$C = \frac{\left(\Delta s_i / \Delta Q_i \right) - \left(\Delta s_{i-1} / \Delta Q_{i-1} \right)}{\Delta Q_{i-1} + \Delta Q_i} \quad (17.1)$$

The delta term in Eq. 17.1 represents increments of drawdown produced by each increase in the rate of pumping (ΔQ). The incremental drawdown, Δs is determined by noting the difference between the water-level at the beginning and at the end of each step keeping the time step Δt to be the same.

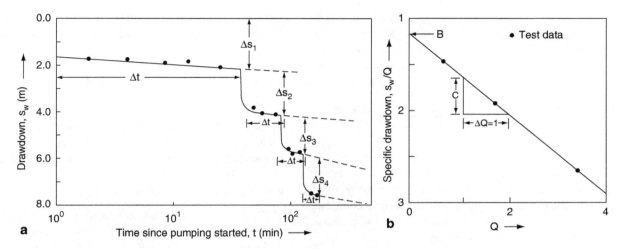

Fig. 17.15 a Time-drawdown plot from step-drawdown test for estimation of C; **b** determination of B and C from graph of s_w/Q vs. Q

For steps 1 and 2, Eq. 17.1 can be written as

$$C = \frac{(\Delta s_2/\Delta Q_2) - (\Delta s_1/\Delta Q_1)}{\Delta Q_1 + \Delta Q_2} \quad (17.2)$$

and for steps 2 and 3

$$C = \frac{(\Delta s_3/\Delta Q_3) - (\Delta s_2/\Delta Q_2)}{\Delta Q_2 + \Delta Q_3} \quad (17.3)$$

The value of C computed from successive steps may not be constant. An increase in the value of C in successive steps indicates well deterioration due to clogging of well screen or other reasons while a decrease in the value of C points to the development of well during pumping.

Specific drawdown, S_w/Q can be expressed as

$$\frac{s_w}{Q} = B + CQ \quad (17.4)$$

Values of specific drawdown (s_w/Q) for each step as determined from Fig. 17.15a, when plotted on arithmetic scale against Q, should give a straight line (Fig. 17.15b). The well loss coefficient, C is given by the slope of the straight line and the formation loss coefficient, B by the intercept Q=O.

In fractured rock aquifers, where turbulent well loss may increase with an increasing pumping rate, the data-plots in Fig. 17.15b would form a parabola rather than a straight line (Kresic 2007).

Well Efficiency (E_w) E_w is the ratio of aquifer loss (BQ) to the total drawdown in the pumped well. It can be expressed as

$$E_w = \frac{BQ}{(BQ + CQ^2)} \times 100 \quad (17.5)$$

The value of well efficiency indicates the state of well and aquifer development. It therefore helps in the maintenance and rehabilitation of the well. No well is 100% efficient. A well efficiency of 70% or more is usually considered acceptable. A progressive decline of well efficiency would indicate clogging of the well screen and surrounding aquifers.

Specific Capacity (S_c) Specific capacity of a well can be estimated from step-drawdown test (see Sect. 9.2.5). Specific capacity (S_c) of a well is the ratio of well yield (Q) to stabilised drawdown (s_w) in the pumping well (dimension L^2T^{-1}). It is a measure of the productivity of both the well and the aquifer; higher the specific capacity better is the well efficiency. The specific capacity (Q/s_w) of a well discharging at a constant rate in a homogenous, isotropic confined aquifer infinite in areal extent can be expressed by Eq. 17.6 which is based on Jacob's Eq. 9.23.

$$\frac{Q}{s_w} = \frac{1}{(2.30/4\pi T)\, log\, (2.25Tt/r_w^2 S)} \quad (17.6)$$

Equation 17.6 assumes that (1) the well has full penetration and receives water from the total saturated thickness of the aquifer, and (2) well loss is negligible. Considering well loss, Eq. 17.6 can be re-written as

$$\frac{Q}{s_w} = \frac{1}{(2.30/4\pi T)\, log\,(2.25Tt/r_w^2 S) + CQ^{n-1}}$$

$$(17.7)$$

Equation 17.7 indicates that specific capacity decreases with Q and t. Specific capacity also varies directly with the logarithm of $1/r_w^2$, i.e. a large increase in the radius of the well results in a comparatively small increase in the values of specific capacity.

Statistical analysis of a large number of specific capacity data from a variety of fractured rocks in Pennsylvania, USA, indicates that the main controlling factors are well diameter and duration of pumping while lithology plays a comparatively smaller role (Knopman and Hollyday 1993).

Considering partial penetration, Jacob's equation can be written as (Kresic 2007)

$$T = \frac{Q}{4\pi\,(s_w - s_L)\left[ln\left(\dfrac{2.25Tt}{r_w^2 S} + 2s_p\right)\right]} \quad (17.8)$$

where s_L is the well loss, s_p is the partial penetration factor defined as

$$s_p = \frac{1 - (L_w/b_a)}{L_w/b_a}\left[ln\left(\frac{b_a}{r_w}\right) - G\left(\frac{L_w}{b_a}\right)\right]$$

$$(17.9)$$

where L_w is the length of the well screen, b_a is the aquifer thickness and G is a function of the ratio of L_w to b_a.

The above equations indicate that in addition to aquifer characteristics and pumping duration, the specific capacity is also a function of well characteristics, such as well radius, degree of penetration and well loss.

Specific capacity decreases with Q and t. Specific capacity also varies directly with the logarithm of $1/r_w^2$, i.e. large increase in the radius of well results in comparatively small increase in the values of specific capacity. A partially penetrating well will have a lower specific capacity than a well penetrating the

full thickness of the aquifer. Kresic (2007) has given Eq. 17.10 for obtaining the corrected value of specific capacity (S'_c).

$$S'_c = S_c\,\frac{s_L}{s_w - s_L} = \frac{Q}{s_w - s_L} \qquad (17.10)$$

In most cases, specific capacity is likely to be overestimated as the drawdown is not stabilised either due to insufficient duration of pumping or aquifer characteristics. Estimation of transmissivity, T from specific capacity data is discussed in Sect. 9.2.6.

Well Productivity Well productivity or specific capacity index (dimension LT^{-1}) is computed by dividing specific capacity by the saturated thickness of the aquifer tapped by the well. Statistical analysis of specific capacity index data from wells screened at different depths in a multi-unit aquifer is useful in determining the relative productivity of different units or variation within the same lithologic unit (Walton 1962, 1970). It is also useful in predicting well yields from a given thickness of the aquifer.

Summary
In hard rock terrains, open dug wells form the main source of water supply. Small diameter bore wells drilled by down-the-hole hammer rig are getting more popular for domestic and irrigation water supplies. Collector wells with radial strainers are successful in tapping the weathered rock (regolith). For large scale municipal water supplies to metropolitan cities, Renney wells are often constructed in the unconsolidated alluvial formations adjacent to the perennial river beds. Location of suitable sites for the installation of wells in fractured rocks is quite a difficult task Integrated geomorphological, structural (fracture and lineament mapping) analysis along with geophysical surveys are found to be very useful for this purpose. In addition to conventional methods of well development used in unconsolidated formations, well yields in hard rocks can be enhanced by hydrofracturing and acidization. Hydraulic performance of wells is expressed in terms of well loss, well efficiency and specific capacity which are estimated from variable rate pumping tests.

Further Reading

Driscoll FG (ed.) (1986) Groundwater and Wells. Johnson Division, Minnesota, 1089 p.

Kresic N (2007) Hydrogeology and Groundwater Modeling. 2nd ed., CRC Press, Boca Raton, FL.

Lloyd JW (1999) Water resources of hard rock aquifers in arid and semi-arid zones. Studies and Reports in Hydrology 58, UNESCO, Paris.

Walton WC (1970) Groundwater Resource Evaluation. McGraw-Hill, New York.

Geothermal Reservoirs and Hot Dry Rock Systems

18.1 Introduction

Most geothermal reservoirs and hot dry rock (HDR) systems are located in fractured low porosity geological formations. In such settings, fractures provide conduits for fluid flow through such rocks. Therefore, fractures either natural or those artificially created are important for the successful operation of geothermal reservoirs and HDR systems.

Geothermal reservoirs are located in areas of youthful mountain building, tectonic plate boundaries and volcanic activity, where subsurface temperatures are higher. They form potential reservoirs for the generation of electrical power. Geothermal fluids are also being used for centuries for domestic and green house heating and for therapeutic and recreation purposes.

There are three types of geothermal energy systems: Low enthalpy or hot water dominated systems (50–150°C), high enthalpy or vapour-dominated systems (150–300°C) and Hot Dry Rock (HDR) systems (50–300°C). In low enthalpy systems, the hot groundwater is used as a source of heat. In high enthalpy systems, steam is the extracted fluid which is used to drive turbines for generating electricity. In Hot Dry Rock (HDR) systems, water is circulated to deeper levels in artificially created fractures where it is heated up and hot water and steam flow back to the ground surface to be used as a source of geothermal energy.

The main requirements for a geothermal system are (1) heat source, (2) sufficient permeability, (3) a source of replenishment, (4) a low permeability bottom boundary, and (5) a low permeability cap rock (Domenico and Schwartz 1998).

Some of the earlier notable power plants generating energy from geothermal resources are in Italy, New Zealand, Japan, western United States, Iceland and Chile. Several new geothermal fields have been discovered in recent years. For example, in the last 20 years, three large geothermal plants have been established in Southeast Asia—Bulalo (426 MW) and Tiwi (330 MW) in the Philippines, and Awibengkok (330 MW) in Indonesia. Each of these geothermal fields is a water-dominated reservoir associated with an andesitic volcano in an island arc setting (Williamson et al. 1996). In the United States, total power production from geothermal energy is about 3000 Mega Watts (MW) and more than 90% of this is generated in California. The Geysers geothermal field in northern California, USA, produces more power than any other geothermal energy field in the world (NRC 1996).

In Russia, the Northern Caucausus, the Volga river area, West Siberia and Far East are identified as the most promising areas for development of geothermal energy. In Switzerland, a geothermal project, Alpine Geothermal Power Production (AGEPP) began in 2006 to generate electricity and heat by drilling 2.5–3 km deep borehole in fractured gneisses. Feasibility studies indicated a reservoir fluid temperature around 100–110°C, a production yield of 50–75 l s^{-1} and weak mineralization. Simulation studies indicate that after about 10–15 years of production, thermal output is likely to decline due to mixing with groundwater (Sonney and Vuataz 2009).

In India, the main geothermal fields are located in the northwestern Himalaya, Narmada valley in central India and along the West Coast. However, as they are mostly low temperature (<150°C) systems, their potentiality for power generation is limited (GSI 1991). Some of the areas, e.g. Narmada geothermal zone in central India is characterized by high heat flow

B. B. S. Singhal, R. P. Gupta, *Applied Hydrogeology of Fractured Rocks*,
DOI 10.1007/978-90-481-8799-7_18, © Springer Science+Business Media B.V. 2010

$(70–300\,\mathrm{mW\,m^{-2}})$ and high thermal gradients $(40–120^{\circ}\mathrm{C\,km^{-1}})$ which can be a potential area for geothermal power development (GSI 1991; Chandrasekharam 1996). Feasibility of binary cycle power plant at some sites in central India and at Puga in Ladakh area in Himalaya has been also established (Pitale and Padhi 1996).

18.2   Hot Dry Rock (HDR) Systems

In some areas with steep geothermal gradients, at depths greater than 3000 m, temperatures are in the range of 200–500°C and rocks are dry. In such areas the attempt is to extract heat from the hot dry rock by circulating water through drill holes. As the geological formations at these places are mostly granites which are impermeable, it is necessary to create fractures for facilitating the movement of fluids by either a suitable array of multiple large yield nuclear explosions or by hydraulic thermal stress fracturing. In the latter case, cold surface water is used to pressurize natural or stimulated fractures. This causes rock shrinkage and development of additional fractures. A second well is drilled later to intersect the hydraulically fractured region thereby a closed loop circulation is established. At Los Almos, New Mexico, USA, two wells were drilled into hot but impermeable granite which were hydraulically fractured to create a connection between the wells. Water is pumped through one of the wells and flows back through the other (production) well (Fig. 18.1). In this process, the circulating water is heated due to hot rocks at depth. Several other areas, viz. Carnmenellis granite in Cornwall, UK, and Soultz-Sous-Forets in the Rhine Graben in France have been also investigated. The data obtained from these studies provide information about the properties of fractures and their effect on flow characteristics with time.

Studies in Carnmenellis granite, UK have been carried out by the researchers of the Camborne School of Mines (McCann et al. 1986; Baria 1990). Four boreholes were drilled approximately to a depth of 300 m. Reservoir development required a reduction in the hydraulic impedance of the borehole walls and an increase in the permeability of the rock mass between the boreholes for efficient extraction of heat. The hydraulic impedance of borehole walls was achieved

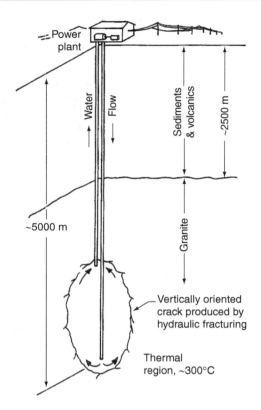

Fig. 18.1 Hot Dry Rock (HDR) geothermal system produced by drilling and hydraulic fracturing. (After Collie 1978)

by explosive stimulation and the bulk permeability of rock mass between the boreholes was increased by hydraulic stimulation. Cross-hole seismic experiments were carried out to delineate the cavity created by explosive stimulation.

An HDR experiment was also performed at Soultz-Sous-Forets in the Rhine Graben in France, where two deep wells of 3590 and 3876 m with a separation of 450 m have been drilled in the granitic basement which is under a 1400 m thick cover of Triassic sedimentary rocks (Gerard et al. 1996). The granite at that depth is highly fractured. Hydraulic performance of wells was improved by hydraulic stimulation, by injecting several thousand cubic metre of water at high flow rates (upto $0.06\,\mathrm{m^3\,s^{-1}}$). The hydraulic performance of one of the deep wells was increased by hydraulic stimulation from $5\times10^{-4}\,\mathrm{m^3\,s^{-1}}$ to $2\times10^{-2}\,\mathrm{m^3\,s^{-1}}$, the water temperature being around 135°C. The pore water in granite was highly saline which is most probably derived from the sedimentary cover.

18.3 Flow Characteristics

During exploitation of geothermal reservoirs, fractures and joints influence the movement of injected water as well as transport of heat from the rock to the circulating fluid. In a closely fractured homogeneous medium, the injected water, which is at a lower temperature, will sweep the heat out of the rock more effectively during its movement. On the other hand, if fractures are isolated and widely spaced, injected water may flow directly through the fractures towards the production well without extracting sufficient heat from the reservoir rock. Therefore, the cold water front will reach the production well in advance.

In a double-porosity reservoir, the matrix blocks will not only influence the transfer of fluids but also the heat from the blocks to the fracture. The hydraulic response of a double-porosity geothermal reservoir is expected to be the same as of a double porosity aquifer. A double porosity framework has been applied to the Geysers Geothermal Reservoir in California.

In geothermal reservoirs, fracture properties and thereby permeability may change with time during the operation of the system. Decrease in pressure due to fluid extraction from the reservoir can cause closure of fractures. On the other hand, fluid injection into the reservoir will cause opening of fractures. Thermal changes during extraction will also affect the fracture permeability with time. Injection of water, which is at a lower temperature, causes contraction resulting in increase in permeability. This is manifested as decrease in flow impedance, i.e. an increase in the rate of injection with time (Fig. 18.2).

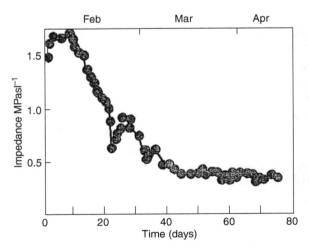

Fig. 18.2 Change in flow impedance in a fracture joining two wells in an HDR experiment. (After Grant et al. 1982)

Injection of water may also lead to changes in permeability due to precipitation and dissolution of minerals.

The change in permeability can be monitored by repeat well discharge tests. As in other fractured aquifers and petroleum reservoirs, the permeability of geothermal reservoirs can also be improved by acid treatment and more often by hydraulic fracturing (see Sect. 17.4.2).

18.4 Reservoir Properties

In geothermal and petroleum reservoirs, the usual practice is to express the transmissivity in terms of intrinsic transmissivity or permeability-thickness (kb) where k is the intrinsic permeability in darcy and b is the saturated thickness of reservoir in metres; kb is expressed in darcy-metre (d-m) (10^2 d-m $= 1$ m^3). Productive geothermal reservoirs usually have kb values of 10 d-m or more and in some cases it is 100 d-m. Values less than 10 d-m indicate poor potentiality of geothermal reservoir.

In addition to permeability-thickness, storativity, S* is also important to know the fluid reserves. Storativity, in the case of geothermal reservoir, is expressed as

$$S^* = \eta\beta b \qquad (18.1)$$

where η is the porosity, β is fluid compressibility and b is the aquifer thickness; S* is expressed in m Pa^{-1}.

It has been discussed in Chap. 8 that in a confined aquifer, storativity depends on the compressibility of the rock matrix and of the fluid. In geothermal reservoirs there are two fluid phases—liquid and vapour, both of which have different compressibilities, therefore, the storativity value of the reservoir will also depend on the relative proportion of the two phases. The compressibility of liquid water is very small (4.7×10^{-10} Pa^{-1}), as compared to that of vapour (1.4×10^{-6} Pa^{-1}). Therefore, even a small amount of vapour in the reservoir will result in a much greater storativity. This implies that a large storativity value in a high temperature reservoir does not always mean a thick high-porosity aquifer. In an inadequate liquid phase system, storativity (S*) can be converted to porosity-thickness (ηb) by simply dividing S* with liquid water compressibility.

The presence of vapour phase also influences the transmission of pressure changes in the geothermal reservoirs. In a low temperature liquid dominated system, the pressure changes are transmitted over long distances in short times but in a high temperature two phase (liquid and vapour) reservoir, the pressure change will move out slowly. Therefore, the effect of boundary, if any, will be reflected only if the test is of longer duration. The fractured nature of the geothermal reservoir also makes the flow mechanism and interpretation of test data more complex. Inspite of these limitations, the hydraulic parameters of geothermal reservoirs can be estimated by well discharge tests as described in Chap. 9.

18.4.1 Single Well Tests

In the single-well test, the effect of discharge (constant or variable) or injection from a single well is observed in the same well. In a drawdown-test, the shut-in is opened to flow. It is assumed that flow rate is constant. The drop in pressure with time is noted. In a pressure build-up test, commonly used in petroleum reservoir, well is produced at a constant rate for a certain period of time and then shut-in. Pressures are measured during the shut-in period as a function of shut-in time. The data is plotted on semi-logarithmic paper and permeability can be estimated by the Horner, and Miller–Dyes–Hutchinson methods (Ramey 1977).

18.4.2 Interference Tests

In a well interference test, one or more wells are discharged simultaneously and the effect is observed in one or several adjacent wells. This is the common method of estimating the hydraulic properties of aquifers (see Chap. 9). Interference tests can also be used when the well production is achieved by injection as is commonly the case in petroleum industry. The interference tests are of greater value as compared to the single well tests because they scan the reservoir properties over a greater area. Interference tests are also useful in identifying boundary conditions but the presence of vapour phase in a geothermal reservoir creates problems for such an analysis. The effect of well-bore storage and skin can also be identified from this type of test data.

The test data can be interpreted by using appropriate flow model of homogeneous, double porosity and discrete fracture flow, as in the case of aquifer tests (Chap. 9). Both type curve and straight line methods of solution are used by plotting pressure change against time, on double logarithmic and semi-logarithmic paper respectively (Grant et al. 1982). McEdwards and Tsang (1977) argued that the type—curve solutions used widely in the field of hydrogeology and petroleum engineering are not applicable to geothermal reservoirs testing where there are two or more producing wells each of which having a unique flow rate history. For such conditions, they suggested a computer based least square fitting technique which is of advantage for accounting effects of two or more flowing wells unique flow rate histories, presence of recharging or discharging boundaries, well-bore storage and skin effects.

As in the case of aquifers, the pressure conditions and levels in wells tapping geothermal reservoirs are also affected by changes in barometric pressure, tides, rainfall and other sources. Therefore, well test data should be corrected for these changes induced by external effects. It is also necessary to have the reliable measuring equipment which can stand the high temperature and salinity of geothermal fluids. Injected water during interference test may cause changes in reservoir permeability due to induced pressure, temperature variation and deposition of precipitates on well face. The reservoir permeability will also change as it is temperature dependent, the decrease in temperature, causes thermal contraction and thereby additional fractures and higher permeability.

Interference test at Wairakei, New Zealand, where the rocks are acid volcanics (rhyolitic tuffs, dacites and andesites) of Pliocene to Recent age, gave permeabilities of the order of 0.05–0.5 darcy (Elder 1981). At Raft River in Idaho, USA, geothermal wells penetrate sedimentary formations and igneous rock (quartz monzonite). Interference tests gave $kb = 70$ d-m whereas single well tests gave kb values of 5–20 d-m, indicating varying permeabilities. Other borehole tests and analysis of barometric and tidal effects, in this area, also indicate varying porosities; the monzonites having low ($\leq 5\%$) while sedimentary formations have high (14–24%) porosities (Grant et al. 1982). A long duration (33 days) well discharge test for geothermal exploration from a 2615 m deep well penetrating Triassic sandstones at Southampton in UK gave a resevoir transmissivity of $6\,m^2\,d^{-1}$ (3.5 d-m). The drawdown pattern indicated bounded nature of the reservoir (Downing et al. 1984).

18.4.3 Tracer Tests

Tracer injection tests are useful for the estimation of reservoir properties and also determination of preferential flow paths and hydraulic interconnectivity of fractures. Both radioactive and non-radioactive tracers are used (see Sect. 10.3). In Broadlands, New Zealand, interference well test data indicated homogeneous reservoir character but tracer tests indicate preferential flow (Grant et al. 1982). Tritium tracer studies at Manikaran geothermal field in India gave velocities of $72\,m\,d^{-1}$ and direction of flow from north to south for geothermal waters in fractured quartzites (Athavale et al. 1992). Tracer tests are also used for siting injection and extraction wells.

18.5 Ground Subsidence

The development of geothermal energy has much less environmental problems as compared with the use of fossil fuels or radioactive waste for energy genera-

tion. It poses problems similar to other areas where subsidence is a result of withdrawal of water or oil. The ground subsidence due to withdrawal of geothermal fluids has been reported from several areas, e.g. Wairakei, New Zealand; Imperial Valley and the Geysers, California, USA. Recently a case of ground subsidence due to inappropriate development of the geothermal field is reported from Germany which is attributed to change of anhydrite to gypsum by water during drilling (Goldscheider and Bechtel 2009).

In geothermal fields, the elastic stresses change due to withdrawal of heat and fluids. Hence, in these areas the subsidence depends both upon fluid pressure and temperature changes, unlike in groundwater and petroleum reservoirs where the changes are largely isothermal. Therefore, in groundwater and petroleum reservoirs, the subsidence can be countered by re-injection of water leading to maintenance of pressure. However, in geothermal reservoirs, the re-injection of fluids will not cure temperature changes. In some cases, the thermal effect may outweigh the pressure effect, e.g. at the Geysers, California, the temperature effects causing subsidence were four times greater than the pressure effects (Grant et al. 1982). Ground subsidence at Wairakei (New Zealand) is well documented where the ground surface has subsided by as much as 5–10 m (Elder 1981). The greatest rate of subsidence is at a distance of 1 km east from the main production area (Fig. 18.3), which is difficult to explain (Grant et al. 1982). Re-injection of thermal

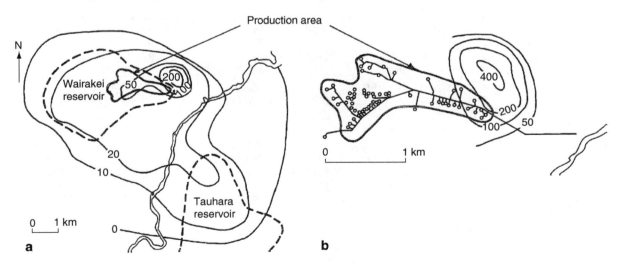

Fig. 18.3 Land subsidence (in mm per year) at Wairakei, New Zealand **a** regional data during 1956–1961, **b** detailed data for production area during 1964–1974. (After Elder 1981)

fluids may also reduce the permeability of the formation, due to the precipitation of salts. The injection of fluids in seismically active areas may also induce seismic activity. Four earthquakes of magnitudes 3.3–4.4 were recorded within 1 month as a result of production of hot water from a reservoir in Texas (Domenico and Schwartz 1998). Further, the reservoir temperature and thereby its energy potential will also be reduced by injection.

18.6 Water Quality

The chemical composition of geothermal waters varies widely (Table 18.1). The composition of waters is mainly controlled by the composition and content of gas, e.g. a large concentration of H_2S will give rise to acid-sulphate waters while systems rich in CO_2 will form $Na–HCO_3$ or $Ca–HCO_3$ waters. Chloride and chloride-sulphate waters are more common. The SiO_2 content in geothermal waters is typically high due to greater solubility of quartz at higher temperatures. Chloride rich waters have higher concentration of silica. SiO_2 and $Na–K$ geothermometry has been used to estimate the base temperature of geothermal

reservoirs and also the extent of mixing with meteoric water (Balmes 1994; Raymahashay 1996; Steinbruch and Merkel 2008).

At places, e.g. in Puga Valley, India, economic deposits of sulphur and borax are associated with hot water springs. Here the B content is in the range of $116–140\,mg\,l^{-1}$ (GSI 1991). Large deposits of calcite, aragonite and travertine are also associated with some thermal springs, e.g. Mammoth Hot springs, Yellowstone National Park, USA. Thermal waters are also well known for their medicinal and curative properties.

18.7 Origin of Geothermal Waters

The origin of geothermal waters has been a matter of great interest. Both geochemical and stable and radioactive isotope data indicate that geothermal waters are mainly of meteoric origin (Elder 1981; Minzi et al. 1988). Plots of δD vs. $\delta^{18}O$ values of thermal fluids from some geothermal fields is given in Fig. 18.4 which shows that the deuterium (D) content is the same as that of local precipitation but the $\delta^{18}O$ is higher. This oxygen isotope shift is attributed to progressive equili-

Table 18.1 Chemical composition of water from some thermal springs (in $mg\,l^{-1}$)

Location	Yellowstone National Park, USA[a]	Puga, India[b]	Manikaran, India[c]	Surdulica, Serbia[d]	Tongonian Geothermal Field, Philippines[e]	Nhambita, Mozambique[f]
Rock type	Volcanic tuff	Ophiolite	Quartzite	Granaodiorite	Andesite	Gneisses
Temperature (°C)	94	55–81	93	78–126	55–81	61.2
pH	9.6	7.4–8.1	8.0	7.5–9.0	7.3–8.2	7.6
TDS	1310	2020–2278	546	1024–1302	–	989
Na	352	530–640	83	18–287	2020–2208	297
K	24	68–80	25	8.6–18.8	172–201	10.5
Ca	0.8	2–17	52	12–32	56–111	32
Mg	0.0	1–5	4	1–7	0.1–0.3	0.27
HCO_3	–	–	195	354–732	–	–
SO_4	23	123–149	50	260–480	81–107	309
Cl	405	370–424	145	43–60	3383–3863	228
SiO_2	363	120–160	90	40–90	245–289	43

[a] Hem (1989)
[b] GSI (1991)
[c] Romani and Singhal (1970)
[d] Hadzisehovic et al. (1995)
[e] Balmes (1994)
[f] Steinbruch and Merkel (2008)

Fig. 18.4 $\delta^{18}O$ and δD variation in geothermal fluids. (Based on data from Craig 1963; GSI 1991; Hadzisehovic et al. 1995; Raymahashay 1996)

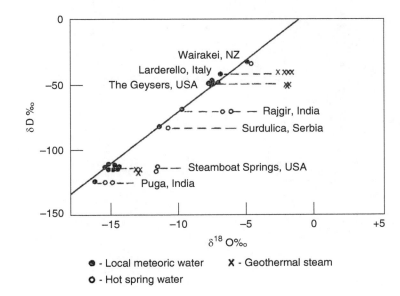

bration of oxygen in the water with silicates and carbonate rocks. δD and $\delta^{18}O$ data from some of the hot springs in the Himalaya indicate recharge from different altitudes (Sukhija et al. 2002).

A comparison of tritium (^3H) in thermal waters with local meteoric water from several geothermal fields also shows large scale mixing with meteoric water (Grant et al. 1982; Raymahashay 1996, Zhiming et al. 1996). Some highly saline thermal waters may also originate due to the mixing of meteoric waters and deep seated brines as is evidenced from the concentration of baron and its isotope, ^{11}B (Balmes 1994; Vengosh et al. 1994).

The age of the geothermal waters at various sites is estimated to be in the range of less than 100–1500 years by tritium and ^{14}C dating (Hadzisehovic et al. 1995; Zhiming et al. 1996). The silica concentration of 43 mg l^{-1} and the ratios of Br/Cl and I/Cl of 2.5×10^{-3} in the hot spring waters in Mozambique suggest that the thermal water comes from approximately 5000 m depth and had a long residence time with silicate rocks. ^{14}C dating suggest groundwater age of 11 000 years (Steinbruch and Merkel 2008). Such variable ages in fractured rocks point towards both shallow and deep flow regimes.

Summary

Low permeability fractured rocks are of interest for the development of geothermal energy where natural thermal gradient is high viz. in areas of recent volcanic activity and at tectonic plate boundaries. Hot dry rock (HDR) systems are developed in low permeability rocks by circulating water through boreholes thereby tapping the subsurface heat. Hydraulic characteristics of fractures are important for the successful functioning of these systems. At places it may be necessary to create more fractures and improve fracture interconnectivity by artificial methods. Careful planning of the withdrawal of geothermal fluids is also required to avoid undesirable environmental effects such as land subsidence.

Further Reading

Domenico PA, Schwartz FW (1998) Physical and Chemical Hydrogeology. 2nd ed., John Wiley & Sons, Inc., New York.
Elder J (1981) Geothermal Systems. Academic Press, London.

Groundwater Modeling

19

19.1 Introduction

Groundwater modeling has become a very useful and established tool for studying problems of groundwater quality and management. Modeling techniques have been used as an aid for: (a) prediction of the effect of groundwater stresses (groundwater recharge and discharge are treated as stresses), (b) groundwater management for sustainable exploitation of groundwater, and (c) to estimate effect of contaminant injection and transport in space and time.

In the earlier times, physical and analog models like sand tank and electric analog were used for simulating groundwater systems; however with the advent of powerful and versatile computers and development of software, computer modeling has replaced other modeling techniques. Presently, computer codes for groundwater modeling provide display and visualization of the dynamic behavior of groundwater flow and pollutant migration in three dimensions. Once a model is validated for a particular type of hydrogeological setting, it can be used for studying different scenarios. Valuable concepts, reviews and methodological details on groundwater modeling are given by many workers (e.g. Narasimhan 1982; Bear and Verruijt 1987; Anderson and Woessner 1992; Indraratna and Ranjith 2001; Lloyd 1991; Pinder 2002; Rushton 2003; Thangarajan 2004; Meijerink et al. 2007).

Limitations Before dealing with the subject matter further, an appreciation of the limitations of modeling techniques is necessary. In spite of all the sophistications in hardware and software, several simplifying assumptions are inescapable such as those related to: (a) estimation of various aquifer parameters of groundwater systems, (b) idealization related to conceptualization, (c) limitations related to the techniques of input data acquisition, and (d) idealization of the groundwater behaviour system etc. Under these constraints, even the most carefully constructed model may at best be an approximation of the reality and not reality itself. Therefore, a model can never be perfectly deterministic for all purposes. However, it can be used to provide useful output for practical groundwater exploration and management purposes.

Modeling Task Overview The following tasks are involved in setting up a groundwater model (Fig. 19.1):

1. Defining the model's purpose.
2. Conceptualization of the system.
3. Setting up the numerical model, including laying out a grid, defining model boundaries, hydrostratigraphic boundaries, aquifer parameters, ground fluxes, and evaluating sink and sources.
4. Model calibration and validation, and
5. Preparation and running simulation for prediction scenarios.

19.2 Defining the Model's Purpose

At the outset, the objective and scale of modeling needs to be clearly defined. For example, one should examine in depth such questions as: what do we need to learn from modeling, on what scale the desired information would be available, is modeling the best way to achieve the objective, do we have sufficient data for modeling implementation, etc. Once the purpose is defined, a flowchart of 'things to do' should evolve.

B. B. S. Singhal, R. P. Gupta, *Applied Hydrogeology of Fractured Rocks,*
DOI 10.1007/978-90-481-8799-7_19, © Springer Science+Business Media B.V. 2010

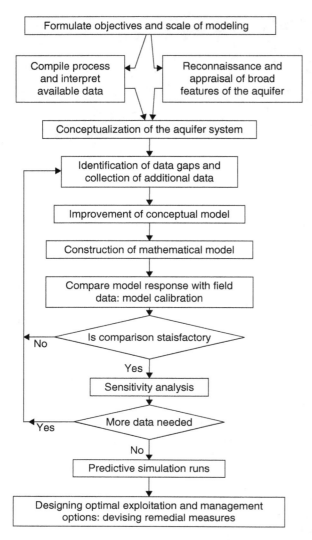

Fig. 19.1 Flow diagram of various activities in groundwater modelling

19.3 Model Conceptualization

Conceptualization of the system to be modeled is the most critical activity in modeling. It consists of understanding the aquifer characteristics and their spatio-temporal quantification. The required details of these parameters will depend upon the particular objective and scale of the model. For example, for a regional study such as regional groundwater resources, small-scale inhomogeneities need not be taken into consideration, while conceptualizing the groundwater system.

Often all the data required for model conceptualization is not available at the beginning. However, one should start by using the existing data supported by judicious understanding of the hydrogeological setting, and at the same time plan for collecting additional data as required to fill the gaps in the database. Useful details of groundwater model conceptualization are given by Mercer and Faust (1981).

Remote sensing data can be extremely useful in setting up the conceptual model, particularly in data-scarce situations where maps and data-bases may not be adequate. Further, strategies can also be conceived to interface numerical modeling (e.g. MODFLOW) with standard GIS (such as ArcGIS). Besides, sophisticated modeling packages such as GMS (EMS 2007) equipped with built-in GIS, allow automatic conversion of data to facilitate implementation of numerical model.

19.4 Setting-up a Numerical Groundwater Model

A variety of data from different sources may be incorporated in the groundwater model.

19.4.1 Model Grid

All numerical models use grid framework and various hydrogeological data are imported and interpolated into the grid. Therefore, the first step is to generate a model grid, which has to be well adjusted to the model domain. Two main types of model grids are used: (a) rectangular grid used in finite-difference method and (b) triangular grid used in finite-element method.

The gridding process leads to subdividing the space into discrete blocks. Modern groundwater modeling packages (i.e. codes) automatically generate model grid for numerical modeling. Frequently, now a days the model grid is generated on a remote sensing image or alternatively on a selected base map (Fig. 19.2). Obviously, the finite difference modeling tool using square grid appear better suited owing to their simplicity, regular structure, easy linkage with standard coordinate systems, and integration with remote sensing data and other raster GIS database that may be required for modeling purposes.

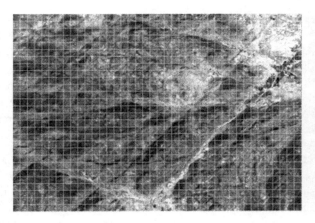

Fig. 19.2 A typical model grid laid over a remote sensing image

The grid size depends upon the type of problem, data availability and the accuracy desired. A single band georeferenced remote sensing image is generally used as a base image. The raster cell size of the image has to be smaller than the cell size of the model grid.

The model grid may not necessarily be uniform throughout. Locally finer grid may be generated for incorporating more details (Fig. 19.3), commensurate with local hydrogeological characteristics, local heterogeneities etc., e.g. to show local karst structures or local seepage zones etc.

19.4.2 Defining Model Boundaries

Defining model boundaries is possibly the most critical step in setting-up of a numerical groundwater model.

This means primarily defining the hydrogeological conditions in the model. For this purpose, the following inputs are generally used:

Surface information and data from various available maps (geological, topographic, soil, hydrogeological etc.) and remote sensing images for defining spatial distribution of impermeable lithological contact, surface water bodies (river, canal, pond, artificial reservoir etc.), springs (groundwater discharge zones), geological faults etc., which may help in defining the no-flow or recharge boundaries.

Subsurface data and information based on geological sections, geophysical surveys, boreholes etc. for defining subsurface hydrogeological characteristics such as depth of weathering, thickness of different aquifers, faults in basement rocks, subsurface salinity variation, buried channels etc. are also highly useful in the above task.

19.4.3 Inputting Model Parameters

Groundwater model parameters are of two types: (a) variable or transient, i.e. those which vary with time such as groundwater-table elevation, piezometric head, rainfall etc. and (b) time-invariant, such as those which are intrinsic and do not vary with time, such as aquifer transmisivity, thickness, fracture density etc. The numerical modeling process creates a grid of 3-D discreet blocks of the space, and numerical values of various parameters are assigned to each unit block or cube.

Fig. 19.3 Comparison of non-uniform grid of node points; **a** nested square grid and **b** conventional grid

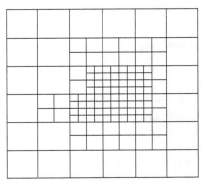

Nested square grid
No. of node points = 129

a

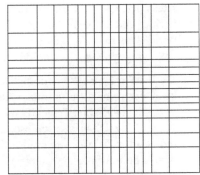

Conventional grid
No. of node points = 210

b

19.4.3.1 Inputting Time Invariant Spatial Data

Each cell of the model grid needs to be assigned time-invariant spatial hydrogeological data which could be of the following types:

Hydrostratigraphic Boundaries It is required that the 3-D space in the form of grid cells is defined in terms of the aquifer characteristics. This implies importing data pertaining to aquifer/impermeable layers, aquifer limits (i.e. aquifer top and bottom etc.). Whereas remote sensing and geologic maps can be used for inputting surface layer characteristics, data for subsurface horizons is derived from lithologic and geophysical well logs, surface geophysical profiling, imaging airborne experiments, GPR etc. Suitable interpolation techniques such as krigging are used to fill in the matrix cells.

Aquifer Hydraulic Characteristics Groundwater modeling invariably needs data on aquifer hydraulic characteristics in each cell in its 3-D cell matrix. This could be, for example:

- Hydraulic conductivity/aquifer transmisivity etc. derived from pumping tests at representing points, MRS (magnetic resonance sounding), laboratory testing of core samples etc.
- Storage coefficient derived from representing pumping tests, laboratory analysis, MRS sounding, gravity satellite data or InSAR techniques

19.4.3.2 Inputting Transient Spatial Data

As mentioned above, this category of data varies with time. Therefore, time is also discretised or split into stress periods (for example recharge period, discharge period) and for each stress period, separate spatio-temporal data matrices are generated. For example, the following type of data could be used:

Hydraulic Head Potentiometric surface or hydraulic head determines the groundwater flow and is a time dependent variable. Typically this exercise implies creation of a 2-D surface representing the water-level in the aquifer. In case there are multiple aquifers, there would be multiple piezometric surfaces, one for each aquifer. Conventionally, the piezometric surface is derived from the reduced water-levels in observation wells drilled in aquifers at various locations. Now differential GPS data are also used for accurate (~1 cm

accuracy) altitude measurements. GPR sounding, geophysical surveys and MRS techniques under suitable conditions can also be used for estimating depth to water-table. DEM derived from stereo-satellite sensors can be used for estimating elevation of spring line/ groundwater discharge sites or appearance of water in ephemeral streams up-to about 1 m accuracy in areas of low to moderate relief.

Groundwater Recharge and Discharge Although it is very important component in groundwater modeling and budgeting, it is rather difficult to estimate the groundwater recharge/discharge (see Sect. 20.2). To some extent the natural discharge can be assessed from interaction of surface water bodies (canals, ponds, rivers etc.) with the top aquifer and various empirical relationships. Spatial distribution of pumping wells and their pumpage can form another basis for estimating groundwater withdrawal.

Evapotranspiration (ET) ET (Evapotranspiration) is an important cause of groundwater discharge and varies with season of the year. Information on ET losses can be best derived from repetitive remote sensing image coverages showing spatial distribution of vegetation and surface water bodies and their variation with time. Using empirical relationships and the 2-D distribution of the surface features, ET can be estimated for groundwater modeling.

It may be re-emphasized that one must keep in view the uncertainties that arise due to the assumptions made during the estimation of these parameters. The correctness of a conceptual model obviously depends upon the accuracy and adequacy of data. However, modeling must be considered as an iterative and dynamic process. A model is progressively refined as more data becomes available and the understanding of the system improves. Here the experience and imagination of the modeller plays an important role and that is why it is said that modeling is both an art and science.

Depending upon the complexity of a groundwater system, various parameters defining its physical framework are highly variable. But it is not possible to estimate them at all the points in an aquifer due to obvious constraints of resources. It therefore becomes imperative to interpolate a parameter at unmeasured locations by using the values of a parameter at the measurement points. It amounts to an effective integration of information on a localized scale to evolve the best regional picture. An alternative approach for interpolation is to utilize the 'structure' i.e., spatial correlation of values over a large area. This method is called 'kriging' which is based on the theory of regionalized variables. It is a method for optimizing the estimation of a variable which is distributed in space and is measured at a set of points. The value at a point is assumed to depend on all the measured values in the neighbourhood.

19.5 Modeling of Homogenous Porous Aquifer

19.5.1 Mathematical Formulation

Homogenous porous isotropic aquifers are typically formed by alluvial sands and sedimentary clastic sequences. Groundwater flow in such media is expressed in terms of basic partial differential equation utilizing the principle of mass conservation as well as momentum conservation (Trescott and Larson 1977) (see Sect. 7.1.3):

$$\frac{\partial}{\partial x}\left(K_x \frac{\partial h}{\partial x}\right) + \frac{\partial}{\partial y}\left(K_y \frac{\partial h}{\partial y}\right)$$
$$+ \frac{\partial}{\partial z}\left(K_z \frac{\partial h}{\partial z}\right) \pm W = S_s \frac{\partial h}{\partial t} \quad (19.1)$$

where, h is the hydraulic head, K_x, K_y and K_z are the principal components of the hydraulic conductivity tensor which are assumed to be colinear with the Cartesian coordinate system, W is the volume flux per unit volume of porous material (a positive sign for the inflow and negative sign for the outflow), S_s is the specific storage, and t is the time.

For movement of a non-reactive solute in groundwater regime, the condition of mass balance of the solute should also be satisfied. The mathematical equation describing this process may be expressed as:

$$\frac{\partial}{\partial x}\left(D_x \frac{\partial C}{\partial x}\right) + \frac{\partial}{\partial y}\left(D_y \frac{\partial C}{\partial y}\right) + \frac{\partial}{\partial z}\left(D_z \frac{\partial C}{\partial z}\right)$$
$$- \frac{\partial}{\partial x}(CV_x) + \frac{\partial}{\partial y}(CV_y) + \frac{\partial}{\partial z}(CV_z)$$
$$- \frac{C_0 W}{\eta_e} = \frac{\partial C}{\partial t} \quad (19.2)$$

where, V_x, V_y and V_z, are the components of velocity of groundwater along the three coordinate axes, C is the concentration of the solute at any point, D_x, D_y, D_z are the principal components of the coefficient of hydrodynamic dispersion, n_e is the effective porosity, and C_0 is the concentration of the solute in the source or sink.

The above mathematical formulation assumes that the medium is a porous continuum in which Darcy's law holds. The partial differential equations may get simplified under different situations. For example, in case the Dupuit approximation is valid, i.e. there is no vertical flow in the system; Eq. 19.1 is reduced to

$$T\left(\frac{\partial^2 h}{\partial x^2} + \frac{\partial^2 h}{\partial y^2}\right) \pm W = S \frac{\partial h}{\partial t} \quad (19.3)$$

where T is the transmissivity and S is the storativity of a confined aquifer.

For a homogeneous isotropic medium, Eq. 19.3 is further simplified to

$$\frac{\partial}{\partial x}\left(T_x \frac{\partial h}{\partial x}\right) + \frac{\partial}{\partial y}\left(T_y \frac{\partial h}{\partial y}\right) \pm W = S \frac{\partial h}{\partial t} \quad (19.4)$$

Similarly the equation for two dimensional solute transport in groundwater gets simplified to

$$\frac{\partial}{\partial x}\left(D_x \frac{\partial C}{\partial x}\right) + \frac{\partial}{\partial y}\left(D_y \frac{\partial C}{\partial y}\right) - \frac{\partial}{\partial x}(CV_x)$$
$$- \frac{\partial}{\partial y}(CV_y) - \frac{C_0 W}{\eta_e} = \frac{\partial C}{\partial t} \quad (19.5)$$

19.5.2 Solution of Mathematical Equation

Sometimes in the literature, the main modeling process is identified with solving of the appropriate partial differential equation to obtain the spatio-temporal distribution of hydraulic heads in case of groundwater flow modeling, or solute concentration in case of mass transport modeling. Exact analytical solution of any of the partial differential equations is possible only for very simple situations. For most of the real life problems, numerical methods are used. In a numerical model, a partial differential equation is approximated by an algebraic equation which may be valid in a small sub-region for a particular time. A number of such equations can be written for all the sub-regions into which the entire domain of study may be suitably divided. Thus, one obtains a set of simultaneous algebraic equations relating all the unknown variables which are characteristic of these sub-regions.

Finite Difference, Integrated Finite Difference and Finite Elements are the main numerical methods for discretizing a partial differential equation. The difference between various methods lies in the process of subdividing the space into discrete blocks. These discrete blocks form a grid which may be either mesh-cantered or block-centered. In the former, the representative nodes are located on the intersection of grid lines whereas in the latter they are centered between grid lines. The time period of interest is also divided into a number of intervals. The closeness of the spatial blocks and time intervals depend upon the gradient of parameter variation, desired accuracy, and availability of data. A sample of non-uniform division of the space is shown in Fig. 19.3 using a conventional or a nested square approach.

In Finite Difference method, the distribution of nodes is regular but the nodal spacing may either be uniform or even non-uniform along an orthogonal coordinate system (Fig. 19.3). The Integrated finite Difference method (IFDM) is more flexible than the Finite Difference method in that the nodes could be distributed in an irregular mesh. Further, the governing equations of IFDM are formulated in an integral form using the basic conservation laws. Despite greater flexibility in the mesh design, larger requirements of data limit the utility of IFDM. The finite Element method (FEM) approximates the partial differential equation by an integral approach. In FEM, a region is divided into sub-regions called elements which could be of any shape. The flexibility in the shape of elements enables a more realistic simulation of the various internal and external boundaries. The value of a variable within an element is obtained by interpolating values of the variables at corner nodes. The integral representation of a partial differential equation in FEM is achieved by using either a weighted residual Galerkin scheme or the Variational method. For further details on use of FEM modeling, the reader is referred to Remson et al. (1971) and Pinder and Grey (1977). In many cases, it

is seen that results of modeling by adopting different methods, do not differ significantly since any sophistication in modeling is largely cancelled out by the inadequacy of field data.

The algebraic equations thus obtained from partial differential equations can be written in a matrix form which can then be solved by either direct or integrative methods. Gaussian elimination procedure or some of its variation have been used as a direct method for solving the matrix equation to find the unknown variables at subsequent time interval. However, on account of large requirements of computer memory, round-off errors, problem of numerical stability in case of a sparse matrix, direct methods have not been popular in solving a matrix equation.

In contrast, an iterative approach avoids the need for storing large matrices and thus saves on memory as well as time resources of a computer. An iterative method starts from a first approximation of the unknown value of the parameter to be computed and its estimate is progressively improved until a sufficiently accurate solution is obtained. The iterative process can be considerably hastened by making a correct initial approximation and use of acceleration factors. The iteration is terminated when the value of the unknown parameter satisfies a pre-set error criterion. Some of the commonly used iterative methods are successive over-relaxation method, alternating direction implicit procedure, iterative alternative direction implicit procedure and strongly implicit procedure. Mercer and Faust (1981) and Javendal et al. (1984), among others, have reviewed the basic principles of these techniques. A number of versatile computer codes have been developed at the US Geological Survey which can be easily used by a hydrogeologist without going deep into the theoretical or mathematical aspects of modeling.

19.5.3 Modeling Software

Modular computer software (MODFLOW) for modeling three-dimensional groundwater flow regime was developed by McDonald and Harbaugh (1988). Later, Pollock (1990) developed software MODPATH to model mass transport in a groundwater system. This is a three dimensional particle-tracking programme which is interfaced with MODFLOW to produce a plot of flow path-lines. Another three dimensional mass transport modeling software MT3D (Zheng 1990) was also interfaced with MODFLOW. These models gained wide acceptance because they are well documented, validated, and verified. All these modeling programmes however, required the input data to be entered as numerical text through standard files which is quite a time-consuming process.

Recognizing the importance of visualization for improving the modeling process, Graphical User Interface (GUI) was incorporated in the computer software which enabled a considerably fast operation. Using this facility, a model grid can be displayed on the monitor screen for graphically inputting the model parameters using menu options and cursor controls. Some of the widely used modeling packages having a GUI are FLOWPATH (Franz and Guiger 1990), Model CAD and Visual MODFLOW. This facility helped in getting rid of the strenuous task of manually creating formatted input data files. Numerical modeling e.g. MODFLOW through preprocessing tools such as PMWIN (processing MODFLOW) can be interfaced with standard GIS (such as Arc GIS). Further, sophisticated modeling packages such as GMS (EMS 2007) equipped with a built-in GIS processor allow automatic conversion of GIS data to facilitate implementation of numerical model.

With these software capabilities, users can now display a model in plan view (layer by layer) as well as in a vertical cross-section during any phase of the model development or prognosis. These software packages are operated on-screen by selecting various options using the mouse. The visualization of simulation results typically include features like transient or steady state equipotential, drawdown contours, water quality values, velocity vectors, path-lines and iso-concentrations. Construction of a groundwater model, its refinement, and prognosis has now become much easier and faster. It is also possible to display an animated migration of contaminants in an aquifer system.

19.5.4 Model Calibration and Validation

Before a deterministic model can be used for prediction purposes, it needs to be calibrated and validated using the available historical data. The computed parameter, say, hydraulic head (in case of groundwater flow model) or solute concentration (in case of mass transport model) should be satisfactorily matched with the field values. The matching criterion will depend upon the objective of modeling and the required accuracy. In case of a mismatch, various parameters should be gradually modified. Essentially, this process is based mostly on a trial-and-error approach and the experience and insight of the modeller plays a significant role in model calibration. After calibration up-to the desired level of accuracy, the model should be validated for prediction by using a new set of field data.

Even after calibration and validation, it may be possible that the model is not an optimised one. Perhaps, a number of combinations of the model parameters may produce the same degree of matching. The model calibration should, therefore, be always optimised with the additional data using suitable optimization software, or manually. Sensitivity analysis of the model can provide a useful insight to plan for additional data collection.

19.5.5 Case Studies

We give here two groundwater case studies to illustrate the potential of groundwater modeling technique.

19.5.5.1 Simulation of Regional Hydrodynamics: Neyveli Lignite Mining Area, India

A semi-confined aquifer comprising sandstone, sands, and gravel/pebbles of Tertiary age occurs below an open cast lignite mine in South India. The aquifer has been very heavily pumped to reduce the upward hydrostatic pressure and thus avoid the flooding of the mine floor. This extensive pumping substantially lowered the hydraulic heads in a large area of about $2300\,km^2$. Besides, stoppage of artesian flows from wells in the area, the decline of hydraulic heads have threatened seawater intrusion in areas which are close to sea. A modeling study was undertaken to quantify future trends of the aquifer response due to the large scale pumping and thereby evolve an optimal groundwater pumping scheme so that the regional lowering of hydraulic heads may be minimal without creating the hazard of flooding of open cast mines (Gupta et al. 1989).

The task of conceptualizing and quantifying the aquifer system was beset with a number of difficulties due to complexity of the system and paucity of data. However, a careful analysis of the available data on the physical framework and the stresses on the groundwater flow regime helped develop a conceptual model of the system.

The aquifer shows an unconfined condition over an area of about 420 km² and is semi-confined in the remaining region of the study area. The input to the unconfined aquifer has mainly been the recharge due to rainfall, whereas in the remaining area there has been a downward leakage from the upper aquifer and an upward feed from the lower confined aquifer. The transverse flows to the aquifer under study were calculated on the basis of difference between the interpolated hydraulic heads in the consecutive aquifer layers, the inferred values of vertical hydraulic conductivity, and thickness of the intervening leaky layers. The available values of the transmissivity at some points, and data on the saturated thickness of the aquifer as obtained from drilling logs were used to workout spatial distribution of transmissivity.

Initially, only a regional model was prepared using a coarse grid of nodes. The entire area was divided into 136 nodes with a spacing of 4 km. The finite difference technique was used in the modeling. Several iterative runs were made to progressively refine the physical

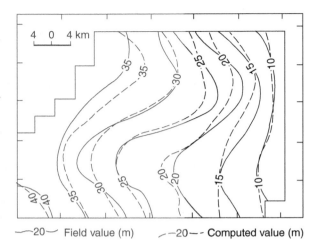

—20— Field value (m)   ⎯⎯20⎯⎯ Computed value (m)

Fig. 19.4 Regional groundwater modelling: contours of initial hydraulic heads

framework as well as the input and output parameters until a satisfactory match between the observed and computed hydraulic heads was obtained both under steady and dynamic states. A comparison of steady state hydraulic heads (observed in the field and computed) is shown in Fig. 19.4. Barring some localised mismatch, the general trend of computed and field piezometric contours agree. The mean annual input and output quantities for the initial steady state as obtained after a number of iterations are illustrated in Fig. 19.5.

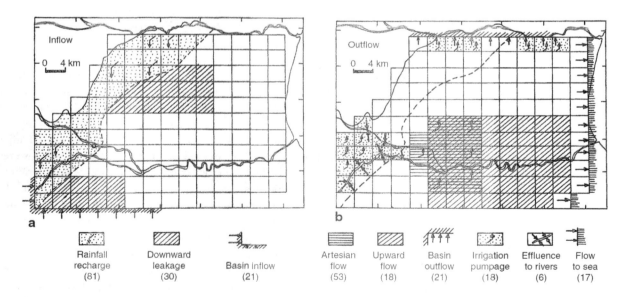

Fig. 19.5 Regional groundwater modelling: **a** input quantities, and **b** output quantities. The data are in million cubic metres per year for the initial state

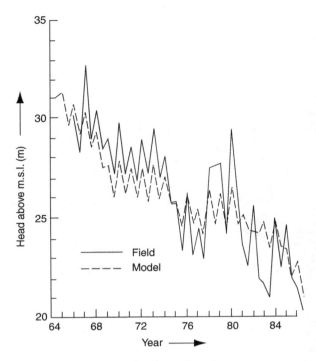

Fig. 19.6 Regional groundwater modelling: comparison of field and computed model hydrographs for a well

The dynamic calibration of the model was carried out for a period of 30 years. A sample hydrograph in a well as observed in the field and that computed from the model is shown in Fig. 19.6.

After obtaining a reasonably representative regional model through this process of calibration, nested square fine grid was adopted in the mine area to evolve a scheme of optimal pumping from mines. There were a number of constraints from the feasibility point of view on the location of pumping wells. These were followed in working out the location of wells and pumping rates. With this scheme, the desired lowering of hydraulic heads beneath the mining area could be achieved while the regional decline of heads was minimized.

19.5.5.2 Industrial Pollution Study

Groundwater in an important tanning centre located in a river basin in South India has been very much polluted. About 650 tanneries have been operating in an area of about 1600 km². These tanneries discharge their effluent (mostly untreated) through unlined channels into the river contaminating the groundwater regime. According

to the available reports, the effluent from the tanneries contain 20 000–30 000 mg l⁻¹ of total dissolved solids (TDS). At some locations, the TDS concentration in groundwater has been as high as 8000 mg l⁻¹. The situation is very precarious as most of the requirements for irrigation, industry and drinking purposes in the basin are being met from groundwater resources.

Main aquifers in the area are weathered and fractured crystalline rocks. However, along the river course, one encounters sand layers and pebbles having high permeability and good groundwater potential. The seepage rate of polluting effluent through the river bed is quite high. Though geohydrological parameters like transmissivity, specific yield, recharge to the aquifer, and groundwater pumpage in the study area were estimated at a few locations, but porosity and dispersivity required for mass transport modeling were not available. The former were interpolated to evolve a spatial distribution but the latter were only guesstimated using the geological information and data from similar studies elsewhere. A uniform porosity of 0.2 was assumed for the entire area. The longitudinal dispersivity α_L was taken as 30 m and transverse dispersivity α_T as 10 m. Historical water-table data monitored every month at a number of wells for the last 12 years was used to calibrate the model (Gupta et al. 1996).

The TDS concentration in groundwater obtained from chemical analysis of water samples from dug-wells represented the cumulative value for the entire depth of the aquifer at a point. However, C_0 which is the source concentration, i.e. actual quantity of TDS reaching the water-table, was not known. It had to be, therefore, inferred only from the TDS content in the effluent discharged at tannery outlets. The TDS load reaching the water-table was assumed to be 12% of that contained in the effluent at the surface.

A finite difference approach coupled with the method of characteristics (Konikow and Bredehoeft 1978) was used to construct groundwater flow and mass transport models. Initially, each node was assumed to be associated with a certain TDS concentration which moved with the groundwater velocity at that point. The numerical solution gave the values of head, h and concentration, C at all the points.

The water-levels obtained from the groundwater flow model were compared with those observed in field. The transmissivity distribution was progressively modified until a satisfactory match was obtained at most of the points. The groundwater flow velocities (Fig. 19.7a)

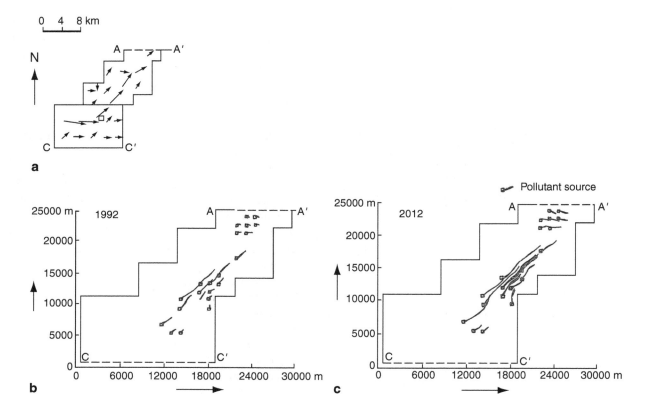

Fig. 19.7 Modelling of groundwater pollution due to industrial waste: **a** graphical view of velocity vectors, **b** modelled path lines of pollutant migration for the initial state (1992), **c** predicted trends after two decades

were computed using FLOWPATH computer code (Franz and Guiger 1990). The solute transport model was calibrated using the historical data for 12 years.

It was seen that the pathlines of pollutant migrtation after two decades were considerably extended (Fig. 19.7c). Actual concentration of TDS at many locations was also found to be higher. Despite lacunae in the data base for modeling of pollutant migration in the aquifer, it is brought out indisputably that if the tannery effluents continue to be discharged at the present level, groundwater pollution will appreciably increase in future. The sensitivity analysis helped to identify major data gaps and the required additional field investigations.

19.6   Modeling of Fractured Rock Aquifer System

As mentioned earlier, crystalline and massive rocks such as granites, quartzites, basalts etc. have little intergranular porosity and in such rocks, secondary poros-

ity in the form of fissures, fractures and joints provide higher permeability and fast movement of water. Thus, most fractured rock systems can be considered to consist of rock blocks bounded by discrete discontinuities comprised of fractures, joints and shear zones, usually occurring in sets with similar geometries (see Sects. 2.2 and 2.3). The fractures may be open, mineral-filled, deformed or any combination thereof. Open fractures may provide conduits for the fast movement of groundwater and contaminants through an otherwise relatively impermeable rock mass. Petroleum industry first developed the theory of fluid flow through fractured rock and the same principles and concepts are used in hydrogeology.

On a local scale, the properties of the fractured rock mass with respect to groundwater flow are extremely heterogeneous. Major factors controlling the groundwater flow through fractured rock include fracture orientation, fracture density (or intensity), effective aperture width and the nature of the rock matrix. Fractures oriented parallel to the hydraulic gradient are more likely to provide effective pathways for flow than fractures oriented perpendicular to the hydraulic gradient.

Fractured rock aquifers may be considered to behave or represent one or more of the following conceptual models: (a) equivalent porous medium, (b) discrete fracture network (DFN), (c) dual porosity medium, (d) triple porosity medium and (e) stochastic continuum (SC) model. Salient features of these models are given in Table 19.1. Out of these, the DFN model has received wider applications and is therefore discussed in more detail here.

19.6.1 Groundwater Flow Equation for Parallel Plate Model

Groundwater flow along discrete fractures can be idealized as that occurring between two parallel plates (parallel plate model):

$$\rho \frac{\partial v}{\partial t} + \rho v \nabla v = \rho g - \nabla P_T + \mu \nabla^2 v \quad (19.6)$$

Table 19.1 Types of models of fractured media

Type of model	Description
1. Equivalent Porous Medium (EPM) Model	When the rock system possesses high intensity of interconnected fractures in various directions, then it can be considered to behave as a statistically continuous porous medium on a regional scale, with equivalent or effective hydraulic properties. The stochastic continuum approach uses the same groundwater flow equations as for ordinary porous medium but with statistical distributions of parameters. The approach assumes that a representative elementary volume (REV) of material characterized by effective hydraulic parameters (hydraulic conductivity, specific storage and porosity) can be defined. When EPM is considered, then standard finite difference method or finite element method may be applied to simulate groundwater flow in a fractured system. It was reported by many research workers that EPM approach might reasonably simulate the behaviour of regional flow system, while it poorly reproduces the local condition
2. Discrete Fracture Network (DFN) Model	This attempts to deal with the typical case of crystalline hard rocks where inter-granular permeability is absent and the water flow system is confined to the inter-connected network of discrete fractures. Initially, the discrete fracture network approach was conceptualized to understand the scale dependence of the effective dispersion parameters for radionuclide transport through fractured rock. Ideally, a DFN model can be applied in domains where the fracture population is homogeneous. Most theoretical developments and applications attempt to define statistical homogeneity in very strict terms of spatially uniform distributions of fracture properties, either univariate (size) or bivariate (orientation). In practice, the required strict statistical homogeneity cannot be met from field condition. Therefore, practical DFN modeling requires a broader definition of statistical homogeneity, based on a statistical description of heterogeneity in the fracture population and characteristics
3. Dual Porosity (DP) Medium Model	If the rock matrix containing the fracture network has some amount of inter-granular permeability also (e.g. sandstones), then the medium will exhibit two flow characteristics—one pertaining to the fractured network and the other to the rock matrix. This is the dual porosity model. The flow through the fractures is accompanied by exchange of water to and from the surrounding porous rock matrix. Exchange between the fracture network and the porous blocks is normally represented by a term that describes the rate of mass transfer. Both the fracture and block physical parameters are needed to defined at each mathematical point in the flow domain. During unsteady (transient) state conditions, the flow phenomenon is characterised by release of water from fractures at the initial stages of the pumping, and as pumping is continued, a differential fluid pressure head is created between the water content in rock matrix and the water flowing fractures, resulting in to the release of water from the rock matrix into the fracture system
4. Triple Porosity (TP) Medium Model	A triple porosity medium model is conceived for karst regions, where three levels of water-bearing porosities and cavities occur: (i) inter-granular, (ii) along fractures and joints (iii) caves, caverns and solution cavities of larger dimensions. Water flow dynamism in each of the above is different and therefore TP may represent the filed hydrogeologic conditions
5. Stochastic Continuum (SC) Models	The stochastic continuum approach (geostatistical approach) for modeling groundwater flow in heterogeneous formations is based on the idea that a formation can be described in terms of physical parameters (conductivity, specific storage, etc.) that vary in space according to spatially varying random functions. The stochastic continuum theory treats the parameter heterogeneity in the context of a statistical (probabilistic) framework. Stochastic continuum (SC) models use blocks of an equivalent porous medium (EPM) to represent blocks of fractured rock

where ρ is fluid density, μ is viscosity of fluid, v is groundwater flow velocity, P_T is total pressure, t is time, and g is the acceleration due to gravity. For the parallel plate situation, the relationship can be simplified in to one-dimensional equation. This is because the aperture is assumed infinite perpendicular to flow. By making use of this condition one can derive transmissivity 'T_f' as a function of aperture width 'a' as given below (Sect. 7.2.1):

$$k = \frac{a^2}{12} \qquad (19.7)$$

$$T_f = \frac{a^3 \rho g}{12\mu} \qquad (19.8)$$

The above equation shows that fracture flux ('T_f') is proportional to the cube of the fracture aperture (a).

19.6.2 Particle Tracking Approach for Solute Transport

Transport of solute in the rock matrix is modeled as the matrix diffusion problem between fractures and rock matrix blocks. In general the particle tracking approach, which considers the concentration of solutes within the fluid (water) by a finite number of discrete particles of equal mass, is used for solving the solute transport equation. Each particle represents a fraction of the total mass of the solute present. Particles are introduced into the modeled region as specified solute sources and removed at encountered sink nodes. At each time step, particles are moved according to a deterministic convective (advective) component and a stochastic dispersive component. The advective component, V_t, is proportional and parallel to the velocity vector at the current particle location. The total distance traveled in a time step can be split into two components, longitudinal and transverse. Prickett et al. (1981) derived the equation for travel distance along the direction of flow (longitudinal) as:

$$X_d = N_L \times (2 \times A_L \times X_c)^{1/2} \qquad (19.9)$$

where X_d=distance traveled along flow (longitudinal) direction (L), X_c=distance traveled due to advection (=V×t), A_L=lateral dispersivity (characteristic length), N_L=a normal random variable with a mean of zero and a variance of one (dimensionless).

Similarly, transverse distance can be described as:

$$Y_d = N_T \times (2 \times A_T \times X_c)^{1/2} \qquad (19.10)$$

where, Y_d=distance traveled normal (transverse) to flow direction (L), X_c=distance traveled due to advection=V×t(L), A_t=transverse dispersivity (characteristic length (L)), N_L=a normal random variable with a mean of zero and a variance of one (dimensionless).

The dispersive component given by Eqs. 19.9 and 19.10 above is proportional to the square root of the advective component. Average concentrations are determined at the end of each time step by summing the particle mass within each fracture element and dividing by the mass of fluid within the element. The equation to determine the particle concentration within specific element is given by:

$$C = N_p \cdot M_p / \rho_w \cdot A \cdot a \qquad (19.11)$$

where, N_p=number of particles in element, M_p=mass of a single particle (M), ρ_w=density of solvent (water), a=fracture aperture width, A=area of fracture element.

Particle movement is based on the velocity field obtained from nodal heads at the end of the current time step. For transient flow simulations, velocities change at each time step, and time steps should be sufficiently small to accommodate changes in the velocity field. For steady state flow simulations, no changes in the velocity field are observed and hence only a single calculation of nodal head is sufficient. Except for changes in element velocities for transient flow simulations, the accuracy of the particle tracking calculations is not affected by the time step size. Each time a particle begins a new time step, new random numbers are drawn to determine the direction and magnitude of the dispersive particle movement.

19.6.3 Discrete Fracture Network (DFN) Model

In fractured rock media, the discrete-fracture network (DFN) models provide a means of explicitly representing the flow path geometry, such that the geometry of interconnection among fractures determines the groundwater movement. The statistical geometry of fractures (orientation, frequency, aperture etc.) can be estimated directly from observations in outcrops and boreholes (see Chap. 2). The applicability of DFN model is limited in terms of scale of the area and volume to be simulated. The DFN models must be used in conjunction with Stochastic Continuum (SC) model. SC model is used to model fluid flow on large scale by making use of probability distribution of fracture properties (i.e., given a definite location, size, transmissivity etc.), as in practice, it will be very difficult to characterise all hydraulically significant fractures in a block of a rock and stochastic modeling can be of much use as uncertainty can be represented explicitly.

The derivation, validation and application of the DFN approach to model flow in discrete fractures have the following broad steps: (i) data analysis, (ii) Monte Carlo simulation, (iii) calibration and validation and (iv) analysis of model predictions (Thangarajan 2004).

19.6.3.1 Data Requirements for DFN Modeling

Direct observation of fracture geometry and properties is limited to a few boreholes and outcrops. Only limited number of major conductive fractures within the rock mass can be directly studied through geophysical methods, well logging, packer tests etc. Therefore, an approach using statistical characterization of fractures is needed. Table 19.2 gives an overview of data requirement to characterize the fracture geometry model and their sources.

Coring During drilling operations, core materials, preferably oriented cores, are collected at different depths to get information on fracture characteristics (Table 2.1; also see Sect. 2.4). Further, data on location of major water bearing fractures, changes in hydraulic head with depth, and changes in the groundwater chemistry are also collected, as feasible. Oblique drilling is carried out to intercept near-vertical fractures. Normally, core data are strongly biased towards the fractures perpendicular to the core, and in that case, Terzaghi's (1965) theoretical correction can be applied to correct the biasedness of core data.

Aquifer Tests Pumping tests (aquifer) can provide valuable information on hydraulic conductivity, anisotropy and average fracture aperture of the fractured formation including heterogeneity and boundary conditions. Cross-hole packer tests are preferred for estimating packer interval transmissivity and storativity (see Sect. 9.2.1).

Tracer Tests Effective porosity, dispersion and matrix diffusion can be obtained through tracer tests, which could be conducted either under natural gradient or stressed gradient conditions. Though these tests will cost more time and money, the above information related to fracture properties can be obtained only through tracer tests (see Sect. 10.3).

Remote Sensing and Geophysical Methods GPR, magnetic surveys, seismic surveys and remote sensing can be used for deducing information on fracture pattern, lineament and dyke etc. (see Chaps. 4 and 5). Borehole geophysical methods include acoustic, electrical resistivity including electrical tomography, calliper and gamma ray logging which may provide more reliable information on fracture intensity. The acoustic televiewer presents a continuous image of the acoustic response of the borehole face, and can detect fracture apertures up to a width of 1 mm. Fracture orientation is also determined by making use of this tool. Recently

Table 19.2 Fracture properties and their source. (After Thangarajan 2004)

Fracture Properties	Data Source
1. Location, orientation and size	Field data, fracture trace maps and borehole logs
2. Conductive fracture intensity	Borehole logs and packer tests
3. Shape	Fracture traces maps, field data
4. Transmissivity	Steady and transient packer tests
5. Dimensionality	Transient packer tests
6. Storativity	Transient packer tests etc.
7. Transmissivity variability	Generic information
8. Fracture hydraulic connectivity	Packer test transient results

down-hole cameras are used to provide in-situ viewing of the fractures in boreholes (see Sect. 2.4).

Borehole Flow Meters with Data Logger Borehole flow meters are used to measure the flow rate during either aquifer tests or packer tests. Three types of flow meters are available viz. Impeller, heat pulse and electromagnetic type. Impeller type is commonly used in the industry and it will be also useful to quantify the water released from fractures to the borehole during pumping.

Packer Test Aquifer tests provide information about fracture intensity, transmissivity and storativity, which can be used to conceptualise DFN model. The packer testing methodology known as Fixed Interval Length (FIL) testing provides the necessary data for statistical derivation of the relevant fracture model parameters. A discrete fracture interpretation of FIL packer test involves three main steps:

- Evaluation of packer test data to estimate interval transmissivity for each test.
- Estimation of variation in bore-hole measurements of fracture transmissivity.
- De-convolution of packer interval transmissivity data to estimate the conductive fracture frequency f_c and the single fracture transmissivity distribution.

The method of packer testing has been described in Sect. 9.2.1 Packer test is usually run and interpreted in conjunction with closed circuit television (CCTV) fracture log to help interpret and identify active (flowing) fracture zones more accurately.

19.6.3.2 Characterization of Fracture Properties

Statistical characterization of fracture properties, i.e. location, orientation, length, size, spacing hydraulic transmissivity etc. is extremely important and critical. Various researchers have proposed different statistical distribution functions to describe these aspects (Table 19.3).

A brief introductory description of characterization of fracture properties is given in the following paragraphs:

1. *Location:* Location of fractures is expressed in terms of probability density function $f_x(x)$ for fracture centers in 3D space. The simplest case is the purely random case, also referred to as the Baecher conceptual model that was introduced for rock mechanics problems.

Table 19.3 Various statistical distribution functions used to describe fracture characteristics. (After Indraratna and Ranjith 2001)

References*	Various Distribution Functions
Joint orientation	
Anderson and Dverstorp* (1987)	Fisher distribution, $f(\theta, \varphi) = \dfrac{Ke^{k \cos \theta} \sin \theta}{4\pi k(\sin h)}$
Samaniego* (1984), Wei et al.* (1995)	Normal distribution, $f(\theta, \varphi) = \dfrac{1}{\sigma\sqrt{2\pi}} e^{-0.5\left(\frac{\theta-\mu}{\sigma}\right)^2}$
Priest* (1993)	Negative exponential distribution, $f(\theta, \varphi) = \dfrac{1}{\mu} e^{-\frac{\theta}{\mu}}$
Joint length/size (l)	
Samaniego* (1984) and Wei et al.* (1985)	Negative exponential distribution
Priest and Hudson* (1981)	Exponential
Bridges* (1976) and McMahon* (1971)	Lognormal
Dershowitz* (1984)	Gamma
Fracture transmissivity	
Snow* (1970)	Lognormal distribution
Spacing (s)	
Priest and Hudson* (1981)	Normal
Sen and Eissa* (1992)	Exponential and Lognormal
Bridges* (1976)	Lognormal, $f(s) = \dfrac{1}{\sqrt{2\pi} s\sigma} exp\left[\left(-0.5\dfrac{1}{s} ln\dfrac{s}{m}\right)\right]$
Wallis and King* (1980)	Exponential

where θ = dip angle, φ = dip direction, k = dispersion parameter about the mean direction, determined by the maximum likelihood estimator (Mardia* 1972), μ = mean, σ = standard deviation, s = spacing and m = median.
*see Indraratna and Ranjith, 2001

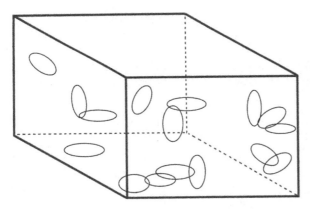

Fig. 19.8 Idealization of circular shape of fractures in a rock mass

2. *Size:* The size of a polygonal fracture is expressed in terms of the equivalent radius r_e of the fracture (Fig. 19.8), i.e. the radius of a circle that has the same area as the polygonal fracture surface area A_f:

$$r_e = \sqrt{(A_f/\pi)} \qquad (19.12)$$

This definition holds good for both terminated (blind) as well as un-terminated fractures. Fracture size (radius) probability density functions are obtained from fracture (and lineament) trace length data. Analytical methods used for deriving this parameter from trace length data are available based upon simplifying assumptions such as normal distribution, uniform orientation distribution etc. (also see Table 19.3).

3. *Orientation:* Orientation data of fractures obtained by sampling from measurements (trace plane, scanline and borehole) needs to be corrected for sampling bias by applying Terzaghi correction (Terzaghi 1965). Fracture orientation is expressed in terms of either fracture pole orientations or fracture dip directions (see Fig. 2.18). Statistical variability of fracture orientation can be expressed in various ways (e.g. Fisher 1953; Table 19.3).

4. *Intensity:* The DFN modeling approach is simplified by modeling only the conductive (flowing) fracture population, which forms only a fraction of all the fractures present. This makes the model to be more computationally efficient and realistic. Conductive fractures can be identified through packer tests and well logging. Surface mapping, scan line surveys along with aerial photographic and remote sensing surveys can generate the input data neces-

sary for describing the fracture distribution. In DFN modeling, it is important to define the 3-D fracture density (intensity) in terms of total fracture area per unit volume (d_3, see Sect. 2.3). This is scale invariant, which implies that it does not depend either on the volume studied or on the assumed fracture size distribution.

5. *Shape:* Fractures are assumed to be repetitive, planar and approximately polygonal in outline or shape. When two planar fractures intersect each other, they generate a linear edge, finally leading to polygonal shape or outline of fractures.

6. *Transmissivity:* Transmissivity at a point ψ on fracture is defined as the constant of proportionality between flux density and hydraulic head gradient. Darcy velocity, also called specific discharge, is expressed as:

$$q_i = -\frac{T(\psi)\delta h}{\delta \psi_i} \qquad (19.13)$$

where, q_i=Component of flux per unit width of plane in the direction ψ_I, $T(\psi)$=transmissivity at the point ψ in the fracture plane, ψ_I=ith component of the local co-ordinate vector ψ, and h=piezometric head. The local transmissivity is assumed to be isotropic in two-dimension and independent of head gradient. In the simplest model, transmissivity is assumed to be constant throughout any given fracture plane. However, this is not true as flow distribution through individual fractures may be irregular due to channelling effects, though it may neglected as a first-order approximation using an average fracture transmissivity value. In a more detailed model, transmissivity is considered to vary as a fractal process within the plane of each fracture such that each fracture is discretized into a number of sub-fractures of approximately equal size possessing a certain transmissivity value.

7. *Storativity:* The storativity at a point ψ on a fracture describes the change in the volume of fluid cowntained per unit area of the fracture, in response to a unit change in pressure:

$$S(\psi) = \lim_{A \to 0} [1/A^* \delta V_w(\psi)/A \delta h(\psi)] \qquad (19.14)$$

where, A=area in the fracture plane, $S(\psi)$=storativity at point ψ in the fracture plane, $V_w(\psi)$=volume of water contained in the fracture plane around the point ψ, and h(ψ)=piezometric head at point (ψ) in the fracture plane.

Storativity is related to the fracture normal stiffness and fluid compressibility. Further, if some infilling is present in the fracture, then it is also related to the porosity and compressibility of the infilling material.

8. *Fracture System Porosity:* The fracture system porosity ϕ_f can be directly estimated as the product of the fracture intensity expressed as fracture area per unit volume d_3 and the storage aperture of the fractures (e) (Dershowitz et al. 1998).

$$\phi_F = \frac{V_F}{V_{cell}} = \frac{\sum (A_F.e)}{V_{cell}} = d_3.e \quad (19.15)$$

where, ϕ_F=fracture system porosity, V_F=fracture system volume, V_{cell}=grid cell volume, A_F=fracture area, "e"=fracture storage aperture, and d_3=fracture area per unit volume (to represent fracture intensity). Obviously, the fracture system porosity depends on the number of fractures per unit volume, fracture size distribution and fracture aperture distribution. Therefore, for every portion of the continuum model where these parameters vary, a different porosity value needs to be calculated. By making use of DFN model parameters, one can compute the fracture system porosity for each grid cell. The major issue in defining the fracture porosity from fracture intensity d_3 is in the selection of an appropriate value for storage aperture "e", which could be based on data such as: (i) aperture derived from transient hydraulic response, (ii) mechanical aperture, (iii) aperture derived from geophysical measurements, or (iv) aperture derived from fracture permeability or transmissivity (cubic law).

9. *Fracture System Permeability (Directional):* The permeability of the fracture system depends on the fracture intensity, distribution of the fracture transmissivity values and inter-connectivity of the fracture network. It can be expressed as a tensor representing the effective fracture flow at each grid cell (Oda 1984).

19.6.3.3   Model Construction and Calibration

Based on the data collected as above, first a conceptual model is evolved, i.e. a hypothesis as to how a system or process operates. The conceptual model is, then, transformed into a numerical model that is tested against data for calibration purposes, and then ultimately used to make predictions. The essential steps in the generation and calibration of a DFN model are as follows:

- Fracture network construction to generate geometric realization of fractures,
- Hydraulic simulation: Verify geometric model against simulated boreholes etc.,
- Assign hydraulic properties based on single borehole flow test, and
- Calibration to large-scale interference tests or other hydraulic events.

19.6.3.4   Merits and Limitations of DFN Model

DFN models are well suited for studying flow and mass transport in small-scale fractured rock mass, explicit representation of flow path geometry (Fig. 19.9), and provide a possibility of modeling fracture zones on various scales. Combination of DFN with SC (stochastic continuum) and DP (double porosity) modeling can enable simulation of flow and solute transport in a larger scale of rock mass. A major limitation is that the DFN approach requires adequate site characterization

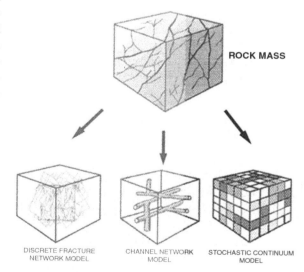

Fig. 19.9 Representation of fracture system in a rock mass through various conceptual models: discrete fracture network model, channel network model, and stochastic continuum model. (After Golder Associates in Thangarajan 2004)

to describe the prevailing fracture systems controlling the hydrogeology. This implies collection of fracture-geometry data at sampling locations well-distributed throughout the region to be modeled including at depth, and requisite interpolation. Also, simplification and idealization of field data is often necessary to simulate large-scale regions. Further, the DFN approach is relatively new, having been developed for specific applications e.g., deep repository or reservoir studies, and hence the numerical modeling tools for groundwater are still under development.

19.6.3.5 Channel Network (CN) Model

A channel network model can be considered as a simplification of the discrete fracture network model and is also based on fracture properties derived from field data. The basic concept of a CN model is that flow within a fracture network is confined to discrete, effectively one-dimensional pathways known as channels produced due to intersection of fractures at various intervals (Fig. 19.9). The CN models are similar to DFN in their recognition of discrete nature of the fracture pathways. A major advantage of CN is that the use of a network composed of effectively 1-D elements can reduce the computational complexities.

19.6.4 Dual Porosity (DP) Model

If the rock matrix containing the fracture network also has some amount of inter-granular permeability (e.g. sandstones), then the medium will exhibit two flow characteristics—one pertaining to the fractured network and the other to the rock matrix. This is the dual porosity or double porosity model (see Sect. 7.2.2). The flow through the fractures is accompanied by exchange of water to and from the surrounding porous rock matrix (Table 19.1). The DP model has certain advantage over DFN regarding model size, speed, and accuracy of multiphase flow. Therefore an integrated approach combining DP and DFN appears powerful option for simulating the heterogeneous fracture aquifer system. The following input parameters could be derived from DFN approach to form as an input in the DP model:

- Fracture system porosity,
- Directional fracture system permeability,
- Optimum grid cell size,
- Matrix-fracture interaction parameters etc.

19.6.5 Stochastic Continuum (SC) Model

The stochastic continuum approach (geostatistical approach) for modeling groundwater flow in heterogeneous formations is based on the idea that a formation can be described in terms of physical parameters (hydraulic conductivity, specific storage, etc.) that vary in space according to spatially varying random functions. The stochastic continuum theory treats the parameter heterogeneity in the context of a statistical (probabilistic) framework (Fig. 19.9). The following data are needed to characterize the flow problem for SC modeling:

- An estimate of minimal rock volume [representative elementary volume (REV)] on which the rock mass can be said to behave as an equivalent porous medium.
- Variability of the effective hydraulic conductivity 'K' of the rock mass.
- An estimate of rock anisotropy in terms of variability of hydraulic conductivity expressed in terms of the ratios of the principal components of the (presumed) hydraulic conductivity tensor (K_x, K_y and K_z) to the average hydraulic conductivity K.
- The form of spatial correlation of rock mass conductivity that results from fracture network effects.
- The relationship between apparent hydraulic conductivities measured by borehole testing and the effective hydraulic conductivities of the rock mass on the scale of blocks used in SC modeling.

19.6.6 Computer Codes

Salient aspects of some of the commercially available computer software to simulate the groundwater flow in fractured media are as follows:

1. FracMan
   FracMan software (developed and marketed by Golder Associates Inc., Redmond, USA) provides an integrated set of tools for simulation of discrete fracture network (DFN) analysis. FracMan includes tools for discrete fracture data analysis, geologic

modeling, spatial analysis, visualization, flow and transport, and geo-mechanics. Its salient features are that:

a. It provides a unique set of tools to transform geological and well test data into quantitative parameters necessary for discrete fracture network modeling.
b. To generate a 3-dimensional realizations of discrete feature geology, including the ability to validate models through simulated exploration.
c. Flow and transport modeling including a dual-porosity concept.
d. Analysis of the flow behavior of discrete fracture network, determining flow pathways.

2. Frac3DVS
   Frac3DVS software (Hydro-geologic Inc., 460 Phillip Street, Suite-101, Waterloo, Ontario, Canada N2L 5J2) is a 3D finite element model for steady/transient, variably saturated flow and advective-dispersive solute transport in porous and or discretely–fractured porous media.

19.6.7 Case Studies

19.6.7.1 Application of EPM in Granitic Rocks of Nalgonda District, India

In the Nalgonda district, Andhra Pradesh, India, weathered and fractured granites commonly form groundwater aquifers. Due to overexploitation, groundwater table has been declining persistently such that many dugwells have gone dry. Accordingly, a study for simulating groundwater flow was taken up with a view to predict the future scenarios and the impact of possible artificial recharge (Prasad 1999).

A distributed modeling for the study area was carried out by designing a grid of 16×18 cells (each of $1\,\mathrm{km} \times 1\,\mathrm{km}$ size) and a finite difference groundwater flow model (MODFLOW) was used for the simulation considering the aquifer as EPM (equivalent porous medium). The results of the distributed model showed that there is an annual recharge of 9.661 MCM while the annual draft is 11.747 MCM. The model was calibrated for the year 1992–1993 (Fig. 19.10a, b) and

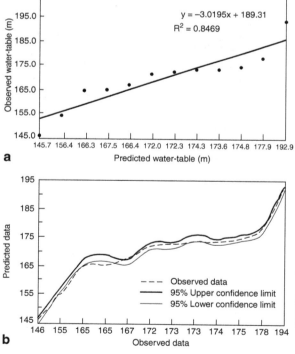

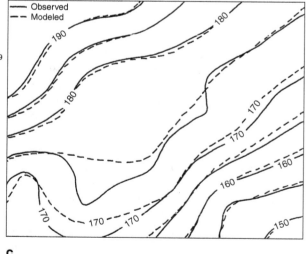

Fig. 19.10 Equivalent porous modeling (EPM) results of weathered and fractured granites in south India. Figures (a) and (b) pertain to 1992 data used for model generation, whereas Figure (c) pertains to 1998 data used for model validation. **a** Scatter plot of observed and predicted results (1992 data), **b** plots of observed data and with 95% confidence limits of predicted data (1992 data), **c** water table contours of observed and modeled (validated) data for No. 1998. (After Prasad 1999)

validated using water-table contour map and water-table hydrographs for the year 1998 (Fig. 19.10c). The model so developed and validated was used to predict scenarios of water-table regime if artificial recharge is implemented in the area.

19.6.7.2 Application of DFN in Limestone of Ireland

The DFN model has been applied to problems in radioactive waste management and hydrocarbon reservoirs. Perhaps the first quantitative study using DFN model to simulate a fractured groundwater aquifer with the purpose to increase well yield was made by Jones et al. (1999). In their study Jones et al. (op cit.) modeled a fractured limestone of Carboniferous age located in SW Ireland. The area is marked by the presence of nearly NE–SW trending major fractures, affected to some extent by shear zones. Brittle fracturing along tensional zones has led to the formation of well-brecciated zones. Detailed geological mapping followed by resistivity surveys helped to define the structural pattern and lineaments. Study of vertical borehole cores and logging yielded data on intensity of fracturing, their spatial distribution and orientation etc. Mapping in local quarries was done to obtain fracture size (trace length) and additional information on fracture statistics. Further, borehole logging and packer tests indicated that the groundwater flow is through discrete fractures in the limestones. A conceptual hydrogeological model was evolved which showed the presence of a major NE–SW trending fracture zone with flow being controlled by discrete fractures (Fig. 19.11). FracMan software was used to characterize and model fracture geometry and fracture properties, which were specified as probability distributions. Hydraulic properties of fractures were estimated from pumping tests and model simulations. The model was conceptualized, calibrated and validated against borehole measurements and pumping test data. Drawdown due to simultaneous pumping of wells was modeled and contoured across fracture surfaces, which was validated against observed drawdown data. The calibrated model was then used to identify optimum borehole orientation and distribution for maximum yield from the fractured aquifer zone.

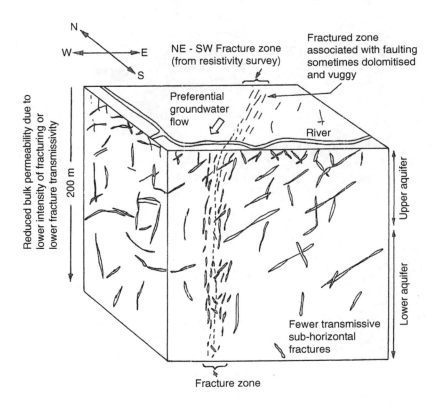

Fig. 19.11 Schematic block diagram of the conceptual model of fractured Carboniferous limestone aquifer in Ireland. (After Jones et al. 1999)

19.7 Concluding Remarks

The development of numerical modeling tools for groundwater applications in fractured rock systems is still in infancy. The task of modeling groundwater flow and solute transport in such media appears to be largely hampered by limitations in site characterization, i.e. due to difficulties in collecting the necessary data to adequately describe the hydrogeological properties of the fractured system. The fracture characteristics are site specific and modeling cannot be a substitute for collection of data. On the other hand, limitations in data collection may be imposed by various factors such as accessibility, time and funding. Innovative approaches in data collection and integration are required to simulate the field conditions to the best approximation for modeling. Further, research is also required to adequately develop tools for double-porosity and triple-porosity models to simulate solute transport phenomenon and regional scale flow in various types of hydrogeological situations.

ant parameters are imported and interpolated into the model grid. Modeling of homogeneous porous aquifer is relatively simple and has been extensively carried out. For fractured and inhomogeneous rock aquifers, several models have been proposed to represent the aquifer conditions. Out of these, the discrete fracture network (DFN) model utilizing the basic parallel plate flow model has been more widely used. However, modeling of fractured rock aquifer system is still in infancy and is largely hampered by inadequate site characterization, as fracture characteristics happen to be site-specific and require extensive field geologic and hydrogeologic characterization.

Summary

A numerical model is a mathematical simplified representation of the field hydrogeological setting. Computer based groundwater modeling is used for studying problems of groundwater resource, quality and management. The task involves defining the purpose of the model, conceptualization of the system, setting up of the numerical model including hydrostratigraphic boundaries, defining model parameters, groundwater fluxes etc., model calibration, validation and running simulation for prediction scenarios. During modeling, the various hydrogeological data of both variable (transient) and time-invari-

Further Reading

Anderson MP, Woessner WW (1992) Applied Groundwater Modeling. Academic Press Inc., San Diego, CA.

Bear J, Verruijt A (1987) Modeling Groundwater Flow and Pollution. Kluwer Academic Publishers, Hingham, MA.

Dershowitz WS, Laponte P, Eiben T, Wei L (1998) Integration of Discrete Fracture Network Models with Conventional simulator approaches. Society of Petroleum Engineers Annual Technical Convention, Louisiana, September 27–30, SPE NO. 49069.

Jones MA, Pringle AB, Fulton IM, O'Neill S (1999) Discrete fracture network modeling applied to groundwater resource exploration in southwest Ireland. Fractures, Fluid Flow and Mineralization. Geological Society of London, London, Spl. Publi. No. 155, pp. 83–103.

Rushton KR (2003) Groundwater Hydrology: Conceptual and Computational Models. John Wiley & Sons Inc., Chichester, UK.

Thangarajan M (2004) Regional Groundwater Modeling. Capital Book Publishing Company, New Delhi.

20.1 Introduction

Assessment and management of groundwater resources is important for their optimum utilization and to avoid any adverse effects. First of all, it is necessary to have a prior knowledge of the distribution as well as hydraulic and geochemical characteristics of aquifers as discussed in earlier chapters. It is also necessary to know the amount of natural recharge and groundwater abstraction. Management of groundwater resources is simple when natural recharge is more than the abstraction. However, when the aquifer approaches full development, the management problems increase significantly. For example, in the past half a century, there has been an explosion in groundwater development for irrigation, domestic and industrial uses, particularly in India and China. Abstraction of water more than natural recharge leads to decline in groundwater levels, which causes uneconomic pumping lifts, land subsidence, deterioration of water quality, sea water encroachment in coastal aquifers and so on. In areas where an aquifer is hydraulically connected with the river, lowering of water table will reduce the groundwater discharge into the stream and may result in induced recharge from the river into the aquifer. On the other hand, extensive use of surface water for irrigation may cause water logging and salinization of water and soil. These problems may arise both in granular rock formations and fractured rocks. Groundwater modelling is a useful tool to study various problems related to effect of various stresses on groundwater regime (Chap. 19). The various technical aspects of groundwater assessment and management are outlined in this chapter. Most of the examples cited are from granular formations as such studies in fractured rocks are limited but the general approach is the same.

20.2 Water Budget and Groundwater Balance

Water budget is a quantitative assessment of the total water resources of a basin over a specific period of time. It helps in evaluating the availability of surface and subsurface water for planning the various utilization patterns and practices, and thereby efficient management of water resources. This involves preparing of (1) hydrologic budget which considers both surface and groundwater resources, and (2) groundwater budget which is more specific for groundwater assessment and management. The various components of groundwater budget are discussed by Sokolov and Chapman (1974), Lloyd (1999), and Scanlon et al. (2002) among others.

20.2.1 Hydrological Balance

The hydrological balance equation can be written as

$$S_m + S_g + S_s = P + I_s + I_g - R - ET - O_s - O_g \pm n \qquad (20.1)$$

where, S_m = change in soil moisture, S_g = change in groundwater storage, S_S = change in surface water storage, P = precipitation, I_S = surface inflow into the basin from outside through canals, rivers etc., I_g = subsurface inflow from adjoining basins, R = runoff, ET = evaporation and evapotranspiration, O_S = surface outflow to other basins, O_g = subsurface outflow to the adjacent basins, and n = errors in estimation or undetermined elements, viz. losses through deep percolation etc.

B. B. S. Singhal, R. P. Gupta, *Applied Hydrogeology of Fractured Rocks*,
DOI 10.1007/978-90-481-8799-7_20, © Springer Science+Business Media B.V. 2010

Changes in soil moisture storage, S_m are negligible if balance is prepared on annual basis. However, for seasonal balance, S_m will be significant. It can be determined from soil moisture profile at the beginning and end of the period. Change in groundwater storage, S_g can be estimated from groundwater level fluctuation at the beginning and end of the period and specific yield of the aquifer. The change in surface water storage, S_s can be estimated from reservoir/lake data. The weighted average depth of precipitation is estimated from rain gauge data usually by constructing Theissen polygons or from isohyetal maps. Surface inflow and outflow, I_S and O_S are estimated from river and canal discharge data. Subsurface inflow and outflow, I_g and O_g can be computed from Darcy's law $(Q = TIL)$ based on data of aquifer transmissivity (T), hydraulic gradient (I) and width of the aquifer (L). Evaporation and transpiration losses (ET) include evaporation from surface water and evapotranspiration from soil, natural vegetation and cropped areas. An idea of evaporation from surface water is obtained from evaporation pan data. ET losses from soil and vegetative cover are more complex which can be estimated using lysimeter and various empirical equations based on types of vegetation and other parameters. In case field hydrological data are not available in an area, remote sensing—GIS techniques can be used to generate surrogate hydrological data (e.g. Meijerink et al. 1994). ET losses should be given due consideration in water-limited environment (WLE) (see Sect. 1.2).

20.2.2   Groundwater Balance

The groundwater balance equation for a given time period can be written as

$$R_p + R_c + R_t + I_r + R_b - ET$$
$$- T_p - S_e - O_g \pm \Delta S \pm n = 0 \quad (20.2)$$

where, R_p = recharge from precipitation, R_c = recharge from canal seepage and field irrigation, R_t = recharge from tanks, I_r = influent seepage from rivers, R_b = subsurface inflow from adjacent basins, ET = evapotranspiration, T_p = draft from groundwater, S_e = effluent seepage to rivers, O_g = outflow to other basins, ΔS = change in groundwater storage, and n = undetermined elements of the balance and errors in

the estimation of various balance elements. ΔS and n can be either positive or negative.

Each item in the groundwater balance Eq. 20.2 is estimated independently. The various items can be expressed either in terms of volume or depth of water using consistent units. The groundwater balance equation can also be used to estimate any one component provided the other items are determined by other methods. The duration of the water balance could be a water year or a shorter period. The span of water year may vary from one country to another based on annual precipitation pattern. In India, water year is taken from 1st Novermber to 31st October of the next year. The water year can be sub-divided into monsoon period (1st June–30th November) and non-monsoon period (1st December–31st May). In USA, water year starts on 1st October and ends on 30th September and is designated by the calendar year in which it ends.

For groundwater budgeting, first, the area of study has to be delineated. For this purpose, the best approach is to select groundwater basin as a unit. In past, the practice in India and in many other countries has been to take the interfluve (*doab*) area or an administrative zone (*taluka* or district) as a unit for such a study. However, such boundaries may not coincide with the groundwater divides. Ideally groundwater balance should be prepared separately for each aquifer unit of a groundwater basin. The reliability of the groundwater balance would depend on the accuracy of the estimation of different parameters. The water balance approach is less accurate in arid and semi-arid regions than in humid areas.

20.2.3   Estimation of Various Components of Groundwater Recharge

Recharge is that portion of infiltered water which reaches the water table (zone of saturation). Recharge estimation is one of the most important and difficult tasks in hydrogeology. Several texts are available on the estimation of groundwater recharge viz. Simmers (1998), Lloyd (1999) and Scanlon et al. (2002).

Recharge is largely controlled by geology (lithology and rock structures), topography, depth to water table, climate (rainfall and evapotranspiration), and nature of vegetative cover. Urbanization usually decreases recharge due to surface impermeabilization but in some other situations, leakage from water supply system may increase the amount of recharge.

In granular aquifers, recharge is usually areally distributed. The water table fluctuations in granular aquifers are also small, and rarely range more than a few metres. By contrast, in fractures recharge is rapid than in the surrounding matrix due to the higher permeability of fractures while in the matrix recharge is less and is delayed. Therefore, in fractured rocks and karst aquifers, water table fluctuations are high and may range up to few tens of meters. This is more true when fracturing is sparse and the rocks are either bare of soil cover or have only a thin overburden. With continuous recharge, an increase in infiltration in fractured rocks is reported which could be explained by the removal of infilling material in fractures (Salve et al. 2008). In the Precambrian rocks of Botswana, recharge varies from 10 mm per year in loamy alluvium to 3 m per year in coarse granular and fractured rocks. Similarly in karst areas in Saudi Arabia, recharge is estimated as 93 mm per year, i.e. 47% of the average annual rainfall (de Vries and Simmers 2002). In mature karst, which have high porosity due to dissolution, recharge rates to the tune of 80% of the precipitation have been reported. As a result of high infiltration in the karst areas of Montenegro, in erstwhile Yugoslavia, the maximum spring discharge is recorded to be about 300 $m^3 s^{-1}$, this being the largest such spring in the world (Scanlon et al. 2002).

20.2.3.1   Recharge from Precipitation (Rp)

The following methods can be used for estimating recharge from precipitation (R_p). It is preferable to use a combination of these methods for this purpose.

Water Table Fluctuations Method

This method is simple and most widely used. It has been applied in areas with different climatic conditions covering areas from tens of square meters to several hundred or thousand square meters (Scanlon et al. 2002). However, it gives only point information about the recharge but can provide dependable values in shallow unconfined aquifers. The magnitude of water level fluctuations would depend on lithology and structural features of the zone of water level fluctuation. Loca-

tion of observation wells is also important. Simulation studies show that all other hydraulic parameters being the same, a well located closer to the discharge zone, such as surface stream, would have a smaller rise in water level than a well located upgradient (Kresic 2007). Schilling (2009) from a study in Iowa, USA, also observed greater recharge in upland areas due to deep water table than in the flood plain areas where water table was at shallow level. As groundwater levels are influenced by other factors also, viz. changes in atmospheric pressure, ET, and pumping etc, necessary corrections in the measured values should be made.

In unconfined aquifers, R_p can be estimated from Eq. 20.3.

$$R_p = S_y h \qquad (20.3)$$

where S_y is the specific yield of the zone of water table fluctuation and h is the weighted average rise in water table. Although, h can be known quite accurately, but large errors in R_p may arise due to unrepresentative value of S_y, especially in fractured rocks. Therefore, recharge estimation from water level fluctuations in fractured rocks may not be dependable. A large variability is reported in the values of S_y, estimated from laboratory and field methods; laboratory values being usually larger than those estimated from field tests (Healy and Cook 2002). S_y can be estimated more accurately from water budget method provided other components of water balance equation are known.

Spring Discharge Method

The springs discharge recession data are also used to estimate recharge as discussed in Sect. 15.5.

Chloride Balance Method

In this method, relative concentration of a stable ion like chloride present in precipitation, groundwater or steady state portion of soil profile and in the dry fallout is measured. Mandel and Shiftan (1981) gave Eq. 20.4 for estimating recharge by salt balance method in well flushed regions where airborne salts are the only source of chloride in groundwater.

$$R_p = P \left(\frac{C_p}{C_g} \right) + \left(\frac{F_d}{C_g} \right) \qquad (20.4)$$

Table 20.1 Rainfall recharge measurements in India Using tritium injection method. (After Rangarajan and Athavale 2000)

Area	Rock type	Rainfall (mm per year)	Recharge		
			Median	Mean (mm per year)	Rain fall (%)
Godavari–Purna basin	Basalt	652	50	56	8.6
Lower Maner basin, A.P.	Sandstone and shale	1250	103	117	9.4
Noyil basin, Tamilnadu	Granite	715	35	69	9.6
Ponnani basin, Tamilnadu	Granite and Gneiss	1320	24	61	4.6

where, R_p = average annual groundwater recharge in m per year, P = average annual rainfall in m per year, C_p = average annual chloride content in rainwater, in $mg\,l^{-1}$, C_g = average chloride (Cl^-) content in groundwater in $mg\,l^{-1}$, F_d = average annual dry fallout of chloride, in $g\,m^{-2}$ per year.

The salt balance method has limitations in the following respect (Mandel and Shiftan 1981): (i) the assumption that the study area is well flushed may not be valid every where, (ii) the airborne chloride data based on limited measurements may not be representative, (iii) estimation of dry fallout component are less satisfactory, and (iv) additional input of chloride from rock weathering and use of fertilizers. In spite of these limitations chloride balance method has provided reliable results from several countries in a variety of hydrogeological environments (Vacher and Ayers 1980; Sharma and Hughes 1985; Sukhija et al. 1988; Gardner et al. 1991; Flint et al. 2002; Abdalla et al. 2009). For example, recharge rates estimated from chloride concentrations range from 0 to 8 mm per year in South Africa, 13 to 100 mm per year in SW Australia, 150 to 800 mm per year in NE Australia (Scanlon et al. 2002), and between 0.1 and 10 mm per year in the volcanic suite of rocks at the Yucca Mountains, Nevada, USA. In the volcanic rocks of the Yucca Mts., the variation in the computed recharge is attributed to differences in the chloride contents of water in fractures and the matrix pore spaces (Flint et al. 2002).

Isotope Techniques

Temperature and several other tracers are being increasingly used for estimating groundwater recharge as well as for understanding the mechanics of recharge. Tritium (^3H) and other environmental isotopes, viz. ^{18}O, ^2H, ^{14}C, ^{32}Si, ^{36}Cl, ^3He and CFCs have been commonly used as tracers and for dating of groundwater (see Sect. 10.5 for details). ^3H, CFCs, and ^3He are used only in areas of relatively shallow (<10 m) water

table and where flow is mainly vertical. An integrated approach using hydrochemical composition and environmental isotope ratios (δ^2H, $\delta^{18}O$ and ^3H) was found useful in determining the recharge due to winter precipitation and glaciated melt-water in the deep groundwater flow system in fractured granites of high altitude Alpine areas (Offerdinger et al. 2004). Recharge estimates from rainfall using tritium in some terrains in India is given in Table 20.1.

Hydrological Balance Method

Equation 20.2 has been commonly used for estimating groundwater recharge (R_p) when all other remaining parameters in the equation are known or computed. For example, this method was used for computing groundwater recharge in the weathered and fractured granodiorite and diorite in the Lee valley of California, USA (Kaehler et al. 1994). This method is however less accurate in arid and semi-arid regions than in humid areas. Further, the recharge term R_p accumulates the errors in all the other terms of the water budget equation (Scanlon and Healy 2002).

Empirical Method

Several empirical equations based on local rainfall characteristics, water level fluctuation and hydrogeological parameters have also been proposed to obtain approximate idea of groundwater recharge (R_p) from precipitation (Karanth 1987). However, these are area-specific and can not be extrapolated to regions of differing hydrogeolgoical setup.

Recharge from Canal Seepage

The seepage losses from canals in alluvial areas can be computed from Eq. 20.5

Fig. 20.1 Processed IRS–LISS II data (NDVI image superimposed over NIR band) showing areas (in *white*) irrigated by large diameter wells in Anantpur district, Central India. Such remote sensing data products alongwith crop inventories are useful for estimating groundwater draft from wells. (After Meijerink et al. 1994)

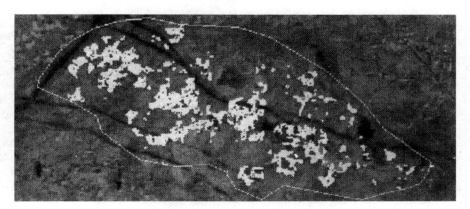

$$\text{Canal losses} \left(m^3 s^{-1} km^{-1} \right) = \frac{C}{200}(B + D)^{\frac{2}{3}}$$

(20.5)

where B and D are the bed-width and depth of the channel respectively in metres, and C is a constant, usually being 1.0 for intermittent running channels and 0.75 for continuously running channels.

Recharge from Irrigated Fields

The magnitude of recharge from irrigated fields depends on source of water (surface or groundwater), method of irrigation, type of crop and nature of soil. When irrigation is from surface water, 35–40% of the applied water recharges the groundwater body while in case of irrigation from groundwater sources, the return flow will be generally between 30% and 35%.

Other Components

Other components of groundwater balance equation, viz. influent and effluent seepage from rivers and flow from/to the adjacent basins can be computed from the Darcy's Law. Leakage across aquifer–aquitard boundaries can be estimated from Eq. 20.6 based on the Darcy's law.

$$Q_v = \frac{K'\Delta hA}{b'}$$

(20.6)

where Q_v is vertical leakage, K' is the vertical hydraulic conductivity of the aquitard, Δh is the head difference between the two aquifers separated by aquitard of thickness b', and A is the plan area of the aquifer through which the vertical leakage occurs. In leaky aquifers, Q_v forms an important component in recharge computations.

Recharge from tanks is estimated by the water balance equation considering surface inflow into the tank as input and outflow through sluices, evaporation and infiltration as output. ET losses are estimated as given in Sect. 1.2.

Groundwater Draft (T_p)

This is mainly through wells. Annual groundwater draft of a well is computed by multiplying its annual working hours and average discharge. The number of working hours can be calculated by the hourly consumption of electrical or diesel energy. The groundwater draft can also be estimated from data on irrigation water requirements of crops in the command area of the well. Remote sensing data can also be used in mapping the irrigated fields (Fig. 20.1), where irrigation is only from groundwater. This information along with cropping calendar and crop-water requirement can be used to estimate the groundwater draft.

20.3 World's Water Resources

Several researchers have attempted assessment of global water resources (e.g. Nace 1971; Shiklomanov 1990, among others). As per the estimates of Shiklomanov (1990), about 96.5% of the total water resources, i.e. 1338 million km^3 occur in the oceans, and only 3.5% i.e. 47 million km^3 occur in the continents (Table 20.2). However, only 5 million km^3 which is about 10% of the total water on the continents, is utilisable as fresh water. The remaining fresh water is stored in glaciers

Table 20.2 Water storage on earth. (After Shiklomanov 1990)

Type of water	Reference area (km² × 10³)	Volume	Share of world reserves (in %)	
			Total water	Fresh water
World oceans	361 300	1338	96.5	–
Total groundwater	134 800	23	1.7	–
Fresh groundwater	134 800	10	0.76	30.1
Soil moisture	82 000	0.017	0.001	0.05
Glaciers and permanent snow pack	16 232	24.0	1.74	68.7
Underground ice in the permafrost zone	21 000	0.300	0.022	0.86
Water reserves in lakes	2058	0.176	0.013	–
Freshwater lakes	1236	0.091	0.007	0.26
Salt-water lakes	822	0.085	0.006	–
Swamps	2683	0.011	0.0008	0.03
Rivers	148 800	0.002	0.0002	0.006
Water in the biosphere	510 000	0.001	0.00007	0.003
Water in atmosphere	510 000	0.012	0.0009	0.04
Total water reserves	510 000	1385	100	–
Fresh water	148 800	35	2.53	100

and polar ice caps and as deep groundwater from where it is difficult to extract. Rivers and fresh water lakes, which are among the main sources of surface water supply, contain only $93 \times 10^3 \, km^3$ of water, which is only 1% of the total amount of fresh groundwater on the earth. Water balance of the continents is given in Table 20.3.

The global estimates do not show regional imbalances in the availability and requirement of water. Due to variations in climatic and hydrogeological conditions, water resources are very irregularly distributed throughout the world, viz. Asia which supports more than half of the world population has only 36% of global water resources, whereas both North America and Australia which have only 8% and <1% of the world's population, have 15% of global water resources (Fig. 20.2). It is estimated that only a few "water-rich" countries enjoy more than 50% of the total world water resources. The water demand also varies due to diverse hydro-economic conditions and population density (Table 20.4).

In the last four decades, the world has seen unprecedented rise in population. In 1960, there were 3 billion people on the planet; in 1987 it rose to 5 billion. The projected increase in world population to 7.2 billion people by the year 2025 will require additional $1000 \, km^3$ of water simply to grow more food. An additional 1.5 billion people should gain access to some form of improved water supply by 2015 (UN 2005).

20.4 Groundwater Resources of India

India has prominently an agricultural economy which makes greater demand of water for irrigation with diverse topography, climate, geology and hydrological conditions. The distribution of available water resources (both surface and groundwater) in the country is highly uneven both in space and time. Rainfall is the main source of groundwater recharge, which

Table 20.3 Water balance of the continents (mm per year). (After Zektser and Loaiciga 1993)

Water balance component	Eurasia	Africa	North America	South America	Australia (without islands)	Earth's land surface
Precipitation	728	686	670	1650	440	834
Evapotranspiration	430	547	383	1065	393	540
Total river run-off	298	139	287	585	47	294
Surface flow	298	91	203	375	40	204
Groundwater contribution	82	48	84	210	7	90

Fig. 20.2 An overview of disparities of water availability with respect to population (in *percentage*) across the continents. (After UN 2005)

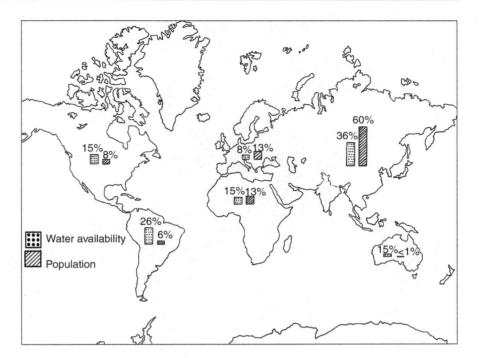

being seasonal, is termed as *monsoon*. The distribution of average annual rainfall on the Indian sub-continent is greatly variable, being maximum (~250 cm) along the Western Ghats (coast) and the NE Himalaya while lowest (<30 cm) is in the western part of the country where arid conditions prevail (Fig. 20.3).

The availability of groundwater is mainly controlled by rainfall distribution and the hydrogeological characteristics of the terrain. The distribution of various groundwater provinces is also shown in Fig. 20.3 and their main hydrogeological characteristics includ-

ing yield potential is outlined in Table 20.5. The vast alluvial plain in the northern part of India forms the most productive region where groundwater has been extensively used for irrigation. Recent studies have indicated the presence of vast reserves of groundwater in the deep alluvial aquifers of the Gangetic plain in North India and fossil groundwater of Mesozoic age in the Lathi aquifers of Mesozoic age in Rajasthan. About 2/3rd of the surface area of the country is covered with hard (crystalline and basaltic) rocks where the availability of groundwater is limited.

Table 20.4 The global water budget by region. (After Dunnivant and Anders 2006)

Region/Country	Total renewable water resources (km³ per year)	Total water withdrawals (m³ per year)	Per capita (m³/person)	Average % of renewable resources	Average % used by agriculture	Average % used by industry
World	43 219	3 414 000	650	–	71	20
Asia (excluding middle east)	11 321	1 516 247	1028	29	79	10
Europe	6590	367 449	503	9	25	48
Middle east and North Africa	518	303 977	754	423	80	5
Sub-saharan Africa	3901	72 556	147	5	66	7
North America	4850	512 440	1720	14	27	58
Central America and Caribbean	1186	105 741	513	9	65	14
South America	12 246	156 948	614	1	76	7
Oceania	1693	16 730	318	1	45	15

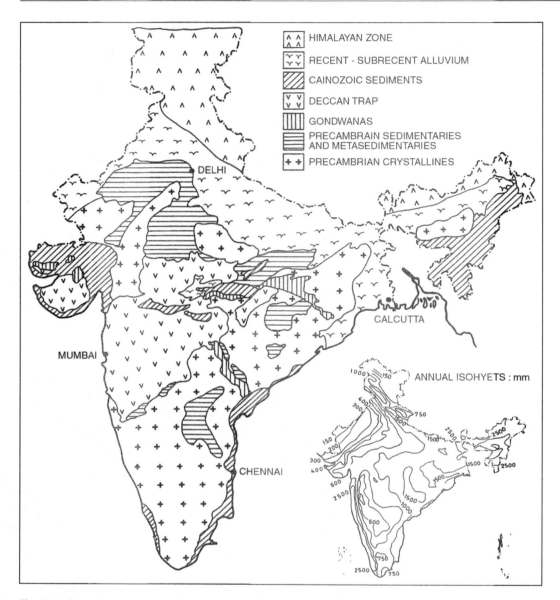

Fig. 20.3 Groundwater provinces of India. (Modified after CGWB 1989)

In the past, several attempts were made to assess the total utilisable water resources of India. Out of a total of 1140 km³ of annual utilisable water, the utilization in 1990 was 572 km³ which comprised 382 km³ from surface water and 190 km³ from groundwater. With increase in population, which crossed 1 billion mark at the end of the last century, and pace of development, there was considerable increase of water requirements for various basic amenities and developmental purposes like drinking and municipal water supplies, irrigation, thermal and hydropower generation, industrial requirement, navigation and maintenance of ecology. In the year 2009, the total utilisation of water was 813 km³ and the projected demand of water for various purposes by 2025 AD is estimated at 1095 km³ (Table 20.6) indicating water scarcity conditions in the next 25–30 years.

Due to increasing population, the per capita water availability in India is also decreasing drastically (Table 20.7).

Table 20.5 Hydrogeological characteristics of major groundwater provinces of India

Groundwater Provinces/ Regions	Location	Lithology (age)	Main features of aquifers	Hydraulic properties T ($m^2 d^{-1}$)	Yield potential
Indo-Gangetic alluvium	Northern India covering an area of more than 1 million km^2	Alluvium with beds of sand, silt and clay with occasional beds of gravel, max. thickness more than 2000 m (Sub-Recent to Recent)	Shalalow aquifers unconfined, deeper ones confined or lf leaky type	$T = 1000–5000$ $S = 10^{-4}–10^{-3}$	0.04–0.11
Cainozoic sedimentary basins	Eastern coast, NE India, and western parts	Unconsolidated to semi-consolidated sandstones, shales and limestones	Deeper aquifer under confined condition; at places flowing wells	$T = 500–5000$ $S = 10^{-5}–10^{-3}$	0.01–0.04
Deccan Traps	Central and western India covering an area of about 5 00 000 km^2	Basaltic lava flows generally flat lying, maximum thickness about 1500 m in the western coast (Upper Cretaceous to Lower Eocene)	Main source of groundwater are (a) Weathered and fractured horizon (b) interflow spaces, (c) intertrappeans (d) vesicular horizons	$T = 10–700$ $S = 10^{-3}–10^{-1}$	0.001–0.03
Gondwana Province	Structurally controlled basins mainly in central India	Semi-consolidated sandstones andshales with coal seams (Carboniferous to Lower Cretaceous)	Shallow aquifers unconfined, deeper ones confined	$T = 50–500$ $S = 10^{-3}–10^{-1}$	0.01–0.10
Precambrian sedimentary basins	Four discrete structural basins (a) Cuddapah, (b) Raipur, (c) Vindhyan, (d) Western Rajasthan	Consolidated sandstones, shales and limestones (Proterozoic)	Intergranular porosity low; fractures are the main source of water in sandstone and shales; solution cavities in limestones	$T = 5–500$ $S = 10^{-3}–10^{-2}$	0.01–0.04
Precambrian crystalline province	It occupies nearly half of the country in Central and South India	Crystalline rocks, viz. granites, gneisses and schists (Precambrian)	Weathered mantle (regolith) is the main source of water supply; fractures and lineaments also facilitate groundwater movement	$T = 5–50$ $S = 10^{-3}–10^{-2}$	0.001–0.005

There is also a large variation in the development of groundwater in the various parts of the country. Vast areas especially those in the drought prone hard rock terrains are either over-exploited or are under critical conditions where groundwater development has exceeded replenishment resulting in the steep decline of water table and in many cases drying of dug wells (Foster and Garduno 2006). The problem is more severe in the hard rock terrains of central and southern parts of the country (Fig. 20.4).

Table 20.6 Present and projected water use in India (in km^3)

Water use	Present (2009 AD)	Projected (2025 AD)
Irrigation	688	910
Domestic	56	73
Industries	12	25
Hydroelectric power	5	15
Other uses	52	72
Total	813	1095

Table 20.7 Decrease in the availability of water over the years

Year	Population in million	Water availability (m^3/year/capita)
1947	400	5000
2000	1000	2000
2025	1390	1500
2050	1600	1000

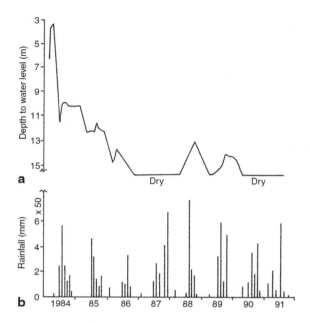

Fig. 20.4 a Hydrograph of an observation well in the granite aquifer in Nalgonda district, South India. **b** Rainfall record note that although rainfall pattern is unchanged, there is a considerable decline in water level. (After Reddy and Raj 1997)

In view of the present scenario of groundwater development, the future emphasis should be on water resources management rather than water resources development. This needs necessary water conservation methods by adopting efficient methods of irrigation, avoiding wasteful use of water and adopting managed aquifer recharge as given in Sect. 20.8.2. The UN has launched the International Decade for Action (2005–2015) "Water for Life" for an integrated approach to the management of the world's water that ensures its sustainable use (www. un.org/waterforlifedecade).

20.5 Groundwater Level Fluctuations

Fluctuations in groundwater levels indicate changes in groundwater storage due to meteorological factors and other environmental impact (Table 20.8). A continuous decline of water levels is observed in areas where there is a greater withdrawal of groundwater than recharge or continuous drought for several years. On the other hand, a rising trend in water levels points towards excessive recharge and less withdrawal. A decreasing trend in water level can cause ground subsidence, especially in unconsolidated formations, while a rising trend is responsible for water logging and water and soil salinization. The main causes of groundwater level fluctuation are the following:

20.5.1 Recharge from Rainfall and Snow-melt

Water levels in wells, both in unconfined and confined aquifers, respond to infiltration due to rainfall and snowmelt. Fluctuations due to snowmelt and rainfall are of seasonal nature. In north Europe and America, highest water levels are observed in late spring and lowest in winter. In India, where the main source of recharge is the *monsoon* rainfall, the highest water levels are observed in September–October and lowest in May–June.

Table 20.8 Common causes of ground water level fluctuations. (Modified after Freeze and Cherry 1979)

Causes of fluctuations	Uncon-fined	Confined	Natural	Man-induced	Short-lived	Diurnal	Seasonal/long-term
Groundwater recharge	✓	✓	✓	–	–	–	✓
Evapotranspiration	✓	–	✓	–	–	✓	–
Tidal effects near oceans	✓	✓	✓	–	–	✓	–
Atmospheric pressure effects	✓	✓	✓	–	–	✓	–
External loading	–	✓	–	✓	✓	–	–
Earthquakes	–	✓	✓	–	✓	–	–
Nuclear explosion	–	✓	–	✓	✓	–	–
Groundwater pumpage	✓	✓	–	✓	–	–	✓
Irrigation and drainage	✓	–	–	✓	–	–	✓
Urbanization	✓	✓	–	✓	–	–	✓

20.5.2 Losses due to Evapotranspiration

In shallow unconfined aquifers, where water table is close to the ground surface, the water level in wells fluctuates in response to losses due to evaporation and transpiration. The loss due to evaporation from water table depends on its depth below the ground surface and the soil structure which controls the capillary rise. For the same depth of water table, losses due to evaporation will be greater in fine grained soils due to higher capillary rise, as compared with coarse grained material. Rate of transpiration also depends on type of vegetation. In shallow water table condition where the plant roots draw water directly from the zone of saturation, the fluctuation is due to direct losses from the water table due to evapotranspiration. Lowering of water table during the day is a result of evapotranspiration and recovery during the night is due to the absence of such losses. These aspects have been discussed in Chap. 1.

20.5.3 Atmospheric Pressure

Changes in atmospheric pressure can produce sizeable fluctuations of water levels in wells tapping confined aquifers. The relationship is inverse, i.e. an increase in atmospheric pressure causes lowering of water levels. This is explained due to the elastic behaviour of confined aquifer (Todd 1980). The barometric efficiency of the aquifer, B is expressed as

$$B = \frac{s_w}{s_p} \qquad (20.7)$$

where s_w is the change in the water level and s_p is the change in barometric pressure expressed in terms of column of water. B is usually in the range of 0.20–0.75.

Wind and cyclones cause reduction in pressure which have similar effect as a vacuum pump causing a sudden reduction of air pressure within the well. This results in quick rise of water level. After the storm passes, the air pressure in the well will rise and the water level will accordingly be lowered.

20.5.4 Ocean Tides

In coastal aquifers and near some lakes and streams, water levels in wells show semidiurnal fluctuations in response to tides. In an unconfined aquifer, the response is due to direct entry of seawater into the aquifer. However, in confined aquifers, which are separated from the ocean water by impervious confining layer, the response in groundwater levels is due to the change in the load on the top of the aquifer due to the weight of the column of ocean water. A rise in the tide will cause additional loading resulting in the compression of the aquifer and thereby, rise in the piezometric surface. This is contrary to the effect and changes in barometric pressure. The ratio of the change in piezometric level, S_w to change in height of tide, S_t is known as the tidal efficiency (C),

$$C = \frac{s_w}{s_t} \qquad (20.8)$$

Equations 20.7 and 20.8 relate the barometric efficiency (B) and the tidal efficiency (C) to the elasticity of a confined aquifer respectively (Ferris et al. 1962).

$$C = \frac{\alpha/\eta\beta}{1 + \alpha/\eta\beta} \qquad (20.9)$$

and

$$B = \frac{1}{1 + \alpha/\eta\beta} \qquad (20.10)$$

where α is the bulk modulus of compression of the solid skeleton of the aquifer, β is the bulk modulus of compression of water and η is the porosity of the aquifer. The sum of the barometric and tidal efficiencies equals unity i.e. $B + C = 1$. The tidal efficiency is a measure of the incompetence of overlying confining layers to resist pressure changes due to loading.

20.5.5 Earthquakes

The amount of water level change is related to the magnitude of the earthquake and its distance from the well (Roeloffs 1998). Generally, wells in limestone aquifers show greater fluctuations as compared with wells in unconsolidated granular material (Ferris et al. 1962). Several mechanisms have been proposed to explain water level response to earthquakes, viz. fracturing of bedrock and earthquake enhanced permeability, consolidation of sediments, and poroelastic strain of aquifers in response to fault displacement. In the near-field, abrupt decrease in water levels is noted, while in the intermediate field changes are more gradual and can persist for days, and at greater distances (the far field), only transient oscillations of the water level have been observed (Roeloffs 1998; Chia et al. 2001; Wang and Chia 2008).

In confined aquifers, an earthquake can lead to small short-lived fluctuations of water levels in wells. These fluctuations are caused by the compression and expansion (dilation) of elastic confined aquifers due to the passage of earthquake waves. In order to adjust to the pressure changes, the water level in the well tapping a confined aquifer first rises and then falls. The amounts of rise and fall of the water level, with respect to the initial position, are approximately the same. However, in cases where earthquake has caused deformation of the aquifer, water levels may not return to the original position.

As the average velocity of earthquake (Rayleigh) waves is approximately $4000\,\mathrm{m\,s^{-1}}$, the effect of earthquake on groundwater levels will be manifested with some time lag at places away from earthquake centres. The large magnitude earthquake can cause widespread fluctuations in water levels. The Alaskan earthquake of March 27, 1964 (magnitude 8.4–8.8 on the Richter scale) produced water level fluctuations all over North America, ranging from about 2.0 to 7.5 m in wells tapping confined aquifers in Pleistocene and late Tertiary strata. None of these wells showed full recovery, suggesting changes in the physical structure of the aquifer and an increase in pore space due to rearrangement of grains or fractures as a result of horizontal extension and/or elastic dilation (Plafker 1969).

Tokunaga (1999) has reported a five fold increase in hydraulic conductivity due to the development of new horizontal fractures as a result of Kobe earthquake in Japan on 17 Feb., 1995. An increase in the HCO_3 content of water was also noted. Studies at other places also show that the changes in hydraulic conductivity, water levels and water quality are usually of heterogeneous nature and are site-specific (Woith et al. 2003).

In connection with the discussion on inter-relationship between earthquakes and groundwater levels here, it may just be mentioned that there are also evidences of triggering of earthquakes due to increase in pore-water pressure as a result of rise of groundwater levels after heavy rainfall and also seepage from reservoirs, indicating a possibility of earthquake prediction based on water level data (Gupta 1992; Heinzl et al. 2006; Chopra et al. 2008).

20.5.6 Underground Nuclear Testing

Underground nuclear tests are carried out in either vertical shafts or horizontal tunnels usually in hard rocks (igneous and metamorphic) in the zone of aeration. In areas, where the water table is shallow, the testing is done below the water table. Abrupt rise in groundwater levels are reported due to detonations of nuclear and also high energy chemical explosions at several sites in China, USA, Canada, Russia (Kazakhstan) and several places in Europe (Matzko 1994). This is explained to be a result of increase in pore water pressure on account of closure of fractures, as the shock waves pass through the dense and low porosity rocks. As the pressure in the explosion cavity is disseminated quickly (within less than 1 min), the rise is followed by a quick drop in water level probably due to the filling of the collapsed cavity. It may take about 10 months for the re-establishment of the original water level due to regional groundwater recharge (Matzko 1994).

Charlie et al. (1994) reported maximum rise of groundwater levels to be 4.8, 0.8, 0.04 and 0.02 m at distances of 650, 2170, 2550 and 3400 m from the detonation site. At Nevada Test Site in Yucca Mts., USA, underground nuclear tests below the water table in welded tuffs, raised water levels by hundreds of metres. The water levels started declining accompanied with land subsidence after the cessation of nuclear testing in 1992 which is attributed to draining of groundwater into the overlying and underlying formations (Halford et al. 2005; Galloway and Hoffmann 2007). The yield of the nuclear test will govern the magnitude of water level fluctuation and the maximum distance up to which such changes will be noticed (Fig. 20.5). This relationship has a potential to determine the yield of underground nuclear detonation from water level data (Charlie et al. 1994).

20.5.7 Urbanization

Increased withdrawal of groundwater for various uses has caused water level decline. For example, in Kol-

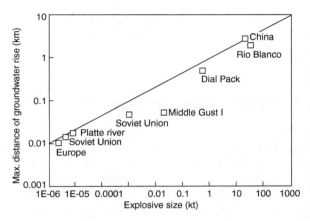

Fig. 20.5 Relation between nuclear explosion size and distance of maximum groundwater rise. (After Charlie et al. 1994, reprinted from Engineering Geology, Vol. 38, 1994, with kind permission of Elsevier Science-NL, Amsterdam, The Netherlands)

kata city, water levels have declined by 4–8 m between 1958 and 1992 due to excessive pumpage (Fig. 20.6). In some cities of north China, over exploitation of groundwater has resulted in a significant (about 60 m)

decline of water levels in several cities in north China. In many other metropolitan cities world over, decline of groundwater levels has resulted in land subsidence (Sect. 20.6). In coastal areas, decline of groundwater level due to excessive pumping has also led to seawater intrusion as described in Sect. 20.7.

In UK, during the late 1800s and early 1900s, due to the heavy withdrawal of groundwater from the Cretaceous Chalk formations in the London Basin and Permo-Triassic sandstones in Birmingham, Liverpool, Manchester and Nothingham areas, the groundwater levels declined by some tens of metres below ground level over large areas. This resulted in the decline of water levels in earlier flowing wells which then ceased to flow. However, in later years, a reverse trend of rise of water levels at the rate of about (1 m per year) is reported (Fig. 20.7). This is attributed to the reduction in groundwater abstraction and additional recharge from sewers, lawns and leaking pipe lines. The rise in groundwater levels has posed problems of flooding of basements, rail tunnels and stability of other underground structures (Wilkinson and Brassington

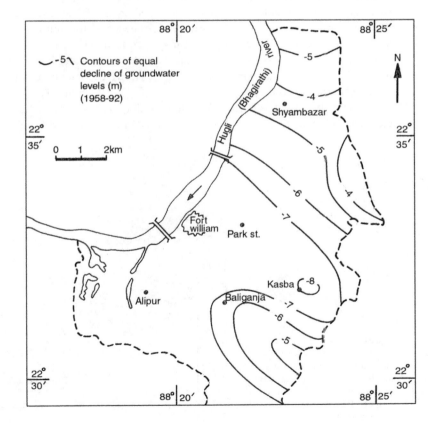

Fig. 20.6 Decline of groundwater levels between 1958 and 1992 in the alluvial aquifer below the city of Kolkata, India. (Redrawn after Goswami 1995)

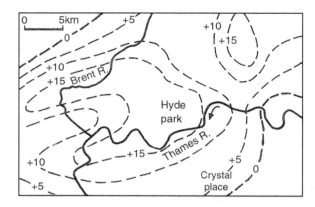

Fig. 20.7 Rise of groundwater levels in the Chalk aquifer below London (1985–1990) (all *contours* in metres). (After Marsh and Davies 1983)

1991). Similarly, in Louisville, Kentucky, USA, increasing groundwater levels have caused problem of structural settlement, drainage to basement floors and disruption of underground tunnels (Hamill and Bell 1986). Preventive measures would include control of groundwater levels by pumping or other remedial engineering works. The disposal of pumped water is likely to pose a problem in heavily urbanised areas like London. It would also be necessary to take necessary care to avoid over-pumping, as it may then lead to problems of subsidence and deterioration in water quality. Therefore, long-term management of groundwater resources is necessary.

In some areas such as Perth, Western Australia, the clearance of vegetation (mainly phreatophytes) for the

development of new housing areas, reduced evapotranspiration losses resulting in water logging (Lerner 1997).

20.5.8 Perennial Irrigation

Excessive seepage losses from unlined irrigation channels and irrigated fields have caused conspicuous rise in groundwater levels, particularly in semi-arid parts of the world, viz. in Pakistan, India, China, Egypt, and USA. This has resulted in water logging and soil salinization thereby reducing the soil fertility. In Pakistan, the rise in water table in Rechna *doab* (interfluve), in between rivers Ravi and Chenab was as much as 28 m from pre-irrigation period (1900) to post irrigation period (1960) (Fig. 20.8). In India, an estimated 23 million hectares is affected by water logging and salinity. In China, water logging is prevalent in the alluvial plain of the Yellow river in the northern part of the country which has semi-arid climate (Chen and Cai 1995). In Egypt, about 8400 km² or 30% of the total arable land is salt effected or water logged (Shata 1982). In the San Joaquin valley of California in USA, where the climate is semi-arid, excessive irrigation caused rise in groundwater levels and thereby higher concentration of salts. Selenium concentration in groundwater reached more than 100 μg l⁻¹ which was beyond the drinking water standards for human beings (<10 μg l⁻¹) and also for aquatic life (<5 μg l⁻¹) resulting in ecological disaster

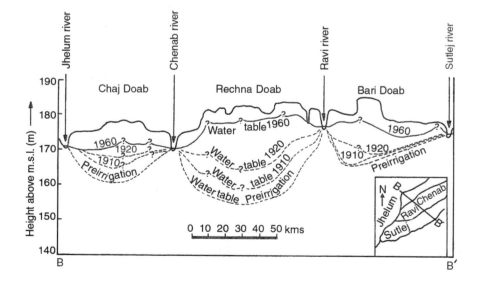

Fig. 20.8 Profile across the northern part of Pakistan showing rise in groundwater levels due to irrigation in various interfluves (*doabs*) of the Punjab alluvial plain. (After Bennett et al. 1967)

(Kehew 2001). Reclamation of water logged areas by surface and subsurface drainage is one of the top priorities in several countries, world over, for augmenting agricultural production.

20.6 Land Subsidence

Over-abstraction of groundwater is a common cause of land subsidence. Land subsidence is also a result of excessive withdrawal of other fluids, e.g. oil, gas and geothermal fluids. A comprehensive account of land subsidence due to withdrawal of fluids is given in Poland (1984) and ASCE (2009). In carbonate rocks, land subsidence can also be a result of rock dissolution and can be a serious geologic hazard causing damage to highways, pipelines, dams and reservoirs.

A fluid saturated porous medium or rock fractures will deform because of either changes in the external load or changes in the internal pore pressure. Heavy withdrawal of groundwater and other fluids causes reduction in pore water pressure which increases the effective stress resulting in the contraction of geological media and thereby an increase in the effective weight of the overburden. The relationship is given by the Eq. 20.11.

$$\sigma' = \sigma - p_w \qquad (20.11)$$

where σ is the total stress, σ' is the effective stress and p_w is the pore water pressure. Therefore, a reduction in pore water pressure results in the compaction of both the aquifer and the interbedded clay layers. The total potential compaction of the confining layer (clay) is much greater as compared with that of sand, as clay has greater compressibility. However, due to lower hydraulic conductivity of clay, the rate of compaction of clay is slower than that of sand aquifer. Therefore, although the compaction of sand aquifer is immediate and elastic but that of clay is much greater but inelastic. The amount of compaction also depends on the clay mineralogy, being about three times greater in montmorillonite as compared with kaolinite while illite has intermediate values (Waltham 1989). There is also a time lag between the withdrawal of fluid and the start of subsidence.

Geodetic techniques (spirit levelling and GPS) have been widely used for measuring land subsidence.

Recently InSAR (SAR interferometry) technique has also been used for monitoring land subsidence and aquifer compaction (Fig. 20.9) (Galloway and Hoffmann 2007).

In general, there exists a fair degree of linearity between the rate of water level decline and the rate of subsidence. The relation between volume of groundwater extraction and rate of subsidence from confined

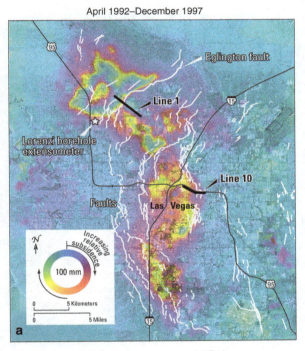

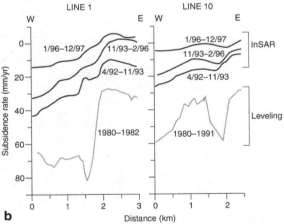

Fig. 20.9 InSAR-derived subsidence, Las Vegas Valley, Nevada. **a** April 1992 to December 1997. **b** Subsidence rates compared to historic levelling at lines 1 and 10 for given periods (month per year). (Figs. a, b: after Galloway and Hoffman 2007)

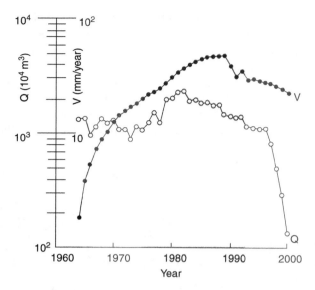

Fig. 20.10 Comparison between the volume of groundwater withdrawn (*Q*) and the land subsidence (*V*) from 1964 to 2000 in Jiaxin City, Zhejiang, China. (After Li, et al. 2006)

aquifer in Jiaxin city, China is shown in Fig. 20.10. It may be noted that although, rate of groundwater withdrawal was considerably reduced after 1996, the rate of subsidence did not respond so quickly, indicating irreversible inelastic deformation of sand and gravel aquifers.

The main effects of land subsidence are: (a) inundation of coastal areas, (b) modification of surface drain-

age gradients, (c) damage to well casings, (d) activation of faults and opening of fissures, and (e) damage to buildings and other surface structures.

There are several examples of land subsidence due to heavy withdrawal of groundwater world over, for example in Santa Clara Valley in California, USA, Tokyo and Osaka in Japan, Bangkok in Thailand, Venice in Italy, and Mexico City in Mexico. In Mexico city, which is located on alluvial fan and lacustrine deposits, heavy withdrawal of groundwater caused land subsidence of about 9 m. In China, over-pumping of groundwater in Beijing, Tianjing and several other cities in the North China Plain has caused significant drawdown which at places exceeded 50 m resulting in land subsidence. In Beijing, the pumping of groundwater has increased by more than 50% during the last 15 years causing a head decline of 20 m and the land subsidence of 59 cm. In Shanghai, the sinking rate is reported to be as much as 24 cm annually during 1949–1965, the maximum subsidence being 2.63 m (Xuanjiang 1994). Large scale extraction of groundwater in the Xian city area, which was the ancient capital of China, has caused the lowering of the groundwater level resulting in ground subsidence (Fig. 20.11). Subsidence has further caused surface rupturing and development of ground fissures damaging buildings and other man made structures (Lee et al. 1996). In Venice, Italy, subsidence due to heavy withdrawal of groundwater and rise in sea level due

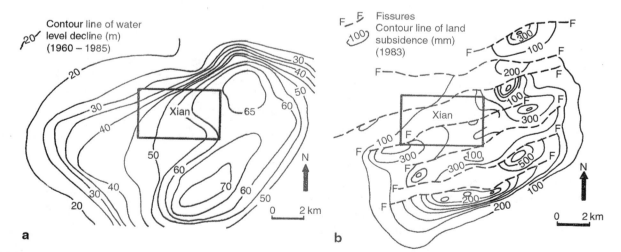

Fig. 20.11 **a** Decline of groundwater levels (1960–1985) in the area around the city of Xian, PR China. **b** Subsidence of the land surface in the same area. (After Lee et al. 1996, reprinted from

Engineering Geology with kind permission of Elsevier Science-NL, Amsterdam, The Netherlands)

to climate change has resulted in inundation of large areas by seawater. Large scale pumping of groundwater to enable open-cast mining of lignite etc. has also resulted in land subsidence (Briechie and Voigt 1980; Alfoldi 1980).

Only a few examples of subsidence in fractured crystalline rocks are available. Zangeri et al. (2003) have reported vertical settlement of about 12 cm in fractured crystalline rocks several hundred meters above a highway tunnel in central Switzerland. Modelling studies indicate that drainage through horizontal joints due to construction of tunnel was mainly responsible for this subsidence.

Extraction of oil and gas can also cause surface subsidence. One of the earliest example is from the sinking of Gaillard Peninsula in Texas, USA where the ground surface sank below the surface of the bay in 1920 as a result of extraction of oil. In the Wilmington Oil Field in California, USA, a subsidence of almost 9 m in the period 1927–1967 is reported. Oil and gas extraction at the Ekofisk Field in the North Sea has caused a subsidence of sea bottom of more than 10 m since the 1970s which necessitated injection of sea water to facilitate drilling operations (Rutqvist and Stephansson 2003). In Japan, heavy withdrawal of gas and brine from poorly consolidated sediments in the coastal city of Niigata, caused sinking of a part of the city below the sea level in 1961 (Marsden and Davis 1967). Subsidence from geothermal fields is described in Sect. 18.5.

The best remedial measures for countering land subsidence will be to reduce groundwater withdrawal and to adopt artificial recharge. Imported water supply will also reduce groundwater demand and thereby reduce subsidence. These measures have proved effective in many areas. For example, in Mexico city the rate of subsidence was reduced from 900 mm per year to around 30 mm per year (Waltham 1989). In China, reduction of groundwater abstraction and artificial recharge reduced annual subsidence in Shanghai from 110 mm in 1961 to 1.5 mm in 1992. Similarly, it was reduced in Tianjin from 86 mm in 1985 to 17 mm in 1991 (Chen and Cai 1995). Artificial recharge through wells proved to be successful in the Wilmington oil field, California, where the subsiding areas reduced from 58 to 8 km^2 and the land surface rebounded as much as 0.3 m. Re-pressuring of aquifer system by artificial recharge also helped in oil production. Artificial recharge by sea water is suggested for the mitigation of subsidence in Venice area, though there are doubts about the use of sea water as it may deteriorate the water quality.

20.7 Fresh–Seawater Relationship in Coastal Areas

20.7.1 Subsurface Groundwater Discharge (SGD) to the Sea

Usually, the groundwater gradient is seaward due to which large amount of fresh groundwater is discharged into the sea as (a) submarine springs, and (b) leakage through the aquifers and semi-confining layers exposed in the sea bottom (Fig. 20.12). SGD is also an important source of nutrients and heavy trace metals to the sea resulting in geochemical reactions which are partly responsible for the formation of Fe–Mn nodules. According to Zektser, about 1300 million t per year of salts are discharged by groundwater into the oceans and seas worldwide (Viventsova and Voronov 2005). Therefore, freshwater outflow to the sea (SGD) is an important resource which should be assessed properly. The following methods have been used for measuring SGD (Simmons 2005; Taniguchi et al. 2007; Kazemi 2008; Lee and Cho 2008):

Calculations based on Darcy's law
Resistivity measurements
Direct measurements with seepage meters
Tracer measurements (e.g. ^{222}Rn, ^{226}Ra, CH_4)
Variable density flow meters
CTD (conductivity, temperature, depth) profiling
Remote Sensing Thermal Imaging Techniques

Moore (1996), based on ^{226}Ra content of coastal waters and sediments in the southeastern part of the Atlantic coast in USA, estimated groundwater flux to oceans to be about 40% of that of river water flux. Based on hydrogeological data, Zektser and Dzhamalov (1988) and Zektsar and Loaiciga (1993) computed the total discharge of groundwater to oceans to be 2400 km^3 per year which includes 1485 km^3 per year from the continents and 915 km^3 per year from islands. Recent studies by Russian and Japanese scientists indicate that the SGD is about 10% of total terrestrial flow to the oceans (Sanford 2007; Taniguchi 2005; Kazemi 2008).

Fig. 20.12 Freshwater flux (surface and subsurface) into seawater along the interface of the Po river plains and the Adriatic sea. Note that the interface extends for a width of about 100 km. (HCMM night thermal IR image dated 25 Nov. 1978; Courtesy NASA)

0   50 km

In order to avoid wasteful discharge of fresh water into the sea, it is necessary to identify such areas for optimum groundwater development without disturbing the fresh–seawater balance.

20.7.2   Seawater Intrusion

Overexploitation of groundwater in coastal aquifers may result in reversal of hydraulic gradient causing sea-water intrusion. The relation between fresh water and seawater in coastal aquifers was first given independently by Ghyben and Herzberg (Eq. 20.12).

$$h_s = \frac{\rho_f}{\rho_s - \rho_f} h_f \qquad (20.12)$$

where, h_f=height of water table above m.s.1, h_s=depth of the fresh–seawater interface below the m.s.1, ρ_s=density of seawater, and ρ_f=density of freshwater (Fig. 20.13). If we assume $\rho_f=1.00\,\mathrm{g\,cm}^{-3}$ and $\rho_s=1.025\,\mathrm{g\,cm}^{-3}$, then h_s will be 40 times of h_f. It means that at any distance from the sea, the depth of a stationary interface below sea level is 40 times the height of the fresh water table above sea level.

As the seawater is not stationary and is moving inland, the head in seawater body in the aquifer will be less than the mean sea level. For such a non-equilibrium condition, where both the saline and fresh waters are in motion, the depth of interface is given by Lusczynski (1961) as

$$h_s = \frac{\rho_f}{\rho_s - \rho_f} h_f \frac{\rho_f}{\rho_s - \rho_f} H_s \qquad (20.13)$$

where H_s is the head difference in seawater wedge above msl (Fig. 20.14). Therefore, the depth of fresh–seawater interface will be greater than that given by the

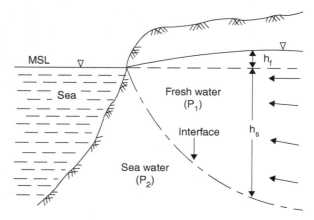

Fig. 20.13 Idealized sketch showing relationship between fresh and saline water in an unconfined coastal aquifer

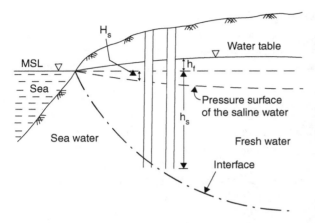

Fig. 20.14 Nonequilibrium conditions in fresh and saline water in an unconfined coastal aquifer

Ghyben–Herzberg relationship. In confined aquifer (Fig. 20.15), the equilibrium position of the intruded seawater wedge is given by

$$L = \frac{1}{2} \frac{\rho_s - \rho_f}{\rho_f} \frac{Kb^2}{q} \qquad (20.14)$$

where L is the length of the seawater wedge, K is the hydraulic conductivity, b is the saturated thickness of the aquifer, and q is the seaward fresh water flow per unit width of the ocean front. Equation 20.14 indicates that the length of the seawater wedge (L) has the following characteristics: (a) It is directly proportional to the hydraulic conductivity of the aquifer; therefore, the most permeable formations, like some of the volcanic rocks, karst and fractured aquifers, will be more affected by seawater intrusion; (b) It is directly related

to the square of the aquifer thickness, b i.e. in thick aquifers and old river channels, filled with gravel and other coarse material, the salt water wedge will penetrate to longer distance; (c) The wedge length, L is inversely proportional to the fresh water flow to the sea, q, therefore greater penetration of wedge is expected in areas with less recharge such as in arid and semi-arid climates or where groundwater basins are of smaller size; heavy withdrawal of groundwater will also reduce freshwater outflow, q to the sea, thereby increasing the length of the seawater wedge. Thus, by pumping from a coastal confined aquifer, the freshwater flow, q will decline and the seawater wedge will advance inland. This phenomenon is known as seawater encroachment or intrusion. In unconfined aquifer also, due to the lowering of water table as a result of overpumping, seawater intrusion can take place (Fig. 20.16).

In multilayered coastal aquifers exposed on the ocean floor, the fresh–seawater wedge will develop in each aquifer separately depending on the piezometric head in each aquifer, its thickness, hydraulic conductivity, and recharge. The other important factor which will influence the length of the wedge is the vertical leakage through the adjacent aquifers. Usually the interface in the lower aquifer is relatively longer than the upper one (Fig. 20.17). Changes in sea level due to melting of glaciers can result in seawater intrusion at different levels in a multi-layered aquifer as is revealed from a study in the coastal parts of Israel. In this area, isotopic data have shown various horizons of seawater intrusion after the last glaciation about 18 000 years ago (Yechieli et al. 2009).

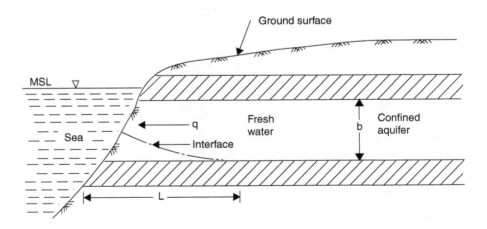

Fig. 20.15 Seawater wedge in a confined coastal aquifer

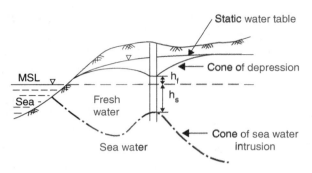

Fig. 20.16 Formation of cone of sea-water intrusion due to pumping in an unconfined coastal aquifer

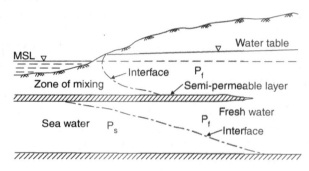

Fig. 20.17 Presence of a semi-permeable layer leading to development of two fresh–seawater interfaces in a coastal aquifer. (After Reilly 1990)

As the fresh water and seawater are miscible fluids, the zone of contact is of transition type. It is characterized by the phenomena of molecular diffusion and especially hydrodynamic dispersion. The thickness of the interface depends on fluctuations of tides and

groundwater level, aquifer permeability, climate and recharge and discharge from wells (Table 20.9). Sea-water intrusion in coastal areas can be monitored from groundwater level and water quality including isotopic data.

India has a coast line of about 7000 km. Sea water intrusion is reported both from the eastern and western coast. The aquifer geometry and fresh–seawater relationship at Digha in the eastern coast of India is illustrated in Fig. 20.18. Electrical resistivity survey and drilling indicated two saline water zones separated by a fresh water aquifer due to clay layers within a depth of 170 m. The shallow saline water (20–30 m) appears to be of connate origin as indicated by hydro-chemical and isotopic data (Singhal 1963; Shivanna et al. 1993). $\delta^{34}S$ data indicate reduction of initial sea water (Sukhija et al. 2002). Further south, in the state of Tamilnadu, the sea water–fresh water interface has moved 2–9 km inland since 1969 due to over-exploitation of groundwater. Based on hydrogeochemistry, radiocarbon and organic biomarkers, Sukhija et al. (1996), have indicated the presence of both palaeomarine and modern sea water in this area. An example of sea water intrusion due to reversal of water table gradient in the coastal parts of Saurashtra in the west is given in Fig. 20.19 which depicts the hydrogeological conditions and changes in the salinity of groundwater.

Increased salinity of groundwater due to seawater intrusion may induce dissolution of certain constituents which were otherwise stable under fresh water environments. Studies in the coastal alluvial aquifers in south Italy indicate high concentration of Hg above

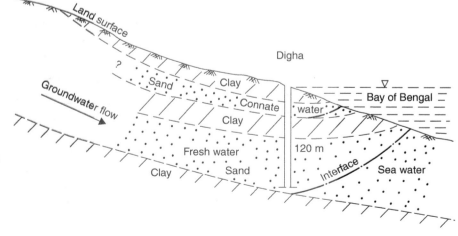

Fig. 20.18 Schematic hydrogeological profile across the coastal town of Digha, West Bengal, India, showing fresh–sea-water relationship. (Simplified after Goswami 1995)

Table 20.9 Factors affecting thickness of freshwater /sea-water interface in coastal aquifers

Factors	Nature of transition zone	Example
Tidal activity		
Negligible	Narrow	Mediterranean area
Intense	Thick	Atlantic area
Hydraulic conductivity		
High	Thick	Hawaii
Low	Narrow	Islands
Climate		
Semi-arid	Thick	Israel
Humid	Narrow	Eastern coast of India

the desired limit in drinking water $(1\,\mu\,gl^{-1})$ due to dissolution of Hg minerals in the aquifer by the intruded sea water (Grassi and Netki 2000).

Sea water encroachment in coastal aquifers can be controlled by reducing groundwater withdrawal and adopting measures such as subsurface barriers (UNESCO 1987). In oceanic islands, radial wells (Maui type), tunnels, galleries and shafts are more suitable depending on topography and rock type to avoid seawater intrusion (Sect. 14.5).

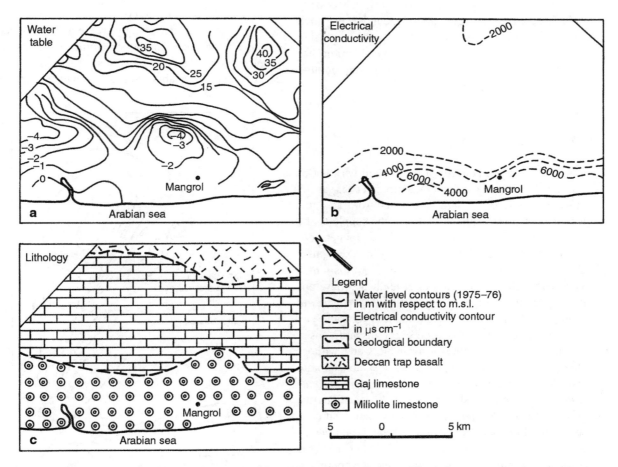

Fig. 20.19 a Water table, and **b** electrical conductivity (EC) contours of groundwater in the coastal part of Saurashtra, Western India (1975–1976); **c** geological map of the area. Note that in the coastal belt, occupied by Miliolite limestone, the water table gradients are inland, due to overpumping, resulting in an increase in EC caused by seawater encroachment. (Redrawn after Desai et al. 1979)

20.8 Groundwater Management and Artificial Recharge

20.8.1 Terminology

Management means planning and implementation of a system to achieve specified goals, without violating technical and non-technical constraints. In this regard, in groundwater management, aquifer is considered as a system that has to be managed (Bear and Zhou 2007). It involves decisions regarding pumping and artificial recharge of the aquifer, including water quality and also its relation with the surface water system.

Groundwater resources should be utilised and managed in a planned manner to avoid any undesirable effects, viz. ground subsidence, seawater intrusion, uneconomic pumping lift etc. It is therefore necessary to define the desirable quantity of groundwater which can be withdrawn safely from a basin without causing any adverse effect. In view of this, several approaches are suggested viz. basin yield, safe yield, water mining and groundwater sustainability.

20.8.1.1 Basin Yield

One of the earliest concept of basin yield was that of *safe yield* which was defined by Meinzer (1923) as the practicable rate of withdrawing water from an aquifer for human use without depleting the supply to the extent that withdrawal at this rate is no longer economically feasible. This would depend on the aquifer characteristics. Other factors that affect estimation of safe yield include economics of groundwater development, and protection of water quality etc. Withdrawal in excess of safe yield is regarded as overdraft. The concept of safe yield has been lately discarded as it considers a fixed quantity of available water without changing hydrological regime and the socioeconomic condition. Some hydrologists prefer the term *groundwater sustainability* and *renewability*. In general terms, *sustainable development* is the management and conservation of the natural resource base and the orientation of technological and institutional change in such a manner as to ensure the attainment and the continued satisfaction of human needs for the present and future generations (Kresic 2007). Accordingly, sustainable yield of an aquifer is the annual quantity of water that

can be withdrawn from the aquifer, year after year, as a constant volume or as one that varies according to some rule, without violating specified constraints, especially that the source will be preserved (quality and quantity) forever (Bear and Zhou 2007). Alternative yield concepts have also been suggested (Todd 1980; ASCE 1987). For example, *perennial* yield is the practicable rate at which groundwater can be withdrawn perennially under specified operating condition without producing any adverse effects. The maximum perennial yield is the maximum quantity of water which can be made available if all possible methods of recharging the basin are adopted. *Mining yield* is the quantity of extractable water which exceeds the recharge. It is non-renewable resource which is like a mineral or petroleum deposit. This can be practiced in areas where groundwater storage is otherwise of no value. This can be used to contribute to the economic development of a water scarcity area, e.g. in the Sahara desert. The degree of mining of groundwater in some countries is excessive as illustrated in Figs. 20.20 and 20.21 from Saudi Arabia and Qatar (Lloyd 1991). Min-

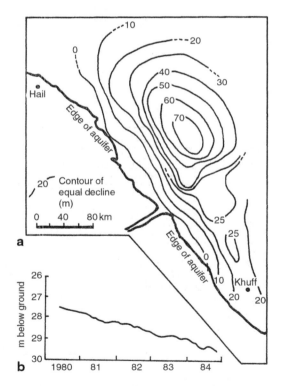

Fig. 20.20 Contours of equal lowering of water-levels in North-Central Saudi Arabia due to groundwater mining. (After Lloyd 1991)

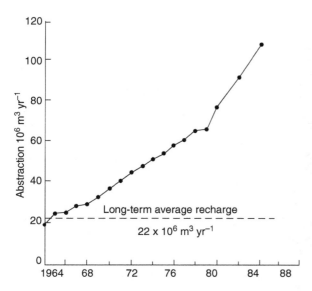

Fig. 20.21 Over-abstraction of groundwater in comparison to long-term average recharge in Qatar. (After Lloyd 1991)

ing yield can be reduced by economic use of water and by finding alternate sources of water supply.

20.8.1.2 Dynamic and Static Resources

Groundwater resources can also be classified as static and dynamic. The *static resource* is the amount of groundwater available in the aquifer below the zone of water level fluctuation. The *dynamic resource* can be defined as the amount of groundwater available in the zone of water level fluctuation. The useable groundwater resource is essentially a dynamic resource which is recharged annually or periodically. The static storage should be utilized during drought periods and the volume to be used from static storage is to be determined considering the need and economics of its exploitation. The groundwater resources of an area can be augmented by adopting artificial recharge, interbasin water transfer and water conservation measures.

20.8.2 Artificial Groundwater Recharge/ Management of Aquifer Recharge

The term *management of aquifer recharge* (MAR) is now being increasingly used for *artificial recharge* to imply additional input of water underground besides natural infiltration (Dillon 2005; BGS 2006). The main purpose of MAR is to augment the groundwater resource. This also helps in checking pollution migration, seawater intrusion in coastal aquifers and land subsidence as mentioned earlier in this chapter. It is also important in groundwater management as it provides natural storage for water for use during dry period. Artificial recharge has an additional benefit of lowering the salinity and temperature of groundwater which helps the industries where water is used for cooling purposes.

The main source of water for artificial groundwater recharge is the storm runoff and river water. Sewage and waste water is also being increasingly used for artificial recharge after necessary treatment. In India, excess monsoon water is used for this purpose.

Techniques of MAR

The various techniques of MAR (Fig. 20.22) are site-specific and the choice of a particular method depends on the nature of aquifers, their hydraulic conductivity and storativity, and quality of recharging water. Surface spreading methods are used in unconfined aquifers where the strata are permeable down to the zone of saturation, and water table is deep to provide adequate storage. Flood plains, alluvial fans and certain glacial deposits like eskers and moraines are suitable for recharge by spreading methods. *Infiltration ponds* are also constructed to recharge unconfined aquifer. These are made by excavation or by building dykes or levees e.g. in the Rhine valley, Germany. A groundwater mound is formed due to infiltration (Fig. 20.22). The shape and size of the mound depends upon basin size and shape, recharge rate and duration, depth of water table and aquifer parameters. The infiltration rate is initially high but usually decreases with time due to accumulation of suspended material, colloidal swelling, and by the growth of algae and bacteria. In order to maintain higher infiltration rates, the basin should be dried periodically to disrupt the growth of algae and bacteria. It also helps in maintaining higher rates of infiltration with the development of shrinkage cracks.

Percolation Tanks

These are a type of infiltration basins built in the ephemeral river bed by providing a low earthen dam

Fig. 20.22 Schematic of types of management of aquifer recharge. (After Dillon 2005)

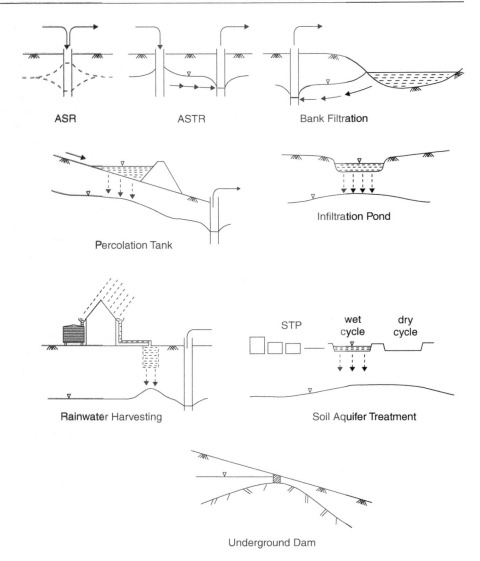

to store rain waters (Fig. 20.22). At places vertical shafts are also constructed in the percolation tanks to enhance recharge. Recharge by percolation (infiltration) tanks is an ancient practice of water conservation in the hard rock formations the world over. The capacity of a percolation tank usually ranges between 0.14 and 3.0 million m^3. Studies in crystalline rocks of south India indicate that the rate of infiltration from percolation tanks varies from 0.5 to 1.5 m d^{-1}; initial rate of infiltration is high but it decreases with time due to silting. In case of non-perennial tanks, the recharge rates are comparatively higher. This is due to the removal of silt by farmers during dry season which is used as manure. The recharge estimates from percolation tanks in Deccan basalts and granitic gneisses

in India are reported to range from 30% to 60% of impounded water depending on bedrock and climatic conditions. At some sites in the crystalline rock terrain in south India, groundwater levels rose by 3–7 m in the vicinity of recharge ponds (BGS 2006). δ^{18}O data from a percolation tank in the basaltic rocks of central India show that the tank water contribution to the nearby wells is 50% and thereafter decreases with distance (Sukhija et al. 2002).

Recharge Through Wells

Recharge through wells is practiced in areas where recharge due to surface spreading and basins is not

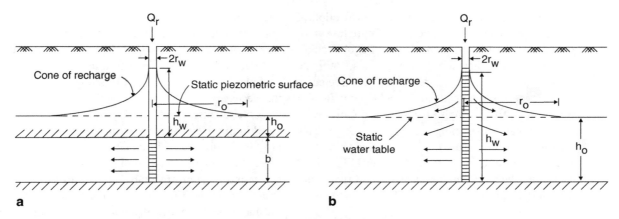

Fig. 20.23 Cone of recharge developed in: **a** confined aquifer, **b** unconfined aquifer

feasible, e.g. in confined aquifers due to the presence of extensive clay layers overlying the aquifer or in metropolitan areas where land values are too high. A cone of recharge is formed which is the reverse of cone of depression developed due to pumping (Fig. 20.23). Under steady state condition, the equation for rate of recharge, Q_r in a confined aquifer can be written as

$$Q_r = \frac{2\pi K b (h_w - h_0)}{\ln(r_0/r_w)} \qquad (20.15)$$

and for an unconfined aquifer,

$$Q_r = \frac{\pi K \left(h_w^2 - h_0^2\right)}{\ln(r_0/r_w)} \qquad (20.16)$$

where, Q_w is the rate of recharge, and the remaining terms are indicated in Fig. 20.23.

As in the case of surface spreading, the recharge rates through wells are initially high but decrease with time unless well rehabilitation and well development methods like surging and pumping are adopted at intervals. Recharge rates from wells vary from 200 to about 2000 $m^3 d^{-1}$ and in exceptional cases, the recharge rates are as high as 51 000 $m^3 d^{-1}$, especially in limestones (Todd 1980).

Recharge through wells has two types of systems: (a) Aquifer storage and recovery (ASR), and (b) Aquifer storage transfer and recovery (ASTR) (Fig. 20.22). In ASR system, during periods of low demand, storm-water or pre-treated wastewater is injected into a well for storage which is recovered from the same well during the periods of greater demand. It is essential that the recharge water is pre-treated for the removal

of suspended material and undesirable chemical constituents to avoid clogging of well and contamination of groundwater depending on its intended use. There are several examples of storage of treated waste water in the shallow Floridan limestone in USA and in Holocene sand dunes in The Netherlands. However, difference in the quality of native water and the injected water has caused increase in As and Ca concentrations due to leaching of aquifer material (Kresic 2007). Barnett et al. (2000) have also described some case studies of aquifer recharge and storage using storm-water and pre-treated sewage effluent in deep limestone and fractured quartzite in the arid regions of South Australia. In the coastal plain of Israel, seasonal storage is created by recharging through wells, with winter rainfall and the stored water is extracted in summer from the same wells. Here imported water from the Jordan river is also used for recharge purposes. Cooling water is used for recharging Chalk aquifer through wells in England. Recharge wells are also extensively used to control sea-water intrusion by creating a fresh water ridge as in Los Angeles and Orange Counties in USA (Todd 1980).

In ASTR, water is injected into a well and is recovered from another well for improving the water quality (Fig. 20.22).

Bank Infiltration (Induced Recharge)

If a well is located adjacent to a stream or lake, the withdrawal of groundwater from the well causes flow from the surface water body to groundwater reservoir (Fig. 20.22). It is effective in cases where there is hydraulic continuity between the aquifer and river bed. Induced recharge also causes self purification of water and regulates its temperature. Withdrawal of groundwater along the river banks will not only augument the groundwater supply but can also mitigate floods. Sometimes collector wells are constructed in the flood plains of river beds for inducing large flow of water from rivers, as along Mississipi and Ohio rivers in USA. In many developing countries also, large scale groundwater withdrawal by induced recharge is being carried out, e.g. in Sudan from the river Nile (Lloyd 1991), from the river Yamuna in India, and river Denube in Serbia (former Yugoslavia) (Sect. 17.1.4).

Underground Dams

The purpose of an underground dam is to check the outflow of groundwater from a sub-basin thereby raising the groundwater storage on the upstream side (Fig. 20.22). Subsurface dams are feasible in narrow and gently sloping valleys where the bedrock occurs at a shallow depth, overlain by valley-fill deposit of 4–8 m thickness. The underground dam consists of an impervious wall built of impervious material like clay, bricks and concrete. Tar felt, resin, polythene sheets and bitumen are also used. The structure usually extends to a depth of 1–3 m below the land surface. Such groundwater dams were also constructed in the ancient times. There are examples of such structures built in Roman times from the islands of Sardinia in Italy and from Tunisia. In recent years, they have gained popularity in many other countries e.g. India, southern and eastern Africa, Brazil and Japan (Sinha and Sharma 1990).

Roof Top Rain Water Harvesting

This method is useful in urban areas where the availability of land for recharge is limited. In this method monsoon runoff from the roof top is diverted into a well provided with sand and gravel filter (Fig. 20.22). The recharged water is pumped for different uses depending on its quality.

Soil Aquifer Treatment (SAT)

In this method (Fig. 20.22) waste water after necessary treatment is applied to shallow basins in porous and permeable formations to remove organic pollutants, nutrients and pathogens as it passes through the unsaturated zone. The infiltered water is recovered by wells for irrigation and industrial uses after residence at some down-gradient point. It is important to assess the necessary residence time underground to ensure the water quality. The feasibility of SAT has been indicated in the river bed of river Sabarmati, near Ahmedabad in Gujarat state of India.

Hydrogeologists have to play an important role for determining the most suitable method of MAR which should be adopted in a given situation for improving the quality and quantity of water supplies. They should also evaluate the possibilities of its adverse effects viz. water logging, changes in water quality, and slope instability etc. Readers may refer to Dillon (2005) and visit IAH website http://www.iah.org/recharge for more information and initiatives on MAR.

Other Methods

Other methods of water conservation include afforestation, evaporation control, reduction in canal losses, and efficient methods of irrigation, viz. drip and sprinkler methods. Desalination of seawater can also solve to some extent the problems of drinking water supply. The various possible methods of water management should be planned in advance so that the feasibility of different alternatives can be assessed by pilot studies. Other aspects which should be considered in the management process include social aspects, environmental impact and economic limitations etc. People's participation in water management programme leads to greater success in their implementation.

20.8.3 Interbasin Water Transfer

A river basin is usually regarded as a basic unit for water resources planning and development. This is based on the concept that each basin has its own characteristics and that development in the upstream part of the basin will affect the downstream regime, both qualitatively and quantitatively. Therefore, there should be necessary coordination between water resources development in different parts of a basin. Further, for the proper utilization of water in a basin, and to meet the requirements in other adjacent basins within the country or in the adjoining countries, interbasin water transfer is advocated. It not only supplies additional surface water but also augments groundwater resources by recharge.

Interbasin water transfer is not a new concept. It was also practiced in the past in several countries. For example, interbasin water transfer projects were implemented in the past in former USSR, China, USA, Mexico and India. A notable scheme of long distance transfer of water in the former USSR is from the Irtysh–Karganda scheme in the central Khazakstan where a 450 km long canal with a maximum capacity of 75 $m^3 s^{-1}$ was built.

In China, interbasin water transfer programmes were implemented as far back as 200 BC. The northern part of China is deficient in water while the south is water rich. The average water resources in the south are more than four times greater than in the north. In view of the water scarcity in the North China Plain in which the cities of Beijing and Tianjing are located, water levels have declined by more than 40 meters in the last 40 years. Therefore, a very ambitious South to North river diversion project involving the construction of three man-made rivers was started in the year 2002 to meet the great demand of water for irrigation and industrial uses by diverting about 20×10^9–$30 \times 10^9 m^3$ of water (Zhengzhou et al. 1996; Liu et al. 2008). It is also reported that China is also planning to dam the river Brahmaputra and divert 200 billion cubic meters of water annually by tunnels to its arid regions in the north–east. However the environmental and ecological impact in these areas is a matter of much concern.

In India, interbasin water transfer has been in practice for the last more than 600 years when in the Moghul times, Western Yamuna Canal and the Agra Canal were built for transferring surplus waters from the Himalayan rivers to other parts in the Gangetic plains. In the 1980's, the Govt. of India mooted the National Perspective Plan for linking various rivers in the country. The National Water Development Agency (NWDA) was set up to carry out the feasibility study but the progress has been very slow. In recent years, a few interbasin transfer programmes have been implemented such as Rajasthan canal, Beas–Sutlej Link Project and Telugu Ganga (IWRS 1996). Proposals for long distance transfer of waters are also under investigations. For example, the Ganga–Cauvery link envisages diversion of surplus waters (1680 $m^3 s^{-1}$) of river Ganga in the north to water

scarcity hard rock areas in the south over a distance of more than 2600 km and over a head of 550 m.

20.9 Transboundary Aquifers

Transboudary aquifers, also known as multinational or shared aquifers, are present in different parts of the world extending from hundreds to thousands of square kilometres. Some of these viz. the Guarani Aquifer System in South America, having an overall area of 1.2 million km^2, can provide huge quantities of groundwater for human use for the next 20 years (Puri and El Naser 2003). Another example of a vast transboundary aquifer is the Nubian sandstone of Palaeozoic to Mesozoic ages in Egypt, Libya, Sudan and Chad in Northern Africa containing about 542 180 km^3 of water capable of supplying water for several tens of decades. In the Indian subcontinent, the Gangetic alluvium of Quaternary age in the northern part, extending from Pakistan in the west to Bangladesh on the east, over a distance of more than 2500 km forms a huge reservoir of groundwater. The concept of transboundary aquifer may also include aquifers of low transmissivity (viz. fractured rock and karst aquifers) as cross-border cooperation of adjoining countries will help in the sustainable cooperative utilisation of water resources (Cobbing et al. 2008).

In view of the importance of sustainable development of transbouhdary aquifers, the International Association of Hydrogeologists (IAH) established a commission (Transboundary Aquifer Resources Management, TARM) and later an International Shared Aquifer Resources Management (ISARM) initiative. A map titled "Groundwater Resources of the World: Transboundary Aquifer Systems" was published in 2006 (Cobbing et al. 2008). A global inventory initiated by the UNESCO has so far documented 90 transboundary aquifers in Western Europe and 60 each in the Americas and Africa. The use, sound management and sustainable development of transboudary aquifers is of importance which has caused several inter-governmental conflicts and need of treaties for sharing waters as between USA and Mexico (Puri and El Naser 2003). The sharing of groundwater between Israel and Palestine from the Western Aquifer Basin has been also a matter of great concern in the peace process between these two countries. This is mainly due to varying hydrogeological conditions and stage of groundwater development on the two sides of the political boundary (MacDoland et al. 2009).

In the last 50 years, approximately 200 treaties have been signed concerning transboundary basins. The UN Intl. Law Commission has taken the initiative to frame a Law on the Use of Treansboundary Aquifers for the African continent. The United Nations has recently adopted the resolution on the law of transboundary aquifers (IAH 2009).

For the management of transboundary aquifers, it is of importance to determine the natural recharge and discharge zones, aquifer delineation, hydraulic parameters, present and planned groundwater development zones, water quality, and pollution vulnerability from human activities. The recharge pattern may also change due to climate change especially where snow packs and glaciers are the source of recharge. It is well known that as a result of climate change over the last 10000 years in regions of North Africa, recharge is substantially reduced since the last pluvial climate. Mathematical modelling has indicated that continued extraction of groundwater in the south-western part of Egypt will result in the spread of cone of depression close to the Sudanese border.

20.10   Impact of Climate Change

The abnormal amount of certain gases viz. CO_2 and CH_4, commonly known as greenhouse gases, cause entrapment of heat from the sun resulting in perceptible changes in the atmosphere and cause increase in temperatures globally. This is known as greenhouse effect or 'greenhouse surprises'. The release of these gases is principally due to human activities.

The Intergovernmental Panel on climate changes (IPCC) has made the following predictions for global climate during this century:

(a) The global surface temperature is predicted to increase by 1.5–4.5°C.
(b) On a regional scale, both increases and decreases in precipitation are expected.
(c) Ice caps and glaciers will retreat and sea ice and polar snow cover will have a decreasing trend.
(d) Greater evapotranspiration in some areas and greater precipitation in others.

The global average sea level is predicted to rise by about 1 m, with regional variation, submerging low-lying island countries and coastal communities around the world viz. Bangladesh, India and elsewhere. Estimates show that for a 1.5 m rise in sea level, 16% of Bangladesh could be inundated.

Rise in sea level, will not only submerge low-lying coastal areas, but will also increase the groundwater salinity due to landward migration of freshwater–sea water interface. There is also likelihood of coastal freshwater springs to turn saline. The climate change will therefore have a significant impact on hydrologic cycle and in turn on water resources.

If the rise in temperature is not limited/controlled, the planet earth will face great havoc. This will include not only flooding of coastal areas but will also cause crop failures, epidemics, severe water scarcity and increase in natural disasters, viz., floods and droughts. In a recent report, the UN has identified 46 countries with 2.7 billion people where climate change and inter-related crises will create "high risk of violent conflict".

Historical and archeological records from Holocene period i.e. the last 10000 years show that the arid areas of the world became drier and desertified by the global warming which led to severe socio-economic crises, resulting in abandonment of urban and agricultural settlements (Issar 2008).

Although much research is focused on the impact of climate change on surface water resources, due attention is also required for studying its effect on groundwater recharge and water quality. The changes in precipitation and evapotranspiration will have a significant impact on groundwater recharge and hence on water availability for various uses. The effect of climate change on groundwater depends on the climate scenario used in the simulation studies (Jyrkama and Sykes 2007). Changes in precipitation are likely to influence the groundwater quality also. Studies in United States show that the total dissolved solids in groundwater generally increases from east to west as a result of varying precipitation and evapotranspiration. Therefore, increase in precipitation should improve the water quality in future. However, due to greater recharge, the aquifer will have greater vulnerability from surface contaminants (Jyrkama and Sykes 2007). Increased recharge, due to climate change, is also likely to increase pore pressure resulting in higher hydraulic conductivities (Reichard and Leap 1998). It will also influence the solute transport characteristics of fractured rocks (Leap and Mai 1992).

There are many evidences of the effect of climate changes on water resources and hydrogeological characteristics of rock formations, in the geological past (Reichard and Leap 1998). Isostatic response due to the unloading of ice sheets in the Holocene period in Fennoscandian countries in Europe, is believed to have caused a rebound effect on the underlying crystalline rocks, resulting in opening of joints, especially sheet joints, and thereby increasing permeabilities (Henriksen 2003). Recent studies in the northern hemisphere also indicate a rise of ground surface due to the melting of glaciers resulting in fall in sea levels. The management of water resources will get more complicated due to uncertainties in the impact of climate on water resources. Therefore, the water users and managers have to adopt integrated and adaptive management approaches based on monitoring and performance evaluation (Sophocleous 2004).

Land use changes due to potential increases in urbanization and greater withdrawal of groundwater to meet the requirements of increasing population in developing countries are likely to have a higher impact on the hydrogeological processes including recharge than the effect of climate change.

20.11 Groundwater Governance

Groundwater has come to be a vehicle for much of rural and industrial development over the last century. It generates employment and livelihood worldwide. However, being treated as a replenishable gift of nature, it is not really valued, monitored and much less governed or managed, so much so that it is on the verge of endangered "water species" threatened from indiscriminate use and quality degradation. Groundwater merits urgent concern for protection and management. It is necessary that appropriate groundwater management policies with legal support be devised to suite local socio-economic conditions. This includes issues like water rights, water pricing, and stakeholders' participation in groundwater management (Romani et al. 2006).

20.11.1 Water Rights

Water has unique features that make it difficult to regulate through laws enacted for primarily land assets. Water is mobile, naturally occurring, has variable supply with respect to time and place, and has multiple uses such as in agriculture, fisheries, hydroelectric generation, industrial etc. Some usages of water are consumptive (e.g. agriculture), some are non-consumptive but need diversion and storage of water (e.g. hydroelectric plants) and still some others are non-consumptive and do not require any diversion (e.g. navigation, boating). Also water is a basic necessity of life. These features and differences render enacting and implementation of water laws difficult. (for details see: http://en.wikipedia.org/wiki/water_law/)

Nature and source of water rights vary in different countries and societies. The ownership rights can be land-based, user-based, or based on ownership of water bodies. (for details see: http://en.wikipedia.org/wiki/water_right/)

(a) *Land-based or riparian rights* are based on the ownership of land: Riparian law states that the owner of the land below which the groundwater occurs, or the bank along which water flows have rights to the water. It is a type of property right. However, over-exploitation of groundwater by a land-owner could affect water-levels in adjoining areas creating disputes in water use.

(b) *Use-based rights* are based on hierarchy of use: The law states that land ownership is not essential, as long as water users have legal access to the water source; the first user has the strongest rights, followed by successively subsequent users.

(c) *Water-rights based on ownership of water bodies:* In some countries such as Finland, water bodies are privately owned. However, this is not the case of most EU countries.

In the US, for example, both riparian water rights and use-based water rights exist in different states, though public trust doctrine is also being recognized more and more. Most countries, including EU, now apply the principle of *water solidarity*, which attempts to reconcile water rights through compromise solutions based on public interest, need and utility. Thus, groundwater is declared as a 'public good' under government guardianship.

20.11.2   Water Pricing

The earlier norms in many countries often used only the area of irrigation or power capacity of the pumping device for water charges and the charges were generally too low, as water has been treated as a naturally occurring replenishable resource and a nature's gift. This did not provide any motivation for its efficient use and resulted in indiscriminate use and wastage of water. It is important to provide incentives to conserve water and encourage responsible use. Therefore, some of the measures being adopted by water utilities are:

(a) Charge higher rates during dry season
(b) Charge higher rates during drought
(c) Differential pricing—low rates initially, and increasing higher rates for larger volumetric amounts of water drawn.

(for details see: http://www.waterencyclopedia.com/)

20.11.3   Groundwater Legislation

In view of the concerns of water scarcity and pollution, legislation has been widely enacted to vest all water resources in the state. Groundwater is declared as a 'public good' with state as the trustee. The intention is to regulate groundwater development and to constrain activities that might adversely affect groundwater quality and quantity. Initially, this was piecemeal but is now being increasingly integrated to include both surface water and groundwater. This permits integrated resource planning at aquifer or river basin level.

A modern groundwater legislation ought to consider the following (for more details, please see http://siteresources.worldbank.org/EXTWAT/Resources/4602122-210186362590/GWM_Briefing_1.pdf):

- Groundwater abstraction and use rights
- Wastewater discharge licensing with 'polluter-pays-principle'
- Penalties for non-compliance
- Controlling well construction activities
- Resource planning at aquifer-level
- Groundwater monitoring and budgeting
- Groundwater artificial recharge
- Conjunctive use of surface water and groundwater

Groundwater legislation has to formulate an administrative set-up for implementation at various levels. This is possible only with political willingness and public awareness. The participation of stakeholders in groundwater management is very important. Experience has shown that involvement of local people who are the real beneficiaries is necessary in the successful implementation of any programme of water management, such as artificial recharge or rainwater harvesting. This is based on the principle of water solidarity.

Further, as groundwater is not limited by political or administrative boundaries, international/national cooperation is necessary for its optimum utilization by the concerned countries and states.

Summary

The assessment and management of groundwater resources is important for its rational utilization. As groundwater is a replenishable resource, there should be a proper balance between recharge and utilization. In this context, preparation of water budget which involves estimation of various components of recharge and discharge for a basin is important. Water level fluctuation method is most commonly used for recharge estimation. Overexploitation of groundwater can lead to several problems, viz. land subsidence, seawater intrusion in coastal aquifers, increasing cost of groundwater withdrawal etc. In areas where these problems have arisen, artificial recharge (managed aquifer recharge) is very useful. Inter-basin water transfer is also practiced in several countries to transport water from water surplus to water deficient areas. However the environmental impact of interbasin water transfer should be assessed in advance.

For the management of groundwater resources, necessary laws depending on socio-economic conditions need to be enacted. Peoples participation at the local level is important. Management of trans-boundary aquifers is also important to avoid any political conflicts between the adjoining states. The possible impact of climate change should also be considered in future planning of groundwater and surface water resources.

Further Reading

Delleur JW (ed) (2007) The Handbook of Groundwater Engineering, 2nd ed., CRC Press, Boca Raton, FL.

IAH (2002) Theme Issue: Groundwater Recharge (eds Scanlon BR, Cook PG) Hydrogeol. J. 10.

IAH (2010) Saltwater and freshwater interactions of coastal aquifers: Theme Issue. Hydrogeol. J. 18(1).

Lloyd JW (1999) Water resources of hard rock aquifers in arid and semi-arid zones. Studies and Reports in Hydrology 58, UNESCO, Paris.

Appendix

English-Metric Unit Conversion Table (from Ground Water, Vol. 33, No. 5, p. 873, reprinted by permission of the Journal of Ground Water). To convert A by C, to convert B to A, divide B by C.

A	B	C	A	B	C
Length			*Hydraulic conductivity/Permeability*		
Inch	metre	2.540E-2	gal/day/ft^2	cm/sec	4.716E-5
Feet	Metre	.3048	gal/day/ft^2	ft/day	0.1337
Yard	metre	.9144	gal/day/ft^2	metre/day	4.075E-2
Mile	kilometre	1.609	gal (UK)/day/ft^2	Metre/day	4.893E-2
Inch	centimetre	2.540			
Area					
sq inch	sq centimetre	6.452	darcy	ft/day	2.433
sq feet	sq metre	9.290E-2	darcy	metre/day	0.7416
sq yard	sq metre	.8361			
sq mile	sq kilometre	2.590	*Transmissivity*		
Acre	sq kilometre	4.047E-3	gal/day/ft	sq metre/day	1.242E-2
			gal (UK)/day/ft	sq metre/day	1.492E-2
			sq ft/sec	sq metre/day	8.027E3
			sq ft/day	sq metre/	9.290E-2
Volume					
cu feet	cu metre	2.832E-2			
cu inch	cu centimetre	1.639E-1			
			Force and pressure		
			pound (f)	newton	4.448
gallon	litre	3.785	pounds/sq. in	pascal	6.895E3
gallon (UK)	litre	4.546	lb/sq./ft	pascal	4.788E1
barrel (petr)	litre	1.590E-2	atmosphere	pascal	1.013E5
acre-feet	cu metre	1.234E-3	psi	kg/cm^2	7.031E2
million gal	cu metre	3.785E-3	ft of H$_2$O (4°C)	Psi	0.4335
gallon (UK)	gallon (US)	1.200			
Velocity and gradient					
feet/sec.	metre/sec	0.3048	*Temperature*		
mile/hour	metre/sec	0.4470	Fahrenheit	Celsius	5 (F-32)/9
feet/mile	metre/km	0.1894	Celsius	Fahrenheit	1.8 (C)+32
Flow rate			Kelvin	Celsius	K-273.2
gal/min	litre/sec	6.309E-2			
gal/min	cu metre/day	5.300			

B. B. S. Singhal, R. P. Gupta, *Applied Hydrogeology of Fractured Rocks*,
DOI 10.1007/978-90-481-8799-7, © Springer Science+Business Media B.V. 2010

A	B	C	A	B	C
gal (UK)/min	litre/sec	7.577E-2			
10^6 gal/day	cu metre/day	3.785E-3			
cu ft/sec (cfs)	litre/sec	2.832E-1			
acre-feet/day	litre/sec	1.458E-1			

Note: 1) The (E) notation indicates exponentiation: $2.540E-2 = 2.540.10^{-2}$. 2) Unless otherwise noted, all gallons are U.S. gallons. 3) The darcy is unit of permeability (k), not of hydraulic conductivity (K). 4) A Newton (force) = $kg.m/s^2$; A pascal (pressure) = $kg/m.s^2$; each in a unit in Sl. 5) Under 'Temperature', entries are formulae, not multipliers.

References

Abdalla OAE (2009) Groundwater recharge/discharge in semi-arid regions interpreted from isotope and chloride considerations in north White Nile Rift, Sudan. Hydrogeol. J. 17(3): 679–92.

Abelin H, Birgersson L (1985) Migration experiments in the Stripa mine, design and instrumentation, in Design and Instrumentation of in-situ Experiments in Underground Laboratories for Radioactive Waste Disposal (eds Come B, Johnston P, Muller A), A.A. Balkema, Rotterdam, pp. 180–90.

Adyalkar PG (1983) Groundwater survey and exploration techniques in karstic regions in Seminar on Assessment, Development and Management of Ground Water Res., Central Ground Water Board, Govt. of India, New Delhi, pp. 45–56.

Adyalkar PG, Mani VVS (1972) An attempt at estimating the transmissibilities of trappean aquifers from specific capacity values. J. Hydrol. 17: 237–41.

Adyalkar PG, Mani VVS (1974) Application of groundwater hydraulics to a basaltic water-table aquifer. J. Hydrol. 21: 211–18.

Adyalkar PG, Rao SS (1979) Hydrodynamic method of assessing groundwater recharge by precipitation in Deccan trap terrain—A case study. J. Geol. Soc. India 20(3): 134–37.

Aggarwal PK, Gat J, Froehlich KFO (eds) (2005) Isotopes in the water cycle. Past, present and future of a developing science. Springer, Dordrecht.

Ahmed F, Andrawis AS, Hagaz YA (1984) Landsat model for groundwater exploration in the Nuba Mountains, Sudan, Adv. Space Res., 4(11): 123-31.

Ahmed S, Jayakumar R, Salih A (eds) (2007) Groundwater dynamics in hard rock aquifers. Capital Publ. Co., New Delhi, p. 251.

Ahn JS (1988) Environment isotope-aided studies on river water and groundwater interaction in the region of Seoul and Taelgu, in Proc. Seminar on Isotope Applications in Hydrology in Asia and the Pacific, IAEA, Beijing, pp. 105–40.

Al-Bassam AM, Awad HD, Al-Alawi JA (1997) Durov plot: A computer program for processing and plotting hydrochemical data. Ground Water 35(2): 362–67.

Alfirevic S (1966) Hydrogeological investigations of submarine springs in the Adriatic. IAH Mem. Belgrade, 6: 225–64.

Alfoldi L (1980) Changes in the physical conditions of aquifers due to withdrawal of large volume of groundwater, in Studies and Reports in Hydrology, UNESCO, Paris, 28, pp. 242–53.

Aller L, Bennett T, Lehr JH, Petty RJ (1987) DRASTIC: A standardised system for evaluating groundwater pollution using hydrogeological settings. United States Environmental Protection Agency, Washington, Report 600/2–85/018.

Al-Saafin AK, Bader AK, Shehata W et al (1990) Groundwater recharge in an arid karst area in Saudi Arbia, in Selected Papers on Hydrogeology (eds Simpson ES, Sharp JM Jr.), IAH/Verlag Heinz Heise, Hannover, Vol. 1: 29–41.

Alsdorf DE, Lettenmaier D, Vorosmarty C (2003) The need for global, satellite-based observations of terrestrial surface waters. EOS Transactions, American Geophysical Union 84(29): 269.

Alumbaugh DL, Morrison HF (1993) Theoretical and practical considerations for crosswell electromagnetic tomography using a cylindrical geometry. Geophysics 60: 846–70.

Amit H, Lyakhovsky V, Katz A et al (2002) Interpretation of spring recession curves. Ground Water 40(5): 543–51.

Anderson MP, Woessner WW (1992) Applied Groundwater Modeling. Academic Press, Inc. San Diego, California.

Angadi KS (1986) Hydrogeological Studies of the Lower Parts of Ghataprabha Basin, Belgaum and Bijapur Districts of Karnataka, India, Ph.D. Thesis, University of Roorkee.

Angino EE, Wilbur CK (1990) The retention of Cd, Pb and Zn by fractured shale units: simulation of reactivity in the unconsolidated zone, in Selected papers on Hydrogeology, Vol. 1 (eds Simpson ES, Sharp JM Jr.), Verlag Heinz Heise, Haunover, pp. 169–77.

Aquilina L, Ladouche B, Doerfliger M et al (2003) Deep water circulation, residence time and chemistry in a karst complex. Ground Water 41(6): 790–805.

Arihood LD (1994) Hydrogeology and paths of flow in the carbonate bedrock aquifer, Northwestern Indiana. Water Res. Bull. 30(2): 205.

Aronoff S (1989) Geographic Information Systems: A Management Perspective. WDL Publications, Ottawa, Canada, p. 294.

American Society of Civil Engineers ASCE (1987) Groundwater management, 3rd ed., Manuals and Report on Engineering Practice, No. 40. Am. Soc of Civil Engrs. New York, p. 263.

American Society of Civil Engineers ASCE (2009) Guidelines for Investigation of Land Subsidence due to Fluid Withdrawal. ASCE/EWRI Standards (Under preparation).

Atekwana EA, Werkena DD (2006) Biogeophysics: The effects of microbial processes on geophysical properties of the shallow subsurface, in Applied Hydrogeophysics (eds Vereecken H et al) Springer, Dordrecht, The Netherlands.

B. B. S. Singhal, R. P. Gupta, *Applied Hydrogeology of Fractured Rocks,*
DOI 10.1007/978-90-481-8799-7, © Springer Science+Business Media B.V. 2010

Athavale RN (1985) Nuclear tracer techniques for measurement of natural recharge in hard rock terrains, in Proc. Intl. Workshop on Rural Hydrogeology and Hydraulics of Fissured Basement Zones (Singhal BBS ed.), University of Roorkee, pp. 71–80.

Athavale RN, Singh VS, Subrahmanyam K (1983) A constant discharge device for aquifer tests on large diameter wells. Ground Water 21, pp. 752–55.

Athavale RN, Rangarajan R, Rao SM (1992) Determination of the flow direction of geothermal waters at Manikaran using the borehole tracer technique. J. Geol. Soc. India 39(4): 329–37.

Atkinson (1977) Diffuse flow and conduit flow. J. Hydrol. 35: 93–110.

Atkinson SF, Thomlinson JR (1994) An examination of groundwater pollution potential through GIS modelling. Technical Papers of the ASPRS/ACSM Annual Meeting, Bethesda, Maryland, pp. 71–80.

Atkinson TC (1977) Diffuse flow and conduit flow in limestone terrain in the Mendip Hills, Somerset (Great Britain). J. Hydrol. 35: 93–110.

Avci CB (1992) Parameter estimation for step-drawdown tests. Ground Water 30(3): 338–42.

Ayers RS (1975) Quality of water for irrigation. Proc. Irrig. Drain. Div. Speciality Conf. ASCE. Logan, Utah, pp. 24–56.

Azeemuddin M, Roegiers JC, Suri P et al (1995) Stress-dependent permeability measurement of rocks in triaxial cell, in Rock Mechanics (eds Daeman and Schultz), A.A. Balkema, Rotterdam, pp. 645–50.

Bacchus ST, Archibald DD, Brook GA, Britton KO, Haines BL, Rathbun SL, Madden M (2003) Near-infrared spectroscopy of a hydroecological indicator: New tool for determining sustainable yield for Floridan aquifer system. Hydrol. Process. 17(9): 1785–809.

Back W (1961) Techniques for mapping hydrochemical facies. US Geol. Surv. Prof. Paper 424-D.

Back W (1966) Hydrochemical facies and groundwater flow patterns in northern part of Atlantic Coastal Plain, US Geol. Surv. Prof. Paper 498-A, p. 42.

Back W, Hanshaw B (1965) Chemical geohydrology, in Advances in Hydroscience (Chow VT ed.), Vol. 2, Academic Press, New York, pp. 49–109.

Bae Dae-soek, Koh Yong-k, Kim KS (2003) The hydrogeological and hydrochemical conditions of deep groundwater system in Yuseong granite area, Korea. IAH Symp. On Groundwater in Fractured Rocks, Prague.

Bajpai RK (2004) Recent advances in the geological disposal of nuclear wastes worldwide and Indian scenario. J. Geol. Soc. India 63(3): 354–56.

Bakalowicz M (2005) Karst groundwater: a challenge for new resources. Hydrogeol. J. 13(1): 148–160.

Baklliwal PC, Grover AK (1988) Signatures and migration of Saraswati river in Thar desert, Western India, Records GSI 116(3–8): 77–86.

Ball TK, Cameron DB, Colman TB, Roberts PD (1991) Behaviour of radon in the geological environment a review. Q. J. Engg. Geol. 24: 169–82.

Balmes CP (1994) The geochemistry of the Mahanagdong sector, Tongonan geothermal field, Philippines. Report No. 2, Geothermal Training Programme, The United Nations University, Reykjavik, Iceland, pp. 31–52.

Banks D, Solbjorg ML, Rohr-Torp E (1992) Permeability of fracture zones in a Precambrian granite. Q. J. Engg. Geol. 25: 377–88.

Barenblatt GE, Zheltov IP, Kochina IN (1960) Basic concepts in the theory of homogeneous liquids in fisssured rocks. J. Appl. Mathematics and Mech. USSR 24(5): 1286–303.

Baria R (ed) (1990) Proc. of the Intl. Conference on Hot Dry Rock Geothermal Energy, Camborne School of Mines, Redruth, Cornwall, UK.

Barker JA (1991) On the discrete kernel method for simulating pumping tests in large diameter wells. J. Hydrol. 124: 177–83.

Barker RD, White CC, Houston JFT (1992) Borehole siting in an African accelerated drought relief, in The Hydrogeology of Crystalline Basement Aquifers in Africa (eds White EP, Burgess WG), Geol. Soc. Spl. Publ. 66, pp. 183–201.

Barnett SR, Howles SR, Martin RR et al (2000) Aquifer storage and recharge: innovation in water resources management. Aust. J. Earth Sci. 47: 13–19.

Barraclough D, Gardner CMK, Wellings SR, Cooper JD (1994) A tracer investigation into the importance of fissure flow in the unsaturated zone of the British Upper Chalk. J. Hydrol. 156: 459–69.

Barton NR (1973) Review of new shear strength criterion for joints. Eng. Geol., 8: 287–332.

Barton CC, Larsen E, Page WR, Howard TM (1987) Characterising fractured rock for fluid-flow, geochemical, and paleostress modelling: Methods and preliminary results from Yucca Mountain, Nevada. U.S. Geol. Survey Open-File Report, 87, p. 36.

Bear J (1979) Hydraulics of Groundwater. McGraw-Hill Int. Book Co., New York, p. 567.

Bear J (1993) Modeling flow and contaminant transport in fractured rocks, in Flow and Contaminant Transport in Fractured Rocks (eds Bear J, Tsang CF, de Marsily G), Academic Press, San Diego, pp. 1–36.

Bear J, Verruijt A (1987) Modeling Groundwater Flow and Pollution. Kluwer Academic Publishers, Hingham, Massachusetts.

Bear J, Zhou Q (2007) Sea water intrusion into coastal aquifers, in The Handbook of Groundwater Engineering, 2nd ed (Delleur JW ed.), CRC Press, Taylor and Francis Group, Boca Raton, FL, pp. 12–1 to 12–29.

Becker MW (2006) Potential for satellite remote sensing of ground water. Ground Water 44(2): 306–18.

Beckie R, Harvey CF (2002) What does a slug test measure: an investigation of instrument response and the effects of heterogeneity. Water Resources Res. 38(12): 1290.

Bedient PB, Rifai H, Nowell CJ (1993) Groundwater Contamination, Transport and Remediation, Prentice Hall Intl. Ltd., London.

Beeson S, Jones CRC, (1988) The combined EM/VES geophysical method for siting boreholes. Ground Water 26(1): 54–63.

Belward AS (1991) Spectral characteristics of vegetation, soil and water in the visible, near infrared and middleinfrared wavelengths, in Remote Sensing Applications and GIS for Resources Management in Developing Countries (eds Belward AS, Valenzuela CR), Kluwer Acad. Publ. Eurocourses, pp. 31–45.

Bennett GD, Rehman A, Sheikh IA, Ali S (1967) Analysis of aquifer tests in the Punjab region of West Pakistan. US Geol. Surv. Water Supply Paper 1608G. Washington, DC, p. 56.

Bentley HW, Fhillips FM, Davis SN et al (1986) Chlorine 36 dating of very old groundwater 1. The great Artesian Basin, Australia. Water Resources Res. 22(13): 1991–2001.

Beres M Jr., Haeni FP (1991) Application of ground-penetrating-radar methods in hydrogeologic studies.

Berglund S, Kautsky U, Lindborg T et al (2009) Integration of hydrogeological and ecological modeling for the assessment of a nuclear waste repository. Hydrogeol. J. 17(1), doi:10.1007/s10040-008-0399-6.

Bernard R, Taconet O, Vidal-Madjar D, Thony JL, Vauclin M, Chapoton A, Wattrelot F, Lebrun A (1984), Comparison of three in-situ surface soil measurements and application to C- band scatterometer calibration. IEEE Trans Geosci Remote Sens GE- 22(4): 388–394.

Berner EK, Berner RA (1987) The Global Water Cycle, Prentice Hall Inc., NJ, p. 383.

Bernstein A, Adar EM, Yakirevich A et al (2007) Dilution tests in low permeability fractured aquifer: matrix diffusion effect. Ground Water 42(3): 363–73.

British Geological Survey BGS (2006) Managed Aquifer Recharge: an assessment of its role and effectiveness in watershed management. Final Report for DFID KAR Project R8169, Augmenting Groundwater Recharge by Artificial Recharge - AGRAR, British Geol Survey, Nottingham, UK, p. 69.

Bhar AK (1996) Mathematical Modelling of Spring Flow, Ph.D. Thesis, University of Roorkee, p. 177.

Bhattacharya PK, Patra HP (1968) Direct current Geoelectric Sounding Principles and Interpretation Elsevier, Amsterdam.

Billings MP (1972) Structural Geology, 3rd ed., Prentice Hall of India, New Delhi, p. 606.

Birkett CM (1998) Contribution of the TOPEX NASA radar altimeter to the global monitoring of large rivers and wetlands. Water Resources Res. 34(5): 1223–239.

Birkett CM, Mertes LAK, Dunne T, Costa MH, Jasinski MJ (2002) Surface water dynamics in the Amazon Basin: Application of satellite radar altimetry. J. Geophys. Res.—Atmospheres 107, no. D20, doi:10.1029/2001JD000609.

Black JH (1987) Flow and flow mechanisms in crystalline rock, in Fluid Flow in Sedimentary Basins and Aquifers (eds Goff JC, Williams BPJ), Geol. Soc., Spl. Publ. No. 34, pp. 185–200.

Black JH, Holmes DC (1985) Hydraulic testing within the cross-hole investigation programme at Stripa, in Design and Instrumentation of In-situ Experiments in Underground Laboratories for Radioactive Waste Disposal (eds Come B, Johnston P, Miiller A), A.A. Balkema, Rotterdam, pp. 116–27.

Blindow N (2006) Ground Penetrating Radar, in Groundwater Geophysics—A tool for Hydrogeology (Kirsch R ed.), Springer, pp. 227–52.

Bobba AG, Bukata RP, Jerome JH (1992) Digitally processed satellite data as a tool in detecting potential groundwater flow systems. J. Hydrol. 131: 25–62.

Bodin J, Delay F, de Marsily G (2003) Solute transport in a single fracture with negligible matrix permeability: 1. fundamental mechanisms. Hydrogeol. J. 11: 418–33.

Boeckh E (1992) An exploration strategy for higher-yield boreholes in the West African Cystalline basement, in The Hydrogeology of Crystalline Basement Aquifers in Africa (eds Wright EP, Burgess WG), The Geol. Society, London, pp. 87–100.

Boehmer WK (1993) Secondary type aquifers and the value of pumping tests for the evaluation of groundwater potential in, Mem. 24th Congress of IAH, Oslo, pp. 159–68.

Boehmer WK, Boonstra J (1987) Analysis of drawdown in the country rock of composite dike aquifers. J. Hydrol. 94: 199–214.

Bogli A (1980) Karst Hydrology and Physical Speleology. Springer-Verlag, Berlin, p. 284.

Bonacci O (1995) Ground water behaviour in karst: example of the Ombla Spring (Croatia). J. Hydrol. 165: 113–34.

Boner F (2006) Complex Conductivity Measurements, in Groundwater Geophysics—A tool for Hydrogeology (Kirsch R ed.), Springer, pp. 119–154.

Boni CF, Bano P, Kovalevsky VS (1984) Evaluation of water resources, in Hydrogeology of Karstic Terrains—Case Histories (eds Burger A, Dubertret L), Vol. 1, pp. 9–17.

Boonstra J (1989) SATEM: Selected Aquifer Test Evaluation Methods: A Microcomputer Program, Publ. 48, Int. Instt. Land Reclamation and Improvement, Wageningen, p. 89.

Boonstra J, Boehmer WK (1986) Analysis of data from aquifer and well tests in intrusive dikes. J. Hydrol. 88: 301–17.

Boonstra J, Boehmer WK (1989) Analysis of data from well tests in dikes and fractures. ILRI Reprint No. 57, in Groundwater Contamination: Use of Models in Decision Making, Kluwer Academic Publishers, pp. 171–80.

Boonstra J, Soppe R (2007) Well hydraulics and aquifer tests, in Handbook of Groundwater Engineering, 2nd ed., (Dlleur JW ed.), CRC Press, Boca Raton.

Bosak P, Ford DC, Glazek J, Horacek I (eds) (1989) Paleokarst. Development in Earth Surface Processess. Elsevier, Amsterdam, p. 725.

Bottomley DJ, Chan LH, Katz A et al (2003) Lithium isotope geochemistry and origin of Canadian shield brines. Ground Water 41(6): 847–56.

Boulton NS (1963) Analysis of data from nonequilibrium pumping tests allowing for delayed yield from storage. Proc. Inst. Civil Engrs. 26: 469–82.

Boulton NS, Streltsova TD (1976) The drawdown near an abstraction well of large diameter under non-steady conditions in an unconfined aquifer. J. Hydrol. 30: 29–46.

Boulton NS, Streltsova TD (1977) Unsteady flow to a pumped well in a fissured water bearing formation. J. Hydrol. 35: 257–69.

Bourdet D, Gringarten AC (1980) Determination of fissure volume and block size in fractured reservoirs by type-curve analysis. Paper SPE 9293 presented at the 1980 SPE Annual Fall Tech. Conf. and Exhibition, Dallas.

Bouwer H (1978) Groundwater Hydrology. McGraw-Hill Book Co., New York, p. 480.

Brace WF (1978) A note on the permeability changes in geologic material due to stress, in Rock Friction and Earthquake Prediction (eds Byerlee JD, Wyss M), Birkhauser, Verlag, Basel, pp. 627–33.

Brace WF (1980) Permeability of argillaceous and crystalline rocks. Int. J. Rock Mech. Sci. & Geomech. Abst. 17: 241–51.

Brace WF (1984) Permeability of crystalline rocks: New in-situ measurements. J. Geophys. Res. 89(B6): 4327–30.

Bredehoeft JD, Papadopulos SS (1980) A method of determining the hydraulic properties of tight formations. Water Resources Res. 16(1): 233–38.

Briechie D, Voigt R (1980) Deep open pit mining and groundwater problems in the Rhenish Lignite District in, Studies and Reports in Hydrology, UNESCO, Paris, 28, pp. 274–90.

Bromley J, Mannstrom B, Nisca D, Jamtlid A (1994) Airborne Geophysics: Application to a groundwater study in Botswana. Ground Water 32(1): 79–90.

Brown RH (1963) Estimating the transmissivity of an artesian aquifer from the specific capacity of a well, in Methods of Determining Permeability, Transmissibility and Drawdown (compiled by R. Bentall), US Gol. Survey Water Supply Paper, 1536-I: 336–38.

Bruce DL, Yechieli Y, Zilberbrand M et al (2007) Delineation of the coastal aquifer of Israel based on repetitive analysis of C14 and tritium. J. Hydrol. 343(1–2): 56–70.

Brunner P, Hendricks-Franssen HJ, Kgotlhang L, Kinzelbach W (2007) Remote sensing in groundwater modelling. J. Hydrol. 15(1): 5–18.

Burdon DJ, Al-Sharhan A (1968) The problem of the palaeo-karstic Damman limestone aquifer in Kuwait. J. Hydrol. 6: 385–404.

Burdon DJ, Mazloum S (1961) Some chemical types of groundwater from Syria in Salinity Problems in the Arid Zones, Arid Zone Res. XIV, UNESCO, pp. 73–90.

Burger A, Dubertret L (eds) (1984) Hydrogeology of Karstic Terrains, IAH Vol. 1, Heise, Hanover, p. 264.

Busenberg E, Plummer LN (1992) Use of chlorofluorocarbons (CCl_3F and CCl_2F_2) as hydrologic tracers and age-dating. Water Resources Res. 28(9): 2257–83.

Butler JJ Jr. (1998) The Design, Performance, and Analysis of Slug Tests. Lewis Publ., New York, p. 252.

Butler JJ Jr., Healey JM (1998) Relationship between pumping-test and slug-test parameters: scale effect or artifact? Ground Water 36(2): 305–13.

Campbell MD, Lehr JH (1973) Water Well Technology, McGraw-Hill Book, Co., New York, p. 681.

Canter LW, Knox RC, Fairchild DM (1987) Ground Water Quality Protection, Lewis Publishers, Inc., Chelsea, Michigan, p. 562.

Canter LW (1985) Methods for the assessment of groundwater pollution potential, in Groundwater Quality (eds Ward CH, Giger W, McCarty PL), John Wiley and Sons, New York, pp. 270–306.

Carlsson A, Olsson T (1977) Variations in hydraulic conductivity in some Swedish rock types. Proc. Ist Int. Symp. Rock Store, Stockholm, pp. 301–7.

Carlsson L, Carlstedt A (1977) Estimation of transmissivity and permeability in Swedish bedrock. Nord. Hydrol. 8: 103–16.

Carruthers RM, Greenbaum D, Jackson PD et al (1993) Geological and geophysical characterisation of lineaments in Southeast Zimbabwe and implications for groundwater exploration. B.G.S. Tech. Report WC/93/7.

Carruthers RM, Smith IF (1992) The use of ground electrical survey methods for siting water supply boreholes in shallow crystalline basement terrains, in The Hydrogeology of Crystalline Basement Aquifers in Africa. (eds Wright EP, Burgess WG), Geological Society of London, Spl. Publ. No. 66, London, pp. 203–20.

Carvalho JM (1993) Mineral and thermal water resources development in the Portugese Hercynian Massif, Mem. 24th Congress of IAH, Oslo pp. 548–61.

Casini S, Martino S, Petirra M et al (2006) A physical analogue model to analyse interactions between tensile stresses and dissolution in carbonate slopes. Hydrogeol. J. 14: 1387–1402.

Castany G, Groba E, Romijn E (eds) (1985) Hydrogeological Mapping in Asia and the Pacific Region in IAH Proc. Workshop, Bandung, 1983, Heise, Hanover, p. 480.

Central Ground Water Board, Govt. of India CGWB (1980) Groundwater studies in the Vedavati river basin-parts of Karnataka and Andhra Pradesh (unpublished report), p. 49.

Central Ground Water Board, Govt. of India CGWB (1989) Hydrogeological Map of India, 2nd ed.

Central Ground Water Board, Govt. of India CGWB (1995b) Hydrogeological atlas of Orissa. Bhubaneswar, p. 41.

Central Ground Water Board, Govt. of India CGWB (1995c) Bhu-Jal News 10(1): 46.

Chachadi AG, Mishra GC, Singhal BBS (1991) Drawdown at a large-diameter observation well. J. Hydrol. 127: 219–33.

Chand R et al (2005) Reliable natural recharge estimates in granitic terrain. Curr. Sci., 88: 821–824.

Chandrasekharam D (1996) Geothermal energy; the future renewable energy source of India, Abstracts 30th IGC, Beijing, Vol. 1, p. 379.

Chandrasekharam D (2005) Arsenic pollution in groundwater of West Bengal, India: where we stand? in Natural Arsenic in Groundwater: Occurrence, Remediation and Management, (eds Bundeschub J et al), A.A. Balkema Publ., pp. 25–30.

Chang KT (2008) Introduction to Geographic Information Systems. McGraw-Hill.

Chapelle EH (1993) Groundwater Microbiology and Geochemistry. John Wiley and Sons, New York.

Charlie WA, Veyera GE, Durnford DS, Doehring DO (1994) Discussion on the paper "Geology of the Chinese nuclear test site near Lop Nor", Xinjiang, Ugyur Autonomous Region, China, by J.R. Matzko. Engg. Geol. 38: 177–79.

Chebotarev II (1950) Genetic types of subterranean waters and their utilisation. Waer and Water Engineering. October 1950, pp. 146–50.

Chen M, Cai Z (1995) Groundwater resources and hydro-environmental problems in China. Episodes 18(1–2): 66–68.

Cheong JY, Hamm SY, Kim HS et al (2008) Estimating hydraulic conductivity using grain-size analyses, aquifer tests, numerical modeling in a riverside alluvial system in South Korea. Hydrogeol. J. 16: 1129–43.

Cherry JA (1989) Hydrogeologic contaminant behaviour in fractured and unfractured clayey deposits in Canada, in Contaminant Transport in Groundwater (eds Kobus HE, Kinzelbach W), A.A. Balkema, Rotterdam, pp. 11–20.

Chia Y, Wang YS, Chin JJ et al (2001) Changes of groundwater level due to the 1999 Chi-Chi earthquake in the Choshui river alluvial fan in Taiwan. Bull. Seismological Soc. Am. 91(5): 1062–68; doi:10.1785/0120000726.

Chiles JP, Marsily Gde (1993). Stochastic models of fracture systems and their use in flow and transport modeling. Flow and Contaminant Transport in Fractured Rocks. Academic Press, pp. 169–231.

Chilton PJ, Lawrence AR, Stuart ME (1995) The impact of tropical agriculture on groundwater quality, in Groundwater

Quality (eds Nash H, McCall GJH), Chapman & Hall, London, pp. 113–22.

Chilton PJ, Foster SSD (1993) Hydrogeological characterisation and water supply potential of basement aquifers in tropical Africa, in Mem 24th Congress of IAH, part 2, pp. 1083–1110.

Chopra S, Madhusudhan Rao K, Sairam B et al (2008) Earthquake swarm activities after rains in peninsular India and a case study from Jamnagar. J. Geol. Soc. India 72(2): 245–52.

Christiansen AV et al (2006) The transient electromagnetic method, in Groundwater Geophysics—A tool for Hydrogeology (Kirsch R ed.), Springer, pp. 179–226.

Clark ID, Fritz P (1997) Environmental Isotopes in Hydrogeology. Lewis Publ., Boca Raton, FL, p. 328.

Clauser C (1992) Permeability of crystalline rocks. EOS Trans. Am. Geophys. Union. 73: 233–38.

Cobbing JE, Hobbs PJ, Meyer R et al (2008) A critical review of transboundary aquifers shared by South Africa. Hydrogeol. J. 16: 1207–14.

Collie MJ (1978) Geothermal Energy: Recent Developments. Noyes Data Corp., Energy Technology Review Series No. 32, Park Ridge, NJ, p. 445.

Collie MJ (ed) (1982) Geothermal Energy—Recent Developments. Noyes Data Corp., NJ, USA, p. 444.

Colwell RN (ed) (1960) Manual of Photographic Interpretation. Am. Soc. Photogramm. Falls Church, VA.

Cook PG (2003) A Guide to Regional Groundwater Flow in Fractured Rock Aquifers. CSIRO Australia, p. 108.

Cook PG et al (2006) Quantifying groundwater discharge to streams using radon. Proc. Joint Congress of 9th Australasian Env. Isotope Conf. And 2nd Austalasian Hydrogeol Reour. Conference, Adelaide, S. Australia, pp. 17–18.

Cooper HH, Bredehoeft JD, Papadopulos IS (1967) Response of finite-diameter well to an instantaneous charge of water. Water Resources Res. 3: 263–69.

Couture RA, Seitz MG, Steindler MJ (1983) Sampling of brines in cores of Precambrian granite from Northern Illinois. J. Geophys. Res. 88(B9): 7331–334.

Craig H (1963) The isotopic geochemistry of water and carbon in geothermal areas, in Nuclear Geology in Geothermal Areas (Tongiorg E ed.), C.N.R., Rome, pp. 17–53.

Cruse K (1979) A review of water well drilling methods. Q. J. Engg. Geol. 12: 79–95.

Csallany SC (1965) The hydraulic properties and yields of dolomite and limestone aquifers. Proc. Dubrovnik Symp. on Hydrogeology of Fractured Rocks 1: 120–38.

Csallany SC, Walton WC (1963) Yield of shallow dolomite wells in Northern Illinois. Illinois State Water Survey Report of Investigation, 46, Urbana, Illinois, p. 43.

Commonwealth Science Council CSC (1984) Water resources of small islands. Proc. Regional Workshop on Water Resources of Small Islands. Part 2, Commonwealth Science Council, London, p. 530.

Custodio E (1985) Low permeability volcanics in the Canary islands (Spain). Mem. IAH. 17(2): 562–73.

Dana ES, Ford WE (1932) A Textbook of Minearology. John Wiley and Sons, Inc., New York, p. 851.

Darko PK, Krasny J (2000) Adequate depth of boreholes in hard rocks: A case study in Ghana. In: O.Sililo et al (eds) Groundwater: Past Achievements and Future Challenges, AA Balkema, Rotterdam, pp. 121–123.

Das S (ed) (2007) Hydrogeological Studies—An anthology of editorials. Geological Soc. India. Spl. Publ, p. 113.

Das S, Kar A, Behera SC (1996) Studies on hydrogeology and groundwater development possibilities in alluvial and fractured basement aquifers in Keonjhar district Orissa, in Proc. Hydro—96, Indian Inst. of Tech., Kanpur.

Davis SN (1969) Porosity and permeability of natural material, in Flow Through Porous Media (DeWiest RJM ed.), Academic Press, New York, pp. 54–90.

Davis SN (1974) Changes of porosity and permeability of basalt with geologic time, in Proc. Symp. on Hydrology of Volcanic Rocks, (unpublished), Lanzarote, Spain.

Davis SN (1989) Use of isotope techniques in the study of the hydrogeology of fractured rock aquifers, in Isotope Techniques in the Study of the Hydrology of Fractured and Fissured Rocks, IAEA, Vienna, pp. 1–4.

Davis SN, DeWiest RJM (1966) Hydrogeology. John Wiley and Sons, Inc., New York, p. 463.

Davis SN, Turk LJ (1964) Optimum depth of wells in crystalline rocks. Ground Water 2: 6–11.

Davison CC (1985) Far-field hydrogeological monitoring at the site of Canada's Underground Research Laboratory, in Design and Instrumentation of In-Situ Experiments in Underground Laboratories for Radioactive Wste Disposal (eds Come B, Johnston P, Muller A), A.A. Balkema, Rotterdam, pp. 142–56.

de Marsily G (1986) Quantitative Hydrogeology–Groundwater Hydrology for Engineers. Academic Press, Orlando, FL.

de Stadelhofen CM (1994) Anwendung geophysikalisher Methoden in der Hydrogeologie. Springer Verlag, Berlin.

de Vries JJ, Simmers I (2002) Groundwater recharge: an overview a processes and challenges. Hydrogeol. J. 10(1): 5–17.

Deike RG (1990) Dolomite dissolution rates and possible Holocene dedolomitization of water-bearing units in the Edwards Aquifer, Southcentral Texas–A summary, in Selected Papers on Hydrogeology (eds Simpson ES, Sharp JM Jr.), Vol. 1, IAH, Verlag Heinz Heise, Hanover, pp. 97–109.

Delleur JW (2007) Elementary groundwater flow and transport processes, in The Handbook of Groundwater Engineering, 2nd ed., (Delleur JW ed.) CRC Press, Boca Raton.

Delleur JW (ed) (2007) The Handbook of Groundwater Engineering, 2nd ed., CRC Press, Boca Raton, Fl, Taylor and Francis Group.

Denny SC, Allen DM, Journeay JM (2007) DRASTIC-Fm: a modified vulnerability mapping method for structurally controlled aquifers in the southern Gulf Islands, British Columbia, Canada. Hydrogeol. J. 15: 483–93.

Deolankar SB (1980) The Deccan basalts of Maharashtra, India—their potential as aquifers. Ground Water 18(5): 434–37.

Dershowitz WS et al (1998) Integration of Discrete Fracture Network Models with Conventional simulator approaches, Society of Petroleum Engineers Annual Technical Convention, Louisiana, September 27–30, SPE NO. 49069.

Desai BI, Gupta SK, Shah MV, Sharma SC (1979) Hydrochemical evidence of sea water intrusion along the Mangrol Chorwad coast of Saurashtra, Gujarat. Hydrol. Sci. Bull. 24(1): 71–82.

DeWiest RJM (1969) Flow Through Porous Media. Academic Press, New York, p. 530.

Dey A, Bhowmick AN, Chakraborty D (2003) Contamination through landfills in NCT, Delhi, in Recent Trends in Hydrogeochemistry (eds Ramanathan A, Ramesh R), Capital Publ. Co., New Delhi, pp. 153–59.

Di Frenna VJ, Price RM, Savabi MR (2008) Identification of a hydrodynamic threshold in karst rocks from the Biscayne Aquifer, South Florida, USA. Hydrogeol. J. 16: 31–42.

Dietvorst EJL, DeVries JJ, Gieske A (1991) Coincidence of well fields and tectonic basins in the Precambrian shield area of south east Botswana. Ground Water 29: 864–77.

Dietz HB (1985) The in-situ test program for site characterization of basalt, in Design and Instrumentation of in-situ Experiments in Underground Laboratories for Radioactive Waste Disposal (eds Come B, Johnston P, Muller A), A.A. Balkema, Rotterdam, pp. 82–92.

Dillon P (2005) Future management of aquifer recharge. Hydrogeol. J. 13(1): 313–16.

Dingman SL (1994) Physical Hydrology. Macmillan Publishing Co., New York, p. 575.

Dissanayake CB (1991) The fluoride problem in the groundwater of Sri Lanka—environmental management and health. Intl. J. Env. Studies. 38: 137–56.

Dobrin MB (1976) Introduction to Geophysical Prospecting. McGraw-Hill Book Co., New York.

Dobrin MB, Savit CH (1988) Introduction to Geophysical Prospecting. McGraw-Hill Inc., New York, p. 639.

Domenico PA (1972) Concepts and Models in Groundwater Hydrology. McGraw-Hill Book Co., p. 405.

Domenico PA, Schwartz FW (1998) Physical and Chemical Hydrogeology, 2nd ed., John Wiley and Sons, Inc., p. 506.

Downing RA, Allen DJ, Barker JA, Burgess WG, Gray DN, Price M, Smith IF (1984) Geothermal exploration at Southampton in the U.K. Energy Exploration and Exploitation, 2(4): 327–42.

Driscoll FG (ed) (1986) Groundwater and Wells. Johnson Division, St. Paul, Minnesota, p. 1089.

Duncan RA, Pyle DG (1988) Rapid eruption of the Deccan flood basalts, Western India in Deccan Flood Basalts (Subbarao KV ed.) Mem. Geol. Soc. India, pp. 1–9.

Dunnivant FM, Anders E (2006) A Basic Introduction to Pollutant Fate and Transport. Wiley-Interscience, p. 479.

Dverstorp B, Anderson J, Nordqvist J (1992) Discrete fracture network interpretation of field tracer migration in sparsely fractured rock. Water Resources Res. 28(9): 2327–343.

Eagon HB, Johe DE (1972) Practical solutions for pumping tests in carbonate rock aquifers. Ground Water 10: 6–13.

Earlougher RC Jr. (1977) Advances in Well Test Analysis. Monograph Series, SPE, Richardson, Texas.

Eaton TT, Anderson MP, Bradbury KR (2007) Fracture control of groundwater flow and water chemistry in a rock aquitard. Ground Water 45(5): 601–15.

Edmunds WM (2005) Contribution of isotopic and nuclear tracers to study of groundwater, in Isotopes in the Water Cycle, (eds Aggarwal PK et al), Springer, pp. 171–92.

Edmunds WM, Savage D (1991) Geochemical characteristics of groundwater in granites and related crystalline rocks, in Applied Groundwater Hydrology (eds Downing RA, Wilkinson WB), Oxford University Press, Oxford, pp. 266–82.

Edwards KB, Jones LC (1993) Modelling pumping tests in weathered glacial till. J. Hydrol. 150: 41–60.

Elachi C, Roth LE, Schaber GG (1984) Spaceborne Radar Subsurface Imaging in Hyperarid Regions. IEEE Trans GE. 22: 382–87.

Elder J (1981) Geothermal Systems. Academic Press, London, p. 508.

El-Kadi AI, Oloufa AA, Eltahan AA, Malik HU (1994) Use of a geographic information system in site-specific ground-water modelling. Ground Water 32(4): 617–25.

Ellefsen KJ, Hsieh, PA, Shapiro AM (2002) Crosswell seismic investigation of hydraulically conductive fractured bedrock near Mirror Lake, New Hampshire. Jr. Appl. Geophys. 50(3): 299–317.

Ellyett CD, Pratt DA (1975) A review of the potential applications of remote sensing techniques to hydrogeological studies in Australia, AWRC, Tech Paper No. 13, p. 147.

EMS-Environmental Modelling System (2007) GMS—Groundwater Modelling Software. http://www.emsi.com/GMS/gms.html.

Entekhabi D, Njoku EG, Houser P, Spencer M, Doiron T, Kim Y, Smith J, Girard R, Belair S, Crow W, Jackson TJ, Kerr YH, Kimball JS, Koster R, McDonald KC, O'Neill PE, Pultz T, Running SW, Shi J, Wood E, van Zyl J (2004) The hydrosphere state (hydros) satellite mission: An earth system pathfinder for global mapping of soil moisture and land freeze/thaw. IEEE Transactions on Geoscience and Remote Sensing 42(10): 2184–95.

Erdelyi M, Galfi G (1988) Surface and Subsurface Mapping in Hydrogeology. John Wiley and Sons, New York, p. 384.

Ernstson K (2006) Magnetic, geothermal, and radioactivity methods, in Groundwater Geophysics—A tool for Hydrogeology, (Kirsch R ed.), Springer, pp. 275–94.

Ernstson K, Kirsch R (2006a) Geoelectrical methods, in Groundwater Geophysics—A tool for Hydrogeology, (Kirsch R ed.), Springer, pp. 85–118.

Ernstson K, Kirsch R (2006b) Aquifer structures: fracture zone and caves, in Groundwater Geophysics—A tool for Hydrogeology, (Kirsch R ed.), Springer, pp. 395–422.

Evans DD, Nicholson TJ (eds) (1987) Flow and Transport Through Unsaturated Fractured Rocks. Monograph 42, American Geophysical Union, Washington, p. 187.

Faillace C (1973) Location of groundwater and the determination of the optimum depth of wells in metamorphic rocks of Karamoja, Uganda. Proc. 2nd Int. Convention on Groundwater, Palermo, Italy.

Falkland AC (1984) Development of groundwater resources on coral atolls; experiences from Tarawa and Christmas islands, Republic of Kiribati, in Water Resources of Samll Island, Republic of Karibati. CSC Tech. Publ. Series No. 154, London, pp. 436–52.

Fayer MJ, Gee GW, Rockhold ML, Freshley, MD, Walters TB (1996) Estimating recharge rates for a groundwater model using a GIS. J. Environ. Qual. 25: 510–18.

Feddes RA, Kabat P, Van Bakel PJT, Bronswijk JJB, Halbertsma J (1988) Modelling soil water dynamics in the unsaturated zone—state of the art. J. Hydrol. 100: 69–111.

Fernandopulle D, Vos J, La Moneda E, Medina L (1974) Groundwater resources of the island of Gran Canaria Paper presented in the Symposium on Hydrogeology of Volcanic Rocks, Lanzarote, Spain, UNESCO, (unpublished).

Ferris JG, Knowles DB, Brown RH, Stallmen RW (1962) Theory of Aquifer Tests, US Geol. Surv. Water Supply Paper 1536-E, p. 174.

Fetter CW (1993) Contaminant Hydrogeology. Macmillan Publishing Company, New York, p. 458.

Fetter CW (1988) Applied Hydrogeology, 2nd ed., Merrill Publ. Co., Columbus, p. 592.

Fisher RA (1953) Dispersion on a sphere. Proc. Royal Society of London, Series A, 217: 295–305.

Fleury P, Bakalowics M, de Marsily G et al (2007) Submarine springs and coastal karst aquifers: A review. J. Hydrol. 339: 79–92.

Fleury P, Bakalowics M, deMarsily G et al (2008) Functioning of a coastal karstic system with a submarine outlet, in southern Spain. Hydrogeol. J. 16: 75–85.

Flint AL, Flint LE, Kwicklis EM et al (2002) Estimating recharge at Yucca Mountain, Nevada, USA: Comparison of methods. Hydrogeol. J. 10: 180–204.

Florea LJ, Vacher HL (2006) Spring flow hydrographs: Eojenetic vs. telogenetic karst. Ground Water 44(3): 352–61.

Florquist BA (1973) Techniques for locating water wells in fractured crystalline rocks. Ground Water 11(3): 26–28.

Fontes JC, Louvat D, Michelot JG (1989) Some constraints on geochemistry and environmental isotopes for the study of low fracture flows in crystalline rocks: The Stripa case. Proc. Advisory Group Meeting. IAEA, Vienna, pp. 29–67.

Ford DC, Williams PW (1989) Karst Geomorphology and Hydrology. Unwin Hyman, London, p. 601.

Ford DC, Williams PW (2007) Karst Hydrology and Geomorphology. Wiley Chichester, 2nd ed., p. 576.

Foster S, Garduno H (2006) Groundwater resources of India: Towards a framework for practical management and effective administration. Proc. 12th National Symposium on Hydrology (eds Romani S et al), Capital Publ. Co., New Delhi, pp. 19–32.

Franz T, Guiger NF (1990) FLOWPATH: Steady State Two-Dimensional Horizontal Aquifer Simulation Model. Waterloo Hydrogeologic Software, ON, Canada.

Freeze RA, Cherry JA (1979) Groundwater. Prentice Hall, Inc., New Jersey, p. 604.

Freeze RA, McWhorter DB (1997) A framework for assessing risk reduction due to DNAPL mass removal from low permeability soils. Ground Water 35(1): 111–23.

Gabrovsek R, Dreybrodt W (2001) A model of early evolution of karst aquifers in limestone in the dimensions of length and depth. J. Hydrol. 240(3–4): 206–224.

Gale JE (1982a) Fundamental hydraulic characteristics of fractures from field and laboratory investigations, in Papers of the Groundwater in Fractured Rock Conference, AWRC, Conference Series No. 5, Conberra, pp. 79–94.

Gale JE, Rouleau A, Witherspoon PA (1982) Hydrogeologic characteristics of a fractured granite, in Papers of the Groundwater in Fractured Rock Conference, AWRC Conference Seris No. 5, Canberra, pp. 95–108.

Gale JE (1982b) Assessing the permeability characteristics of fractured rocks, in Recent Trends in Hydrogeology (Narasimhan TN ed.), Geol. Soc. Am. Special Paper 1989, pp. 163–81.

Galloway DL, Hudnut KW, Ingebritsen SE et al (1998) Detection of aquifer system compaction and land subsidence using interferometric synthetic aperture radar, Antelope Valley,

Mojave Desert, California. Water Resources Res. 34(10): 2573–85.

Galloway DL, Hoffmann J (2007) The application of satellite differential SAR interferometry—derived ground displacements in hydrogeology. Hydrogeol. J. 15: 133–54, doi:10.1007/s10040-006-0121-5.

Gardner CMK, Bell JP, Cooper JD et al (1991) Groundwater recharge and water movement in the unsaturated zone, in Applied Groundwater Hydrology (eds Downing RA, Wilkinson WB), Oxford University Press, Oxford, p. 54.

Garnier, JM, et al (1985) Traçage par 13C, 2H, I – et uranine dans la nappe de la craie sénonienne en écoulement radial convergent (Béthune, France). J. Hydrol. 78: 379–392.

Garrels RM, Christ CL (1965) Solutions, Minerals and Equilibria. Harper and Row, New York, p. 450.

Gascoyne M, Ross JD, Watson RL (1996) Highly saline pore fluid in the rock matrix of a granitic batholith in the Canadian shield, Abstracts 30th IGC, Beijing, Vol. 3, p. 269.

Gascoyne M, Kamineni DC (1993) The hydrogeochemistry of fractured plutonic rocks in the Canadian shield, in Mem. 24th Congress of IAH, Oslo, pp. 440–49.

Gerard A, Baumgartner J, Baria R (1996) European Hot Dry Rock Research project at Soultz-sous-Forets. 30th IGC, Abstracts, Beijing, Vol. 1, p. 385.

Germain D, Frind EO (1989) Modelling of contaminant migration in fracture networks: Effects of matrix diffusion, in Contaminant Transport in Groundwater (eds Kobus HE, Kinzelbach W), A.A. Balkema, Rotterdam, pp. 267–74.

Ghosh A (1990) Role of Radial Cracks, Free face and Natural Discontinuities of Fragmentation from Bench Blasting, Ph.D. Thesis, University of Arizona.

Giusti EV (1977) Hydrogeology and "Geoesthetics" applied to land use planning in the Puerto Rican Karst, in Karst Hydrogeology (eds Tolson JS, Doyle FL), Mem. 12th Congress of IAH, Alabama, pp. 149–68.

Glynn PD, Plummer LN (2005) Geochemistry and the understanding of groundwater systems. Hydrogeol. J. 13(1): 263–87.

Goldscheider N, Bechtel TD (2009) Editor's message: The housing crisis from underground—damage to a historic town by geothermal drillings through anhydrite, Staufen, Germany. Hydrogeol. J. 17(3): 491–93.

Goody DC, Kinniburgh DG, Barker JA (2007) A rapid method for determining apparent diffusion coefficients in Chalk and other consolidated porous media. J. Hydrol. 343(1–2): 97–103.

Goswami AB (1995) A Critical Study on Water Resources of West Bengal, Ph.D. Thesis (unpublished), Jadavpur University, p. 309.

Grant MA, Donaldson IG, Bixley PF (1982). Geothermal Reservoir Engineering. Academic Press, New York, p. 367.

Grassi S, Netki R (2000) Sea water intrusion and mercury pollution of some coastal aquifers in the province of Grossete, in southern Italy. J. Hydrol. 237(3–4): 198–211.

Graziadei W, Zotl JG (1984) Karstwater gallery and hydroelectric power plant Muhau—the water supply of Innsbruck, Austria, in Hydrogeology of Karstic Terrains (eds Burger A, Dubertret L) IAH Vol. 1, pp. 113–16.

Greenbaum D (1992) Structural influences on the occurrence of groundwater in SE Zimbabwe, in The Hydrogeology of Crystalline Basement Aquifers in Africa (eds Wright EP,

Burgess WG), Geol. Soc. Spl. Publ. No. 66, London, pp. 77–85.

Greene EA, Rahn PH (1995) Localized anisotropic transmissivity in a karst aquifer. Ground Water 33(5): 806–16.

Greenman DW, Swarzenski WV, Bennett GD (1967) Groundwater Hydrology of the Punjab, West Pakistan with Emphasis on Problems Caused by Canal Irrigation. US Geol. Surv. Water Supply Paper. 1608-H, Washington, p. 66.

Gregory KJ et al (1973) Drainage Basin Form and Process—A Geomorphic Approach. Edward Arnold (Publ.) Ltd., London, p. 458.

Griffiths DH, Turnbill JH (1985) A multi-electrode array for resistivity surveying. First Break. 3: 16–20.

Gringarten AC (1982) Flow-test evaluation of fractured reservoirs, in Recent Trends in Hydrogeology (Narasimhan TN ed.), Geol. Soc. Am. Special Paper 189: 237–63.

Grisak GE, Pickens JF (1980) Solute transport through fractured media 1. The effect of matrix diffusion. Water Resources Res. 16(4): 719–30.

Grisak GE, Pikens JF, Cherry JA (1980) Solute transport through fractured media 2. Column study of fractured till. Water Resources Res. 16(4): 731–39.

Ground Water 29, pp. 375–86.

Geological Survey of India GSI (1969) Geohydrological Map of India. Geol. Surv. India. Calcutta.

Geological Survey of India GSI (1991) Geothermal Atlas of India. Geol. Surv. India, Calcutta. Spl. Publ. 19, p. 144.

Gudmundsson A, Gjesdal O, Brenner Sl et al (2003) Effects of linking up of discontinuities on fracture growth and groundwater transport. Hydrogeol. J. 11: 84–99. doi:10.1007/s10040-002-0238-0.

Guo Q, Wang Y, Gao X, Teng M (2007) A new model (DRARCH) for assessing groundwater vulnerability to arsenic contamination at basin scale: A case study in Taiyuan basin, northern China. Environ. Geol. 52: 923–32.

Gupta CP, Thangarajan M, Guranadharao VVS et al (1996) Groundwater pollution in the upper palar basin, Tamilnadu. Proc. Workshop on Challenges in Groundwater Development, Central Board of Irrigation and Power, New Delhi, pp. 200–13.

Gupta CP, Thangarajan M, Gurunadha Rao VVS et al (1989) Simulation of regional hydrodynamics in the Neyveli aquifer IGW. Vol. II, pp. 607–18.

Gupta HK (1992) Reservoir Induced Earthquakes. Elsevier Press, Amsterdam, p. 369.

Gupta RP (1991) Remote Sensing Geology. Springer-Verlag, Berlin, p. 356.

Gupta RP (2003) Remote Sensing Geology, 2nd ed., Springer, Verlag, p. 655.

Gupta SK (1996) Evaluation of Fracture Characteristics for Hydrogeological Modelling, M.Tech. Dissertation (unpublished), University of Roorkee, Roorkee.

Gustafson G, Krasny J (1994) Crystalline rock aquifers: Their occurrence, use and importance. Hydrogeol. J. 2: 64–75.

Gustafsson P (1993) SPOT satellite data for exploration of fractured aquifers in a semi-arid area in southern. Botswana Mem. of the 24th Congress of IAH, Oslo, pp. 562–75.

Hadzisehovic M, Dangic A, Miljevic N et al (1995) Geochemical-water characteristics of the Surdulica aquifer. Ground Water 33(1): 112–23.

Haeni FP, Lane JW Jr., Lieblich DA (1993) Use of surface-geophysical and borehole-radar methods to detect fractures in crystalline rocks, Mirror Lake Area, Grafton County, Near Hampshine. Mem. of the 24th Congress of IAH, Oslo 1993, pp. 577–87.

Haitjema HM, Mitchell-Bruker S (2005) Are water tables a subdued replica of the topography? Ground Water 43(6): 781–796.

Halford KJ, Laczniak RJ, Galloway DL (2005) Hydraulic characterization of overpressured tuffs in central Yucca Flat, Nevada Test Site, Nye County, Nevada. USGS Sci Invest Rep 2005–5211. http:pubs.usgs.gov/sir/2005/5211/. Cited 29 Sept 2006.

Halliday SL, Wolfe ML (1991) Assessing groundwater pollution potential from nitrogen fertilizer using a Geographic Information System. Water Resources Bull. 27(2): 237–45.

Hamill L, Bell FG (1986) Groundwater Resource Development. Butterworths, London, p. 344.

Handa BK (1988) Fluoride occurrence in natural waters in India and its significance. Bhu-Jal News. Central Ground Water Board, Govt. of India 3(2): 31–37.

Handa BK (1994) Ground water contamination in India, in Proc. Regional Workshop on Env. Aspects of Groundwater Dev., (Singhal DC ed.), Ind. Assoc. Hydrol., Kurukshetra, pp. 1–33.

Hansmann BC, Meijerink AMJ, Kodituwakku KAN (1992) An inductive approach for groundwater exploration. ITC J. 1992–93: 269–76.

Hantush MS (1956) Analysis of data from pumping test in leaky aquifers. Trans. Amer Geophys. Union. 37: 702–14.

Hantush MS (1964) Hydraulics of Wells, in Advances in Hydroscience, (Chow VT ed.), Vol. 1, Academic Press, New York, pp. 281–432.

Hantush MS, Jacob CE (1955) Non-steady radial flow in an infinite leaky aquifer. Trans. Amer. Geophys. Union. 36: 95–100.

Hao Y, Yeh TCJ, Xiang J et al (2008) Hydraulic tomography for detecting fracture zone connectivity. Ground Water 46(2): 183–92.

Harvey F (2008) A Primer of GIS: Fundamental Geographic and Cartographic Concepts. Guilford Press.

Hasan SE (1996) Geology and Hazardous. Waste Management. Prentice Hall, New Jersey, p. 387.

Hazell JRT, Cratchley CR, Jones CRC (1992) The hydrogeology of crystalline aquifers in northern Nigeria and geophysical techniques used in their exploration, in The Hydrogeology of Crystalline Basement Aquifers in Africa. (eds Wright EP, Burgess WG), Geological Society of London, Spl. Publ. No. 66, London, pp. 155–82.

Hazen A (1893) Some physical properties of sands and gravels with special reference to their use in filtration. Massachusetts State Board of Health, 24th Ann. Report, Boston, pp. 541–56.

Healy RW, Cook PG (2002) Using groundwater levels to estimate recharge. Hydrogeol. J. 10: 91–109.

Heath MJ (1985) Solute migration experiments in fractured granite, South West England, in Design and Instrumentation of in-situ Experiments, in Underground Laboratories for Radioactive Waste Disposal (eds Come B, Johnston P, Muller A), A.A. Balkema, Rotterdam, pp. 191–200.

Heath RC (1976) Design of ground-water level observation-well programs. Ground Water 14(2): 71–77.

Heilman JL, Moore DG (1982) Evaluating depth to shallow groundwater using Heat-Capacity Mapping Mission (HCMM) data. Photogramm. Eng. Rem. Sens. 48(12): 1903–6.

Heilweil VM, Solomon DK, Gingerich SB et al (2009) Oxygen, hydrogen, and helium isotopes for investigating groundwater systems of the Cape Verde Islands, West Africa. Hydrogeol. J. 17: 1157–74. doi:10.1007/s10040-009-0434-2.

Heinzl S, Kraft T, Wassermann J et al (2006) Evidence of rain-triggered earthquake activity. Geophys. Res. Letters. 33, L 19303.

Hem JD (1989) Study and Interpretation of the Chemical Characteristics of Natural Water, 3rd ed., US Geol. Surv. Water Supply Paper 2254, p. 263.

Hendry MJ (1982) Hydraulic conductivity of a glacial till in Alberta. Ground Water 20: 162–69.

Henriksen H (1995) Relation between topography and well yield in boreholes in crystalline rocks, Sognag Fjordane, Norway. Ground Water 33(4): 635–43.

Henriksen H (2003) The role of some regional factors in the assessment of well yields from Hard-rock aquifers of Fennoscandia. Hydrogeol. J. 11: 628–45.

Herbert R, Kitching R (1981) Determination of aquifer parameters from large-diameter dug well pumping tests. Ground Water 19(6): 593–99.

Herbert RB Jr. (1992) Evaluating the effectiveness of a mine tailing cover. Nord. Hydrol. 23: 193–208.

Herwanger JV, Worthington MH, Lubbe R, Binley A, Khazanehdari J (2004) A comparison of cross-hole electrical and seismic data in fractured rock. Geol. Prospecting 52(2): 109–121.

Hodgkinson D, Benabderrahmane H, Elert M et al (2009) An overview of Task 6 of the Aspo Task Force: Modelling groundwater and solute transport: Improved understanding radionuclide transport in fractured rock. Hydrogeol. J. 17(5) doi:10.1007/s10040-008-0416-9.

Hoffmann J, Galloway DL, Zebker HA (2003) Inverse modeling of interbed storage parameters using land subsidence observations, Antelope Valley, California. Water Resources Res. 39(2): 13.

Hoffmann J, Zebker HA, Galloway DL, Amelung F (2001) Seasonal subsidence and rebound in Las Vegas Valley, Nevada, observed by synthetic aperture radar interferometry. Water Resources Res. 37(6): 1551–66.

Hook SJ, Dmochowski JE, Howard KA, Rowan LC, Karlstrom KE, Stock JM (2005) Mapping variations in weight percent silica measured from multispectral thermal infrared imagery–examples from Hiller Mountains, Nevada, USA and Tres Virgenes-La Reforma, Baja California Sur, Mexico. Remote Sensing of Environment 95: 273–289.

Horne RN (1990) Modern Well Test Analysis—A Computer Aided Approach. Petroway Inc., Palto Alto, Clif., p. 185.

Horta J (2005) Saline waters. In Isotopes in the water cycle: past, present and future of a developing sciences (eds Aggarwal PK et al), pp. 271–87.

Houlihan MF, Berman MH (2007) Remediation of contaminated groundwater, in The Handbook of Groundwater Engineering, 2nd ed., (Delleur JW ed.), CRC Press, Taylor and Francis, pp. 36–1 to 36–48.

Houlihan MF, Botek MH (2007) Remediation of contaminated groundwater, in The handbook of Groundwater Engineering, 2nd ed., (Delleur JW ed.), CRC press, Boca Raton, FL, pp. 36–1 to 36–48.

Houston J (1992) Rural water supplies: comparative case histories from Nigeria and Zimbabwe, in The Hydrogeology of Crystalline Basement Aquifers in Africa (eds Wright EP, Burgess WG), Geol. Soc. Spl. Publ. No. 66, London, pp. 243–57.

Hsieh PA, Tiedeman CR (2001) Evaluation of forced-gradient tracer tests in fractured rock aquifers, Canadian Symp.

Humphreys WF (2009) Hydrogeology and groundwater ecology. Does each inform the other? J. Hydrol. 17(1), doi:10.1007/s10040-008-0349-3.

Huntley D, Normenson R, Steffey D (1992) The use of specific capacity to assess transmissivity in fractured rock aquifers. Ground Water 30(3): 396–402.

Huntoon PW (1992a) Hydrogeologic characteristics and deforestation of the stone forest karst aquifers of South China. Ground Water 30(2): 167–76.

Huntoon PW (1992b) Exploration and development of ground water from the Stone Forest karst aquifers of South China. Ground Water 30(3): 324–30.

International Atomic Energy Agency IAEA (1983) Guidebook on Nuclear Techniques in Hydrology. Tech. Reports Series No. 91.

International Atomic Energy Agency IAEA (1989) Isotope techniques in the study of the hydrology of fractured and fissured rocks. Proc. of Advisory Group Meeting. IAEA, Vienna, 1989, p. 306.

International Association of Hydrogeologists IAH (2002) Theme Issue: Groundwater recharge, (eds Scanlon BR, Cook PG) Hydrogeol. J. 10(1): 237.

International Association of Hydrogeologists IAH (2005) The Future of Hydrogeology, in J. Hydrol. (Voss C ed.), 13(1): 349.

International Association of Hydrogeologists IAH (2009) Codification of Law of Transboundary Aquifers, in News and Information, Issue D31, IAH, p. 8.

Ilman WA, Neuman SP (2001) Type-curve interpretation of a cross-hole pneumatic injection test in unsaturated fractured tuff. Water Resources Res. 37(3): 583–603.

Ilman WA, Neuman SP (2003) Steady-state analysis of cross-hole pneumatic tests in unsaturated fractured tuff. J. Hydrol. 281(1–2): 36–54.

Ilman WA, Tartakovsky DM (2005) Asymptotic analysis of three-dimensional pressure interference tests: A point source solution. Water Resources Res. 41(1): W01002.

Ilman WA, Tartakovsky DM (2006) Asymptotic analysis of cross-hole hydraulic tests in fractured granite. Ground Water 44(4): 555–63.

Inamdar PM, Kumar D (1994) On the origin of bole beds in Deccan traps. J. Geol. Soc. India 44(3): 331–34.

Indraratna B, Ranjith P (2001) Hydromechanical Aspects and Unsaturated flow in Jointed Rock. AA Balkema Publ., Tokyo, p. 286.

Inseson J, Downing RA (1964) The groundwater component of river discharge and its relationship to hydrogeology. J. Inst. Water Engrs. 18: 519–41.

Irmay S (1954) On the hydraulic conductivity of unsaturated soils. Trans. Am. Geophys. Union. 35: 463–67.

Issar AS (2008) Progressive development in arid environments: adapting to concept of sustainable development to a changing world. Hydrogeol. J. 16: 1229–31.

IWRS (1996) Interbasin Transfer of Water for National Development: Problems and Prospects. Indian Water Resources Society, Roorkee, p. 35.

Jacobson G (1982) Groundwater in fractured rock aquifers of the Australian Capital Territory—A valuable resource or a major problem, in Papers of the Groundwater in Fractured Rock Conference 1982, AWRS, Series No. 5, Canberra, pp. 109–18.

Jacobus J et al (2002) Flood frequency studies on ungaged urban watersheds using remote sensing data, in Proc. of the Natural Symposium on Urban Hydrology, University of Kentucky, pp. 3–9.

Jagannathan V (1993) Deep seated fractured aquifers in crystalline rocks. J. Geol. Soc. India 41(3): 280–82.

Jalludin M, Razack M (1993) Groundwater occurrence in weathered and fractured basalts, Republic of Djibouti, in Mem. 24th Congress of IAH, Oslo, Part 1, pp. 248–58.

Javendal I, Doughty C, Tsang CF (1984) Groundwater Transport: Hand Book of Mathematical Models. Water Resources Monograph Series 10, American Geophysical Union, p. 228.

Jayasena HAH, Singh BK, Dissanayake CB (1986) Groundwater occurrences in the hard rock terrains of Sri Lanka—A case study. Aqua 4: 214–19.

Jensen JR (1986) Introductory Digital Image Processing. Prentice Hall, Englewood Cliffs, New Jersey, p. 379.

Jezersky Z (2007) Hydrogeochemistry of a deep gas–storage cavern, Czech Republic. Hydrogeol. J. 15: 599–614.

Jha MK, Chowdary VM (2007) Challenges of using remote sensing and GIS in developing nations. Hydrogeol. J. Vol. 15(1): 197–200.

Jiao JJ, Zheng C (1995) The different characteristics of aquifer parameters and their implications on pumping-test analysis. Ground Water 35(1): 25–29.

Jones IC, Banner JL (2000) Estimating recharge in a tropical karst aquifer. Water Resources Res. 36(5): 1289–99.

Jones L, Lemar T, Tsai C (1992) Results of two pumping tests in Wisconsin age weathered till in Iowa. Ground Water 30(4): 529–38.

Jones MA, Pringle AB, Fulton IM, O'Neill S (1999) Discrete fracture network modeling applied to groundwater resource exploration in southwest Ireland. Fractures, Fluid Flow and Mineralization, Geological Society of London, Spl. Publ. No. 155, pp. 83–103.

Jorgensen DG (1991) Estimating geohydrologic properties from borehole geophysical logs. Groundwater Monitoring and review, Summer 1991, pp. 123–29.

Jorgensen PR, Broholm K, Sonnenborg TO et al (2003) Monitoring well interception with fractures in clayey till. Ground Water 41(6): 772–79.

Jouanna P (1993) A summary of field test methods in fractured rocks, in Flow and Contaminant Transport in Fractured Rock (eds Bear J, Tsang CF, de Marsily G), Academic Press, Inc., San Diego, pp. 437–543.

Jyrkama MI, Sykes J F (2007) The impact of climate change on groundwater, in The Handbook of Groundwater Engineering, 2nd ed., (Delleur JW ed.), pp. 28–1 to 28–42.

Kaehler CA, Hsieh PA (1994) Hydraulic Properties of a Fractured Rock Aquifer, Lee Valley, San Diego County, California. US Goel. Surv. Water Supply Paper 2394, p. 64.

Kaila KL (1988) Mapping the thickness of Deccan Trap flows in India from DSS studies and inferences about a hidden Mesozoic basin in the Narmada—Tapti region, in Deccan Flood Basalts, (Subbarao KV ed.), Geol. Soc. Ind., Mem. 10: 91–116.

Kakar YP (1988) Ground water quality monitoring in India. Jalvigyan Sameeksha, NIH, Roorkee, Vol. III, pp. 52–63.

Kakar YP (1990) Groundwater pollution due to industrial effluents. Bhu-Jal News, Central Ground Water Board, Govt. of India 2: 1–12.

Kalinski RJ, Kelly WE, Bogardi I, Ehrman RL, Yamamoto PO (1994) Correlation between DRASTIC vulnerabilities and incidents of VOC contamination of municipal wells in Nebraska. Ground Water 32(1): 31–34.

Kalkoff SJ (1993) Using Geographic Information Systems to determine the relation between stream quality and geology of the Roberts Creek watershed, Clayton County, Iowa. Water Resources Bull. (AWRA), 29(6): 989–96.

Karanga FK, Hansman BKG, Meijerink AMJ (1990) Use of remote sensing and GIS for the district water plan, Samburu District, Kenya. Intern. Symp. Remote Sensing and Water Resources, Enschede, The Netherlands, IAH & Neth. Soc. RS, pp. 835–48.

Karanth KR (1987) Ground Water Assessment, Development and Management. Tata McGraw-Hill Publ. Co. Ltd., New Delhi, p. 720.

Karanth KR (1992) Pegmatites: A potential source of siting high yielding wells. J. Geol. Soc. India 39: 77–81.

Karlsson F (1989) The Swedish research programme on radioactive waste disposal: Experiences gained with the use of isotope techniques in fractured rock. Proc. Advisory Group Meeting. IAEA, Vienna, pp. 69–86.

Karundu AJ (1993) Hydrogeology of fractured bedrock systems of Nyabisheki catchment in SW Uganda. Mem. 24th Congress of IAH, Oslo, pp. 90–102.

Kazemi GA (2008) Editor's message: Submarine groundwater discharge studies and the absence of hydrogeologists. Hydrogeol. J. 16(2): 201–4.

Kazi A, Sen Z (1985) Volumetric RQD: an index of rock quality, in Int. Symp. on Fundamentals of Rock Joints, Bjorkliden, pp. 95–102.

Kearey P, Brooks M, Hill I (2002). An Introduction to Geophysical Exploration. Wiley-Blackwell Science.

Kehew AE (2001) Applied Chemical Hydrogeology. Prentice Hall, New Jersey, p. 368.

Kehew AE, Straw WT, Steinmann WK et al (1996) Groundwater quality and flow in a shallow glacio-fluvial aquifer impacted by agricultural contamination. Ground Water 34(3): 491–500.

Keister BA, Repetto PC (2007) Landfills, in The Handbook of Groundwater Engineering, 2nd edn., (Delleur JW ed.), pp. 33–1 to 33–31.

Keppler A, Drost W, Wohnlich S (1996) Comparison of different radioactive tracer within a single fracture of granitic rock at the Grimsel Test Site, in Hardrock Hydrogeology of the Bohemian Massif (eds Krasny J, Mls J), Acta Universitatis Caroline, Geologica, Univerzita Karlova, Czech Republic, Vol. 40, 247–55.

Kessels W, Kuck J (1995) Hydraulic communication in crystalline rock between the two boreholes of the Continental Deep Drilling Project in Germany. Int. J. Rock Mech. Min. Sc. & Geomech. Abstr. 32(1): 37–47.

Khair K, Haddad (1993) The fractured carbonate rocks of Lebanon and their seasonal springs, in Hydrogeology of Hard Rocks, Mem. 24th congress of IAH, Oslo, Part 2, pp. 1135–44.

Khan SHA, Raja J (1989) Management of groundwater resources in complex aquifer system of Rajmahal traps of Sahibganj district, Bihar, in Proc. International Workshop on Appropriate Methodlogies for Development and Management of Ground Water Resources in Developing Countries, NGRI, India. (Gupta CP ed.), pp. 649–60.

Khare MC (1981) Sedimentological and Hydrogeological Studies of parts of Durg and Raipur Districts, M.P., India, Ph.D Thesis (Unbpulished), University of Roorkee, p. 351.

Kimblin RT (1995) The chemistry and origin of groundwater in Triassic sandstone and Quaternary deposits: northwest England and some UK comparisons. J. Hydrol. 172: 293–311.

Kiraly L (1975) Report on the present knowledge on the physical characteristics of karstic rocks, in Hydrogeology of Karstic Terrains, (eds Burger A, Dubertret L), IAH—Paris, Ser. B. No. 3, pp. 53–67.

Kirsch R (2006) Petrophysical properties of permeable and low-permeable rocks, in Groundwater Geophysics—A tool for Hydrogeology (Kirsch R ed.), Springer, pp. 1–22.

Kittu N (1990) High yielding bore wells in basaltic terrain at Yerangaon, Nagpur district, Maharashtra. Bhu-Jal News, 5(4), Coverstory.

Kittu N, Mehta M (1990) Numerical technique for evaluation of aquifer parameters on large diameter wells in hard rocks. Proc. All India Seminar on Modern Techniques of Rain Water Harvesting, Water Conservation and Artificial Recharge for Drinking Water, Pune, pp. 259–62.

Klint KE, Rosenbaum (2001) A new approach to analysis of mechanical fracture aperture in rocks, in Proc. Conf. Fractured Rock 2001 EPA, (eds Keuper BH et al), Ontario, p. 5.

Klitten K (1991) Hydrofracturing of hard rock borewells. Lecture notes (unpublished) Regional Training Course on Groundwater Exploration and Assessment, University of Roorkee, Vol. 2, pp. 129–57.

Kloska M, Ostrowski L, Pusch G, Yerby Y (1989) Gas pulse-test-a new method for low permeability formations, in Rock at Great Depth, Vol. 1 (eds Maury V, Fourmaintraux D), A.A. Balkema, Rotterdam, pp. 219–26.

Knopman DS, Hollyday EF (1993) Variation in specific capacity in fractured rocks, Pennsylvania. Ground Water 31(1): 135–45.

Koefoed O (1979) Geosounding Principles, (1) Resistivity Sounding Measurements, Elsevier, Amsterdam, p. 276.

Koh Dong-Chan et al (2007) Evidence for terrigenic SF6 in ground water from basaltic aquifers Jeju Island, Korea. J. Hydrol. 339: 93–104.

Komatina M (1975) Development conditions and regionalization of karst, in Hydrogeology of Karstic Terrains (eds Burger A, Dubertret L), IAH, Paris, pp. 21–29.

Konikow LF, Bredehoeft JD (1978) Computer model of two dimensional solute transport and dispersion in groundwater, Techniques of Water Resources Investigation, US Geol. Surv., Chapter C-2, p. 90.

Korkmaz N (1990) The estimation of groundwater recharge from spring hydrographs. J. Hydrol. Sci. 35(2): 209–17.

Krampe K (1983) Hydrogeological requirements for rural water supply in Central Java, Indonesia, in Proc. Int. Symp, on Ground Water in Water Resources Planning,

Vol. 1, UNESCO—IAH—IAHS, Kolbanz, Germany, pp. 201–7.

Krasny J (1996) State of the art of hydrogeological investigations in hard rocks: The Czech Republic, in Hardrock Hydrogeology of the Bohemain Massif (eds Krasny J, Mls J), Acta Universitatis Carolinac, Geologica, Univerzita Karlova, Vol. 40, pp. 89–101.

Krasny J, Sharp JM (eds) (2008) Groundwater in Fractured Rocks. IAH Selected Papers on Hydrogeology, Vol. 9. CRC Press, Taylor and Francis Group, The Netherlands, p. 654.

Krasny J (1999) Hard-rock hydrogeology in the Czech Republic. Hydrol. 2: 25–38.

Kresic N (2007) Hydrogeology and Groundwater Modeling, 2nd ed., CRC Press, Boca Raton, FL, p. 805.

Krishnamoorthy TM, Nair RN, Sarma TP (1992) Migration of radionuclides from a granite repository. Water Resources Res. 28(7): 1927–34.

Krishnan MS (1949) Geology of India and Burma, The Madras Law Journal Office, Madras, p. 544.

Krishnaswamy VS (2008) The geological environment of some ancient caves of India:their optimum utilization for speleological exploration and hydrogeological research. J. Geol. Soc. India 71(5): 630–50.

Krumbein WC, Monk GD (1942) Permeability as a function of the size parameters of unconsolidated sand. Am. Inst. Mining Engg. Tech. Publ., Littleton, Co., pp. 153–63.

Kruseman GP, de Ridder NA (1970) Analysis and Evaluation of Pumping Test Data. Intl. Inst. for Land Reclamation and Improvement, Bulletin 11, Wageningen, p. 200.

Kruseman GP, de Ridder NA (1990) Analysis and Evaluation of Pumping Test Data, 2nd ed., Intl. Inst. for Land Reclamation and Improvement, Publication 47, Wageningen, p. 377.

Kueper BH, McWhorter DB (1992) The behavior of dense non-aqueous phase liquids in fractured clay and rock. Ground Water 29(5): 716–28.

Kukillaya JP, Kunhi AM, Abdul Rahman AK (1992) Basic and ultrabasic dykes as potential aquifers in hard rock terrains. Bhu-Jal News. 7(4): 14–19.

Kuo MCT, Fan K, Kuochen H et al (2006) A mechanism for anomalous decline in radon precursory to an earthquake. Ground Water 44(5): 642–47.

LaMoreaux PE (1986) Hydrology of Limestone Terranes. IAH Intl. Contributions to Hydrogeology, Vol. 2, Verlag Heinz Heise, Hannover, p. 342.

LaMoreaux PE, LeGrand HE, Stringfield VT (1975) Progress of knowledge about hydrology of carbonate terrains, in Hydrogeology of Karstic Terrains (eds Burger A, Dubertret L), IAH, Paris, pp. 41–52.

LaRiccia MP, Rauch HW (1977) Water well productivity related to photo-lineaments in carbonates of Fredrick Valley, Maryland, in Hydrologic Problems in Karst Regions (eds Dilamarter RR, Csallany SC), Western Kentucky University, Bowling Green, Kentucky, pp. 228–34.

Laslett GM (1982) Censoring and edge effects in areal and line transect sampling of rock traces. J. Int. Assoc. Math. Geol. 14: 125–40.

Lasserre F, Razack M, Banton O (1999) A GIS-linked model for the assessment of nitrate contamination in groundwater. J. Hydrol. 224: 81–90.

Lattman LH, Parizek RR (1964) Relationship between fracture traces and the occurrence of groundwater in carbonate rocks. J. Hydrol. 2: 73–91.

Laubach S (1992) Fracture networks in selected Cretaceous sandstones of the Green River and San Juan Basins, Wyoming, New Mexico, and Colorado. Geological studies relevant to horizontal drilling: Examples from Western North America (eds Schmoker JM, Coalson EB, Brown CA), Rocky Mountain Assoc. of Geologists, pp. 61–73.

Lawrence A, Stuart M, Cheny C et al (2006) Investigating the scale of structural controls on chlorinated hydrocarbon distributions in the fractured-porous unsaturated zone of a sandstone aquifer in the UK. Hydrogeol. J. 14: 1470–82.

Lawrence CR (1975) Geology, Hydrodynamics and Hydrochemistry of the Southern Murray Basin, Vol. 1, Geol. Surv. of Victoria, Mem. 30, Mines Dept., Victoria, p. 357.

Leap DI, Mai PA (1992) Influence of pore pressure on apparent dispersivity of a fissured dolomitic aquifer. Ground Water 30(1): 87–95.

Leblanc M, Casiot C, Elbaz- Poulichert F et al (2002) Arsenic removal by oxidizing bacteria in a heavily arsenic-contaminated acid mine drainage system (Carnoules, France), in Mine Water Hydrogeology and Geochemistry (eds Younger PL, Robins NS), Geol. Soc. London, Spl. Publ. No. 189, pp. 267–74.

Lee CF, Zhang JM and Zhang YX (1996) Evaluation and origin of the ground fissures in Xian. China. Engg. Geol. 43(1): 45–55.

Lee CH, Deng BW, Chang JL (1995) A continuum approach for estimating permeability in naturally fractured rocks. Engg. Geol. 39: 71–85.

Lee CH, Farmer I (1993) Fluid Flow in Discontinuous Rocks. Chapman and Hall, London, p. 169.

Lee JY, Cho BW (2008) Submarine groundwater discharge into the coast revealed by water chemistry of man-made underwater liquefied petroleum gas caverns. J. Hydrol. 360(1–4): 195–206.

Lee K (1969) Infrared exploration for shoreline springs of Mono Lake, California test site. Proc. Sixth Int. Symp., on Remote Sensing of Evnironment, Ann Arbor, Michigan, pp. 1275–87.

Legchenko A, Baltassat JM, Bobachev A et al (2004) Magnetic resonance sounding applied to aquifer characterization. Ground Water 42(3): 363–73.

LeGrand HE (1964) System for evaluation of contamination potential of some waste disposal sites. J. Am. Water Work Assoc. 56: 959–74.

LeGrand HE (1967) Groundwater of the Piedmont and Blue Ridge Provinces of the Southeastern States. US Geol. Surv. Circular. 238: 11.

LeGrand HE (1970) Movement of agricultural pollutants with groundwater, in Agricultural Practices and Water Quality, Iowa State University Press, Ames, Iowa, pp. 303–13.

Lerner D, Stelle A (2001) In-situ measurement of fracture apertures using NAPL injection, in Conf. Proc. Fractured Rock 2001 (eds Keuper BH et al), EPA; Ontario.

Lerner DN (1997) Groundwater recharge, in Geochemical processes, weathering and groundwater recharge in catchments (eds Saether OM, de Caritat P), AA Balkema, Rotterdam, pp. 109–50.

Levens RL, Williams RE, Rushton KR (1994) Hydrogeologic role of geologic structures. Part 2: Analytical models. J. Hydrol. 156: 245–63.

Li C, Tang X, Ma T (2006) Land subsidence caused by groundwater exploitation in Hangzhou–Jiaxing- Huzhon Plain, China. Hydrogeol. J. 14(8): 1652–1665.

Li W, Englert A, Cirpka OA et al (2007) Two dimensional characterization of hydraulic heterogeneity by multiple pumping tests. Water Resources Res. 43(4): W04433.

Lillesand T, Kiefer RW, Chipman J (2007) Remote Sensing and Image Interpretation, 6th ed., Wiley, Chicester, p. 804.

Linsley RK, Kohler MA, Paulhus JLH (1982) Hydrology for Engineers, 3rd ed., McGraw-Hill Book Co., New York.

Liu J, Zhaeng C, Zheng L et al (2008) Ground water sustainability: Methodology and application to the North China Plain. Ground Water 46(6): 897–909.

Lloyd JW (1991) Hydrogeology in developing countries, in Geosciences in Development (eds Stow DAV, Laming DJC), A.A. Balkema, Rotterdam, pp. 107–14.

Lloyd JW (1999) Water resources of hard rock aquifers in arid and semi-arid zones. Studies and Reports in Hydrology. UNESCO, Paris, 58.

Lloyd JW, Hearthcote JA (1985) Natural Inorganic Hydrochemistry in Relation to Ground Water—An Introduction Clarendon Press, Oxford, p. 296.

Loiselle M, Evans D (1995) Fracture density distributions and well yields in coastal Maine. Ground Water 33(2): 190–96.

Long A (1995) Lecture notes in Training workshop on Environmental Isotope Techniques in Hydrology. Roorkee, India, p. 232.

Long AJ, Sawyer JF, Putnam LD (2008) Environmental tracers as indicators of karst conduits in groundwater in South Dakota, USA. Hydrogeol. J. 16(2): 263–80. doi:10.1007/s10040-007-0232-7.

Long JCS, Remer JS, Wilson CR, Witherspoon PA (1982) Porous media equivalents for networks of discontinuous fractures. Water Resources Res. 18(3): 645–58.

Long JCS, Witherspoon PA (1985) The relationship of degree of interconnection to permeability in fracture networks. J. Geophys. Res. 90(B4): 3087–98.

Loosli HH, Purtschert R (2005), Rare Gases. In: Aggarwal PK et al (eds) (2005) Isotopes in the Water Cycle: Past, Present and Future of a Developing Science. Springer, Dordrecht.

Louis C (1974) Rock hydraulics, in Rock Mechanics, (Muller L ed.), Springer, Verlag, Wien, pp. 299–387.

Louis C (1977) Suggested methods for determining hydraulic parameters and characteristics of rock masses. Intl. Soc. of Rock Mechanics, unpublished report, p. 137.

Louis C (1984) An introduction to rock hydraulics, lecture notes (unpublished), SIMESCOL, Paris, p. 16.

Lu Z, Danskin WR (2001) In SAR analysis of natural recharge to define structure of a ground-water basin, San Bernardino, California. Geophys. Res. Lett. 28(13): 2661–64.

Lubczynski MW (2009) The hydrogeological role of trees in water-limited environments. Hydrogeol. J. 17(1): 247–59. doi:10.1007/s10040-008-0357-3.

Lubczynski MW, Gurwin J (2005) Integration of various data sources for transient groundwater modeling with spatio-temporally variable fluxes—Sardon study case, Spain. J. Hydrol. 306: 71–96.

Lusczynski NJ (1961) Head and flow of ground water of variable density. J. Geophys. Res. 66: 4247–56.

Mabee SB, Hardcastle KC (1997) Analyzing outcrop-scale fracture features to supplement investigations of bedrock aquifers. Hydrogeol. J. 5(4): 21–36.

MacDoland AM, Dochartaigh BEO, Calow RC (2009) Mapping of groundwater development costs for the transboundary

Western Aquifer Basin, Palestine/Israel. Hydrogeol. J. 17, doi:10.1007/s10040-009-0471-x.

Mace RE (1999) Estimation of hydraulic conductivity in large-diameter hand dug wells using slug test methods. J. Hydrol. 219: 34–45.

Mace, RE (1997) Determination of transmissivity from specific capacity tests in karst aquifers. Ground Water 35(5): 738–742.

Mackie CD (1982) Multi-rate testing in fractured rock aquifers, in Proc. AWRC Conference on Groundwater in Fractured Rocks, Canberra, pp. 139–50.

Maini T, Hocking G (1977) An examination of the feasibility of hydrologic isolation of an high level waste repository in crystalline rocks. Invited paper Geologic Disposal of High Radioactive Waste Session, Ann. Meet Goel. Soc. Am. Seattle, Washington.

Mallik SB, Bhattacharya DC, Nag SK (1983) Behaviour of fractures in hard rocks—A study by surface geology and radial VES method. Geophysics. 21: 181–89.

Maloszewski P, Zuber A (1992) On the calibration and validation of mathematical models for the interpretation of tracer experiments in groundwater. Adv. Water Resour. 15: 47–62.

Mandel S, Shiftan ZL (1981) Groundwater Resources—Investigations and Development. Academic Press, New York, p. 269.

Marechal JC, Ladouche B, Dorflinger N et al (2008) Interpretation of pumping tests in a mixed flow karst system. Water Resources Res. 44, W05401: 1–18.

Marsden SS, Davis SN (1967) Geological subsidence. Sci Am. 216(6): 93–100.

Marsh TJ, Davies PA (1983) The decline and partial recovery of groundwater levels below London. Proc. Inst. Civil Engrs., Part 1, 74: 263–76.

Maslia ML, Prowell DC (1990) Effect of faults on fluid flow and chloride contamination in a carbonate aquifer system. J. Hydrol. 115(1): 1–49.

Mather JD (1993) Pollution of shallow hard rock systems an overview, in Mem. 24th Congress of IAH, Olso.

Mather JD (1995) Preventing groundwater pollution from landfilled waste—is engineer containment an acceptable solution, in Ground Water Quality (eds Nash H, McCall GJH), Chapman and Hall, London, pp. 191–95.

Mather PM (2004) Computer Processing of Remotely Sensed Images–An Introduction, 3rd ed., Wiley, Chicester, p. 442.

Mathers S, Zalasiewicz J (1994) A Guide to the Sedimentology of Unconsolidated Sedimentary Aquifers, Tech. Report WC/93/32, BGS, p. 117.

Matter JM, Goldberg DS, Morin RH et al (2006) Contact zone permeability at intrusion boundaries: new results from hydraulic testing and geophysical logging in the Newark Rift Basin, NY, USA. J. Hydrol. 14: 689–699.

Matthess G (1982) The Properties of Groundwater. John Wiley and Sons, New York, p. 406.

Matthews RA (1985) High elevation aquifer-potential sources of water and electrical supply. Mem. 17 of IAH, Tucson, pp. 720–28.

Matzko JR (1994) Geology of the Chinese nuclear test site near Lop Nor, Xinjiang Ugyur Autonomous Region, China. Engg. Geol. 36: 173–81.

Maxey GB (1964) Geology: Part I, Hydrogeology, in Handbook of Applied Hydrology (Chow VT ed.), McGraw-Hill Book Co., New York, pp. 4–38.

Mazor E (1991) Chemical and isotopic groundwater hydrology, 2nd ed., Marcel Dekker, Inc.

McCann DM, Baria R, Jackson PD, Green ASP (1986) Application of cross-hole seismic measurements in site investigation surveys. Geophysics. 51(4): 914–29.

McCauley JF, Schaber GC, Breed CS et al (1982) Subsurface valleys and geoarcheology of the eastern Sahara revealed by shuttle radar. Science 218: 1004–19.

McConnell CL (1993) Double porosity well testing in the fractured carbonate rocks of the Ozarks. Ground Water 31: 75–83.

McDonald MG, Harbaugh AW (1988) A modular three dimensional finite difference groundwater flow model. Techniques of Water Resources Investigations, Chapter A1, US Geol. Surv, p. 586.

McDowell PW (1979) Geophysical mapping of water filled fracture zones in rocks. Int. Asso. Eng. Geol. Bull. 19: 258–64.

McEdwards, DG, Tsang CF (1977) Variable rate multiple well testing analysis, in Proc. Int. Well-Testing Symposium, Berkeley, pp. 92–99.

McFarlane MJ, Chilton PJ, Lewis MA (1992) Geomorphological controls on borehole yields; a statistical study in an area of basement rocks in central Malawi, in The Hydrogeology of Crystalline Basement Aquifers in Africa (eds Wright EP, Burgess WG), Geol. Soc. Spl. Publ. No. 66, pp. 131–54.

Mehta NS, Rajawat AS, Bahuguna IM et al (1994) Remote sensing for mineral exploration: Geological potential of ERS-I SAR data covering parts of Aravalli hills and Thar desert (Rajasthan, India), Project Report, SAC/RSA/RSAG/ERS-1/GEO/PR/01/94, p. 67.

Meijerink AMJ, de Brouwer HAM, Mannacrts CM, Valenzuela CR (1994) Introduction to the Use of Geographic Information Systems for Practical Hydrology. ITC Publ. No. 23, Enschede.

Meijerink AMJ, Bannert D, Batelaan O, Lubcyznski MW, Pointet T (2007), Remote Sensing Applications to Groundwater. IHP-VI, Series on Groundwater No. 16, UNESCO, Paris, p. 312.

Meinzer DE (1923), Outline of Groundwater Hydrology, with Definitions. US Geol. Surv. Water Supply Paper 494.

Mekel JFM (1978) ITC textbook of photo-interpretation. Chap. VIII. The use of aerial photographs and other images in geological mapping. ITC, Enchede.

Mercer JW, Faust CR (1981) Groundwater Modelling. National Water Well Association (NWWA), Worthington, Ohio, p. 60.

Merin IS (1992) Conceptual model of ground water flow in fractured siltstone based on analysis of rock cores, borehole geophysics, and thin sections. GWHR, Fall. 1992, pp. 118–25.

Meyer PD, Valocchi AJ, Eheart JW (1994) Monitoring network design to provide initial detection of groundwater contamination. Water Resources Res. 30(9): 2647–59.

Meyer RR (1963) A chart relating well diameter, specific capacity, and the coefficient of transmissibility and storage, in Methods of Determining Permeability and Drawdown (compiled by R. Bentall) US Geol. Surv. Water Supply Paper, 1536–I, pp. 338–41.

Michael HA, Voss CI (2009) Controls on groundwater flow in the Bengal basin of India and Bangladesh: Regional modeling analysis. Hydrogeol. J. 17(7), doi:10.1007/s10040-008-0429-4.

Michalski A (1990) Hydrogeology of the Brunswick (Passaic), Formation and implications for groundwater monitoring practice. GWMR, Fall. 1990, pp. 134–43.

Michalski A, Britton R (1997) The role of bedding fractures in the hydrogeology of sedimentary bedrock-evidence from the Newark basin, New Jersey. Groundwater. 35(2): 319–27.

Milanovic PT (1981) Karst Hydrology. Water Resources Publ. Colorado.

Miller DW (1985) Chemical contamination of ground water, in Ground Water Quality (eds Ward CH, Giger W, McCarty PL), John Wiley and Sons, New York, pp. 39–52.

Miller VC, Miller CF (1961). Photogeology. McGraw-Hill, New York.

Minzi S, Fagao H, Ruifen L, Baoling N (1988) Application of oxygen and hydrogen isotopes of waters in Tengchong hydrothermal systems of China in, Isotope Applications in Hydrology in Asia and the Pacific, Beijing, IAEA, pp. 23–34.

Mishra SK, Singhal BBS, Singhal DC (1989) Analysis of pumping test data from borewells in fractured rocks of Karnataka, in Proc. International Workshop on Appropriate Methodologies for Development and Management of Ground Water Resources in Developing Countries, NGRI, India. (Gupta CP ed.), Vol. I, pp. 15–25.

Misra KS (2002) Arterial system of lava tubes and channels within the Deccan volcanics of western India. J. Geol. Soc. India 59(2): 115–24.

Moench AF (1984) Double porosity model for a fissured ground water reservoir with fracture skin. Water Resources Res. 20(7): 831–46.

Mojtabai N, Cetintas A, Farmer IW, Savely J (1989) In-place and excavated block size distributions, 30th US Symp. Rock Mechanics, West Virginia University, pp. 537–44.

Mondal NC, Singh VS (2004) A new approach to delineate the groundwater recharge zone in hard rock terrain. Curr. Sci., 87: 658–662.

Moore RB, Schwartz GE, Clark SF Jr. et al (2002) Factors related to well yield in the fractured- bedrock aquifer of New Hampshire. USGS Professional Paper 1660, USGS, Denver, Co, USA.

Moore WS (1996) Large groundwater inputs to coastal waters revealed by ^{226}Ra enrichments, Nature 380: 612–14.

Morland G, Strand T, Furuhaug H et al (1998) Radon in Quaternary aquifers related to underlying bedrock geology. Ground Water 36(1): 143–46.

Morris DA, Johnson AI (1967) Summary of Hydrologic and Physical Properties of Rock and Soil Materials, as Analyzed by the Hydrologic Laboratory of the U.S. Geological Survey 1948–1960. US Geol. Surv. Water Supply Paper, 1839-D, p. 42.

Morrison RJ, Prasad RA, Brodie JE (1984) Chemical hydrology of small tropical islands, CSC Tech. Publ. Science, 154: 211–23.

Motz LH (1982) Lower Hawthorn Aquifer on Sanibel Island, Florida. Ground Water 20(2): 170–78.

Nace RL (ed) (1971) Scientific Framework of World Water Balance UNESCO Tech. Papers Hydrol. 7: 27.

Nair AR, Sinha, UK, Joseph, TB (1993) Application of environmental radioisotopes for studying groundwater recharge process in arid areas, in Proc. 2nd National Symp. on Environment, Jodhpur, India, pp. 188–90.

Nair VN (1981) Sedimentological, Hydrogeological Studies of Chorward. Madhavpur area, Junagadh District, Gujarat, India, Ph.D Thesis, (unpublished) University of Roorkee, p. 328.

Narasimhan TN (1982) Numerical modeling in hydrogeology, in Recent Trends in Hydrogeology (Narasimhan TN ed.), Special Paper 189, Geological Society of America, Boulder, Colorado, pp. 273–96.

Nath B, Chakraborty S, Jana J et al (2006) Arsenic in the groundwater of the Bengal delta plain: hydrogeochemical studies from the districts of Nadia and South 24 Parganas, West Bengal, India (Abst), in Intl. Conference on Groundwater for Sustainable Dev. (eds Ramanathan AL et al), Delhi.

Nativ R, Adar E, Assaf L et al (2003) Characterization of the hydraulic properties of fractures in chalk. Ground Water 41(4): 532–43.

Navada SV, Nair AR, Rao SM et al (1993) Groundwater recharge studies in arid region of Jalore, Rajasthan using isotope techniques. J. Arid Environ. 24: 125–33.

Neerdael B, Wemaere I, Thury M et al (1996) Characteristion of argillaceous formations as potential host rock for geological disposal in Belgium and in Switzerland. Abstract Vol. 2, 30th IGC, Beijing.

Neuman SP (1972) Theory of flow in unconfined aquifers considering delayed response of the water table. Water Resources Res. 8: 1031–49.

Neuman SP (1975) Analysis of pumping test data from anisotropic unconfined aquifers considering delayed gravity response. Water Resources Res. 11(2): 329–42.

Neuman SP (2005) Trends, prospects and challenges in quantifying flow and transport through fractured rocks. Hydrogeol. J. 13(1): 124–47.

Neuzil CE (1982) On conducting the modified slug test in tight formations. Water Resources Res. 18(2): 439–41.

Neuzil CE (1986) Groundwater flow in low permeability environments. Water Resources Res. 22(8): 1163–95.

Neves MA, Morales NC (2007) Well productivity controlling factors in crystalline terrains of south eastern Brazil. Hydrogeol. J. 15: 471–82.

Niedzielski H (1993) Productivity of the basalt aquifer of the eastern part of Chichinautzin Group, Mexico, in Mem. 24th Congress of IAH, Oslo, Part I, pp. 281–89.

Nilsson B, Sidle RC (1996) Flow and solute transport in a contaminated clay till in Denmark, Abstracts 30th IGC, Beijing, Vol. 3, p. 270.

Ninomiya Y, Fu B, Cudahy TJ (2005) Detecting lithology with advanced spaceborne thermal emission and reflection radiometer (ASTER) multispectral thermal infrared 'radiance-at-sensor' data. Remote Sensing of Environment, 99, pp. 127–139.

Novakowski KS et al (2007) Groundwater flow and solute transport in fractured media, in The Handbook of Groundwater Engineering, 2nd edition, (Delleur JW ed.), CRC Press, Boca Raton, pp. 20–1 to 20–43.

National Research Council, USA NRC (1996) Rock Fractures and Fluid Flow- Contemporary Understanding and Applications. National Academic Press, Washington, DC, p. 551.

Oda M (1984) Permeability Tensor for Discontinuous Rock Masses. Geotechnique 35: 483–485.

Oda M, Saitoo T, Kamemura K (1989) Permeability of rock masses at great depth, in Proc. Symp. Rock at Great Depth (eds Maury V, Fourmaintrant D), A.A. Balkema, Rotherdam, pp. 449–56.

Oden M, Niemi A, Tsang CF et al (2008) Regionalized channel transport in fractured media with matrix diffusion and linear sorption. Water Resources Res. 44(2): W02421.

Offerdinger US, Balderer W, Loew S et al (2004) Environmental isotopes as indicator for ground water recharge to fractured granite. Ground Water 42(960): 868–79.

O'Leary DW, Friedman JD, Pohn HA (1976) Lineament, linear and lineation: some proposed new standards for old terms. Geol. Soc. Am. Bull. 87: 1463–69.

Ollier C (1969) Volcanoes. The MIT Press, Cambridge, USA, p. 177.

Olsson O, Falk L, Forslund O, Lundmark L, Sandberg E (1992) Borehole Radar Applied to the Characterization of Geophysical Prospecting 40: 109–42.

Otkun G (1977) More about Palaeocene karst aquifer in Saudi Arabia, in Karst Hydrogeology (eds Tolson JS, Doyle FL), Mem. 12th Congress, IAH, UAH Press, Alabama, pp. 25–38.

Owen M (1981) The Thames groundwater scheme, in Case Studies in Groundwater Resources Evaluation, (Lloyd JW ed.), Clarendon Press, Oxford, pp. 186–202.

Owoade A (1993) Some field observations on the groundwater potential in tropical hard rock catchments, in Mem. 24th Congress of IAH, Oslo Part 2, pp. 1161–65.

Padadopulos SS, Bredehoeft JD, Cooper HH (1973) On the analysis of slug test data. Water Resources Res. 9: 1087–89.

Pahl PJ (1981) Estimating the mean lenth of discontinuity trace. Int. J. Rock Mech. Min. Sci. & Geomech. Abstr. 18: 221–28.

Paloc H, Back W (eds) (1992) Hydrogeology of Selected Karst Regions, IAH contributions to Hydrogeology, Vol. 13, Verlag-Heinz, Hannover, p. 514.

Panda BB, Kulatilake HSW (1995) Study of the effect of joint geometry parameters on the permeability of jointed rocks. In: Daemen and Schultz (eds), Rock Mechanics, AABalkema, Rotterdam, pp. 273–78.

Papadopulos IS, Cooper HH (1967) Drawdown in a well of large diameter. Water Resour. Res. 3(1): 241–44.

Parasnis DS (1997) Principles of Applied Geophysics. 5th ed, Chapman & Hall, 413 p.

Parizek RR (1976) Lineaments and groundwater, in Interdisciplinary application and interpretations of EREP data within the Susquehanna River Basin (eds McMurthy GT and Peterson GW), Pennsylvania State Univ., pp. 4–59 to 4–86.

Patel SC, Mishra GC (1983) Analysis of flow to large-diameter well by discrete kernel approach. Ground Water. 21(5): 573–76.

Payne BN (1967) Isotope techniques in groundwater hydrology, in Methods and Techniques of Groundwater Investigation and Development. Water Resources Series, No. 33, UNESCO, pp. 107–13.

Pearce BR (1982) Fractured rock aquifers in central Queensland, in Papers of the Groundwater in Fractured Rock Conference 1982, AWRC Conference Series No. 5, pp. 161–72.

Perumal A (1990) Hydromorphogeological Investigations in parts of Athur Valley, Tamilnadu, India, using remote sensing data, Ph.D. Thesis (unpublished), University of Roorkee.

Peterson FL (1972) Water development on tropic volcanic islands—type example: Hawaii. Ground Water 10(5): 18–23.

Peterson FL (1984) Hydrogeology of high oceanic islands. Proc. Regional Workshop on Water Resources of Samll Islands, Part 2, CSC, London, pp. 431–35.

Philip G, Singhal BBS (1992) Importance of geomorphology for hydrogeological study in hard rock terrain, an example from Bihar Plateau through remote sensing. Ind. J. Earth Sci. 19(4): 177–88.

Pinder GF (2002) Groundwater Modeling using Geographical Information System. John Wiley and Sons, Inc., New York, p. 233.

Pinder GF, Grey WG (1977) Finite Element Simulation in Surface and Subsruface Hydrology. Academic Press, New York, p. 295.

Piper AM (1953) A graphic procedure in the geochemical interpretation of water analyses. Ground water Notes No. 12, USGS, pp. 14.

Pitale U, Padhi R (1996) Geothermal environment and energy potential of mobile crustal belt and intraplate tectonic zones in India, Abstracts 30th IGC, Beijing, Vol. 1, p. 386.

Plafker G (1969) Tectonics of the March 27, 1964, Alaska Earthquake, US Goel. Surv. Prof. Paper 543-I, p. 174.

Pointet T (1989) Hydrogeological prospecting of hard rock aquifers in semi-arid climates for small scale irrigation. Abstracts 28th IGC, Washington, Vol. 2, p. 620.

Poland JF (ed) (1984) Guidebook to Studies of Land Subsidence due to Ground-Water Withdrawal. Studies and Reports in Hydrology. UNESCO, 40, p. 305.

Pollock DW (1990) Documentation of Computer Programs to Compute and Display Path Lines Using Results from the US Geol. Surv. Modular Three Dimensional Finite Diffeence Groundwater Flow Model. US Geol. Surv. Open-File Report 89 pp. 381–88.

Pool DR, Eychaner JH (1995) Measurements of aquifer-storage change and specific yield using gravity surveys. Ground Water 33(3): 425–32.

Powell RL (1977) Joint patterns and solution channel evolution in Indiana, in Karst Hydrogeology (eds Tolson JS, Doyle FL), UAH Press, Huntsville, Alabama, pp. 255–69.

Prasad RD (1999) Analysis of groundwater flow for central part of Paler subbasin, Andhra Pradesh. M.Tech Dissertation (unpublished), University of Roorkee, Roorkee, p. 71.

Price NJ, Cosgrove J (1990) Analysis of Geological Structures, Cambridge University Press, p. 502.

Prickett TA (1965) Type curve solution to aquifer tests under water-table conditions. Ground Water 3(3): 5–14.

Prickett, TA, Naymik TG, Lonnquist CG (1981) A "Random Walk" Solute Transport Model for Selected Groundwater Quality Evaluations, Illinois State Water Survey Bulletin No. 65.

Priest SD (1993) Discontinuity Analysis for Rock Engineering. Chapman & Hall, London, p. 473.

Pronk M, Godscheider N, Zopfi J (2009) Microbial communities in karst groundwater and their potential use for biomonitoring. Hydrogeol. J. 17, doi:10.1007/s10040-008-0350-x.

Pruess K, Wang JSY (1987) Numerical modelling of isothermal and nonisothermal flow in unsaturated fractured rock—A review. Geophysical Monograph 42, Am. Geophys. Union, pp. 11–21.

Puri S, El Naser H (2003) Intensive use of groundwater in trans-boundary aquifers, in Intensive Use of Groundwater (eds Llmas R, Custodio E) A.A. Balkama Publ., Tokyo, pp. 415–38.

Put M, Ortiz L (1996) Large scale in-situ measurement of the hydraulic parameters of the Boom clay formation, Abstract 30th IGC, Beijing, Vol. 3, p. 268.

Rabbel W (2006) Seismic methods, in Groundwater Geophysics—A tool for Hydrogeology, (Kirsch R ed.), Springer, pp. 23–84.

Radhakrishna I, Athavale RM, Singh VS, Govil PK (1976) Hydrogeolgical and hydrochemical studies in Lower Maner Basin. Technical Report No. GH3-HG1, Indo-German Collaboration Project on Exploration and Management of Groundwater Resources, NGRI, Hyderabad.

Radhakrishna I et al (1979) Evaluation of Shallow Aquifers in Lower Maner Basin. Unpublished Report, NGRI, Hyderabad, p. 115.

Rai KL (1994) Environmental aspects of base-metal mining in India: selected case studies, in Mineral Development and Environment (eds Rai KL, Singh G), Recent Researches in Geology, Hindustan Publishing Corporation (India), New Delhi. Vol. 15, pp. 325–40.

Ramakrishna DM, Viraraghavan T, Jin YC (2006) Iron oxide coated sand for arsenic removal: Investigation of Coating Parameters using factorial design approach. Prentice Periodical of Hazardous, Toxic and Radioactive Waste Management. ASCE, 10(4): 198–206.

Ramey HJ Jr. (1977) Petroleum engineering well test analysis—State of the art, in Proc. Invitational Well Testing Symp., Lawrence Berkeley Lab, California, pp. 5–13.

Randall-Roberts JA (1993) Geophysical prospecting for water in Precambrian Gneiss, Oaxaca, Mexico, Mem. of the 24th Congress of IAH, Oslo, 1993, pp. 648–54.

Ranganai RT, Ebinger CJ (2008) Aeromagnetic and Landsat TM structural interpretation for identifying regional groundwater exploration targets, south-central Zimbabwe Craton. J. Appl. Geophys. 65(2): 73–83.

Rangarajan R, Athavale RN (2000) Annual replenishable ground water potential of India-an estimate based on injected tritium studies. J. Hydrol. 234(1–2): 38–53.

Rao SH (1975) Hydrogeology of parts of the Deccan Traps, Ph.D. Thesis (unpublished), Poona University, India, p. 190.

Rasmuson A, Neretnieks I (1981) Migration of radionuclides in fissured rock: the influence of micropore diffusion and longitudinal dispersion. J. Geoph. Res. 86(B5): 3749–58.

Raven KG, Novakowski KS, Lapsevic PA (1988) Interpretation of field tracer tests of a single fracture using a transient solute storage model. Water Resources Res. 24(12): 2019–32.

Raymahashay BC (1996) Geochemistry for Hydrologists. Allied Publishers Ltd., New Delhi, p. 190.

Razack M, Huntley D (1991) Assessing transmissivity from specific capacity in a large and heterogeneous alluvial aquifer. Ground Water. 29(6): 856–61.

Read MD, Meredith PG and Murrell SAF et al (1989) Permeability measurement techniques under hydrostatic and deviatoric stress conditions, in Rock at Great Depth (eds Maury V, Fourmaintraux D) Vol. 1, A.A. Balkema, Rotterdam, pp. 211–17.

Read RE (1982) Estimation of optimum drilling depth in fractured rock, in Papers of the Groundwater in Fractured Rocks, AWRC Conference, Series No. 5, Canberra, pp. 191–97.

Reboucas A, daC (1993) Groundwater development in the Precambrian shield of South America and west side of Africa, in Hydrogeology of Hard Rocks. Mem. 24th Congress IAH, Oslo, PT. 2, pp. 1101–15.

Rector JW (ed) (1995) Cross Well Methods. Geophysics, special issue, Vol. 60, No. 3.

Reddy DV, Sukhija BS, Nagabhushanam P et al (2006) Rn-emanometry: A tool for delineation of fractures for groundwater in granitic terrains. J. Hydrol. 329: 186–93.

Reddy TN, Raj P (1997) Hydrogeological conditions and optimum well discharges in granitic terrain in parts Nalgonda district, Andhra Pradesh, India. J. Geol. Soc. Ind. 49(1): 61–74.

Reichard JS, Leap DL (1998) The effects of pore pressure on the conductivity of fractured aquifers. Ground Water 36(3): 450–56.

Reilly TE (1990) Simulation of dispersion in layered coastal aquifer systems. J. Hydrol. 114: 211–28.

Remson I, Hornberger GM, Molz FJ (1971) Numerical Methods in Subsurface Hydrology. Wiley-Interscience, New York, p. 389.

Richards CJ, Roaza H, Roaza RM (1993) Integrating Geographic information systems and Modflow for Groundwater resources assessment. Water Resour. Bull. (AWRA) 29(5): 847–53.

Rifai HS, Hendricks LA, Kilborn K, Bedient PB (1993) A geographic information system (GIS) user interface for delineating wellhead protection areas. Ground Water 31: 480–88.

Ritzi RF Jr., Andolsek RH (1992) Relation between anisotropic transmissivity and azimuthal resistivity surveys in shallow fractured carbonate flow systems. Ground Water 30(5): 774–780.

Roaza H, Roaza RM, Wagner JR (1993) Integrating geographic information systems in groundwater applications using numerical modelling techniques. Water Resources Bull. (AWRA), 29(6): 981–88.

Rodell M, Famiglietti JS (2002) The potential for satellite-based monitoring of groundwater storage changes using GRACE: The high plains aquifer, Central US. J. Hydrol. 263(1–4): 245–56.

Roeloffs EA (1998) Persistent water level changes in a well near Parkfield, California due to local and distant earthquakes. J. Geophys. Resources 103B1: 869–89.

Rogbeer G (1984) Mauritius—Water resources, in Water Resources of Small Islands, CSC Tech. Publ. Series No. 182, pp. 114–39.

Rogers RB, Kean WF (1981) Monitoring groundwater contamination at a flash disposal site using surface electrical resistivity methods. Ground Water 18(5): 472–78.

Romani S, Sharma KD, Ghosh NC et al (eds) (2006) Groundwater Governance—Ownership of groundwater and its pricing. Proc. 12th National Symp. on Hydrology. Capital Publ. Co., New Delhi, p. 514.

Romani S, Singhal BBS (1970) A study of some of the thermal springs of Kulu district, H.P. Indian. Geohydrology 6: 57–68.

Ronka E, Niini H, Suokko T (eds) (2005) Proceedings of the Fennoscandian 3rd Regional Workshop on Hardrock Hydrogeology. Helsinki, Finland. Finnish Environment Institute, Helsinki, Finland, p. 121.

Rosenbom AE et al (2001) Migration and distribution of NAPL in fractured clayey till, in Proc. Conference Fractured Rock 2001 (eds Kueper BH et al).

Rotzoll K, El-Kadi AI (2008) Estimating hydraulic conductivity from specific capacity for Hawaii aquifers. Hydrogeol. J. 16(5): 969–79.

Rovey CWII, Cherkauer DS (1995) Scale dependency of hydraulic conductivity measurements. Ground Water 33(5): 769–80.

Rowan LC, Hook SJ, Abrams MJ, Mars JC (2003) Mapping hydrothermal altered rocks at Cuprite, Nevada: the advanced spaceborne thermal emission and reflection radiometer (ASTER), a new satellite imaging system. Economic Geology 98(5): 1019–1028.

Rowe RK, Booker JR, Hammoud A (1989) The effect of multi-directional matrix diffusion on contaminant transport through fractured systems, in Contaminant Transport in Groundwater (eds Kobus HE, Kinzelbach W), A.A. Balkema, Rotterdam, pp. 259–66.

Roy J, Lubczynski M (2003) The magnetic resonance sounding technique and its use for groundwater investigations. Hydrogeol. J. 11(4): 455–65.

Ruden F (1993) A revival of the qanat principle for rural water supply in a third world context, in Mem. 24th Congress of IAH, Uslo, pp. 1183–90.

Rugh DF, Burbey T (2008) Using saline tracers to evaluate preferential recharge in fractured rocks, Floyd County, Virginia, USA. Hydrogeol. J. 16: 251–62.

Ruhland M (1973) Me'thode d'etude de la fracturation naturelle des roches associee'e a' divers mode'les structureaux. Bulletin des Sciences Ge'ologiques 26: 91–113.

Ruland WW, Cherry JA, Feenstra S (1991) The depth of fractures and active groundwater flow in a clayey till plain in Southern Ontario. Ground Water 29(3): 405–17.

Ruprecht JK, Schofield NJ (1991) Effects of partial deforestation on hydrology and salinity in high salt storage landscapes. J. Hydrol. 12(9): 19–38.

Rushton KR (2003) Groundwater Hydrology: Conceptual and Computational Models. John Wiley and Sons, Chichester, U.K.

Rushton KR, Holt SM (1981) Estimating aquifer parameters for large-diameter wells. Ground Water 19: 505–9.

Rushton KR, Raghava Rao SV (1988) Groundwater flow through a Miliolite limestone aquifer. J. Hydrol. Sci. 33: 449–64.

Rushton KR, Redshaw SC (1979) Seepage and Ground Water Flow. John Wiley and Sons, New York, p. 339.

Rushton KR, Singh VS (1983) Drawdown in large-diameter wells due to decreasing abstraction rates. Ground Water 21(6): 670–77.

Rushton KR, Singh VS (1987) Pumping test analysis in large diameter wells with a seepage face by kernel function technique. Ground Water 25(1): 81–90.

Russell CE, Hess JW, and Tyler SW et al (1987) Hydrogeologic investigations of flow in fractured tuffs, Rainier Mesa, Nevada Test Site (eds Evans DD, Nicholson TJ), American Geophysical Union Monograph 42: 43–50.

Rutqvist J, Stephansson O (2003) The role of hydromechanical coupling in fractured rock engineering. Hydrogeol. J. 11: pp. 7–40.

Sabins FF (1983) Geologic interpretation of Space Shuttle radar images of Indonesia. Am. Assoc. Petro. Geol. Bull. 67: 2076–99.

Sabins FF (1997) Remote Sensing–Principles and Interpretation, 3rd ed., Freeman, New York.

Sakthivadivel R, Rushton KR (1989) Numerical analysis of large diameter wells with a seepage face. J. Hydrol. 107: 43–57.

Salama RB, Tapley I, Ishii T, Hawkes G (1994) Identification of areas of recharge and discharge using landsat-tm satellite imagery and aerial-photography mapping techniques. J. Hydrol. 162(1–2): 119–41.

Salve R, Ghezzchi TA, Jones R (2008) Infiltration into fractured bed rocks. Water Resources Res. 44(1), W01434.

Sami K (1996) Evaluation of the variations in borehole yield from a fractured Karoo aquifer, South Africa. Ground Water 34(1): 114–20.

Sanchez-Vila X, Carrera J, Girardi JP (1996) Scale effects in transmissivity. J. Hydrol. 183(1): 1–22.

Sanchez-Vila X, Meier PM, Carrera J (1999) Pumping tests in heterogeneous aquifers: An analytical study of what can be obtained from their interpolations using Jacob's method. Water Resource Res. 35(4): 943–52.

Sander P, Minor TB, Chesley MM (1997) Ground-water exploration based on lineament analysis and reproducibility tests. Ground Water 35(5): 888–99.

Sanford WE, Cook PG, Dighton JC (2002) Analysis of vertical dipole tracer test in highly fractured rock. Ground Water 40(5): 535–42.

Sanford W et al (eds) (2007) A new focus on groundwater-seawater interactions. IAHS Publ. 312, p. 344.

Saraf AK, Choudhury PR (1998) Integrated remote sensing and GIS for groundwater exploration and identification of artificial recharge sites. Int. J. Remote Sens. 19(10): 1825–41.

Scanlon BR, Healy RW, Cook PG (2002) Choosing appropriate techniques for quantifying groundwater recharge. Hydrogeol. J. 10: 18–39.

Schilling KE (2009) Investigating local variation in groundwater recharge along a topographic gradient, Walnut Creek, Iowa, USA. Hydrogeol. J. 17(2): 397–407.

Schmidt DA, Burgmann R (2003) Time-dependent land uplift and subsidence in the Santa Clara valley, California, from a large interferometric synthetic aperture radar data set. J. Geophys. Res.—Solid Earth. 108, no. B9. doi:10.1029/2002JB002267.

Schoeller H (1959) Arid Zone Hydrology: Recent Developments. UNESCO, Paris, p. 125.

Schwartz FW, Zhang H (2003) Fundamentals of Ground Water. John Wiley and Sons, Inc., New York, pp. 583.

Screaton EJ, Martin JB, Gim B et al (2004) Conduit properties and karstification in the unconfined Floridan aquifer. Ground Water 42(3): 338–46.

Sekhar M, Kumar SM, Sridharan K (1993) A leaky aquifer model for hard rock aquifers, in IAH Mem. 24th Concretess of IAH, Oslo, Part 1, pp. 338–48.

Shaban A, Khawlie M, Abdalla C (2006) Use of GIS and remote sensing to determine recharge potential zones: the case of occidental Linanon. Hydrogeol. J. 14(4): 433–43.

Shahin M (1985) Hydrology of the Nile Basin, in Development in Water Science 21, Elsevier, Amsterdam, p. 575.

Shapiro AM, Hsieh P (1998) How good are estimates of transmissivity from slug tests in fractured rocks? Ground Water 36(1): 37–48.

Shapiro AM (2003) The effect of scale on the magnitude of the formation properties governing fluid flow movement and chemical transport in fractured rocks. Proc. Symposium Groundwater in Fractured Rocks, Prague, pp. 13–14.

Shapiro AM (2007) Fractured-rock aquifers: Understanding an increasingly important source of water. USGS Fact Sheet 112–02, p. 3.

Sharma ML, Hughes MW (1985) Groundwater recharge estimation using chloride, deuterium and oxygen-18 profiles in the deep coastal sands of Western Australia. J. Hydrol. 81: 93–109.

Sharp JM Jr. (1993) Fractured aquifers/reservoirs: approaches, problems and opportunities. Mem. 24th Congress of IAH, Oslo, Part 1, pp. 23–38.

Sharp JM Jr., Boulton RCS, Fuller CM (1993) Permeability and fracture patterns in a weathered and variably welded tuff: implications for flow and transport. Mem. of 24th Congress of IAH, Oslo, Pat 1, pp. 103–14.

Shata AA (1982) Hydrogeology of the Great Nubian Sandstone basin Egypt. Q. J. Engg. Geol. 15: 127–33.

Sheila BB, Banks D (eds) (1993) Hydrogeology of Hard Rocks. Mem. 24th Congress, IAH, Oslo, Norway.

Shelton ML (1982) Ground water management in basalts. Ground Water 20: 86–93.

Shestopalov VM, Rudenko YF, Bohuslavsky AS et al (2006) Chernobyl-born radionuclides: Groundwater protectability with respect to preferential flow zones, in Applied Hydrogeophysics. (eds Vereecken H et al) Springer, Dordrecht, The Netherlands, pp. 341–76.

Shiklomanov IA (1990) Global water resources. Nature and Resources. UNESCO, Paris, 26(3): 34–43.

Shivanna K et al (1993) Isotope and geochemical evidence of past seawater salinity in Midnapore groundwaters, in Int. Symp. on Application of Isotope Techniques in Studying Past and Current Env. Changes in the Hydrosphere (Unpublished) IAEA/UNESCO, Vienna.

Sibanda T, Nonner JC, Uhlenbrook S (2009) Comparison of groundwater recharge estimation methods for the semi-arid Nyamandhlovu area, Zimbabwe. Hydrogeol. J. 17 doi:10.1007/s10040-009-0445-z.

Siddiqui SH, Parizek RR (1971) Geohydrological factors influencing well yields in folded and faulted rocks in Central Pennsylvania. Water Resources Res. 7(5): 1295–312.

Sidle WC, Lee PY (1995) Estimating local groundwater flow conditions in a granitoid: preliminary assessments in the Waldoboro Pluton Complex, Maine. Ground Water 33(2): 291–303.

Siegal BS, Abrams MJ (1976) Geologic mapping using Landsat data, Photogramm. Eng. Remote Sensing 42: 325–37.

Siemon B (2006) Electromagnetic methods—frequency domain, in Groundwater Geophysics—A tool for Hydrogeology, (Kirsch R ed.), Springer, pp. 155–78.

Silar J (1996) Groundwater residence time in crystalline rocks, in Hardrock Hydrogeology of the Bohemian Massif (eds Krasny J, Mls J), Acta Universitatis Carolinae Geologica, Univerzita Karlova, Czech Republic, Vol. 40: 279–88.

Simmers I (ed) (1988) Estimation of natural groundwater recharge. NATO ASI Ser C 222. Reidel, Dordrecht, pp. 510.

Simmons CT (2005) Variable density ground water flow: from current challenges to future possibilities. Hydrogeol. J. 13(1): 116–19.

Simmons GR (1985) In situ experiment in granite in underground laboratories—A review, in Design and Instrumentation of In-situ Experiment in underground Laboratories for Radioactive Waste Disposal (eds Come B, Johnston P, Muller A), A Balkema, Rotterdam, pp. 56–81.

Singh G (1988) Impact of coal mining on mine water quality. Int. J. Mine Waters. 7(3): 49–59.

Singh VS and Gupta CP (1986) Hydrogeological parameter estimation from pump test on a large diameter well. J. Hydrol. 87, pp. 223–32.

Singhal BBS (1963) Occurrence and geochemistry of ground water in the coastal region of Midnapur, West Bengal, India. Econ. Geol. 58, pp. 419–33.

Singhal BBS (1973) Some observations on the occurrence, utilisation and management of groundwater in the Deccan trap areas of Central Maharashtra, India, Proc. Int. Symp. on Development of Ground Water Resources, Vol. 3, Madras, India, pp. 75–81.

Singhal BBS, Singhal DC, Awasthi AK (1987) Study of borewells and dugwells in the state of Karnataka, India, Tech. Report prepared for NABARD, Bombay (Unpublished), p. 223.

Singhal BBS, Singhal DC (1990) Evaluation of aquifer parameters and well characteristics in fractured rock formations of Karnataka, India, in Selected Papers on Hydrogeology (eds Simpson ES, Sharp JM, Jr.), IAH, Vol. 1, pp. 951–62.

Sinha BPC, Kakar YP (1974) Geochemistry of fluoride in groundwater of north-Central Rajasthan, India. IAH Mem. 10, pp. 130–32.

Sinha BPC, Sharma S K (1990) Groundwater dams—Concepts and case histories. Bhu-Jal News 5: 3–13.

Skinner A Ch (1985) Groundwater protection in fissured rocks, in International Contributions to Hydrogeology 6 IAH, Hannover, pp. 123–43.

Skjernaa L, Jorgensen NO (1993) Detection of local fracture systems by azimuthal resistivity survey: examples from south Norway. Mem. of the 24th Congress of IAH, OSLo, 1993, pp. 662–71.

Smart PL, Hobbs SL (1986) Characterization of carbonate aquifer: A conceptual base, in Proc. of the Env. Problems in karst terranes and their solutions conference, Dublin, Ohio: National Water Well Associations, pp. 1–14.

Smedley PL (1992) Relationship between trace elements in water and health, with special reference to developing countries. Interim Report 1992. BGS Tech. Report WD/92/39, p. 33.

Smeldley P, West J (1995) Arsenic in water-implications for health. Earthworks 7, BGS, Nottingham, p. 3.

Smellie JAT, Laaksoharju M, Wikberg P (1995) Aspo, SE, Sweden a natural groundwater flow model derived from hydrogeochemical observations. J. Hydrol. 172: 147–69.

Smith L, Schwartz FW (1993) Solute transport through fracture network, in Flow and Contaminant Transport in Fractured Rock (eds Bear J, Tsang CF, de Marsily G), Academic Press, San Diego, pp. 129–66.

Smith LC (2002) Emerging applications of interferometric synthetic aperture radar (InSAR) in geomorphology and hydrology. Ann. Assoc. Am. Geogr. 92(3): 385–98.

Snow DT (1968a) Rock fracture spacings, openings and porosities. J. Soil Mech. Found. Div., ASCE, 94(SMI), pp. 73–91.

Snow DT (1968b) Hydraulic character of fractured metamorphic rocks of the front-range and implications to the Rocky Mountain Arsenal Well, Qr. of Colorado School of Mines, Denver 63(1): 167–99.

Sokolov AA, Chapman TG (eds) (1974) Methods of water balance computations: An international guide for research and practice. Studies and Reports in Hydrology, No. 7 UNESCO Press, Paris, p. 127.

Sonney R, Vuataz FD (2009) Numerical modeling of Alpine deep flow systems: a management and production tool for an exploited geothermal reservoir (Lavey-les-Basin, Switzerland). Hydrogeol. J. 17(2): 601–16.

Sophocleous M (2004) Climate change: why should water professional care?. Ground Water 42(5): 637.

Spies BR, Habashy TM (1995) Sensitivity analysis of crosswell electromagnetics. Geophysics 60: 834–45.

Springer A E, Stevens LE (2009) Spheres of discharge of springs. Hydrogeol. J. 17(1) doi:10.1007/s10040-008-0341-y.

Srivastava N (1993) Technique for Processing and Interpreting Groundwater Data Using GIS. M.Tech. Thesis, Department of Earth Sciences, University of Roorkee, Roorkee, India (unpublished).

Sridharan K, Sekhar M, Mohan Kumar MS (1990) Analysis of aquifer-water table-aquitard system. J. Hydrol. 114: 175–89.

Star J, Estes J (1990) Geographic Information Systems: An Introduction. Prentice Hall, Englewood Cliffs, NJ.

Stasko, S, Tarka, R. (1996) Hydraulic parameters of hard rocks based on long-term field experiment in the Polish Sudetes, in Acta Universitis Carolinae, Geologica (eds Krasny J, Mls J), 40(2): 167–78.

Stearns HT (1942) Hydrology of volcanic terraines, in Hydrology (Meinzer OE ed.), Dover Publ., Inc., New York, pp. 678–703.

Steinbruch F, Merkel BJ (2008) Characterization of a Pleistocene spring in Mozambique. Hydrogeol. J. 16(8): 1655–68.

Stober I, Bucher K (2005) Deep fluids: Neptune meets Pluto. Hydrogeol. J. 13(1): 112–15.

Straface S, Yeh T CJ, Zhu J et al (2007) Sequential aquifer tests at a well field, Montalto Ufffugo Scalo, Italy. Water Resources Res. 43(7): 1–13 W07432., doi:10.1029/2006WR005287.

Strahler AN (1964) Geology: Part II, Quantitative geomorphology of drainage basins and channel networks, in Handbook of Applied Hydrology (Chow VT ed.), McGraw-Hill Book Co., New York, pp. 4–39 to 4–70.

Strassberg G, Maidment DR, Jones NL (2007) A geopgraphic data model for representing groundwater systems. Ground Water Vol. 45(4): 515–18.

Streltsova TD (1976a) Hydrodynamics of groundwater flow in fractured formation. Water Resources Res. 12(3): 405–14.

Streltsova TD (1977) Storage properties of fractured formations, in Hydrologic Problems in Karst Regions (eds Dilamarter RR, Csallany SC), Kentucky University, pp. 188–92.

Streltsova TD (1976b) Advances and uncertainties in the study of groundwater flow in fissured rocks, in Advances in Groundwater Hydrology (Saleem ZA ed.), Am. Water Resources Assoc., pp. 18–56.

Streltsova-Adams TD (1978) Well hydraulics in heterogeneous aquifer formations, in Advances in Hydroscience (Chow VT ed.), Vol. 11, Academic Press, New York, pp. 357–423.

Struckmeier WF (1993) Hydrogeological mapping in hard rock terrains. Mem. of the 24th Congress of IAH, Oslo.

Subramanian PR (1993) High yielding borewell at Khairi village in basaltic terrain in Maharashtra. Bhu-Jal Nfews, 8(3–4): 35–36.

Subramanian PR (1992) Hydrogeology and its variations in the granites and associated rock formations in India. Proc. Workshop on Artificial Recharge of Groundwater in Granitic Terrains, Central Ground Water Board, Govt. of India, Bangalore, pp. 1–13.

Sudicky EA, Frind E O (1982) Contaminant transport in fractured porous media: Analytical solutions for a system of parallel fractures. Water Resources Res. 18(6): 1632–42.

Sukhija BS, Nagabhushanam P, Reddy DV (1996) Groundwater recharge in semi-arid regions of India: An overview of results obtained using tracers. Hydrogeol. J. 4(3): 50–71.

Sukhija BS, Nagabhushanam P, Reddy DV (2002) Use of stable isotopes in groundwater hydrology. Indian J. Geochem. 17: 89–119.

Sukhija BS, Rao AA (1983) Environmental tritium and radiocarbon studies in the Vedavati river basin, Karnataka and Andhra pradesh, India. J. Hydrol. 60: 185–96.

Sukhija BS, Reddy DV, Nagabhushanam P, Chand R (1988) Validity of the environmental chloride method for recharge evaluation of coastal aquifers, India. J. Hydrol. 99: 349–66.

Swenson S, Wahr J, Milly PCD (2003) Estimated accuracies of regional water storage variations inferred from the Gravity Recovery and Climate Experiment (GRACE). Water Resources Res. 39(8): 1223.

Talbot JC, Buckley DK, Herbert R (1993) Hydraulic Fracturing: Further Investigations on the use on Low Yielding Boreholes in the Basement Rocks of Zimbabwe, Tech. Report WD/93/16, BGS, Keyworth, UK.

Tam VT, De Smedt F, Batelaan O, Dassargues A (2004) Study on the relationship between lineaments and borehole specific capacity in a fractured and karstified limestone area in Vietnam. Hydrogeol. J. 12(6): 662–73, doi:10.1007/s10040-004-0329-1.

Taniguchi M, Ishitobi T, Saeki K (2005) Evaluation of time-space distributions of submarine ground water discharge. Ground Water 43(3): 336–42.

Taniguchi, M, Ishitobi T, Burnett WC (2007) Evaluating groundwater—sea water interactions via resistivity and seepage-meters. Ground Water 45(6): 729–35.

Tanwar BS (1983) Groundwater Studies of the Alluvial Belt of Yamuna and Ghaggar Doab in Haryana, Ph.D. Thesis (unpublished), Kurukshetra University, Kurukshetra, India, p. 205.

Taylor GC Jr. (1959) Groundwater provinces of India. Econ. Geol. 54: 683–97.

Taylor RW, Fleming AH (1988) Characterizing jointed systems by azimuthal resistivity surveys. Ground Water 26: 464–74.

Telford WM, Geldart LP, Sheriff RE (1999) Applied Geophysics. Cambridge University Press.

Terzaghi R (1965) Sources of error in joint surveys. Geotechnique 15: 287–304.

Thangarajan M (2004) Regional Groundwater Modeling. Capital Book Publishing Company, New Delhi.

Theis CV (1935) Relation between the lowering of the piezometric surface and the rate and duration of discharge of a well using ground-water storage. Trans. Am. Geoph. Union Pt. 2: 519–24.

Thornthwaite CW (1948) An approach toward a rational classification of climate. Geographical Review 38: 55–94.

Thorstenson DC, Fisher DW, Croft MG (1979) The geochemistry of Fox Hills—Basal Hill Creek aquifer in southwestern

North Dakota and northwestern South Dakota. Water Resources Res. 15(6): 1479–98.

Thrailkill J (1985) Flow in a limestone aquifer as determined from water tracing and water levels in wells. J. Hydrol. 78: 123–36.

Tickell SJ (1994) Dryland Salinity Hazard Map—The Northern Territory. Northern Territory of Australia. Darwin, Power and Water Authority, Report 54/94 D, p. 43.

Tiedeman CR, Hsieh PA (2004) Evaluation of longitudinal dispersivity estimates from saturated forced- and natural-gradient tracer tests in heterogeneous aquifers. Water Resource Res. 40, W01512.

Tiwary RK, Dhakate R, Rao VA et al (2005) Assessment and prediction of contaminant migration in groundwater from chromite waste dump. Env. Geol. 48(4–5): 420–29.

Todd DK (1980) Groundwater Hydrology, 2nd ed., John Wiley and Sons, New York, p. 535.

Todd DK, Mays LW (2005) Groundwater Hydrology, 3rd ed., John Wiley and Sons, NJ.

Tokunaga T (1999) Modeling of earthquake-induced hydrological changes and possible permeability enhancement due to the 17 January, 1995 Kobe Earthquake, Japan. J. Hydrol. 223: 221–29.

Toth J (1966) Groundwater geology, movement, chemistry, and resources near Olds, Alberta. Res. Council of Alberta, Canada, Geol. Div., Bull. 17.

Travers C, Martin T, Ruby M et al (2001) Geologic controls on DAPL migration in a fractured bed-rock system: Avtex Fibers Superfund site, Part I. Proc. Intl. Conf. Fractured Rock, EPA, ON, Canada.

Trescott PC, Larson SP (1977) Solution of three dimensional groundwater flow equations using the strongly implicit procedure. J. Hydrol. 35: 49–60.

Trevett JW (1986) Imaging Radar for Resources Surveys. Chapman and Hall, London, 313.

Trimmer D, Bonner B, Heard HC, Duba A (1980) Effect of pressure and stress on water transport in intact and fractured gabbro and granite. J. Geophys. Res. 85(B12): 7059–71.

Tsai TT, Kao CM, Yeh TY et al (2008) Chemical oxidation of chlorinated solvents in contaminated groundwater: Review. Practice Periodical of Hazardous, Toxic, and Radioactive Waste Management, ASCE/EWRI. 12(2): 116–26.

Tsang YW (1999) Usage of 'equivalent apertures' for rock fractures as derived from hydraulic and tracer tests. Water Resources Res. 28(5): 1451–53.

Tsang YW, Tsang CF (1987) Channel flow through fractured media. Water Resources Res. 23: 467–79.

Tsang YW, Witherspoon PA (1985) Effects of fracture roughness on fluid flow through a single deformable fracture, in Mem. 17 of IAH, Tuscon, pp. 683–94.

Tweed SO, Leblanc M, Webb JA, Lubczynski MW (2007) Remote Sensing and GIS for mapping groundwater recharge and discharge areas in salinity prone catchment, south-eastern Australia. Hydrogeol. J. 15(1): 75–96.

Tyagi NK (1994) Prevention and amelioration of water logging and salinity in India: An overview, in Proc. Regional Workshop on Environmental Aspects of Groundwater Development, (Singhal DC ed.), Kurukshetra, pp. III–1–III–10.

Uhl VW (1979) Occurrence of groundwater in the Satpura Hills region of Central India. J. Hydrol. 41: 124–41.

Uma KO, Egboka BCE, Onuoha KM (1989) New statistical grain size method for estimating the hydraulic conductivity of sandy aquifers. J. Hydrol. 108: 343–66.

United Nations UN (2005) Water for Life Decade 2005–2015, UN Deptt. of Public Information New York, p. 17.

United Nations Educational, Scientific and Cultural Organisation UNESCO (1972) Ground Water Studies (eds Brown RH, Konoplyantsev AA, Ineson J, Kovalevsky VS), UNESCO, Paris.

United Nations Educational, Scientific and Cultural Organisation UNESCO (1975) Analytical and investigational techniques for fissured and fractured rocks, in Ground Water Studies Studies (eds Brown RH et al), Studies and Reports in Hydrology (Chapter 14), Supplement 2, UNESCO, Paris.

United Nations Educational, Scientific and Cultural Organisation UNESCO (1977) Planning and design of groundwater level networks, in Groundwater Studies, Chapter 7.6, Studies and reports in hydrology 7, UNESCO, Paris.

United Nations Educational, Scientific and Cultural Organisation UNESCO (1979) Water in Crystalline Rocks, SC-77/WS/71, UNESCO, Paris, p. 344.

United Nations Educational, Scientific and Cultural Organisation UNESCO (1980) Aquifer Contamination and Protection, studies and reports in Hydrology 30, p. 440.

United Nations Educational, Scientific and Cultural Organisation UNESCO (1983) International Legend for Hydrogeological Maps, Tech. Document SC-84/WS/7, UNESCO Paris, p. 51.

United Nations Educational, Scientific and Cultural Organisation UNESCO (1984a) Groundwater in Hard Rocks, Studies and Reports in Hydrology 33, UNESCO, Paris.

United Nations Educational, Scientific and Cultural Organisation UNESCO (1984b) Guide to the Hydrology of Carbonate Rocks, Studies and reports in hydrology 41, p. 345.

United Nations Educational, Scientific and Cultural Organisation UNESCO (1987) Groundwater Problems in Coastal Areas. Studies and reports in hydrology 45, UNESCO, Paris, p. 596.

United Nations Educational, Scientific and Cultural Organisation UNESCO/IAHS (1967) Proc. Dubrovnik Symposium on Hydrogeology of Fractured Rocks Vol. I and II, UNESCO, Louvain, Belgium.

United Nations Educational, Scientific and Cultural Organisation UNESCO/UNEP (1989) Geology and the Environment, Vol. II (Vartanyan GS ed.), UNESCO, Paris, p. 201.

Vacher HL, Ayers JF (1980) Hydrology of small oceanic islands–utility of an estimate of recharge inferred from the chloride concentration of the fresh water lenses. J. Hydrol. 45: 21–37.

Van Golf-Racht TD (1982) Fundamentals of Fractured Reservoir Engineering, Elsevier Scientific Publ. Co., Amsterdam, p. 710.

Veeger A, Davis S, Long A, Cuslodio E (1989) Groundwater chemistry related to sources of recharge on the Island of La Palma, Canary Islands, Spain, Abstracts Vol. 3, 28th IGC, Washington, pp. 288–89.

Veeger, AI, Rudereman. NC (1998) Hydrogeologic controls on Radon-222 in a buried valley fractured bedrock aquifer system. Ground Water 36(4): 596–604.

Vengosh A, Rosenthal E (1994) Saline groundwater in Israel: Its bearing on the water crisis in the country. J. Hydrol. 156: 389–430.

Vengosh A, Starinsky A, Kolodny Y, Chivas AR (1994), Boron isotope geochemistry of thermal springs from the northern rift valley, Israel. J. Hydrol. 162: 155–69.

Venkateswarlu J, Sen AK, Dubey JC et al (1990) Water 2000 AD The Scenario for Arid Rajasthan, Central Arid Zone Res. Inst., Jodhpur, India, p. 49.

Verma MN, Jolly PB (1992) Hydrogeology of Helen Springs—Explanatory Notes for 1:250 000 Scale Map, Power and Water Authority, Report 50, Darwin.

Verma MN, Qureshi H (1982) Groundwater occurrence in fractured rocks in Darwin Rural Area, N.T., in AWRC Conference on Groundwater in Fractured Rocks, Series No. 5, pp. 229–39.

Versey HR, Singh BK (1982) Groundwater in Deccan basalts of the Betwa Basin, India. J. Hydrol. 58: 279–306.

Vivenstsova E, Voronov A (2005) The interaction of seawater and fissured water in the basement of the Gulf of Finland basin in Proc. Fennoscandian 3rd Regional Workshop on Hardrock Hydrogeology, Helsinki, pp. 106–8.

Vogel P, Giesel W (1989) Propogation of dissolved substances in rocks: Theoretical consideration of the relationship between systems of parallel fractures and homogeneous aquifers, in Contaminant Transport in Groundwater (eds Kobus HE, Kinzelbach W), A.A. Bolkema, Rotterdam, pp. 275–80.

Viventsova EA, Voronov AN (2005) Radon in groundwater in magmatic rocks. Proc. of the Fennoscandian 3rd Regional Workshop on Hardrock Hydrogeology. Helsinki, pp. 103–5.

Vouillamoz JM, Legchenko A, Albouy Y et al (2003) Localization of saturated karst aquifer with magnetic resonance sounding and resistivity imagery. Ground Water 41(5): 578–81.

Walker DD, Gylling B, Strom A et al (2001) Hydrogeologic studies for nuclear- waste disposal in Sweden. Hydrogeol. J. 9(4): 419–31.

Walker GPL (1973) Length of lava flows. Philosophical Trans. of the Royal Soc. of London, 274(B): 107–18.

Walker WH, Walton WC (1961) Groundwater development in three areas of Central Illinois, Report of Investigation 41, Illinois State Water Surv., Urbana, p. 43.

Waltham AC (1989) Ground Subsidence. Blackie, Glasgow & London, p. 202.

Walton WC (1962) Selected Analytical Methods for Well and Aquifer Evaluation, Bull. 49, Illnois State Water Survey, Urbana, p. 81.

Walton WC (1970) Groundwater Resource Evaluation. McGraw-Hill, New York, p. 664.

Walton WC (1979) Review of leaky artesian aquifer test evaluation methods. Ground Water 17(3): 270–83.

Walton WC, Csallany S (1962) Yields of Deep Sandstone Wells in Northern Illinois, Illinois State Water Survey, Report of Inv. 43, p. 47.

Walton WC, Stewart JW (1961) Aquifer tests in the Snake River basalts. Am. Soc. Civil Engrs., Trans. 126, part III: 612–32.

Wang C, Chia Y (2008) Mechanism of water level changes during earthquakes: Near field versus intermediate field. Geophys. Res. Lett. 35, L12402. doi:10.1029/2008GL034227.

Wang JSY, Narasimhan TN (1993) Unsaturated flow in fractured porous media, in Flow and Contaminant Transport in Fractured Rock (eds Bear J, Tsang CF, de Marsily G), Academic Press, Inc., San Diego, pp. 325–95.

Wang Y, Merkel BJ, Li Y, Ye H, Fu S, Ihm D (2007) Vulnerability of groundwater in Quaternary aquifers to organic contaminants: A case study in Wuhan City, China. Environ. Geol. 53: 479–84.

Warwick D, Hartopp PG, Vilijoen RP (1979) Application of thermal infrared linescanning technique to engineering mapping in South Africa. Q. L. Engg. Geol., 12, 159–79.

Water Res. Board (1972) The Hydrogeology of the London Basin, Water Res. Board, Reading, UK p. 139.

Waters P, Greenbaum D, Smart PL, Osmaston H (1990) Applications of remote sensing to groundwater hydrology. Remote Sensing Reviews. 4(2): 223–64.

Watson I, Burnett AD (1993) Hydrology—An Environmental Approach, Buchanan Books, Cambridge, p. 702.

Watson KM, Bock Y, Sandwell DT (2002) Satellite interferometric observations of displacements associated with seasonal groundwater in the Los Angeles basin. J. Geophys. Res.—Solid Earth 107, no. B4. doi:10.1029/2001JB000470.

Wealthall GP, Kueper BH, Lerner DN (2001) Fractured rockmass characterization for predicting the fate of DNAPLs, in Proc. Conference Fractured Rock 2001 (eds Kueper BH et al).

Weatherill D, Simmons CT, Cook PG (2006) Using dipole tracer tests to predict solute transport in fractured rock. Proc. Joint Congress of 9th Australasian Env. Isotope Conf. And 2nd Australasian Hydrogeol. Res. Conf., Adelaide, South Australia, pp. 165–66.

Weeks EP (1987) Effect of topography on gas flow in unsaturated fractured rock: Concepts and observations. Geophysical Monograph 42, Am. Geophys. Union, pp. 165–70.

Welch AH Stollenwerk KG (eds) (2003) Arsenic in Ground Water. Kluwer Academic Publ., Dordrecht, p. 475.

Wels C, Smith L (1994) Retardation of sorbing solutes in fractured media. Water Resources Res. 30(9): 2547–63.

Wenzel LK (1942) Methods for Determining Permeability of Water Bearing Materials US Geol. Surv. Water Supply Paper 887, p. 192.

Wesslen A, Gustafson G, Maripuu P (1977) Groundwater and storage in rock caverns, pumping tests as an investigation method. Proc. Ist Intl. Symp. Rock Store, Stockholm, pp. 359–66.

White DE (1957) Magmatic, connate, and metamorphic waters. Bull. Geol. Geol. Am. 68: 659–82.

White DE, Hem JD, Waring GA (1963) Chemical composition of sub-surface waters, in Data of Geochmistry, 6th ed., (Fleischer M ed.), US Geol. Surv. Prof. Paper 440F, p. 67.

White RB, Gainer RB (1985) Control of groundwater contamination at an active uranium mill. Ground Water Monit. Rev. 5(2): 75–82.

White WB (2002) Karst hydrology: recent developments and open questions. Engg. Geol. 65(2–3): 85–105.

White WB (2007) Groundwater Flow and Transport in Karst, in The Handbook of Groundwater Engineering (Delleur JW ed.), CRC Press, Boca Raton, FL, pp. 18–1 to 18–36.

World Health Organisation WHO (1984) Guidelines for Drinking Water Quality Vol. 1, 2 and 3, WHO, Geneva.

Wilcox LV (1955) Classification and Use of Irrigation Waters. US Deptt. Agri. Circ. 969, Washington, DC, p. 19.

Wilkins A, Subbarao KV, Ingram G, Walsh JN (1994) Weathering regimes within the Deccan Basalts, in Volcanism (ed Subbarao KV), Wiley Eastern Ltd., pp. 217–31.

Wilkinson WB, Brassington FC (1991) Rising groundwater levels—an international problem, in Applied Groundwater Hydrology (eds Downing RA, Wilkinson WB), Oxford University Press, Oxford, pp. 35–53.

Williamson KH, Fernandez JC, Aunzo ZP, Sussman D, Prabowo HT (1996) Large scale geothermal development in Asia—UNDCAL's experience. Abstracts, 30th IGC, Beijing Vol. 1, p. 384.

Williamson WH (1982) The use of hydraulic techniques to improve the yield of bores in fractured rocks. Paper of the Groundwater in Fractured Rock Conference, AWRC conference series No. 5, Canberra, pp. 273–83.

Wilson CR, Witherspoon PA (1985) Steady state flow in rigid networks of fractures. Water Resources Res. 10(2): 328–35.

Wilt MJ et al (1995) Crosswell electromagnetic tomography: system design considerations and field results. Geophys. 60, pp. 871–85.

Winter TC, Harvey JW, Franke QL, Alley WM (1998) Ground water and surface water: A single resource. USGS circular 1139. Denver, Colorado: USGS.

Witherspoon PA, Wang JSY, Iwai K et al (1980) Validity of cubic law for fluid flow in deformable rock fracture. Water Resources Res. 16(6): 1016–24.

Witherspoon PA, Bodvarsson GS (eds) (2006) Geological challenges in radioactive waste isolation: Fourth worldwide review, Report 59808, Lawrence Berkeley National Laboratory, Berkeley, California, p. 283.

Woith H, Wang R, Milkeveit C et al (2003) Heterogeneous response of hydrogeological systems to the Izmit and Duzce (Turkey) earthquakes of 1999. Hydrogeol. J. 11(1): 113–21.

Wolf PR (1983) Elements of Photogrammetry, 2nd ed., McGraw-Hill, New York.

Woolley DR (1982) Depth—yield relationships of bores in fractured rocks in New South Wales, in AWRC Conference on Groundwater in Fractured Rock Conference Ser. No. 5, Canbera, pp. 283–92.

Worthington MH (1984) An introduction to geophysical tomography. First Break 2: 20–26.

Wright EP (1984) Drilling for groundwater in the Pacific region, in Water Resources of Small Islands, Tech. Proc. (Part 2) of the Regional Workshop on Water Resources of Small Islands, CSC, London, pp. 525–29.

Wright EP (1985) Groundwater in the Third World, in Hydrogeology in the Service of Man, Mem. IAH, Cambridge, UK, pp. 53–64.

Wright EP (1992) The hydrogeology of crystalline basement aquifers in Africa, in The Hydrogeology of Crystalline Basement Aquifers in Africa (eds Wright EP, Burgess WG), Geol. Soc. Special Publ. No. 66, The Geological Society, London, pp. 1–27.

Wright EP, Burgess WG (eds) (1992) The Hydrogeology of Crystalline Basement Aquifers in Africa Geol. Soc. Special Publ. No. 66, The Geological Society. London, p. 264.

Xuanjiang H (1994) The mutual relationship between urban development, surface settlement and its counter measure. Proc. Int. Conference on Geoscience in Urban Development (LANDPLAN IV), China Ocean Press, Beijing, China, pp. 271–76.

Yang L (2008) Phytoremediation: An accurate technology for treating contaminated sites. Practice Periodical of Hazardous, Toxic and Radioactive Waste Management, ASCE/EWRI, 12(4): 290–98.

Yashpal, Sahai B, Sood RK et al (1980) Remote sensing of the lost Saraswati river, Proc. Ind. Acad. Sc. (Earth Planet Sc.). Vol. 89.

Yechieli Y, Kafri U, Sivan O (2009) The inter-relationship between coastal sub-aquifers and the Medterranean sea, deduced from radioactive isotope analysis. 17(2): 265–74.

Yeh T-CJ, Liu S (2000) Hydraulic tomography: Development of a new aquifer test method. Water Resources Res. 36(8): 2095–2105.

Yeh T-CJ, Lee C-H. (2007) Time to change the way we collect and analyze data for aquifer characterization. Ground Water 45(2): 116–18.

Yin Zhi-Yong, Brook GA (1992) The topographic approach to locating high–yielding wells in crystalline rocks: Does it work? Ground Water 30: 96–102.

Yuan D (ed) (1996) Karst Processes and the Carbon Cycle. Newsletter, Project 379, Institute of Karst Geology, Guilin, China, p. 188.

Yuan D, Drogue C, Dai A et al (1990) Hydrology of the karst aquifer at the experimental site of Guilin in Southern China. J. Hydrol. 115(2): 285–96.

Zangeri C, Eberhardt E, Loew S et al (2003) Ground settlements above tunnels in fractured crystalline rocks: numerical analysis of coupled hydromechanical mechanisms. Hydrogeol. J. 11: 162–73.

Zaporozec A (1972) Graphical interpretation of water quality data. Ground Water 10(2): 32–43.

Zeil P, Volk P, Saradeth S (1991). Geophysical method of lineament studies in ground water exploration: A case history from SE Botswana. Geoexploration 27: 165–77.

Zeizel AJ, Walton WC, Prickett TA et al (1962) Groundwater Resources of Dupage County, Illinois, Ground Water Report No. 2, Illinois State Water Survey and Geol. Surv.

Zektser IS, Dzhamalov RG (1988) Role of Groundwater in the Hydrological Cycle and in Continental Water Balance. IHP III Project, 2.3, UNESCO, Paris, p. 133.

Zektser IS, Loaiciga HA (1993) Groundwater fluxes in the global hydrological cycle past, present and future. J. Hydrol. 144(2): 405–27.

Zhao J, Brown ET (1992) Thermal cracking induced by water flow through joints in heated granite. Int. J. Rock Mech. Min. Sc. and Geomech. Abstr. 29(1): 77–82.

Zheng C (1990) A Modular Three Dimensional Transport Model for Simulation of Advection, Dispersion and Chemical Reactions, of Contaminants in Groundwater Systems., Environmental Protection Agency, Oklahoma, USA, R.S. Ken Environmental Research Laboratory.

Zhengzhou Y, Zhenmin H, Shimin X, Junxiang X (1996) Field Trip Guide, T337, Experiment, Exploitation and Utilization of Groundwater in the North China Plain. 30th IGC, Geol. Publ. House, Beijing, p. 29.

Zhiming W, Shokang H, Feng X et al (1996) Isotope-geochemical studies of geothermal water in Chenrian uranium mine, in Abstracts of Uranium Geology Theses, Bureau of Geology, China National Nuclear Corporation, Beijing.

Zuber A, Motyka J (1994) Matrix porosity as the most important parameter of fissured rocks for solute transport at large scales. J. Hydrol. 158: 19–46.

Index

2-D resistivity imaging, 78
3-D resistivity imaging, 79

A
Acidizing, 310
Active remote sensing, 46
Addition image, 70
Advection, 131
Age dating, 202, 203
Air rotary method, 308
Albedo, 53
Anisotropy, 14, 151
Aperture, 30
    hydraulic, 146
    mechanical, 146
Apparent resistivity figure, 80
Applachian basin, 296
Aquiclude, 8
Aquifer
    confined, 9
    defined, 8
    double porosity, 9
    fractured rock, 2, 334
    granular rock, 2
    leaky, 9
    perched, 9
    unconfined, 9
    water-table, 9
Aquitard, 8
Archie's law, 75
Areal fracture density, 27
Areal surveys, 32
Arithmetic overlay, 102
ARS, 80
Arsenic, 218, 223, 224, 226
Artificial recharge, 364–370
Aspo Hard Rock Laboratory, 229
ASR, 369
ASTER, 51
ASTR, 369
Atmospheric
    absorption, 46
    interaction, 54

scattering, 46
windows, 47
Attribute data, 97
Average velocity, 117
Azimuthal resistivity survey, 80
Azimuthal seismic method, 85

B
Bank infiltration, 370
Bank storage, 5
Barometric efficiency, 355
Basalt, 257, 259
Baseflow, 5
Basin yield, 366
Bedding, 13, 63
Berea sandstone, 294
Bioremediation, 234
Blasting, 311
Block size, 29
Bole, 261
Boolean logic, 102
Boom clay, 296
Bore fluid log, 92
Borehole televiewer, 93
Borewells, 300
Boulton's delay index, 144
Boulton's method, 177
Bourdet – Gringarten method, 187
Bouwer and Rice method, 162
Breakthrough curve, 136, 197, 198
Bredehoeft and Papadopulos method, 165
Brittle deformation, 16, 17
Brunswick shale, 296
Bundelkhand granite, 126
Buried pediment, 238
Buried valleys, 291

C
Cable tool method, 307, 308
Caliper log, 91
Camera system, 47
Canary islands, 266

B. B. S. Singhal, R. P. Gupta, *Applied Hydrogeology of Fractured Rocks,*
DOI 10.1007/978-90-481-8799-7, © Springer Science+Business Media B.V. 2010